W0246163

Semiconductor Alloys

Physics and Materials Engineering

MICRODEVICES
Physics and Fabrication Technologies

Series Editors: Ivor Brodie and Arden Sher
SRI International
Menlo Park, California

COMPOUND AND JOSEPHSON HIGH-SPEED DEVICES
Edited by Takahiko Misugi and Akihiro Shibatomi

ELECTRON AND ION OPTICS
Miklos Szilagyi

ELECTRON BEAM TESTING TECHNOLOGY
Edited by John T. L. Thong

GaAs DEVICES AND CIRCUITS
Michael Shur

ORIENTED CRYSTALLIZATION ON AMORPHOUS SUBSTRATES
E. I. Givargizov

PHYSICS OF HIGH-SPEED TRANSISTORS
Juras Požela

THE PHYSICS OF MICRO/NANO-FABRICATION
Ivor Brodie and Julius J. Muray

PHYSICS OF SUBMICRON DEVICES
David K. Ferry and Robert O. Grondin

THE PHYSICS OF SUBMICRON LITHOGRAPHY
Kamil A Valiev

SEMICONDUCTOR ALLOYS
Physics and Materials Engineering
An-Ban Chen and Arden Sher

SEMICONDUCTOR LITHOGRAPHY
Principles, Practices, and Materials
Wayne M. Moreau

SEMICONDUCTOR PHYSICAL ELECTRONICS
Sheng S. Li

A Continuation Order Plan is available for this series. A continuation order will bring delivery of each new volume immediately upon publication. Volumes are billed only upon actual shipment. For further information please contact the publisher.

Semiconductor Alloys

Physics and Materials Engineering

An-Ban Chen
Auburn University
Auburn, Alabama

Arden Sher
SRI International
Menlo Park, California

Plenum Press • New York and London

Library of Congress Cataloging-in-Publication Data

Chen, An-Ban.
Semiconductor alloys : physics and materials engineering / An-Ban Chen, Arden Sher.
p. cm. -- (Microdevices)
ISBN 978-1-4613-7994-2
1. Semiconductors--Materials. 2. Alloys. 3. Free electron theory of metals. I. Sher, Arden. II. Title. III. Series.
QC611.8.A44C48 1995
537.6'22--dc20 95-41549
CIP

Front cover: Lattice arrangement of $Ga_{0.5}In_{0.5}As$. This is the relaxed atom arrangement in a cation-substituted disordered alloy, $Ga_{0.5}In_{0.5}As$. The cations are located near their face-centered cubic sublattice sites, but the anions assume relaxed positions determined by their nearest-neighbor cation site occupancy. Other bond length mismatched alloys have similar arrangements.

ISBN 978-1-4613-7994-2 e-ISBN-13: 978-1-4613-0317-6
DOI: 10.1007/978-1-4613-0317-6

Softcover reprint of the hardcover 1st edition 1995
A Division of Plenum Publishing Corporation
233 Spring Street, New York, N. Y. 10013

10 9 8 7 6 5 4 3 2 1

To our wives Mayurase Chen and Lois Sher

Preface

Semiconductor alloys, with tailorability of their electronic structures and other material properties, have found wide applications in optic and electronic devices. Literature on the fundamental properties of these materials is growing rapidly. Many major discoveries and advances have been made in the past decade. However, owing to difficulties in treating disorder in these materials, it probably will take another decade for semiconductor alloy theory to reach the same level of maturity as that for current crystalline semiconductors. Against this background, there is a need for a book which provides simple, yet practical, theories linking the properties of semiconductor alloys to their constituent compounds. The main purpose of this book is to provide such a linkage. Often, even now, it is possible to do this at a level of accuracy capable of serving as engineering design tools.

The topics treated in this book include crystal structures, bonding, elastic properties, phase diagrams, band structures, and transport. We begin with the constituent compounds and gradually pass to alloys. A fair fraction of the subject matter selected reflects our previous work, but a broader view is adopted whenever possible. *Ab initio* theories are discussed to provide an appreciation of the status of current theory. However, semiempirical theories are emphasized. The idea is to assemble thought-provoking models and simple computational tools that still correlate with experiments. Extensive tables and figures are provided so that the book can serve as a useful reference for workers in the field. This is particularly true in Chapter 7, where detailed quantitative band structures of all common III–V and II–VI compounds and alloys are compiled. At the end of each chapter, we comment on future developments. Problem sets are also included so that the book can be used as a text.

Readers with a background in introductory solid-state physics and quantum mechanics should be able to understand the majority of the material, although some treatments require a knowledge of these topics at the graduate level.

ACKNOWLEDGMENTS

Many results discussed in this book originated from collaborative research with Srini Krishnamurthy, Mark van Schilfgaarde, Marcy Berding, Chin-Yu Yeh, and Tony Paxton. Walt Harrison and Tom Casselman provided valuable insights for improving the content and presentation of many sections of the book. We are deeply in debt to Joyce Garbut, Robin Burns, and Julie Kirkpatrick, without whose skilled and patient work on the manuscript this book would never have materialized. Some financial aid to An-Ban Chen for manuscript preparation was supplied by ONR. We also acknowledge AFOSR, ONR, ARPA, and NASA for helping support aspects of treatments presented here that, in several instances, were never previously published. Finally, but not least, we thank our wives for their continuous support and understanding.

An-Ban Chen
Arden Sher

Contents

CHAPTER 4. Alloy Statistics and Phase Diagrams

CHAPTER 5. Band Structure Theory

CHAPTER 6. Transport

CHAPTER 7. Band Structures of Selected Semiconductors and Their Alloys

PROBLEMS

1

Crystal Structures

The crystal structures of semiconductors are well known, but those of their alloys are not. In this chapter the three most elementary crystal structures in which most semiconductors form are introduced. Then the structures of nearly random bulk alloys revealed by extended x-ray absorption fine-structure spectroscopy (EXAFS) are discussed. The features of three additional structures related to long-range ordered alloys are also described.

1.1. DIAMOND, ZINC BLENDE, AND WURTZITE STRUCTURES

The diamond, zinc blende, and wurtzite structures are the three most common crystal structures of semiconductors grown for practical applications. The diamond structure consists of two interpenetrating face-centered-cubic (fcc) sublattices displaced from each other by $(1/4,1/4,1/4)a$, where a is the cube edge length, often called the lattice constant. Besides diamond (C), the two elementary semiconductors Si and Ge, and the semimetal gray tin (Sn) assume this structure. The zinc blende structure also has the same sublattices, except that the anions (the nonmetal elements) occupy one sublattice, while the cations (the metal atoms) sit on the other. Both structures have an fcc lattice with two atoms per unit cell. Figure 1.1 shows the arrangement of the atoms inside a cube of the zinc blende structure. Many important III–V compound semiconductors, such as GaAs and InP, and II–VI compounds, such as CdTe, have this crystal structure.

Wurtzite is similar to the zinc blende structure, both are fourfold coordinated, except that the former has a hexagonal close-packed (hcp) Bravais lattice rather than an fcc. The difference between wurtzite and zinc blende can be described most easily by examining the stacking sequence of planes of close-packed spheres of the same radius along the (111) direction. The fcc lattice repeats a characteristic stacking sequence every three layers, while the hcp repeats the stacking sequence every two layers. After the one sublattice is specified, say the anion sublattice, between every two anion planes a plane of cations can be inserted such that each cation is surrounded by four anions in a tetrahedral arrangement, and vice versa. Thus, as far as the local tetrahedral bonding is concerned, there is no difference between zinc blende and wurtzite structures. The difference comes when the second and farther neighboring atoms are considered. The other difference is that there are four atoms per unit cell in a wurtzite semiconductor, because each hcp unit cell contains two nonequivalent atomic sites. It should be noted that while the ideal hcp has a hexagon height c to edge

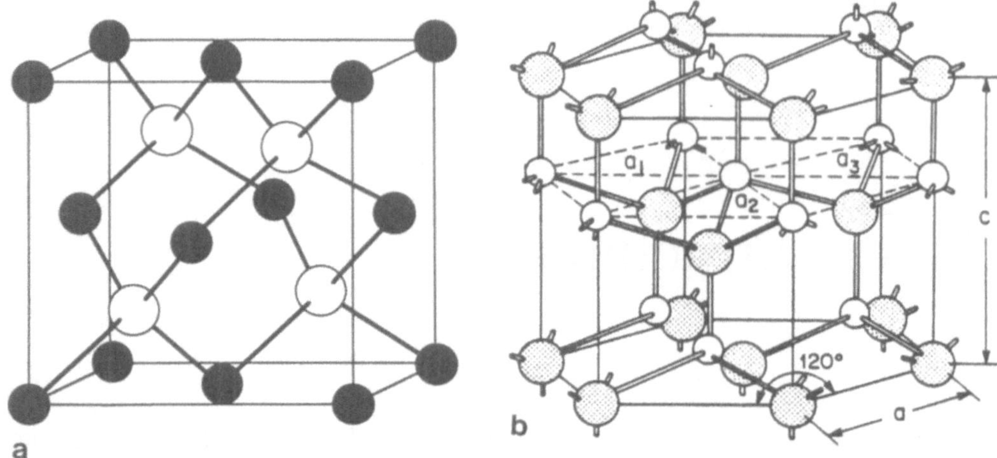

FIGURE 1.1. Atomic positions in (a) zinc blende and (b) wurtzite crystal (Sze, 1981).

length a (c/a) ratio of $(8/3)^{1/2} = 1.633$, most wurtzite semiconductors have slightly different c/a ratios. Unlike the zinc blende structure, which only requires a single parameter (the lattice constant a) to specify the crystal, the wurtzite structure requires three: the lattice constant a, the c/a ratio, and the ratio, γ, of the bond length along the c direction to the other three (equal) bond lengths. Table 1.1 is a collection of the lattice parameters and crystal structures for a number of crystalline semiconductors belonging to the three structures just mentioned.

One special feature in these structures is their tetrahedral bonding (Harrison, 1980). The tetrahedral coordination is perfect for the diamond and zinc blende (zb) structures and is nearly ideal for most wurtzite crystals. Thus, the bond length d is a parameter often used to scale various physical quantities in semiconductors. Table 1.1 also lists the experimental values of d for the diamond and zb semiconductors and the average d for wurtzite.

Pauling (1960) suggested a set of empirical radii for atoms, the so-called covalent radii, to predict the bond length in the tetrahedral crystals from the expression $d = r_A + r_C$, where r_A and r_C are the covalent radii for the anion and the cation, respectively. The values of these covalent radii are listed in Table 1.2. These radii generally predict d accurately to 1% or better, although there are exceptions (e.g., AlSb). However, one should not take these radii as having unique values. As pointed out by Phillips (1973), one can add a constant value to the radii for the group III elements and subtract the same value from those of the group V elements without changing the results of the bond length of the III–V compounds. A similar argument also holds for the II–VI and I–VII compounds.

In addition to the three crystal structures considered above, there are more complicated tetrahedral semiconductors, such as the ternary (Berger and Prochukhan, 1969) and multinary (Deb and Zunger, 1987) diamond-like semiconductors. There are also many other semiconductors with different coordination numbers and crystal structures (Phillips, 1973). These more complicated materials have not found nearly as many applications as the simple diamond, zinc blende, and wurtzite semiconductors and consequently will not be emphasized. However, as shown later, some ordered alloys grown epitaxially have these complex structures. They will be treated in some detail in Section 1.4.

TABLE 1.1. Crystal Structures, Lattice Parameters a, c/a, the Bond Ratio γ, and the Bond Lengths d Defined in the Text for the Common Tetrahedral Semiconductors (all lengths in Å)

	Zinc blende or diamond		Wurtzite			
	a	d	a	c/a	γ	d
C	3.567	1.543	2.52	1.635	—	1.54
Si	5.431	2.352	3.80	1.653	—	2.34
Ge	5.657	2.450	—	—	—	—
Sn	6.489	2.810	—	—	—	—
SiC	4.357	1.887	3.079	1.641	1.008	1.889
BN	3.615	1.565	2.55	1.647	—	1.57
AlN	—	—	3.111	1.601	1.007	1.893
GaN	—	—	3.190	1.627	1.004	1.951
InN	—	—	3.533	1.611	—	2.154
BP	4.538	1.965	3.562	1.656	—	2.192
AlP	5.467	2.367	—	—	—	—
GaP	5.447	2.359	—	—	—	—
InP	5.869	2.541	—	—	—	—
BAs	4.777	2.069	—	—	—	—
AlAs	5.639	2.442	—	—	—	—
GaAs	5.654	2.448	—	—	—	—
InAs	6.058	2.623	4.274	1.638	—	2.620
AlSb	6.136	2.657	—	—	—	—
GaSb	6.094	2.638	—	—	—	—
InSb	6.479	2.805	—	—	—	—
ZnO	—	—	3.253	1.603	1.008	1.980
ZnS	5.406	2.341	3.811	1.636	—	2.335
CdS	5.835	2.527	4.137	1.623	1.005	2.528
HgS	5.872	2.543	—	—	—	—
ZnSe	5.669	2.455	4.003	1.637	—	2.453
CdSe	6.05	2.62	4.30	1.631	1.006	2.63
HgSe	6.085	2.635	—	—	—	—
ZnTe	6.103	2.643	4.310	1.645	—	2.646
CdTe	6.478	2.805	4.572	1.637	—	2.802
HgTe	6.460	2.797	—	—	—	—
CuCl	5.416	2.345	3.91	1.642	—	2.40
CuBr	5.691	2.464	4.06	1.640	—	2.49
CuI	6.055	2.822	4.31	1.645	—	2.65
AgI	6.486	2.809	4.592	1.635	1.001	2.813

After R. Zallen (1982).

TABLE 1.2. Pauling's Covalent Radii (in Å)

	Be 1.06	B 0.88	C 0.77	N 0.70	O 0.56	F 0.64
	Mg 1.40	Al 1.26	Si 1.17	P 1.10	S 1.04	Cl 0.99
Cu 1.35	Zn 1.31	Ga 1.26	Ge 1.22	As 1.18	Se 1.14	Br 1.11
Ag 1.52	Cd 1.48	In 1.44	Sn 1.40	Sb 1.36	Te 1.32	I 1.28
	Hg 1.48	Tl 1.47	Pb 1.46	Bi 1.45		

After Pauling (1960).

1.2. BULK ALLOYS

The most common semiconductor alloys are the binary IV–IV mixtures $A_{1-x}B_x$ (e.g., $Si_{1-x}Ge_x$) and the pseudobinary alloys of the form $A_{1-x}B_xC$ (e.g., $Ga_{1-x}In_xAs$) and $CA_{1-x}B_x$ (e.g., $GaP_{1-x}As_x$), where x is the percent concentration. The form $CA_{1-x}B_x$ indicates an anion substituted alloy, which may be thought of as resulting from replacing a fraction x of the anions A in the CA compound by B atoms from the same column in the periodic table: examples are $GaP_{1-x}As_x$, $ZnS_{1-x}Se_x$ and $CuCl_{1-x}Br_x$. Similarly $A_{1-x}B_xC$ is a cation-substituted alloy with the atoms A and B distributed on the cation sublattice, such as $Ga_{1-x}In_xAs$ and $Hg_{1-x}Cd_xTe$. Although there is an attempt to establish a notation that takes the A atoms to be lighter than the B atoms, this rule will not be followed in this book because several contrary examples, such as $Hg_{1-x}Cd_xTe$, have been used so often in the past that it would be confusing to change.

If we look only at gross structural features, these alloys are similar to their constituent crystals. For example the x-ray diffraction pattern of a homogeneous pseudobinary alloy is similar to those of the pure compounds, except that the intensity profile for alloys may have a broader width, as shown in Fig. 1.2 for GaAs, InAs, and the alloys of the two compounds. The lattice parameters a (cube edge length) of these alloys determined from x-ray diffraction are found to be well approximated by the concentration weighted average of those of the constituents, which is usually referred to as Vegard's law (1921),

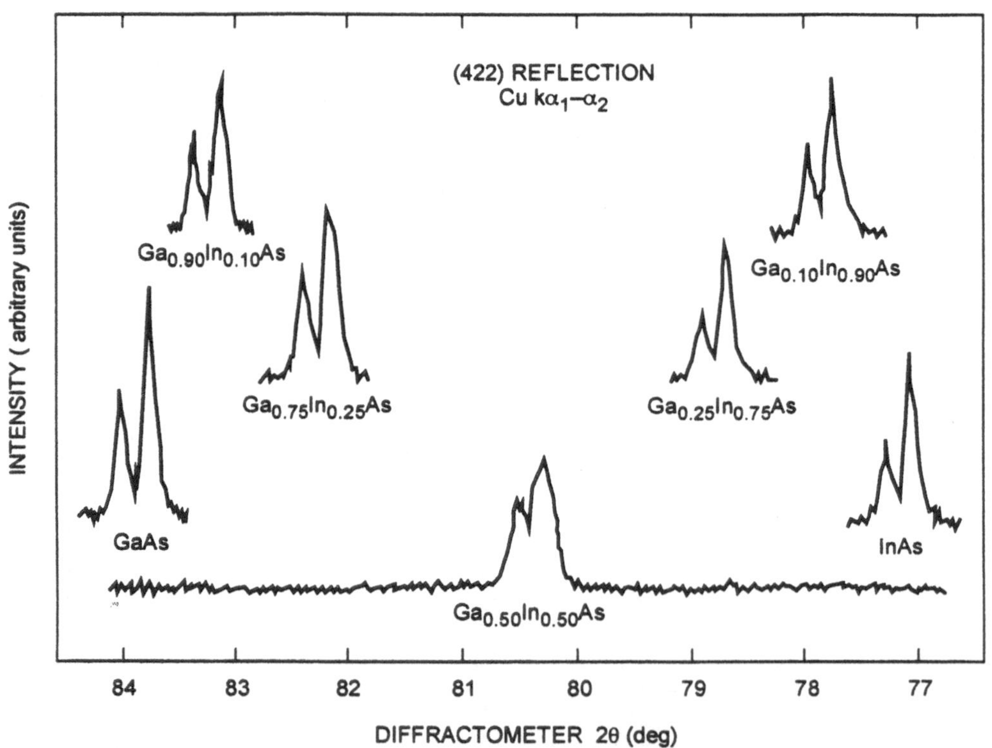

FIGURE 1.2. X-ray diffraction spectro of $Ga_{1-x}In_xAs$ alloys (Mikkelsen and Boyce, 1983).

$$a = (1 - x)a_{AC} + xa_{BC} \tag{1.2.1}$$

Because many properties of pseudobinary alloys are also approximately equal to the average values of the constituent compounds, the virtual crystal approximation (VCA)* has been used frequently to describe these alloys. VCA implicitly assumes that the substituted atoms A and B share one regular fcc sublattice, and the third kind of atoms C sit on the other fcc sublattice in zinc blende pseudobinary alloys. This, if taken literally, implies that the AC and BC bonds have the same bond length. However, if Pauling's covalent radii are correct, the local bond lengths for the bonds AC and BC in the alloy ought to be different and should preserve their respective AC and BC crystal values. Extended x-ray absorption fine-structure spectroscopy measurements (Mikkelsen and Boyce, 1982, 1983) showed that the local bond lengths in pseudobinaries are indeed closer to pure-crystal values than their VCA values.

1.3. ALLOY STRUCTURE DETERMINED BY EXAFS

EXAFS is one of the very few techniques capable of measuring each separate local atomic distance in a material. Normal x-ray diffraction methods are only sensitive to average values. The bond-length information comes from analyzing the fine structure of the frequency dependence of the x-ray transmission above the absorption edge. The results discussed in this section are based on the work of Mikkelsen and Boyce (1983). Figure 1.3a shows the 10.37-keV K-edge absorption spectra of the Ga atom in GaAs as a function of photon energy. Figure 1.3b displays the oscillatory part above the edge after removal of the main absorption structure. The spectra are plotted as a function of an effective electron wave vector $k = \sqrt{2m(\hbar\omega - E_0)}/\hbar$, where $\hbar\omega$ is the photon energy and E_0 is the absorption edge. The oscillation in the spectra comes from the interference between the outgoing wave of the photoexcited core electrons with a small fraction of the electron wave scattered back from nearby atoms (Hayes and Boyes, 1982). A Fourier transform of the spectra in Fig. 1.3b to real space is shown in Fig. 1.3c. It exhibits well-defined peaks located in the vicinity of the first few near-neighbor distances from a Ga atom. The same analysis can also be done on alloys. Figure 1.4 shows a comparison for the real-space Ga K-edge EXAFS spectra in GaAs and two GaInAs alloys, from which the bond-length changes in the alloy can be deduced. After detailed analysis of all the K-edges of Ga, In, and As in their compounds GaAs and InAs, and in their alloys, Mikkelsen and Boyce arrived at the following conclusions:

(1) The individual nearest-neighbor (NN) Ga–As and In–As bond lengths d in the alloy are closer to the pure crystal values than the concentration weighted average value,

$$d(x) = (1 - x)d^0_{AC} + xd^0_{BC} \tag{1.3.1}$$

where d^0_{AC} and d^0_{BC} are, respectively, the bond lengths of the constituent AC and BC compounds. However, the average lattice constant agrees with Vegard's law. Figure 1.5 summarizes the results for the NN bond lengths. If one defines relaxation parameters $\Gamma_{AC}(x)$ and $\Gamma_{BC}(x)$ as

*In the context of band structures, VCA means an approximation that treats the alloy potential as the average of those of the constituent crystals. This concept is addressed in Chapter 5.

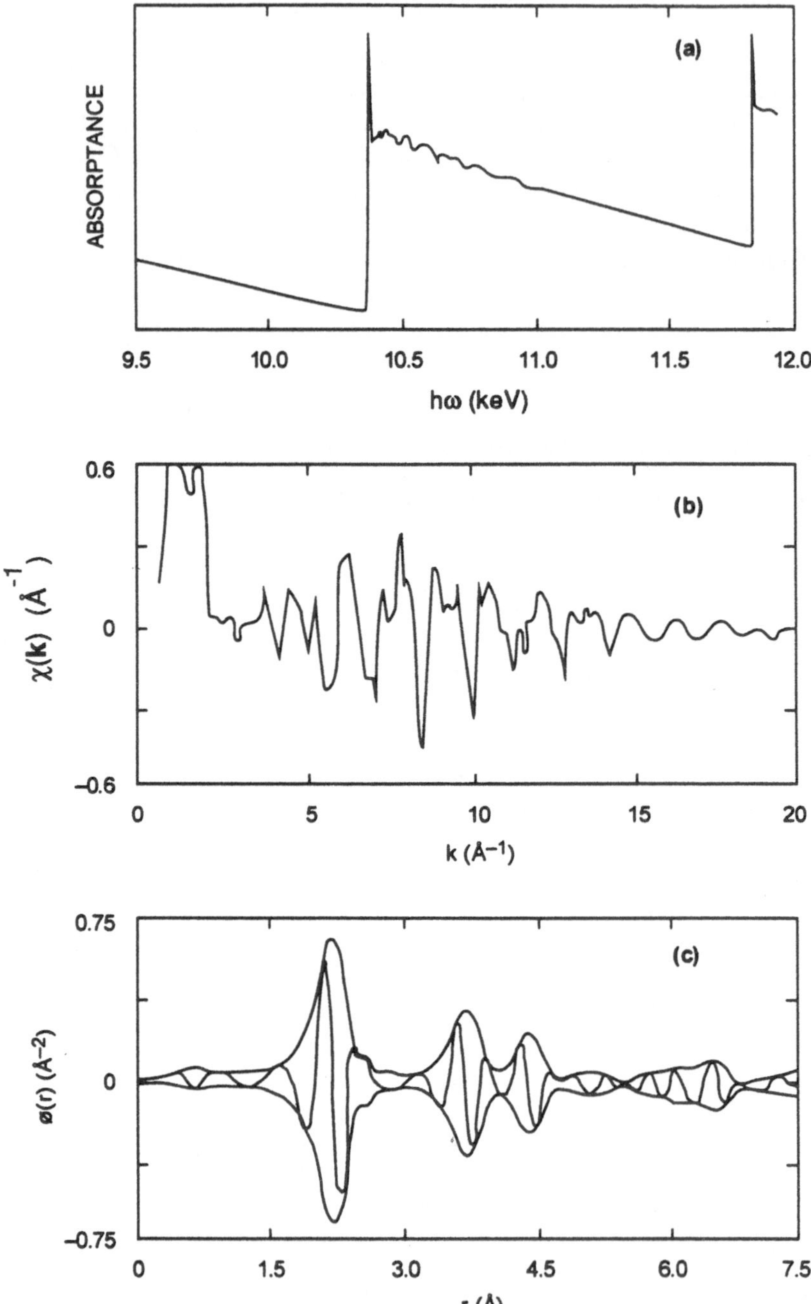

FIGURE 1.3. (a) Absorption as a function of photon energy about the Ga K edge at $\hbar\omega = 10.37$ keV in GaAs at 77 K. (b) Ga K-edge EXAFS oscillations, $\chi(k)$, as a function of k after removal of the background absorption. (c) Fourier transform of (b) to real space. The transform window is 3.76 – 18 Å^{-1}, broadened by a Gaussian of width 0.7Å^{-1} (Mikkelsen and Boyce, 1983).

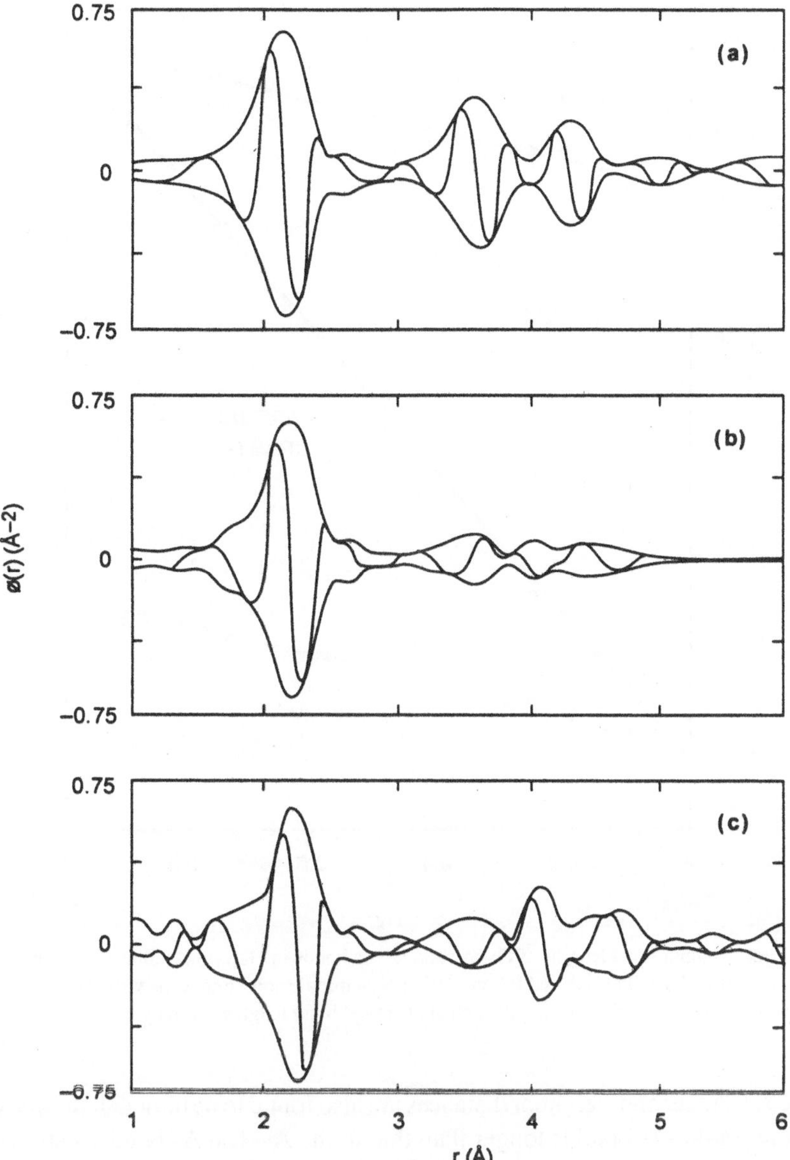

FIGURE 1.4. Ga K-edge EXAFS in real space transformed using a window of $k = 3.56 - 15.65$ Å^{-1} broadened by a Gaussian of width 0.7 Å^{-1} for (a) pure GaAs, (b) $Ga_{0.5}In_{0.5}As$, and (c) $Ga_{0.1}In_{0.9}As$. Note the similarity of the first-neighbor As peak position, width, and amplitude. The second-neighbor peaks, on the other hand, are quite different (Mikkelsen and Boyce, 1982).

$$\Gamma_{AC}(x) = [d_{AC}(x) - \bar{d}(x)]/[d^0_{AC}(0) - \bar{d}(x)] \tag{1.3.2}$$

and similarly for $\Gamma_{BC}(x)$, then Γ is about 75% and the values are nearly the same for all x. In Eq. (1.3.2), $d_{AC}(x)$ and $d_{BC}(x)$ are the NN bond lengths in the alloy and

$$\bar{d} = (1 - x)d_{AC}(x) + xd_{BC}(x) \tag{1.3.3}$$

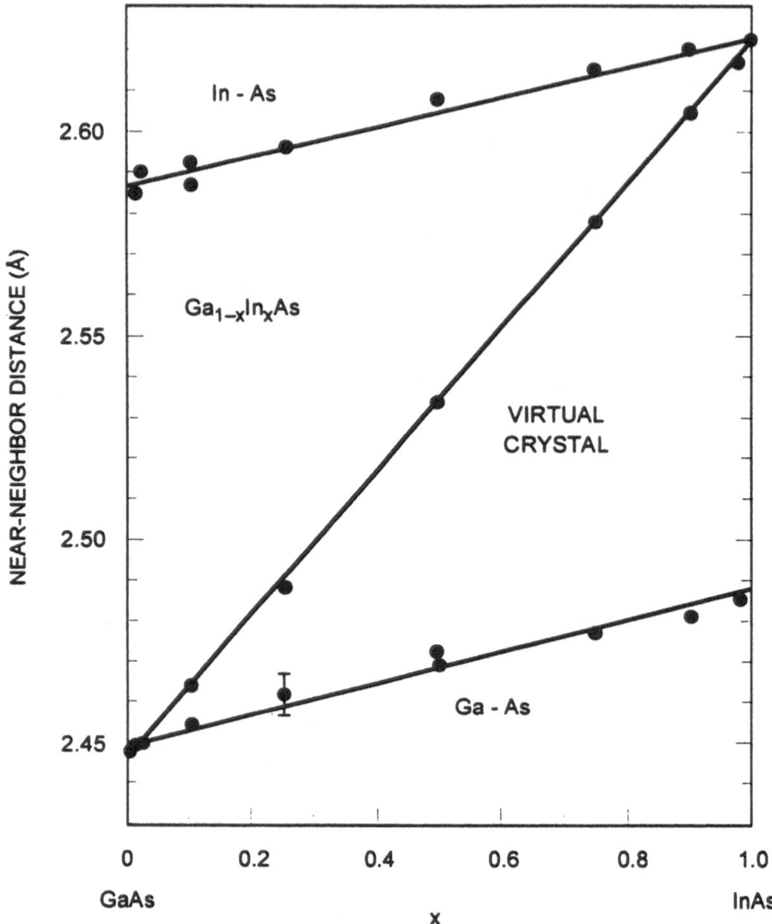

FIGURE 1.5. First-neighbor bond lengths of Ga–As and In–As bonds in $Ga_{1-x}In_xAs$ alloys as a function of alloy concentration x measured by Mikkelsen and Boyce (1982). The middle solid line is the virtual crystal value given by Eq. (1.3.1) and the dots along this line are the actual averaged bond lengths given by Eq. (1.3.3).

(2) The As–As second-neighbor distances are also found to be bimodal: the alloy As–As distance in the As–In–As bond is longer than that in the As–Ga–As bond, as shown in Fig. 1.6. This figure also shows that the As–As bond distances are nearly 100% relaxed; that is, they preserve their pure crystal values.

(3) The Ga–In, Ga–Ga, and In–Ga second-neighbor distances are closer to the VCA values than the pure crystal values, as shown in Fig. 1.7.

EXAFS experiments have been performed on many pseudobinary semiconductor alloys. Table 1.3 summarizes the results for the NN bond lengths in terms of the Γ parameters defined in Eq. (1.3.2). Also listed are the Γ values for the ordered alloys to be considered shortly. Except for $Ga_{1-x}In_xSb$, the measured relaxation parameters for the III–V and II–VI alloys, defined in Eq. (1.3.2), are between 70% and 80%, and relaxations for the A–C and B–C bonds are about equal. For the two I–VII systems listed in Table 1.3, while (I, Br)Cu is not too different from the rest, (Br,Cl) shows a nearly 100% relaxation. The reason for such a large difference is still not known.

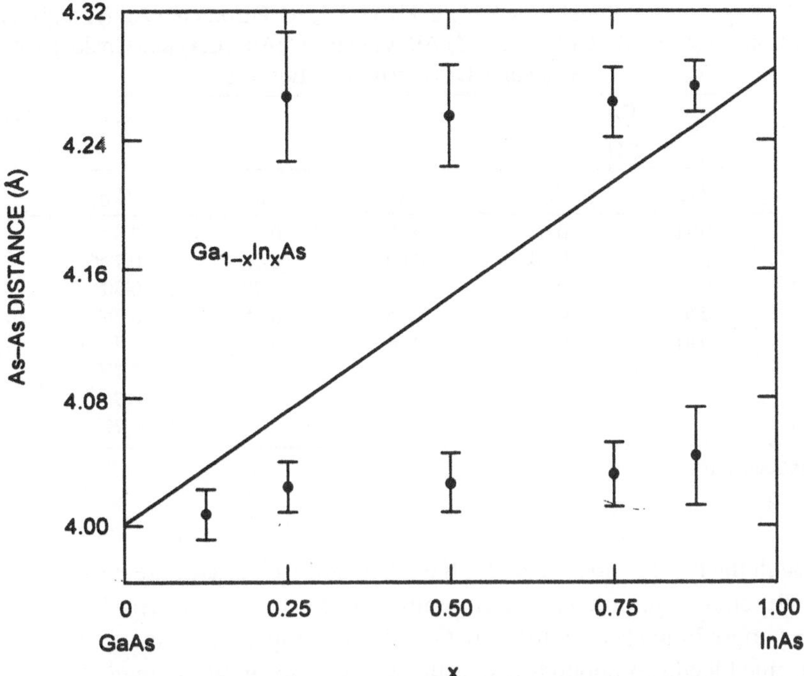

FIGURE 1.6. As–As second-neighbor distance for $Ga_{1-x}In_xAs$ as a function of composition. Two As–As distances are observed, the shorter one corresponding to As–Ga–As bonds and the longer one corresponding to As–In–As bonds. The middle curve is the VCA As–As distance (Mikkelsen and Boyce, 1982).

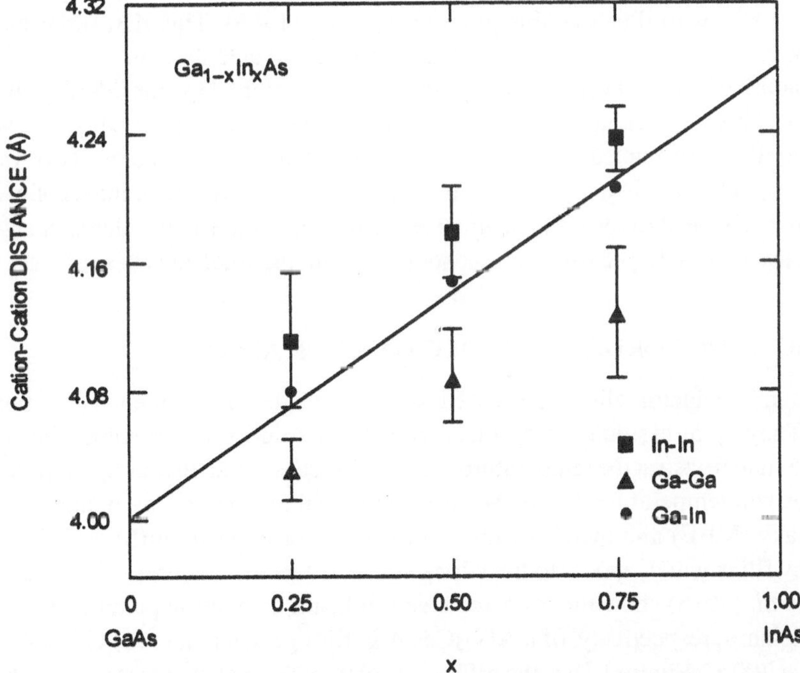

FIGURE 1.7. Ga–Ga, In–In, and Ga–In second-neighbor distance for 25, 50, and 75 mole % GaAs. The cation–cation distances are seen to approach the VCA values, the solid line (Mikkelsen and Boyce, 1982).

TABLE 1.3. Experimental Ratios Γ Defined in Eq. (1.3.2) for Disordered Alloys and Comparison with Calculated Values for Ordered ABC_2 Alloys in the CuAuI (CA) and Chalcopyrite (CH) Structures Discussed in Section 1.4

(A,B)C	CA ETB		CH ETB		Disordered alloys Experiments	
	Γ_{AC}	Γ_{BC}	Γ_{AC}	Γ_{BC}	Γ_{AC}	Γ_{BC}
(In,Ga)Sb	0.71	0.64	0.80	0.76	0.89[a]	0.88[a]
(In,Ga)As	0.71	0.66	0.80	0.78	0.77[a]	0.80[a]
(In,Ga)P	0.74	0.67	0.83	0.79	0.80[a]	0.76[a]
(As,P)Ga	0.59	0.57	0.75	0.75	0.76[a]	0.75[a]
(Hg,Zn)Te	0.81	0.76	0.86	0.84	0.72[b]	0.73[b]
(Te,Se)Zn					0.78[a]	0.80[a]
(Br,Cl)Cu					0.94[a]	0.97[a]
(I,Br)Cu					0.79[a]	0.70[a]

[a]Boyce and Mikkelsen (1987).
[b]Balzarotti (1987).

Although the local bond lengths for the 50–50 alloy discussed are very similar to those found in the long-range ordered chalcopyrite structure to be considered shortly, the x-ray diffraction pattern indicates that the bulk $Ga_{1-x}In_xAs$ alloy has a lattice symmetry similar to that of the zinc blende compounds across the whole concentration range.

These EXAFS results give us a rough picture of the way a bulk semiconductor alloy fits together to fill space; the substituted atoms retain their fcc arrangement with a VCA lattice constant, while the atoms on the other sublattice arrange themselves in their local environment to minimize the strain energy caused by bond distortions. A simple spring model in Problem 1.1 serves to illustrate this point (Shih *et al.*, 1985). This first-order picture still needs to be refined with more accurate measurements and calculations.

The nature of the next order refinements can be envisioned by considering clusters of a C atom and its four near neighbors. There are five types of clusters, those in which the neighboring sites are occupied by four A atoms, three A atoms and one B atom, two A atoms and two B atoms, and so on. In the first-order picture presented above, the volumes of each cluster are the same. The next order refinement takes into account the small volume changes of the different cluster types. Higher-order corrections deal with modifications to still larger clusters.

1.4. LONG-RANGE ORDERED SEMICONDUCTOR ALLOYS

The semiconductor alloys discussed so far have been assumed to be bulk-grown systems. They were assumed to be disordered solid solutions. For some alloys this is a metastable state frozen at the temperature when atoms cease to diffuse as the sample is cooled from its growth temperature. However, recent epitaxial techniques, particularly molecular-beam epitaxy (MBE) and metal-organic chemical vapor deposition (MOCVD), have produced alloy films which appear to have long-range order. For example, Fig. 1.8a,b reported by Kuan *et al.* (1985) show the reflected electron-beam diffraction patterns from the (110) and (001) planes, respectively, of a $Al_{0.75}Ga_{0.25}As$ film grown in a (110) GaAs substrate by MOCVD at 700°C. Figure 1.8a is the diffraction pattern from an incident electron beam (200

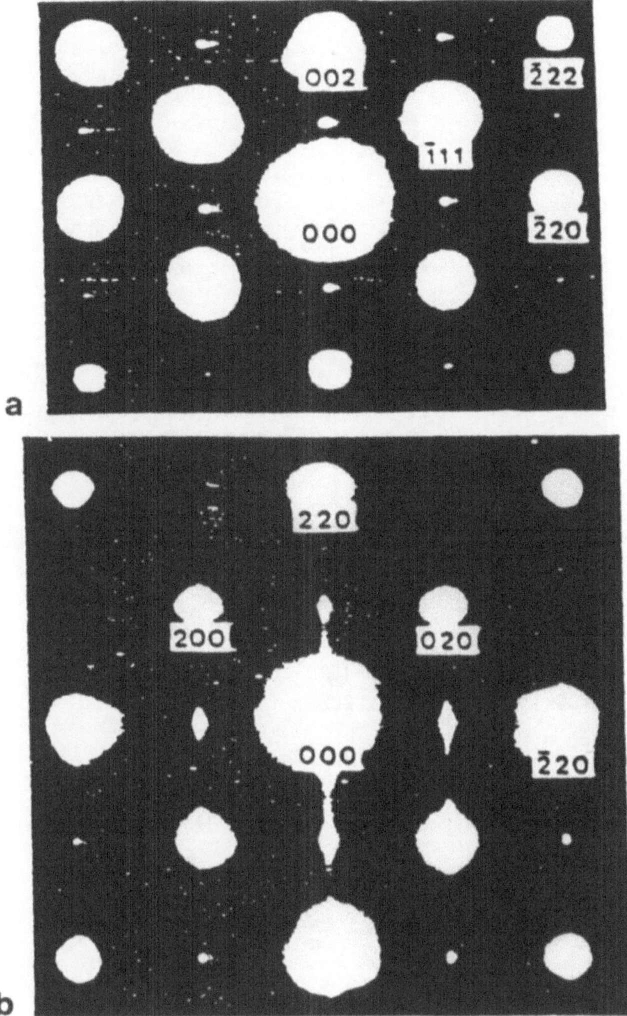

FIGURE 1.8. (a) (110) diffraction pattern and (b) (001) pattern from a $Al_{0.25}Ga_{0.25}As$ film grown on a (110) GaAs substrate by MOCVD at 700°C (Kuan *et al.*, 1985).

keV in energy) which is parallel to the (110) growth direction, and Fig. 1.8b corresponds to the case where the electron beam is perpendicular to (110). If Ga and Al atoms are distributed randomly on the cation sublattice, then the reflection patterns should be that of a zinc blende structure, namely those having the mixed even and odd indices (100), (110), (210), (211), and so on, are not allowed. However, Figs. 1.8a,b show a set of the forbidden indices [e.g., (001), ($\bar{1}$10), ($\bar{1}$12), (110), (130), etc.], indicating that the alloy has some long-range order. These diffraction patterns were mapped into a reciprocal lattice shown in Fig. 1.9a. This reciprocal lattice structure corresponds to an ordered $GaAlAs_2$ structure with Ga atoms sitting on the (0, 0, 0)*a* and (1/2, 1/2, 0)*a* sites and the Al atoms sitting on the (1/2, 0, 1/2)*a* and (0, 1/2, 1/2)*a* sites of each unit cell. This ordered structure consists of alternating AlAs

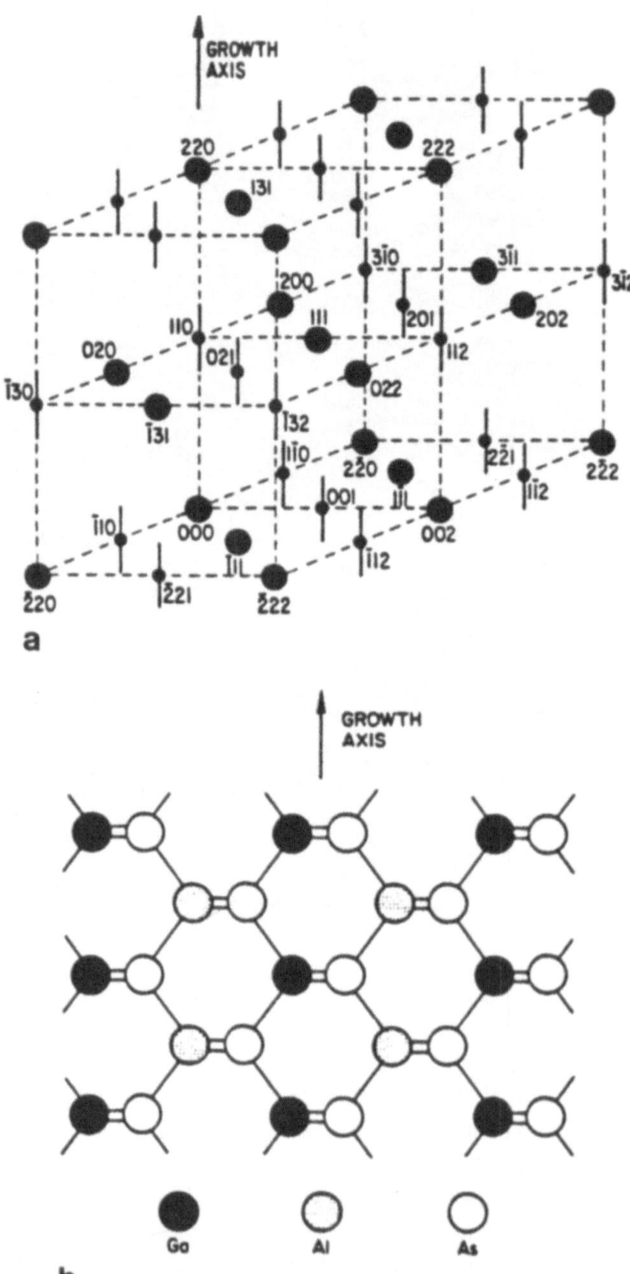

FIGURE 1.9. (a) Reciprocal lattice of an ordered $Al_xGa_{1-x}As$ crystal grown along the (110) direction. Large closed circles represent the Bragg reflections, and small closed circles are superstructure reflections. (b) Schematic diagram of a perfect long-range order in $Al_{0.5}Ga_{0.5}As$. After Kuan *et al.* (1985).

TABLE 1.4. Intensities of Superstructure Reflections of $Ga_{1-x}Al_xAs$ Thin Crystals Grown on GaAs by MOCVD Observed by Kuan *et al.* (1985)

Substrate orientation	x	Growth Temperature			
		600°C	650°C	700°C	800°C
(110)	0.25	—	Weak	—	—
	0.50	Weak	Weak	Medium	Weak
	0.75	Weak	Medium	Strong	Weak
(100)	0.25	—	—	Weak	—
	0.50	—	—	—	—
	0.75	—	—	—	Weak

and GaAs monolayers when viewed along either the (110) growth direction or the (100) direction normal to the growth axis, as depicted in Fig. 1.9b. This type of crystal structure is referred to as the CuAuI (copper–gold–one) structure.

The ordering of these samples is sensitive to the growth conditions. Ordering occurs only in a very narrow temperature range, as noted by Kuan *et al.* (1985). While a minimum temperature is needed for surface atoms to diffuse to reach their lower free-energy state, higher temperatures would tend to drive the system into the disordered phase. Table 1.4 summarizes their findings of the effects of concentration, temperature, and substrate on the intensities of the superstructure, a quantity directly related to the degree of ordering of the samples. The degree of ordering can be defined in terms of the probabilities that certain atoms occupy specified preferred sites in an ordered alloy. Let r_{Ga} be the fraction of the preferred Ga sites that are occupied by Ga atoms, and similarly let r_{Al} be the fraction of the preferred Al sites occupied by Al atoms. Then an order parameter S can be defined as

$$S = r_{Ga} + r_{Al} - 1 \quad (1.4.1)$$

TABLE 1.5 Ordered III–V Semiconductor Alloys Identified Experimentally

Alloys	Structure	Growth method	Substrate	Substrate T (°C)	Reference
$AlGaAs_2$	CA	MOCVD & MBE	GaAs (100) & (110)	600–800	Kuan *et al.* (1985)
$AlInAs_2$	CP	MOCVD	InP (001)	600	Norman *et al.* (1987)
$InGaAs_2$	Famatinite	LPE	InP (110)	630	Nakayama and Fujita (1986)
	Luzonite	MBE	InP (001)	400	Matsui *et al.* (1986)
	CA	MBE	InP (110)	500	Kuan *et al.* (1987)
	CP	VLE	InP (001)	650–660	Shahid *et al.* (1987)
$INAlP_2$	CP	MOCVD	GaAs (001)	650–700	Yasuami *et al.* (1988)
$GaInP_2$	CP	MOCVD	GaAs (001)	650	Gomyo *et al.* (1987, 1988)
	CP	MOCVD	GaAs (001)	640	Bellon *et al.* (1988)
	CP	MOCVD	GaAs (001)	650–700	Yasuami *et al.* (1988)
	CP	MOCVD	GaAs (001)	600–630	Ueda *et al.* (1987)
	CP	MOCVD	GaAs (001)	600–700	Kondow *et al.* (1988)
Ga_2AsSb	CA	MOCVD	InP (100)	550–680	Jen *et al.* (1986)
	CH	MOCVD	InP (100)	600	Jen *et al.* (1986)
	CP	MBE		540	Murgatroyd *et al.* (1985, 1986)

A high value, $S = 1$, corresponds to a perfectly ordered crystal, and a low value, $S = 0$, means that the alloy is disordered. The strongest intensity for the 700°C sample in Table 1.4 was estimated to have an S value of 0.5. The rest of the samples have smaller values, $S \leq 0.3$. These results show that the ordering is far from perfect in these alloys. (It is likely that these samples consist of domains of ordered material surrounded by regions of nearly disordered material, i.e., material in a correlation state with only short-range order.)

There have been several other semiconductor alloys grown by MOCVD and MBE which were found to have similar long-range order. Table 1.5 is a list of several ordered III–V alloys that have been grown, along with the growth conditions and ordered structure. A great majority of these ordered alloys form ABC_2 compounds in one or more of the three crystal structures, CuPt (CP), CuAuI (CA), and chalcopyrite (CH), shown in Fig. 1.10. However, a

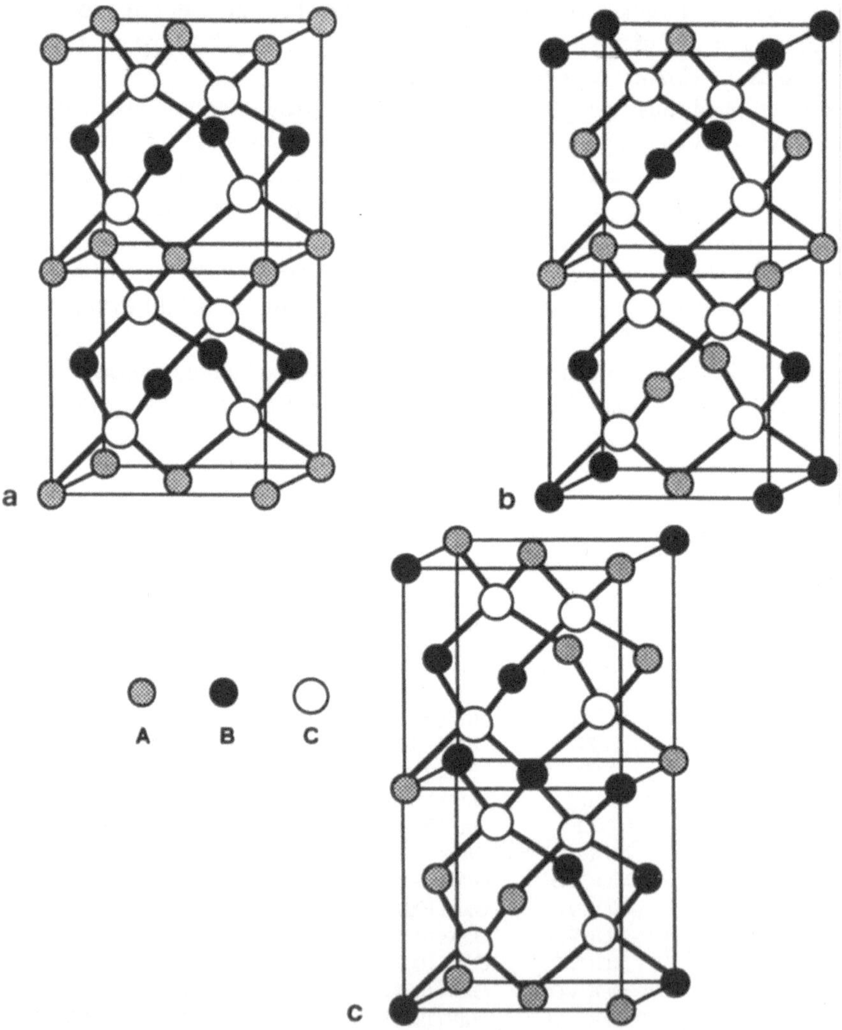

FIGURE 1.10. Three ABC_2 structures: (a) CuAuI ordered in (100) direction, (b) chalcopyrite structure ordered in (201) direction, and (c) CuPt structure ordered in (111) direction.

few alloys are ordered A_3BC_4 compounds in the farmatinite or luzonite structure (Wei and Zunger, 1989). The substrate temperatures for the ordering to occur range from 400 to 800°C, and the ordering directions are not necessarily the same as the growth direction. For example, $AlInAs_2$ (Norman *et al.*, 1987) was found to order in the (111) direction when grown on an InP (110) substrate. There is one difference between the other alloys in the list and $GaAlAs_2$; the others all have appreciable bond-length mismatches. Therefore, mechanisms that drive the systems to order may also be different. Finally, in addition to the III–V alloys mentioned above, some weak ordering has also been seen in epitaxial films of $Si_{1-x}Ge_x$ alloys (Ourmazd and Beam, 1975).

1.4.1. ABC_2 Structures

The actual atomic positions in the ordered ABC_2 alloys deviate slightly from those shown in Fig. 1.10 because internal bond-length relaxation occurs. We summarize the atomic positions in these alloys.

1.4.1.1. CuAuI Structure The ABC_2 semiconductor alloy in the CuAuI-type structure in Fig. 1.10a forms a layered structure ACBCACBC . . . along the (001) direction. After

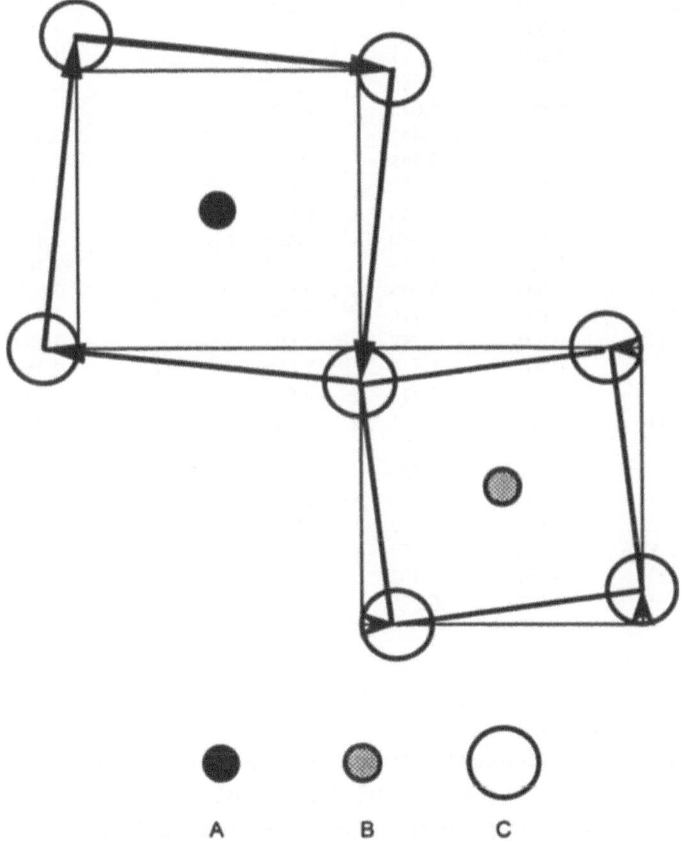

FIGURE 1.11. Atomic arrangement in ABC_2 chalcopyrite structure with internal distortion parameter $\delta \neq 0$, as viewed along the crystal *c*-axis. Note the clockwise and counterclockwise rotations of the C atoms in the *xy* plane around A and B atoms respectively to reach the final positions when the bond length d_{AC} is larger than d_{BC}.

TABLE 1.6. Calculated Equilibrium Bond Lengths d_{AC} and d_{BC} and Equilibrium Average Lattice Constants a (all in Å) for Ordered Alloys in the CuAuI (CA), Chalcopyrite (CH), and CuPt (CP) Structures (In CP structures, the first values d_{AC} and d_{BC} are for those bonds along the (111) direction, and the second values are for those in the other three directions.)

$(A,B)C_2$		c/a	a	d_{AC}		d_{BC}	
$(Al,Ga)As_2$	CA	1.000	5.658	2.450		2.449	
	CH	1.000	5.6566	2.450		2.449	
	CP	0.998	5.660	2.451	2.449	2.444	2.451
$(Al,Ga)P_2$	CA	1.000	5.457	2.365		2.362	
	CH	1.000	5.459	2.366		2.361	
	CP	0.998	5.461	2.371	2.362	2.354	2.366
$(In,Ga)Sb_2$	CA	1.009	6.272	2.781		2.669	
	CH	0.995	6.298	2.788		2.660	
	CP	1.001	6.295	2.788	2.762	2.650	2.699
$(In,Al)As_2$	CA	1.013	5.831	2.597		2.479	
	CH	0.992	5.870	2.604		2.469	
	CP	1.001	5.861	2.599	2.578	2.456	2.512
$(In,Ga)As_2$	CA	1.102	5.834	2.597		2.478	
	CH	0.993	5.866	2.605		2.468	
	CP	1.002	5.859	2.602	2.577	2.457	2.510
$(In,Al)P_2$	CA	1.007	5.658	2.520		2.395	
	CH	0.996	5.674	2.527		2.384	
	CP	1.001	5.677	2.516	2.504	2.381	2.425
$(In,Ga)P_2$	CA	1.013	5.640	2.518		2.390	
	CH	0.992	5.672	2.526		2.379	
	CP	1.003	5.665	2.520	2.499	2.368	2.425
$(As,P)Ga_2$	CA	1.001	5.549	2.430		2.379	
	CH	0.999	5.552	2.437		2.371	
	CP	0.998	5.556	2.450	2.422	2.354	2.390
$(Sb,As)Ga_2$	CA	1.003	5.880	2.601		2.499	
	CH	0.998	5.884	2.613		2.482	
	CP	0.995	5.905	2.637	2.585	2.439	2.531
$(Sb,P)Ga_2$	CA	1.008	5.776	2.586		2.435	
	CH	0.994	5.787	2.603		2.407	
	CP	0.993	5.822	2.635	2.567	2.344	2.486
$(Cd,Hg)Te_2$	CA	1.000	6.471	2.805		2.799	
	CH	1.000	6,471	2.806		2.798	
	CP	0.999	6.473	2.817	2.798	2.788	2.805
$(Hg,Zn)Te_2$	CA	1.010	6.256	2.783		2.656	
	CH	0.995	6.286	2.787		2.650	
	CP	1.003	6.277	2.771	2.770	2.659	2.679
$(Cd,Zn)Te_2$	CA	1.016	6.252	2.790		2.657	
	CH	0.991	6.302	2.795		2.650	
	CP	1.007	6.275	2.787	2.771	2.647	2.685

relaxation, if $d^0_{AC} > d^0_{BC}$, the C layers will move toward B layers to reach the equilibrium positions. The basic lattice vectors can be chosen as $\mathbf{a}_1 = (1/2, 1/2, 0)a$, $\mathbf{a}_2 = (1/2, -1/2, 0)a$, and $\mathbf{a}_3 = (0, 0, \beta)a$, where β is the c/a ratio with a being the lattice constant. The ideal β value is 1. There are four atoms per unit cell: one A atom at (0, 0, 0), one B atom at $(1/2, 0, \beta/2)a$, and two C atoms at $(1, 1, \beta + \delta)a/4$ and $(3, 1, 3\beta - \delta)a/4$, where δ is the internal distortion parameter for the C atoms. There are only two different nearest-neighbor bond lengths in the alloy: $d_{AC} = (a/4)\sqrt{2 + (\beta + \delta)^2}$ and $d_{BC} = (a/4)\sqrt{2 + (\beta - \delta)^2}$.

1.4.1.2. Chalcopyrite Structure The ABC_2 semiconductor alloy in the CH structure (Fig. 1.10b) forms a layered structure ACACBCBCACACBCBC... along the (012) direction. The basic lattice vectors can be chosen as $\mathbf{a}_1 = (1,1,-2\beta)a/2$, $\mathbf{a}_2 = (-1,1,2\beta)a/2$, and $\mathbf{a}_3 = (1,-1,2\beta)a/2$, where β again is the c/a ratio. There are now eight atoms per unit cell: two A atoms at (0,0,0) and $(0,1,\beta)a/2$, two B atoms at $(1,0,\beta)a/2$ and $(1,1,0)a/2$, and four C atoms at $(1+\delta, 1, \beta)a/4$, $(1, 3+\delta, 3\beta)a/4$, $(3, 1-\delta, 3\beta)a/4$, and $(3-\delta, 3, \beta)a/4$, where δ is the internal distortion parameter for the C atoms. If $d^0_{AC} > d^0_{BC}$, the A atoms are displaced clockwise while the B atoms are displaced counterclockwise about the C atoms on the xy plane to reach equilibrium positions as shown in Fig. 1.11. Again there are only two different nearest-neighbor bond lengths in the alloy: $d_{AC} = (a/4)\sqrt{1+(1+\delta)^2+\beta^2}$ and $d_{BC} = (a/4)\sqrt{1+(1-\delta)^2+\beta^2}$.

1.4.1.3. CuPt Structure In the CuPt-type structure (Fig. 1.10c), the alloy forms a (111) superlattice ACBCACBC. . . . Because of lack of reflection symmetry about any of these planes, the B layer does not need to be located exactly midway between the two successive A layers. Also the distance between two closest atoms from two different A layers need not correlate with that between two A atoms on the same plane. Thus, there are a total of five independent parameters required to describe the crystal structure: the lateral lattice constant a for the layers, the spacing D between two successive A layers, and the three spacing parameters for the three layers (one B and two C) inside D. These relaxations can be described in terms of the small displacement parameters Δ, δ_1, δ_2, and δ shown in Fig. 1.12. The basic lattice vectors can be chosen as $\mathbf{a}_1 = (0, -1, 1+2\Delta/3)a/2$, $\mathbf{a}_2 = (-1, 0, 1+2\Delta/3)a/2$, and $\mathbf{a}_3 = (1, 1, 2+2\Delta/3)a/2$. There are four atoms per unit cell: one A atom at (0,0,0), one B atom at $(\Delta/3+\delta/4, 1/2+\Delta/3+\delta/4, 1/2+\Delta/3+\delta/4)a$, and two C atoms at $(-1+\Delta/3+\delta_1, 1+\Delta/3+\delta_1, 1+\Delta/3+\delta_1)a/4$, and $(-1-\Delta-\delta_2, -1-\Delta-\delta_2, -1-\Delta-\delta_2)a/4$, each atom coming from a different layer. There are four different bond lengths in the crystal, two for the AC and two for the BC bonds, given by

$$d^S_{AC} = (a/4)\sqrt{3}(1+\Delta+\delta_2)$$

$$d^T_{AC} = (a/4)\sqrt{(1-\Delta/3-\delta_1)^2+2(1+\Delta/3+\delta_1)^2}$$

$$d^S_{BC} = (a/4)\sqrt{3}(1+\Delta-\delta_1+\delta)$$

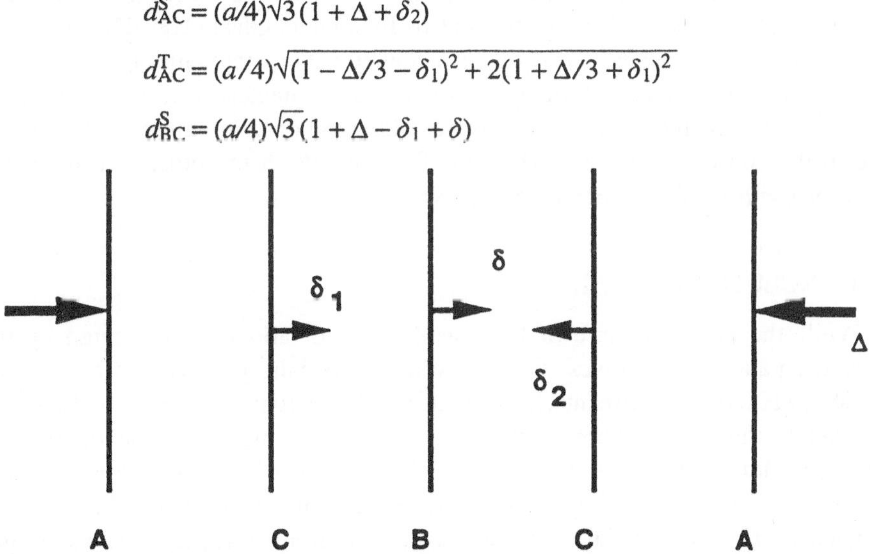

FIGURE 1.12. Schematic diagram indicating the displacement parameters δ_1, δ_2, δ, and Δ for the CuPt layer structures ACBCA. . . along the (111) direction.

$$d^{T}_{BC} = (a/4)\sqrt{(1 - \Delta/3 + \delta_2 + \delta)^2 + 2(1 + \Delta/3 - \delta_2 - \delta)^2}$$

where the superscript S means singlet and T means triplet bonds.

1.4.2. Bond Lengths

There has not been a detailed measurement performed on the bond lengths of these ordered alloys. The results from a detailed calculation (Yeh *et al.*, 1991) to be discussed in Chapter 3 are listed in Table 1.6. The values listed are the c/a ratios and the first-neighbor bond lengths for a collection of ABC_2 alloys in the three structures mentioned above. These results are useful for checking against future experiments such as EXAFS that may be performed on these alloys. The c/a ratios are nearly unity for all alloys tabulated. The notation in the table chooses the equilibrium bond lengths of the constituent crystals in the order $d^0_{AC} > d^0_{BC}$. The bond lengths in the CuAuI (CA) and chalcopyrite (CH) structures are in general bimodal, with $d_{AC} > d_{BC}$. This result is similar to that found in disordered bulk alloys. There are four different bond lengths in the CuPt (CP) structure: d^T_{AC}, d^S_{AC}, d^T_{BC}, and d^S_{BC} where again the superscripts T and S mean the triplet and singlet bonds, respectively. However, the alloy lattice constants a are quite close to the mean value $\bar{a}$ of the constituent compounds. A more sensitive measure of the bimodal distribution is the relaxation parameter $\Gamma_{AC} = (d_{AC} - \bar{d})/(d^0_{AC} - \bar{d})$ and a similar expression for Γ_{BC}, as was done in Section 1.3 for disordered alloys. The calculated Γ ratios in the CA and CH structures are compared with experimental values for the disordered bulk alloys in Table 1.3. Note that a value of $\Gamma = 1$ corresponds to the totally relaxed case where there is no bond stretching, whereas $\Gamma = 0$ corresponds to a rigid virtual crystal where all the atoms are on the zinc blende crystal sites. The values of Γ in Table 1.3 show that CH structures are more relaxed than the CA structures. The calculated values of Γ for $GaInAs_2$ and $HgZnTe_2$ do correlate well with the experimental results. While all the calculated values of Γ in this table range from 0.6 to 0.9 for the lattice-mismatched alloys, this trend does not hold for the lattice-matched alloys. For example, Table 1.5 shows that Hg–Te and Cd–Te bond lengths in $HgCdTe_2$ in both CH and CA structures nearly retain their respective constituent crystal values (i.e., $\Gamma \cong 1$). Finally, the bond length for the CuPt-type (CP) structure are characteristically different from those in CA and CH structures. The singlet bonds along the ordering direction (111) tend to be close to the constituent values, while the triple bonds in the other directions have less relaxation (with Γ values around 0.5 or less).

1.5. CONCLUDING REMARKS

While the gross structure of bulk semiconductor alloys, as measured by the x-ray diffraction patterns, resembles virtual (average) crystals, the detailed bond lengths, as revealed by EXAFS experiments, have significant deviations from the virtual crystal values. For example, the first-neighbor bond lengths in an alloy exhibit a bimodal distribution when the two constituent compounds have a significant difference in their lattice constants. A rough view of an $A_{1-x}B_xC$ bulk alloy is one where the substituted atoms A and B occupy (nearly at random) perfect fcc sublattice sites while the C atoms assume positions that minimize the local strain energy. Refined structures still await the next generation of experiments. They also can be determined theoretically using the bonding models of Chapters 2 and 3 in a large-scale molecular dynamics or Monte Carlo simulation. A detailed treatment of the atomic distribution in disordered semiconductor alloys is discussed in Chapter 4.

While bulk semiconductor alloys are disordered, many epitaxially grown alloys exhibit a partial long-range ordering into CuAuI, chalcopyrite, CuPt, or more complicated structures. The bulk alloy bond lengths, when compared with those calculated for the ordered alloys, are closer to the chalcopyrite than the CuAuI or the CuPt structure. However, the real structure of a semiconductor alloy is neither totally disordered nor perfectly ordered and, to a great extent, is influenced by growth and annealing processes. In the absence of understanding the real mechanism driving ordering, particularly in lattice-matched alloys, a more systematic correlation of the states of ordering to growth conditions would help control the growth of ordered alloys.

REFERENCES

Balzarotti, A. (1987), in *Ternary and Multinary Compounds—Proceedings of the 7th International Conference on Ternary and Multinary Compounds*, eds. S.K. Deb and A. Zunger (Materials Research Society, Pittsburgh, PA), p. 333.

Bellon, P., J.P. Chevalier, G.P. Martin, E. Dupont-Nivet, C. Thiebaut, and J.P. Andre (1988), *Appl. Phys. Lett.* **52**, 567.

Berger, L.I., and V.D. Prochukhan (1969), *Ternary Diamond-Like Semiconductors* (Consultants Bureau, New York).

Boyce, J.B., and J.C. Mikkelsen, Jr., (1987), in *Ternary and Multinary Compounds—Proceedings of the 7th International Conference on Ternary and Multinary Compounds*, eds. S.K. Deb and A. Zunger (Materials Research Society, Pittsburgh, PA), p. 333.

Deb, S., and A. Zunger (1987), *Ternary and Multinary Compounds—Proceedings of the 7th International Conference on Ternary and Multinary Compounds* (Materials Research Society, Pittsburgh, PA).

Gomyo, A., T. Suzuki, K. Kobayashi, S. Kawata, I. Hino, and T. Yuasa (1987), *Appl. Phys. Lett.* **50**, 673.

Gomyo, A., T. Suzuki, and S. Iijima (1988), *Phys. Rev. Lett.* **60**, 2645.

Harrison, W.A., (1980), *Electronic Structure and Properties of Solids* (W.H. Freeman, San Francisco).

Hayes, T.M., and J.B. Boyce (1982), *Solid State Phys.* **37**, 173.

Kondow, M., H. Kakibayashi, and S. Minagawa (1988), *J. Cryst. Growth* **88**, 291.

Jen, H.R., M.D. Cherng, and G.B. Stringfellow (1986), *Appl. Phys. Lett.* **48**, 1603.

Kuan, T.S., T.F. Keuch, W.I. Wang, and E.L. Wilkie (1985), *Phys. Rev. Lett.* **54**, 201.

Kuan, T.S., W.I. Wang, and E.L. Wilkie (1987), *Appl. Phys. Lett.* **51**, 51.

Matsui, Y., H. Hayashi, and K. Yoshida (1986), *Appl. Phys. Lett.* **48**, 1060.

Mikkelsen, J.C., Jr., and J.B. Boyce (1982), *Phys. Rev. Lett.* **49**, 1412.

Mikkelsen, J.C., Jr., and J.B. Boyce, (1983), *Phys. Rev. B***28**, 7130.

Murgatroyd, I.J., A.G. Norman, and G.R. Booker (1985), Paper presented at Inst. Phys. Solid State Phys. Conf. Univ. Reading.

Murgatroyd, I.J., A.G. Norman, and G.R. Booker (1986), in *Materials Characterization Symposium, April 15–17*, MRS Symposia Proc. vol. 69, eds. N. Cheung, M.A. Nicolet, Palo Alto, CA.

Nakayama, H., and H. Fujita (1986), in *Proceedings of the 12th International Symposium on GaAs and Related Compounds*, 1985, Inst. Phys. Conf. Ser. vol. 79, ed. M. Fujimoto (IOP, London), p. 289.

Norman, A.G., R.E. Mallard, I.J. Murgatroyd, G.R. Booker, A.H. Moore, and M.D. Scott (1987), in *Microscopy of Semiconducting Materials*, Inst. Phys. Conf. Ser. vol. 87, eds. A.G. Cullis and P.A. Adjustus (IOP, London), p. 77.

Ourmazd, A., and J.C. Bean (1985), *Phys. Rev. Lett.* **55**, 762.

Pauling, L. (1960), *The Nature of the Chemical Bond* (Cornell University Press, Ithaca, NY).

Phillips, J.C. (1973), *Bonds and Bands in Semiconductors* (Academic Press, New York).

Shahid, A., S. Mahajan, D.E. Laughlin, and H.M. Cox (1987), *Phys. Rev. Lett.* **58**, 2567. [Note that these authors used the "vapor levitation epitaxy" (VLE) method.]

Shih, K., W.E. Spicer, W.A. Harrison, and A. Sher (1985), *Phys. Rev. B* **31**, 1139.

Sze, S.M. (1981), *Physics of Semiconductor Devices* (Wiley, New York).

Ueda, O., M. Takikawa, J. Komeno, and I. Umebu (1987), *Jpn. J. Appl. Phys.* **26**, L1824.
Wei, S.-H., and A. Zunger (1989), *Phys. Rev. B* **39**, 3279.
Yasuame, S., C. Nozaki, and Y. Ohba (1988), *Appl. Phys. Lett.* **52**, 2031.
Yeh, C.-Y., A.-B. Chen, and A. Sher (1991), *Phys. Rev. B* **43**, 9138.
Zallen, R. (1982), in *Handbook on Semiconductors*, ed. T.S. Moss (North-Holland, Amsterdam), Chap. 1, Vol. 1.

2

Bonding in Ordered Structures

This chapter deals with the binding energies of semiconductors and their ordered alloys. Disordered alloys will be treated in Chapter 4. The cohesive energy and alloy excess energy will be defined and their values presented. Several important theoretical models for the total energy calculation, including density functional theory, the tight-binding approach, and Harrison's bond-orbital model, will be introduced. The concept of polarity will be discussed within the tight-binding approach and compared with the Phillips (1973) ionicity factor. Finally, we will show that the excess energies for long-range ordered alloys are mostly positive, implying they are nonequilibrium states formed during growth.

2.1. COHESIVE ENERGY IN THE BORN–OPPENHEIMER ADIABATIC APPROXIMATION

The energy (Hamiltonian) of a solid consists of five parts: the kinetic energies of the electrons and ions, K_e and K_I, respectively; the ion–ion potential energies, U_{II}; the electron–electron potential energies, U_{ee}; and electron–ion potential energies, U_{eI}. The standard way to treat this Hamiltonian is to use the Born–Oppenheimer (1927) adiabatic approximation. In this approximation, one freezes the ionic motion (because the speed of ions in a solid is much smaller than that of the electrons) in a set of ionic positions $\{R\}$. Then the many-body Schrödinger equation is solved for the part of the Hamiltonian that involves the set of electronic coordinates $\{r_i\}$:

$$H\psi_\gamma(\{r_i\}, \{R\}) = E_\gamma(\{R\})\psi_\gamma(\{r_i\}, \{R\}) \tag{2.1.1}$$

where $H = K_e + U_{ee} + U_{eI}$. The eigenenergy E_γ is a function of the ionic positions $\{R\}$. The lowest-energy curve of the sum of E_γ and U_{II} is the ground state curve E_g, which serves as the potential energy for the ionic motion. A Taylor expansion of E_g about its minimum value E_0, and retaining only terms quadratic in atomic displacements η from their equilibrated positions produces the expression

$$E_g(\{R\}) \cong E_0 + \frac{1}{2} \sum_{m\alpha} \sum_{n\beta} A_{nm}^{\alpha\beta} \eta_{n\alpha} \eta_{m\beta} \tag{2.1.2}$$

In Eq. (2.1.2) the minimum energy E_0 corresponds to the value in the equilibrium ionic configuration $\{R^0\}$, $\eta_{n\alpha}$ is the α-component of the small displacement $\mathbf{R}_n - \mathbf{R}_n^0$ of the nth ion, and $A_{nm}^{\alpha\beta}$ are second derivatives of E_g with respect to the components of the ionic coordinates evaluated at the equilibrium positions. A canonical transform of the second term leads to the normal-mode vibrations for the lattice—the phonons. Finally, the total crystal energy E_T at zero temperature is E_0 plus the minimum lattice vibrational energy, i.e. the sum of zero-point energies of the phonons. The difference between the average free-atom energy and E_T per atom in a solid is the cohesive energy.

The ability to compute E_T accurately allows determination of the structure or the phase a system will have at zero temperature. The dynamic matrix $A_{nm}^{\alpha\beta}$ also enables us to calculate the elastic and lattice vibrational properties.

2.2. DENSITY FUNCTIONAL THEORY

Equation (2.1.1) is computationally the most difficult part of the problem because it deals with on the order of 10^{23} electrons that are interacting with each other and with the ions and have to obey Pauli's exclusion principle. Self-consistent density functional theory (DFT) (Hohenberg and Kohn, 1964; Kohn and Sham, 1965) represents the simplest and most effective approximation to treat this difficulty.

In DFT the ground state energy of a solid is assumed to be

$$E_g = K_e + U_{ee} + U_{eI} + U_{xc} + U_{II} \tag{2.2.1}$$

where U_{xc} denotes the many-body electron–electron exchange and correlation energies. The kinetic energy K_e is the sum of the single-particle kinetic energies of the occupied states ϕ_v:

$$K_e = \sum_v \int \phi_v^*(\mathbf{r}) \frac{p^2}{2m} \phi_v(\mathbf{r})\, d^3r \tag{2.2.2}$$

The terms U_{ee}, U_{eI}, and U_{xc} in Eq. (2.2.1) are functionals of the electronic density ρ:

$$\rho(\mathbf{r}) = \sum_v \phi_v^*(\mathbf{r})\phi_v(\mathbf{r}) \tag{2.2.3}$$

They are given by

$$U_{ee} = \frac{1}{2}e^2 \int\int \frac{\rho(\mathbf{r})\rho(\mathbf{r}')}{|\mathbf{r}-\mathbf{r}'|}\, d^3r\, d^3r' \tag{2.2.4}$$

$$U_{eI} = -e^2 \sum_n Z_n \int \frac{\rho(\mathbf{r})}{|\mathbf{r}-\mathbf{R}_n|}\, d^3r \tag{2.2.5}$$

$$U_{xc} = \int \rho(\mathbf{r})\, \varepsilon_{xc}[\rho(\mathbf{r})]\, d^3r \tag{2.2.6}$$

where Z_n is the ionic charge. The exchange correlation energy $\varepsilon_{xc}[\rho]$ as a function of ρ is taken to be the same as that derived for a uniform electron density. $\varepsilon_{xc} = \varepsilon_x + \varepsilon_c$ contains the exchange energy, $\varepsilon_x \propto \rho^{1/3}$, and the correlation energy ε_c contains the remaining many-body corrections. Several approximate forms of ε_c have been derived (e.g., Hedin and Lundquist, 1971; Ceperley and Alder, 1980; Perdew and Zunger, 1981). The use of this form of ε_{xc} is called the local density approximation (LDA). It has proven to be remarkably accurate in its predictions of solids' total energies.

To find the ground state energy E_g, a variational principle can be used to minimize E_g with respect to the wave functions ϕ_v under the constraint that the total number of electrons in the solid is a constant:

$$\delta\left[E_g - \varepsilon \sum_{v} \int |\phi_v(r)|^2 \, d^3r\right] = 0 \tag{2.2.7}$$

where ε is a Lagrange multiplier. Equation (2.2.7) leads to a single-particle Schrödinger equation for ϕ:

$$\left(\frac{p^2}{2m} + V\right)\phi_v = \varepsilon_v \phi_v \tag{2.2.8}$$

The effective one-electron potential contains three parts:

$$V = V_{ee} + V_{eI} + V_{xc} \tag{2.2.9}$$

where

$$V_{ee}(\mathbf{r}) = e^2 \int \frac{\rho(\mathbf{r}')}{|\mathbf{r} - \mathbf{r}'|} \, d^3r' \tag{2.2.10}$$

$$V_{eI} = -e^2 \sum_{n} \frac{Z_n}{|\mathbf{r} - \mathbf{R}_n|} \tag{2.2.11}$$

$$V_{xc} = \frac{\partial}{\partial \rho}(\rho \varepsilon_{xc}) = \varepsilon_{xc} + \rho \frac{\partial}{\partial \rho} \varepsilon_{xc} \tag{2.2.12}$$

Equations (2.2.10) through (2.2.12) and Eq. (2.2.3) form a repeated loop, $\rho \rightarrow V \rightarrow \phi \rightarrow \rho$, which must be iterated to achieve self-consistency. Once the self-consistency is established, the ground state energy E_g of the solid can be calculated from Eqs. (2.2.1) through (2.2.6). It is often useful to write E_g as a band structure term E_{bs} plus a repulsive term U_R, where E_{bs} is the sum of the single-electron eigenenergies ε_v in Eq. (2.2.8) of the occupied states and U_R is the remainder:

$$\begin{aligned} E_g &= E_{bs} + U_R \\ &= \sum_{v} \varepsilon_v + \left[-U_{ee} + U_{II} + \int (\varepsilon_{xc} - V_{xc})\rho \, d^3r\right] \end{aligned} \tag{2.2.13}$$

From here the determination of E_T in LDA involves heavy computations. Various properties predicted by LDA will be discussed here and in Chapters 3 and 4. The underlying band theory will be considered in Chapter 5.

2.3. BONDS AND BANDS FROM LOCAL DENSITY FUNCTIONAL THEORY

When atoms form into crystals, the outer electrons of the atoms (the valence electrons) redistribute in space to minimize the total free energy. There are on average eight valence

FIGURE 2.1. Experimental and calculated valence charge density in the (110) plane of Si (in units of electrons per cell) (Zunger, 1980).

electrons per atomic pair in the tetrahedral semiconductors (e.g., zinc blende and wurtzite): in group IV elementary semiconductors each atom in the pair contributes four electrons, in the III–V compound systems three electrons come from the cation and five from the anion, and in II–VI compounds two come from the cation and six from the anion, etc. It is evident from both experimental results and the LDA calculations as shown in Fig. 2.1 (Zunger, 1980) that the valence electron density in the group IV semiconductors tends to concentrate along the line connecting the nearest-neighboring atoms by covalent bonds, with a symmetrical peak at the center of the bond. For the III–V, II–VI, and even I–VII compounds, the electron charge distribution along the bond becomes asymmetric, with a tendency for electrons to transfer from the cation to the lower-energy anion site as shown in Fig. 2.2 (Wang and Klein, 1981). This is the reason why the elementary semiconductors Si and Ge are called homopolar semiconductors and the compound semiconductors are called polar semiconductors.

In a polar semiconductor, there is no unique way of assigning charge to each atom, but there are some logical ways to proceed. Once a decomposition method is formally established, the resulting partially ionic bond becomes a convenient extensive unit for many physical quantities. The concepts of polarity (Harrison, 1980) and ionicity (Phillips, 1973),

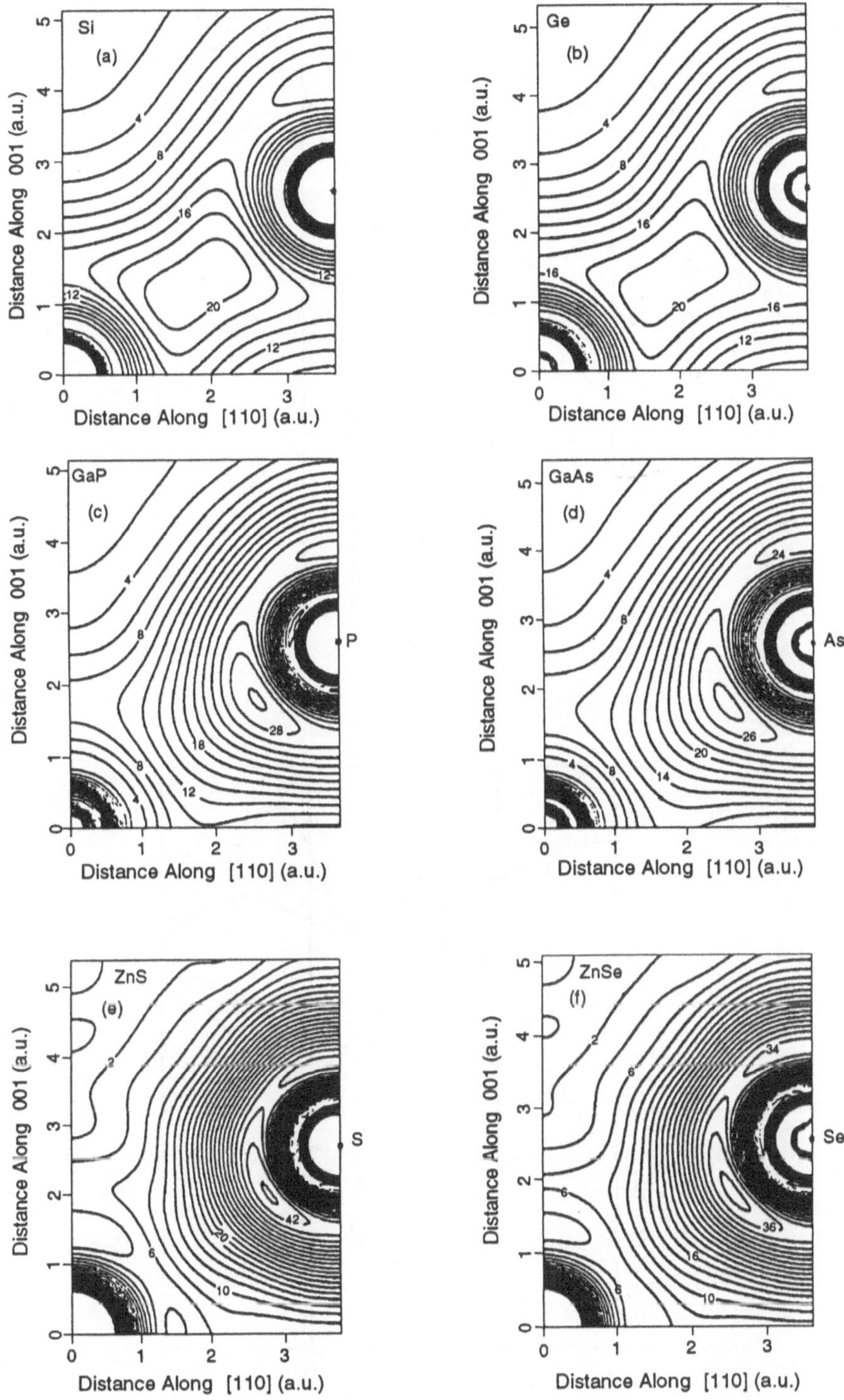

FIGURE 2.2. Comparison between the valence charge densities of diamond and zinc blende semiconductors in the $(1\bar{1}0)$ plane calculated from the SCDFT (Wang and Klein, 1981).

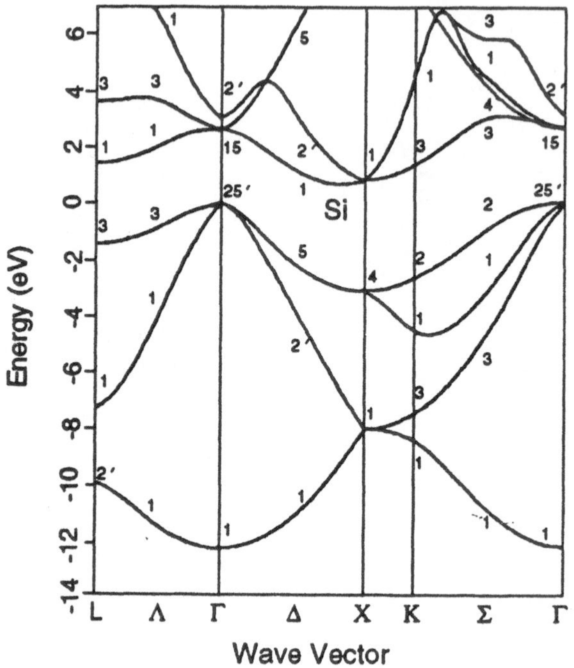

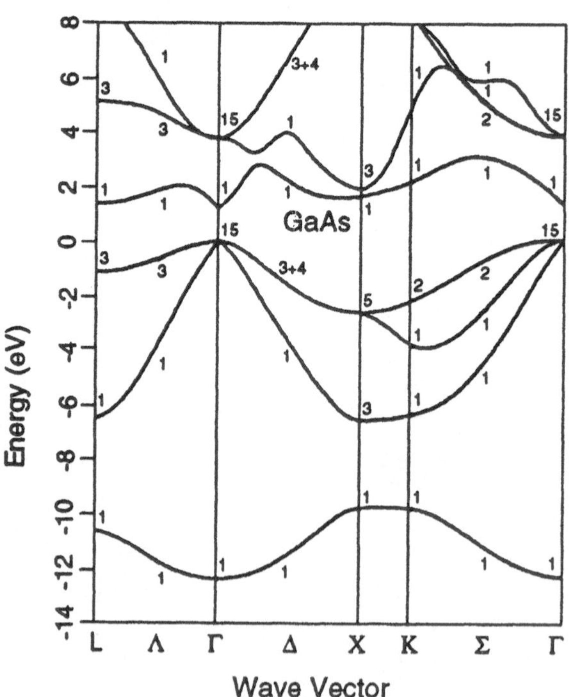

FIGURE 2.3. Band structures of Si and GaAs calculated from SCDFT by Wang and Klein (1981).

factors to be considered in Section 2.6, represent special ways to accomplish this charge decomposition.

The actual calculational procedure in DFT is quite involved, particularly in the solution of the single particle eigenvalue problem in Eq. (2.2.8). In fact in its most general form it has never been solved, but it has been done in the LDA. However, for crystalline solids (including semiconductors), by taking advantage of crystal periodicity in the form of the Bloch theorem, techniques have almost been perfected so that uncertainties in the calculated results can now be attributed to basic assumptions in the LDA rather than to numerical errors. Figure 2.3 shows the band structures obtained in an LDA calculation for Si and GaAs, respectively (Wang and Klein, 1981). These are plots of the eigenvalues of Eq. (2.2.8) as a function of their quantum numbers, which are the wave vectors **k** inside the first fcc Brillouin zone. Note that although these bands resemble the band structures normally used in the interpretation of optical transitions, they should not be used for such purposes. Remember that the eigenenergies here are merely the Lagrange multiplier in the total energy minimization, so they should not be treated as the excitation energies. The energy gaps calculated with DFT in LDA are usually smaller than the experimental values. However, when corrections in the many-body excitation self-energy are made using the so-called GW approximation (Hybertsen and Louie, 1985), the band gaps are properly predicted. The cohesive energies calculated from LDA are in general several tenths of an electron volt larger than the experimental values, primarily due to errors in the predicted ground state energies of free atoms. However, very accurate results have been obtained for the lattice constants and the elastic constants (see Tables 3.1, 3.2, and 3.4) in addition to the charge distribution already presented.

2.4. TIGHT-BINDING APPROACH

Because of calculational complexity in self-consistent density functional theory, semiempirical tight-binding approaches provide a convenient way of calculating structural properties and gaining insight into the underlying basic physics. In such an approach, short-range interactions involving only several neighbors are used to represent both the band structure energy contribution and the repulsive term in the total energy expression in Eq. (2.2.13). For a crystalline semiconductor, the total energy may be rewritten as

$$E_{\mathrm{T}} = E_{\mathrm{bs}} + U_{\mathrm{R}} = \sum_{\gamma}\sum_{k} \varepsilon_{\mathrm{v}}(\mathbf{k}) + \sum\sum_{i>j} u_{ij} \tag{2.4.1}$$

where the band energies $\varepsilon_{\mathrm{v}}(\mathbf{k})$ are obtained from a tight-binding (TB) Hamiltonian which contains only the term values of the atoms and a handful of interaction parameters between orbitals of the nearest-neighboring atoms. Similarly, the repulsive energies, u_{ij}, are assumed to involve only the nearest pairs (Chadi, 1979).

The simplest TB Hamiltonian contains only the s and p atomic term values ε_s and ε_p for both cations and anions, and the nearest-neighbor two-center interactions $V_{ss\sigma}$, $V_{sp\sigma}$, $V_{pp\sigma}$, and $V_{pp\pi}$. These two-center interactions are schematically sketched in Fig. 2.4 and are discussed in detail by Harrison (1980). To be more explicit, the 8×8 **k**-dependent Hamiltonian for a zinc blende or diamond semiconductor contains the term values as the

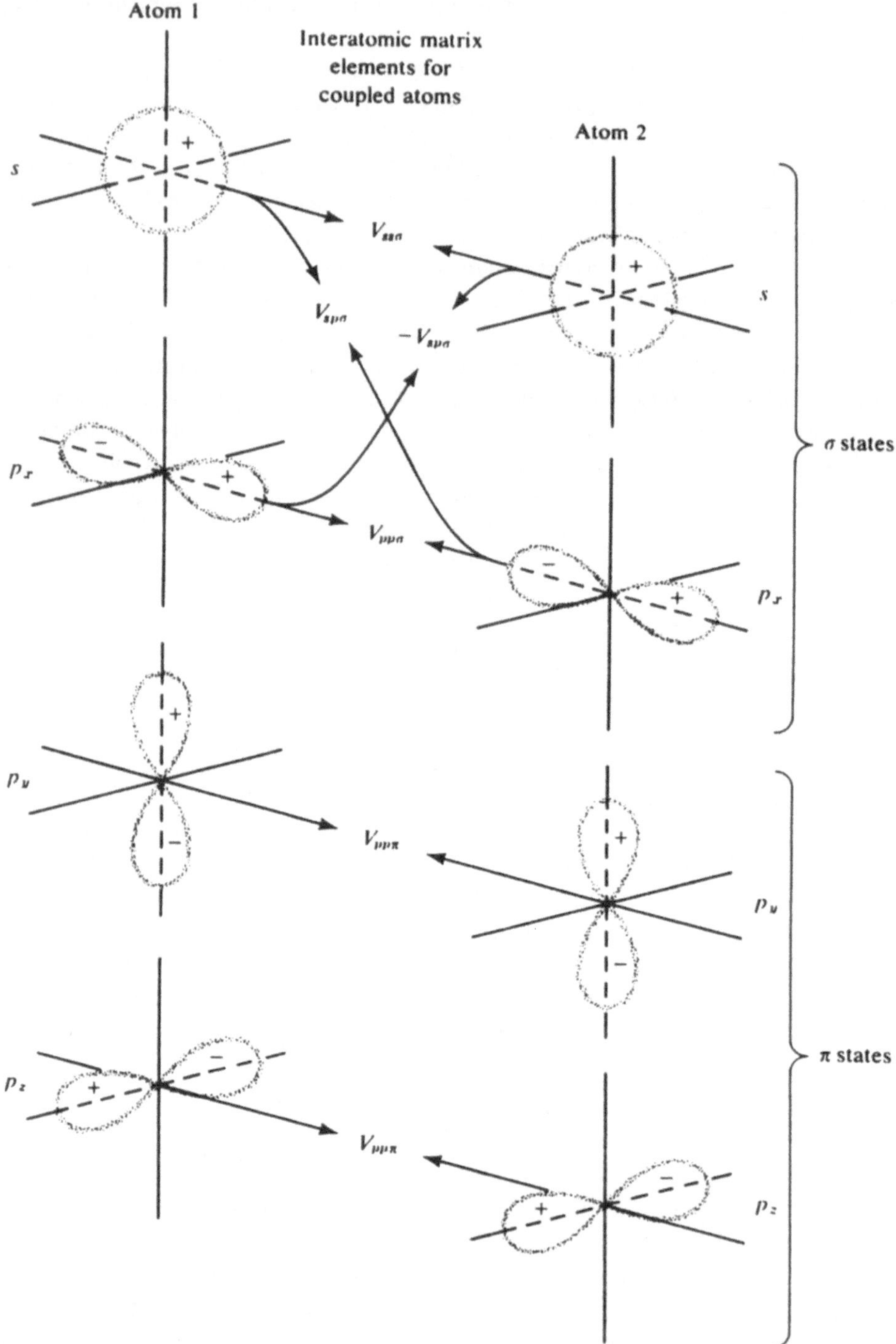

FIGURE 2.4. The coupling of atomic orbitals in lithium-row diatomic molecules, and the resultant bond disignations (at right) (Harrison, 1980).

diagonal matrix elements, while the off-diagonal matrix elements between the cation and anion orbitals are given by*

$$H_{\gamma\gamma'}(\mathbf{k}) = \sum_{d} e^{i\mathbf{k}\cdot\mathbf{d}} h_{\gamma\gamma'}(\mathbf{d}) \tag{2.4.2}$$

where the sum runs over the four first-neighbor atoms specified by the nearest-neighbor bond vector **d**. The γ's are the orbital indices standing for s, p_x, p_y, and p_z. The local TB matrix elements $h_{\gamma\gamma'}$ are related to the two-center interactions by the Slater–Koster (1954) relations:

$$h_{ss} = V_{ss\sigma} \tag{2.4.3}$$

$$h_{sx} = \xi_1 V_{sp\sigma} \tag{2.4.4}$$

$$h_{xx} = \xi_1^2 V_{pp\sigma} + (1 - \xi_1) V_{pp\pi} \tag{2.4.5}$$

$$h_{xy} = \xi_1 \xi_2 (V_{pp\sigma} - V_{pp\pi}) \tag{2.4.6}$$

where the ξ_i's are the directional cosines of **d**, and the two-center interactions are only functions of the bond length d.

Once the values of these TB parameters and their dependence on the bond length are known, the Hamiltonian at each **k** inside the Brillouin zone (BZ) can be evaluated and the summation over **k** carried out to obtain the band structure energy, which, when added to the repulsive energy, gives the total energy of any specified geometry. In this manner, it can also be effectively used to calculate various quantities, such as elastic constants, by examining how E_T changes under prescribed distortions. Conversely, one can adjust the interaction parameters so that these important structural properties are produced correctly for the pure compounds. Then these adjusted parameters can be used to interpolate the properties of alloys. In actual applications, the TB calculation is either carried out using full band structures in quantitative studies or approximated by simpler local theory, such as Harrison's bond-orbital model (BOM) for comparative studies ranging across different systems.

2.5. THE BOND-ORBITAL MODEL

A simple model of bonding in semiconductors based on the TB approach has been proposed by Harrison (1980, 1983). This model contains all the essential features of semiconductor bonding: covalent, polar and metallization contributions. It also provides an excellent picture and language that assist understanding of the underlying physics, thereby lending insight into many properties. Although the results of the original model are often only semiquantitative, the trends are correct and therefore can be trusted in comparative studies. The essence of the model is reviewed here. Details can be found in Harrison (1980) and elsewhere (1983).

The model is based on a tight-binding description. The basis functions for the valence electrons are the atomic orbitals: each atom is assigned one s-type, $|s\rangle$, and three p-type, $|p_x\rangle$, $|p_y\rangle$, and $|p_z\rangle$, orbitals. The only Hamiltonian matrix elements retained in the model are the atomic term values ε_s and ε_p of each atom and the nearest-neighbor interactions. Harrison

*An explicit expression for this 8×8 Hamiltonian can be found in Harrison (1980, p. 77).

used a set of universal TB parameters that apply to all the semiconductors. Based on a comparison with the free-electron bandwidth (Froyen and Harrison, 1979) and with empirical TB parameters (Chadi, 1979), Harrison (1983) deduced the following set of universal two-center interactions:

$$V_{\alpha\alpha'} = \eta_{\alpha\alpha'} \hbar^2/md^2 \tag{2.5.1}$$

with $\eta_{ss\sigma} = -1.32$, $\eta_{sp\sigma} = 1.42$, $\eta_{pp\sigma} = 2.22$, and $\eta_{pp\pi} = -0.63$, where m is the free-electron mass and d is the bond length. Again we have used the traditional Slater–Koster (1954) notation, and the connections between the two-subscript and three-subscript notation in made in Fig. 2.4. In units where d is in Å and V in eV, $V_{\alpha\alpha'} = 7.62\eta_{\alpha\alpha'}/d^2$.

The total energy of the crystal per bond is calculated in the following four steps, with Fig. 2.5 depicting the interactions and energy levels involved.

Step 1. Construct from the s- and p-orbitals four hybrid orbitals, $|h\rangle$, directed toward the neighboring atoms. For example, the hybrid orbital with its principal lobe in the (111) direction is given by

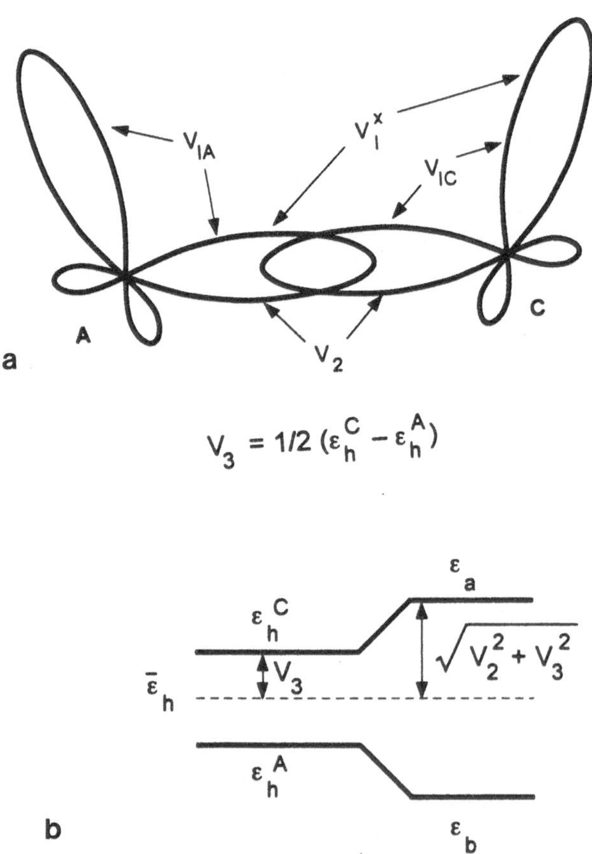

FIGURE 2.5. Schematic diagrams for (a) sp^3 hybrid orbitals and their interaction parameters, and (b) energy levels of the hybrids and the molecular bonding and antibonding states.

$$|h\rangle = (|s\rangle + |p_x\rangle + |p_y\rangle + |p_z\rangle)/2 \tag{2.5.2}$$

For a hybrid in the (1 −1 −1) direction the signs in front of $|p_y\rangle$ and $|p_z\rangle$ become negative. The hybrid energy is then given by $\varepsilon_h = (\varepsilon_s + 3\varepsilon_p)/4$, the weighted average of the s and p energies. Note that the hybrid energy for an anion $\varepsilon_h^{\rm A}$ is different from that for a cation $\varepsilon_h^{\rm C}$.

Step 2. Construct the bonding and antibonding molecular orbitals, $|b\rangle$ and $|a\rangle$ respectively, from the two hybrids directed toward each other along the same bond:

$$|b\rangle = U_b^{\rm A}|h^{\rm A}\rangle + U_b^{\rm C}|h^{\rm C}\rangle$$

$$|a\rangle = U_a^{\rm A}|h^{\rm A}\rangle + U_a^{\rm C}|h^{\rm C}\rangle \tag{2.5.3}$$

where A and C denote the anion and cation, respectively. The 2×2 molecular Hamiltonian in the basis $|h^{\rm A}\rangle$ and $|h^{\rm C}\rangle$ contains the two hybrid energies as the diagonal elements and the interaction V_2, called the covalent energy, as the off-diagonal elements. V_2 can be evaluated by expressing the hybrid orbitals in terms of the atomic orbitals as in Eq. (2.5.2), using the expressions for the various matrix elements in Eqs. (2.4.3) through (2.4.6) and corresponding values of $V_{\alpha\alpha'}$ in Eq. (2.5.1). The result is $V_2 = -24.5/d^2$. The diagonalization of this 2×2 matrix yields the bonding and antibonding energies $\varepsilon_{\rm b}$ and $\varepsilon_{\rm a}$, respectively:

$$\varepsilon_{\left\{\begin{smallmatrix}a\\b\end{smallmatrix}\right\}} = \bar{\varepsilon}_{\rm h} \pm \sqrt{V_2^2 + V_3^2} \tag{2.5.4}$$

where the minus sign is for $\varepsilon_{\rm b}$. In Eq. (2.5.4), $\bar{\varepsilon}_{\rm h}$ is the average hybrid energy $\bar{\varepsilon}_{\rm h} = (\varepsilon_{\rm h}^{\rm C} + \varepsilon_{\rm h}^{\rm A})/2$ and V_3 is given by $V_3 = (\varepsilon_{\rm h}^{\rm C} - \varepsilon_{\rm h}^{\rm A})/2$, which is called the polar energy by Harrison. The probability amplitudes in Eq. (2.5.3) are given by

$$U_{\rm b}^{\rm A} = \sqrt{(1+\alpha_{\rm p})/2} = -U_{\rm a}^{\rm C}$$

$$U_{\rm b}^{\rm C} = \sqrt{(1-\alpha_{\rm p})/2} = U_{\rm a}^{\rm A} \tag{2.5.5}$$

where $\alpha_{\rm p}$, called the polarity, is given by

$$\alpha_{\rm p} = V_3/\sqrt{V_2^2 + V_3^2} \tag{2.5.6}$$

The polarity, varying between the limits $0 \le \alpha_{\rm p} \le 1$, is a measure of the difference in the fractions of electrons occupying the anion and cation sites. When $\alpha_{\rm p} = 0$, the bonds are completely covalent, and when $\alpha_{\rm p} = 1$, the bonds are completely ionic.

Step 3. Compute the correction to the bond energy in a solid. Expanded in the basis of bonding and antibonding states, the solid Hamiltonian has no matrix element between the bonding and antibonding states of the same bond, but interbond matrix elements, $\langle b|H|a'\rangle$, $\langle b|H|b'\rangle$, and $\langle a|H|a'\rangle$, between neighboring bonds are not zero. Here the prime indicates the states of different bonds. The $\langle b|H|b'\rangle$ terms spread the bonding levels into the valence bands, but do not change the center of gravity of the occupied states, which is $\varepsilon_{\rm b}$. However, the bonding–antibonding interactions result in a repulsion between them, thereby lowering the average energy of the occupied states. This can be handled with perturbation theory, which yields a shift in $\varepsilon_{\rm b}$ by

$$\Delta\varepsilon_{\rm b} = \sum_{a'} \frac{|\langle b|H|a'\rangle|^2}{\varepsilon_{\rm b} - \varepsilon_{\rm a'}} \tag{2.5.7}$$

Harrison calls this term the metallization energy because it is a measure of the delocalization energy of the valence electrons responsible for the bonding of metals. If only interactions between the nearest bonds are retained, the matrix elements $\langle b|H|a'\rangle$ become $U_b^A U_a^A V_{1A}$ or $U_b^C U_a^C V_{1C}$, depending on whether $|b\rangle$ and $|a'\rangle$ join through a common anion or cation. The quantities V_{1C} and V_{1A} are called the metallic energies and are given by

$$V_1 = (\varepsilon_s - \varepsilon_p)/4 \tag{2.5.8}$$

Step 4. Include the repulsive energy term. The electronic energy per bond (for two electrons) $E_b^0 = 2(\varepsilon_b + \Delta\varepsilon_b)$ obtained so far in this model is negative, and its magnitude increases as d decreases. To prevent a crystal from collapsing, there must be a repulsive term u_0 added. Harrison argues that this repulsive term arises mainly from the fact that the atomic orbitals on different sites are not orthogonal. He therefore related u_0 to the overlap integral between two hybrids that form the bond, which can be expressed in terms of the average hybrid energy ε_h and V_2 by $1.41V_2^2/|\varepsilon_h|$. Thus $u_0 = C/d^4$, with C being a constant. In practice, C is adjusted so that the net bond energy has its minimum energy at the equilibrium bond length. The bond energy (B.E.) is then the difference between the energy per bond in the solid and the average energy per two electrons in the free atoms:

$$\text{B.E.} = 2\varepsilon_b + 2\Delta\varepsilon_b + u_0 - 2\bar{\varepsilon}_{\text{atom}} \tag{2.5.9}$$

Note that the cohesive energy per bond ε_{CH} is –B.E., if the zero-point vibrational energy is neglected.

TABLE 2.1. The s- and p-State Term Values (in eV) Deduced from Extraction Energies of Neutral Atoms[a]

	Be	B	C	N	O	F
	9.320	14.003	19.814	26.081	28.551	36.229
	5.412	8.300	11.260	14.540	13.613	17.484
	5.412	8.300	11.260	14.540	13.610	17.420
	Mg	Al	Si	P	S	Cl
	7.640	11.780	15.027	19.620	21.163	25.812
	3.926	5.980	8.150	10.610	10.449	13.136
	3.926	5.980	8.150	10.550	10.360	13.010
Cu	Zn	Ga	Ge	As	Se	Br
7.720	9.390	13.230	16.390	20.015	21.412	24.949
2.991	4.237	6.000	7.880	10.146	11.188	12.353
2.965	4.011	5.850	7.694	9.810	9.750	11.840
Ag	Cd	In	Sn	Sb	Te	I
7.570	8.990	12.032	14.525	17.560	19.120	21.631
3.647	4.313	5.780	7.340	9.391	9.951	11.470
3.487	4.097	5.453	6.879	8.640	9.010	10.450
Au	Hg	Tl	Pb	Bi		
9.220	10.430	13.048	15.250	19.949		
4.349	4.998	6.110	7.410	10.407		
3.688	4.031	4.741	5.979	7.290		

[a]These energies are adjusted to have the correct first ionization energies. The top entry is the s-state, the second the $p_{1/2}$-state, and the third the $p_{3/2}$-state.

TABLE 2.2. Various Energies Entering the Bond Model, the Polarities and the Calculated and Experimental Bonding Energies (in eV)

	V_1^C	V_1^A	V_2	V_3	α_p	U_0	B.E.	exp
C	−2.14	−2.14	−10.33	0.0	0.0	10.0	−7.05	−3.68
Si	−1.72	−1.72	−4.44	0.0	0.0	3.94	−2.50	−2.32
Ge	−2.16	−2.16	−4.12	0.0	0.0	3.26	−2.35	−1.94
Sn	−1.80	−1.80	−3.12	0.0	0.0	2.35	−1.86	−1.56
SiC	−1.72	−2.14	−6.93	1.76	0.25	6.41	−4.77	−3.17
BN	−1.42	−2.91	−9.94	3.82	0.36	9.08	−6.99	−3.34
BP	−1.42	−2.26	−6.31	1.55	0.24	5.81	−3.72	−2.52
AlN	−1.45	−2.89	−6.86	5.00	0.59	5.56	−5.54	−2.88
AlP	−1.45	−2.26	−4.40	2.70	0.52	3.68	−2.55	−2.13
AlAs	−1.45	−2.52	−4.15	2.51	0.52	3.46	−2.24	−1.89
AlSb	−1.45	−2.20	−3.46	1.77	0.45	2.88	−1.61	−1.76
GaN	−1.83	−2.89	−6.51	4.85	0.60	5.25	−4.80	−2.24
GaP	−1.83	−2.26	−4.40	2.55	0.50	3.69	−2.15	−1.78
GaAs	−1.83	−2.52	−4.08	2.36	0.50	3.39	−1.84	−1.63
GaSb	−1.83	−2.20	−3.49	1.62	0.42	2.86	−1.40	−1.48
InN	−1.62	−2.89	−5.30	5.12	0.70	3.94	−4.63	−1.93
InP	−1.56	−2.26	−3.80	2.74	0.59	3.09	−2.05	−1.74
InAs	−1.56	−2.52	−3.60	2.56	0.58	2.94	−1.76	−1.55
InSb	−1.56	−2.20	−3.10	1.81	0.50	2.55	−1.27	−1.40
BeO	−0.98	−3.74	−9.00	5.48	0.52	7.61	−5.42	−3.06
ZnO	−1.32	−3.74	−6.25	5.96	0.69	4.67	−3.50	−1.89
ZnS	−1.32	−2.69	−4.47	3.83	0.65	3.49	−1.79	−1.59
ZnSe	−1.32	−2.88	−4.08	3.68	0.67	3.16	−1.48	−1.29
ZnTe	−1.32	−2.45	−3.52	3.18	0.67	2.73	−1.04	−1.20
CdS	−1.20	−2.69	−3.83	3.85	0.71	2.85	−1.60	−1.42
CdSe	−1.20	−2.88	−3.54	3.70	0.72	2.64	−1.34	−1.21
CdTe	−1.20	−2.45	−3.10	3.20	0.72	2.33	−0.96	−1.10
HgS	−1.52	−2.69	−3.80	3.61	0.69	2.91	−1.06	−1.02
HgSe	−1.52	−2.88	−3.52	3.45	0.70	2.70	−0.80	−0.85
HgTe	−1.52	−2.45	−3.12	2.95	0.69	2.43	−0.48	−0.82
CuCl	−1.19	−3.19	−4.47	6.04	0.80	2.85	−1.84	−1.58
CuBr	−1.19	−3.23	−3.95	5.55	0.81	2.51	−1.53	−1.45
CuI	−1.19	−2.79	−3.57	4.72	0.80	2.35	−1.24	−1.33
AgI	−1.01	−2.79	−3.12	4.52	0.82	1.98	−1.17	−1.18

All the terms can now be grouped to yield an explicit expression for B.E. It takes the following simple form for a homopolar semiconductor:

$$\text{B.E.} = V_2(1 - \alpha_m + 9\alpha_m^2/16) \tag{2.5.10}$$

where α_m is called the metallicity and is defined as $2V_1/V_2$. The expression for a polar semiconductor is slightly more complicated:

$$\text{B.E.} = 2\bar{\varepsilon}_h - 2\bar{\varepsilon}_{\text{atom}} - 2\sqrt{V_2^2 + V_3^2}\left[1 - \frac{1}{2}\alpha_c^2 + \frac{9}{16}\alpha_c^4 \frac{V_{1C}^2 + V_{1A}^2}{V_2^2 + V_3^2}\right] \tag{2.5.11}$$

where $\alpha_c = \sqrt{1 - \alpha_p^2}$ is called the covalency, and V_{1C} and V_{1A} are the metallic energies for the cation and the anion, respectively.

To calculate the above quantities, one needs to input the atomic term values. Table 2.1 lists a set of these values obtained from a total-energy calculation based on so-called norm-conserved pseudopotentials (Hamann *et al.*, 1979; Bachelet *et al.*, 1982). These values have been rigidly shifted to give the correct first atomic ionization energies. The *p*-term values include the spin-orbit interaction. The nonrelativistic value is taken to be the weighted average $\varepsilon_p = (\varepsilon_{p1/2} + 2\varepsilon_{p3/2})/3$. Although most of the values in the table are comparable to the Hartree–Fock term values used in Harrison's (1980) original table, there are some differences. The discrepancy is mainly due to the scalar relativistic shift included in Table 2.1.

Table 2.2 lists the values of interaction parameters, V's, and calculated α_p, u_0, and B.E., and the experimental values for the B.E. of many semiconductors. We note that this simple model produces reasonable results for most systems except for those with small and large bond lengths. One can improve the quantitative agreement by improving the scaling rules for the covalent energy V_2 and the effective repulsive energy u_0. Examples of such applications will be considered in Section 2.7 and in Chapter 3.

2.6. POLARITY AND IONICITY

As mentioned earlier, there is no unique way to allocate charges between the anion and cation sites in a semiconductor. However, the concept of polarity that serves as a measure of this division is a useful one, and the polarity can be defined unambiguously in a tight-binding representation with an orthonormal basis set. Let the valence-band electron eigenfunctions in a semiconductor be expanded in these local orbitals:

$$\psi_v(\mathbf{r}) = \sum_n \sum_\alpha C^v_{n\alpha}\phi_\alpha(\mathbf{r} - \mathbf{R}_n) \tag{2.6.1}$$

where $\phi_\alpha(\mathbf{r} - \mathbf{R}_n)$ is an orbital of the α-symmetry located at site n specified by the position vector $\mathbf{R}_n$. Then $\Sigma_\alpha |C^v_{n\alpha}|^2$ is the probability that an electron resides on site n when it is in state v. By summing up this probability for all the occupied states, the total number of electrons residing on site n can be computed as

$$Z(n) = \sum_v \sum_\alpha |C^v_{n\alpha}|^2 \tag{2.6.2}$$

In this way the number of electrons on an anion site Z(A) and on a cation site Z(C) can be defined. Then a logical definition of polarity (Berding *et al.*, 1987) is

$$\alpha_p = [Z(A) - Z(C)]/[Z(A) + Z(C)] \tag{2.6.3}$$

If the interactions among bonds such as the metallization effect in Eq. (2.5.7) are neglected in the bond-orbital model, Z(C) is simply the sum of the probability contributions from the four bonds attached to a cation; i.e., $Z(C) = 8|U^C_b|^2 = 4(1 - \alpha_p)$ and similarly $Z(A) = 4(1 + \alpha_p)$, with α_p given by Eq. (2.5.6). A more precise α_p can be computed by diagonalizing the tight-binding Hamiltonian $H(\mathbf{k})$ of Eq. (2.4.2) and then calculating Z(A) and Z(C) from Eq. (2.6.2). In bond-orbital terms this band structure polarity not only contains the metallization corrections but all long-range terms as well. The values of polarity obtained from this band structure calculation are denoted α_p(BS) and are compared with BOM values α_p(BOM) in Table 2.3. The BOM always overestimates the polarity.

TABLE 2.3. Comparison of Polarities in Semiconductor Compounds[a]

Compound	α_p (BOM)	α_p (BS)	f_p	α_p (exp)
AlP	0.52	0.44	0.31	0.33
GaP	0.50	0.40	0.33	0.31
InP	0.60	0.48	0.42	0.32
AlAs	0.52	0.41	0.27	0.31
GaAs	0.50	0.37	0.31	0.30
InAs	0.59	0.44	0.36	0.30
AlSb	0.46	0.34	0.25	0.30
GaSb	0.43	0.30	0.26	0.29
InSb	0.53	0.37	0.32	0.29
ZnS	0.65	0.56	0.62	0.60
CdS	0.71	0.61	0.69	
HgS	0.69	0.57	0.79	
ZnSe	0.67	0.56	0.63	0.59
CdSe	0.72	0.61	0.70	
HgSe	0.70	0.55	0.68	
ZnTe	0.67	0.55	0.61	0.57
CdTe	0.72	0.59	0.72	0.58
HgTe	0.69	0.53	0.65	0.53

[a]All symbols are defined in the text except α_p (exp), which are the experimentally deduced values by Falter *et al.* (1984).

It should be noted that the polarity defined in Eq. (2.6.2) is what Coulson *et al.* (1962) and Falter *et al.* (1984) refer to as ionicity. However, the often-referenced Phillips (1973) spectroscopic ionicity f_p is defined differently:

$$f_p = C^2/(C^2 + E_h^2) \tag{2.6.4}$$

where E_h and C are respectively the symmetrical and antisymmetrical combination of the atomic pseudopotentials of the constituent atoms. Qualitatively C behaves like V_3 and E_h like V_2 when compared to Harrison's BOM, so Phillips's ionicity resembles the square of polarity; that is,

$$f_p \cong \alpha_p^2 \tag{2.6.5}$$

Table 2.3 shows that α_p(BS) is smaller than α_p(BOM) even though the same Hamiltonian is used. α_p(BS) compares reasonably well with the experimental values for the II–VI systems deduced by Falter *et al.*, but are consistently larger than the experimental values for the III–V compounds. Although Phillips's f_p values are very close to the experimental numbers, this resemblance is fortuitous.

In summary, while polarity and ionicity are both parameters intended to quantify the charge transferred from cations to anions in compound semiconductors, they are not identical, nor even simply related. This has been a source of confusion in the literature. This section is intended to clarify this point.

2.7. EXCESS ENERGIES OF ORDERED ALLOYS

It is noteworthy that the long-range ordered semiconductor alloys mentioned in Section 1.5 are found only in epitaxially grown samples. It is therefore interesting to ask whether these alloys are in their thermal equilibrium state at the growth temperature. Although the answer to this question requires detailed examination of the free energies of different phases as a function of temperature, which will be discussed in Chapter 4, the essence of the matter can be deduced if one knows the energetics at zero temperature. The quantity to be examined is the excess energy ΔE. We consider the three most important structures for the ordered ABC_2 systems. The excess energy per four atoms may be defined as

$$\Delta E = E(ABC_2) - E(AC) - E(BC) \tag{2.7.1}$$

where $E(ABC_2)$ is the energy per ABC_2 formula unit in the alloy, and E(AB) and E(BC) are the energies per two atoms in the pure AC and BC compounds respectively. If ΔE is positive, then the ordered structure is thermally unstable at any temperature. If ΔE is negative, phase stability depends on the magnitude of ΔE and the temperature.

Because the structures and local bonding of these ordered alloys are very similar to the constituent compounds, the tight-binding model can be used as an interpolation scheme for a systematic and quantitative calculation. Since the internal distortions and strain energy play an important role in the alloy excess energy, the tight-binding parameters controlling the elastic properties have to be represented accurately in the model. A simple way to achieve this end is to keep these parameters close to Harrison's original form but to free them from the $1/d^2$ and $1/d^4$ scaling rules for $V_{\alpha\alpha'}$ and u respectively. This can be done if one assumes the following forms:

$$V_{\alpha\alpha'}(d) = V^0_{\alpha\alpha'}(d_0/d)^n \tag{2.7.2}$$

and

$$u(d) = u_0(d_0/d)^m \tag{2.7.3}$$

where the superscript and subscript 0 indicate the values evaluated at the equilibrium bond length d_0. The values of $V^0_{\alpha\alpha'}$ are taken to be Harrison's universal forms in Eq. (2.5.1) scaled by a factor f:

$$V^0_{\alpha\alpha'} = f\, V^{\rm Harr}_{\alpha\alpha'} \tag{2.7.4}$$

There are now four parameters to be adjusted—the scaling parameter f, the powers n and m, and the repulsive energy parameter u_0. These parameters can be determined by requiring that the model produce the correct experimental values for the bonding energy B.E., the equilibrium bond length d_0, the bulk modulus B, and the shear elastic constant C_{11}–C_{12}. These parameters and the way they are determined will be discussed in Chapter 3 (see Table 3.5) after methods for calculating the elastic constants are introduced. This empirical tight-binding (ETB) model not only produces the quantities used in the fitting, it also predicts other elastic quantities very well, e.g., C_{44} and optical phonon frequencies.

The bulk excess energies ΔE for the ABC_2 ordered alloys are then calculated from this parameterized Hamiltonian using the band structure approach described in Section 2.4. The values of ΔE in the three structures CuAuI (CA), chalcopyrite (CH), and CuPt (CP) are listed in Table 2.4 and are compared with those from self-consistent LDA calculations. For the lattice-mismatched alloys, all the ΔE values from ETB are positive. For the lattice-matched systems $GaAlAs_2$ and $HgCdTe_2$, the ΔE values are nearly zero. Although there are detailed

TABLE 2.4. Excess Energies ΔE (in meV per four atoms) for the Three Ordered Semiconductor Alloy Structures from Tight-Binding Calculation and Comparison with Results from Local Density Functional Approximation (LDA)

	CA		CH		CP	
Alloys	TB	LDA	TB	LDA	TB	LDA
$AlGaAs_2$	0.6	10.8[a]	0.6	11.4[c]	0.8	7.5[b]
		11.5[b]		9.8[b]		
		15.1[c]				
		13.5[j]				
		35[k]				
$AlGaP_2$	−2.4		−2.8		−2.6	
$InGaSb_2$	57.4		37.4		85.8	
$InAlAs_2$	68.8	35.0[l]	45.8	−15.0[l]	107.0	
$InGaAs_2$	73.2	60.1[e]	48.2	16.5[f]	113.0	108.5[f]
		83.6[a]				
		66.7[d]				
$InAlP^2$	69.0	43.0[l]	44.0	−21.0[l]	114.2	97.0[d]
$InGaP_2$	88.4	115.6[a]	57.2	19.0[g]	139.4	155.4[f]
		91.0[g]				
		54.4[i]				
Ga_2AsP	30.0	26.6[f]	19.4	6.5[f]	33.6	37.2[f]
Ga_2AsSb	113.0	129.2[a]	69.8	52.0[b]	128.0	132.0[b]
		114.8[h]				
		115.0[b]				
Ga_2SbP	260.2		67.4		290.8	
$HgCdTe_2$	−2.3	12.1[f]	−2.7	11.3[f]	−2.7	9.8[f]
$HgZnTe_2$	29.7	42.5[f]	21.0	11.4[f]	54.9	103.3[f]
$CdZnTe_2$	34.3	54.2[f]	22.4	19.2[f]	65.3	103.5[f]

[a]Buguslawski and Baldereschi (1989)
[b]Bernard *et al.* (1988)
[c]Bylander and Kleinman (1986, 1987)
[d]Bernard *et al.* (1990)
[e]Ohno (1988)
[f]Wei *et al.* (1990)
[g]Srivastava *et al.* (1988)
[h]Qteish *et al.* (1989)
[i]Nelson and Batra (1989)
[j]Min *et al.* (1988)
[k]Ciraci and Batra (1987a,b)
[l]Dandrea *et al.* (1990)

quantitative differences between the results calculated from ETB and LDA, both models give the same trend $\Delta E_{CH} < \Delta E_{CA} < \Delta E_{CP}$ among the three structures, as shown in Fig. 2.6.

The most important conclusion drawn from these results is that these ordered bulk alloys are not thermal equilibrium states at the experimental growth temperatures shown in Table 1.4. For this to happen, the ΔE value has to be negative, and its magnitude, as discussed in Chapter 4, has to be greater than 200 meV per four atoms. This required ΔE value is far lower than all the calculated values and lies beyond the uncertainties of our ETB model and the LDA calculations listed above. These results imply that the ordering of these alloys must be driven by dynamic processes encountered in epitaxial growth.

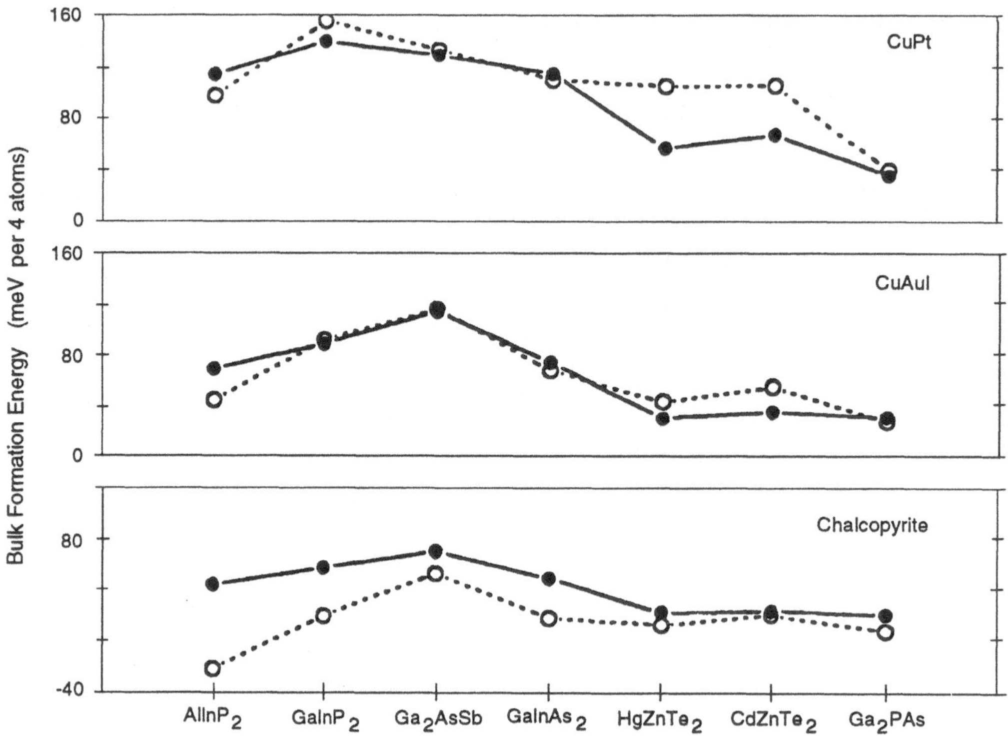

FIGURE 2.6. Excess energies for three ordered alloys from tight-binding (Yeh *et al.*, 1991, solid circles) and LDA (Wei *et al.*, 1990, open circles) calculations.

2.8. CONCLUDING REMARKS

Density functional theory in LDA has proven to be remarkably accurate in predicting, with no adjustable parameters, structural properties of all classes of solids, including the semiconductors emphasized in this book. Bond lengths and elastic constants are predicted to within a few percent of measured values (see Table 3.2). Predicted cohesive energies are generally 20% lower than experiment, but most of the error can be ascribed to inaccuracy in the values calculated for the free atoms which are mostly eliminated by the addition of gradient corrections (Van Schilfgaarde and Berding, 1995). These errors in the absolute energies do not affect most structural properties. One drawback of LDA is that it is computationally intensive, which so far has limited LDA to dealing only with ordered solids involving few atoms. However, a newly developed order-N method could revolutionize LDA and allow it to deal with large systems, as needed for simulating disordered alloys. This new method (Daw, 1993; Lee *et al.*, 1993) manipulates the density matrix directly, bypassing the direct diagonalization of the Hamiltonian matrix and thus reduces the calculational steps from N^3 to N, where N is the number of atoms per unit cell.

The tight-binding method is much less computationally demanding, and when there is sufficient experimental information to determine the parameters it is also accurate (Chadi, 1984). This is the case for the full-band structure TB model for the structural properties treated here. After fitting the constituents' bond energies, bond lengths, and elastic constants, this TB model predicts accurate results for the alloys, as evidenced by the bond lengths calculated in Chapter 1, the energetics of ordered alloys in this chapter, disordered alloys in Chapter 4, and elastic constants in Chapter 3. The reason this TB model works is that the alloy structures have nearly retained the basic tetrahedral coordination. We do not expect this model to work well for cases in which large deviations from the tetrahedral coordination occur, for example, on surfaces and in amorphous states.

Of the three bonding theories considered, the bond orbital method of Harrison is the simplest. Its greatest virtues are in analytical studies which may be only semiquantitative but often give great insight into mechanisms responsible for the properties studied. This is an outstanding tool to help experimentalists as well as theorists develop their intuition.

When applied to epitaxially prepared ordered alloys, LDA and TB both suggest that these materials do not result from thermal equilibrium processes at the growth temperature. For most of the alloys that form ordered structures, the calculated mixing energy has the wrong sign to account for the observations, and for those few with the proper sign the magnitude is far too small to account for the observed ordering. Although there are indications that surface reconstruction plays an important role (Zunger and Mahajan, 1993), realistic simulations are still needed to bring about a convincing picture.

REFERENCES

Bachelet, G.B., D.R. Hamman, and M. Schluter (1982), *Phys. Rev. B* **26**, 4199.
Berding, M.A., A. Sher, and A.-B. Chen (1987), *Phys. Rev. B* **36**, 7433.
Bernard, J.E., R.G. Dandrea, L.G. Ferreira, S. Froyen, S.-H. Wei, and A. Zunger (1990), *Appl. Phys. Lett.* **56**, 731.
Bernard, J.E., L.G. Ferreira, S.-H. Wei, and A. Zunger (1988), *Phys. Rev. B* **38**, 6338.
Born, M., and J.R. Oppenheimer (1927), *Ann. Phys.* (4) **84**, 457.
Boguslawski, P., and A. Baldereschi (1989), *Phys. Rev. B* **39**, 8055.
Bylander, D.M., and L. Kleinman (1986), *Phys. Rev. B* **34**, 5280.
Bylander, D.M., and L. Kleinman (1987), *Phys. Rev. B* **36**, 3229.
Ceperley, D.M., and B.J. Alder (1980), *Phys. Rev. Lett.* **45**, 566.
Chadi, D.J. (1979), *Phys. Rev. B* **19**, 2074.
Chadi, D.J. (1984), *Phys. Rev. B* **29**, 785.
Ciraci, S.C., and I.P. Batra (1987a), *Phys. Rev. Lett.* **58**, 14; (1987b), *Phys. Rev. B* **36**, 1225.
Coulson, C.A., L.B. Redei, and D. Stocker (1962), *Proc. R. Soc. (London)* **270**, 352.
Dandrea, R.G., J.E. Bernard, S.-H. Wei, and A. Zunger (1990), *Phys. Rev. Lett.* **64**, 36.
Daw, M.S. (1993), *Phys. Rev. B* **47**, 10895.
Falter, C., W. Ludwig, M. Selmke, and W. Zierau (1984), *Phys. Lett.* **105A**, 139.
Froyen, S., and W. Harrison (1979), *Phys. Rev. B* **20**, 2420.
Hamann, D.R., M. Schluter, and C. Chiang (1979), *Phys. Rev. Lett.* **43**, 1494.
Harrison, W.A. (1980), *Electronic Structure and Properties of Solids* (W.H. Freeman, San Francisco, CA).
Harrison, W.A. (1983), *Phys. Rev. B* **27**, 3592.
Hedin, L., and B.I. Lundquist (1971), *J. Phys. C* **4**, 2064.
Hohnberg, P., and W. Kohn (1964), *Phys. Rev.* **126**, B864.
Hybertsen, M.S., and S.G. Louis (1985), *Phys. Rev. Lett.* **55**, 1418.
Kohn, W., and L.J. Sham (1965), *Phys. Rev.* **140**, A1133.
Lee, X-P, R.W. Nunes, and D. Vanderbilt (1993), *Phys. Rev. B* **47**, 10891.
Min, B.I., S. Massidda, and A.J. Freeman (1988), *Phys. Rev. B* **38**, 1970.
Nelson, J.S., and I.P. Batra (1989), *Phys. Rev. B* **39**, 3250.
Ohno, T. (1988), *Phys. Rev. B* **38**, 13191.
Perdew, J.P., and A. Zunger (1981), *Phys. Rev. B* **23**, 5048.
Phillips, J.C. (1973), *Bonds and Bands in Semiconductors* (Academic Press, New York).
Qteish, A., N. Motta, and A. Balzarotti (1989), *Phys. Rev. B* **39**, 5987.

Slater, J.C., and G.F. Koster (1954), *Phys. Rev.* **94**, 1498.
Srivastava, G.P., J.L. Martins, and A. Zuner (1988), *Phys. B* **38**, 12694.
van Schilfgaarde, M., and M.A. Berding (1995), private communication.
Wang, C.S., and B.M. Klein (1981), *Phys. Rev. B* **24**, 3393.
Wei, S.-W., L.G. Ferreira, and A. Zunger (1990), *Phys. Rev. B* **41**, 8240.
Yeh, C.-Y., A.-B. Chen, and A. Sher (1991), *Phys. Rev. B* **43**, 9138.
Zunger, A. (1980), *Phys. Rev. B* **21**, 4785.
Zunger, A., and S. Mahanjan (1993), *Handbook on Semiconductors*, Vol. 3, 2nd ed. (Elsevier, Amsterdam).

3

Elasticity

The objective of this chapter is to review the elastic constants of semiconductors and their alloys and to present theoretical models for treating them. After a brief review of notation and traditional distortion modes, we describe the status of modern *ab initio* calculations. Then a number of elastic constant models of varying complexity are introduced. These include the valence-force-field (VFF) model, Harrison's bond-orbital model (BOM), and empirical tight-binding (ETB) calculations. The elastic constants for a large number of semiconductors evaluated from these models are analyzed and compared with experiment. These models are then used to study the bulk moduli of alloys. We show that, although the calculated bulk moduli of alloys tend to be slightly smaller than the average of the constituents' values, experimental results do not always support this trend.

3.1. DEFINITIONS AND ANALYSIS

In the regime of linear elasticity theory, deformations are assumed to be infinitesimal. The relative displacement $\mathbf{x}'$ between two points in a deformed solid under stress is related to the corresponding vector $\mathbf{x}$ in the unstressed solid through a nine-component strain tensor, by the following equation written in component form:

$$x_\alpha' = x_\alpha + \sum_\beta \eta_{\alpha\beta} x_\beta \tag{3.1.1}$$

The change in the internal energy associated with η is also small and will be denoted UV, where V is the equilibrium volume of a solid and U is an energy density. The strain tensor η, thought of as a thermodynamic parameter, is a generalization of the pressure. If only mechanical forces are present, U is a function only of η and is quadratic in η:

$$U = \frac{1}{2} \sum_{\alpha\beta\mu\nu} \eta_{\alpha\beta} c_{\alpha\beta\mu\nu} \eta_{\mu\nu} \tag{3.1.2}$$

where $c_{\alpha\beta\mu\nu}$ are the elastic stiffness coefficients, which are characteristic properties of the solid. From this definition, $c_{\alpha\beta\mu\nu} = c_{\mu\nu\alpha\beta}$ is required for U to be an analytic function of η. Moreover, $c_{\alpha\beta\mu\nu} = c_{\beta\alpha\mu\nu} = c_{\alpha\beta\nu\mu}$ is also required to ensure that U is zero under any infinitesimal rigid rotation. In light of these properties, the energy can be expressed in terms of a symmetrical strain tensor ε as

$$U=\frac{1}{2}\sum_{\alpha\beta\mu\nu}\varepsilon_{\alpha\beta}c_{\alpha\beta\mu\nu}\varepsilon_{\mu\nu} \tag{3.1.3}$$

where $\varepsilon_{\alpha\beta}$ is defined to be

$$\varepsilon_{\alpha\beta}=\frac{1}{2}(\eta_{\alpha\beta}+\eta_{\beta\alpha}) \tag{3.1.4}$$

The conjugate variable to ε is the stress tensor σ, with its components given by

$$\sigma_{\alpha\beta}=\frac{\partial U}{\partial\varepsilon_{\alpha\beta}}=\sum_{\mu\nu}c_{\alpha\beta\mu\nu}\,\varepsilon_{\mu\nu} \tag{3.1.5}$$

It is clear that σ is also a symmetrical tensor.

The most frequently used notation is the engineering convention, in which the strain tensor **e** is related to ε of Eq. (3.1.4) by $e_{\alpha\alpha}=\varepsilon_{\alpha\alpha}$ for the diagonal components, but $e_{\alpha\beta}=2\varepsilon_{\alpha\beta}$ for $\alpha\neq\beta$. Furthermore, since **e** has at most six independent components, it is treated as a six-component vector, with the vector components 1 to 6 standing respectively for the tensor components *xx, yy, zz, yz, xz*, and *xy*. For a more complete discussion see Hirth and Lothe (1982). In terms of **e** the strain energy density is written as

$$U=\frac{1}{2}\sum_{ij}C_{ij}\,e_i\,e_j \tag{3.1.6}$$

where C_{ij} can be identified as $c_{\alpha\beta\mu\nu}$ with $i=\alpha\beta$ and $j=\mu\nu$. Because C_{ij} is symmetrical, it has at most 21 independent components for any crystal. Crystal symmetries reduce this number further. For a cubic lattice, to which the zinc blende and diamond semiconductors belong, there are only three independent components, namely C_{11}, C_{12}, and C_{44} (Love, 1944; Ashcroft and Mermin, 1976).

The three independent elastic constants of the diamond and zinc blende (zb) semiconductors can be calculated by considering the following three strains:

(i) Under a uniform expansion, which changes a displacement **x** into $\mathbf{x}'=(1+e)\mathbf{x}$, $e_1=e_2=e_3=e$ and the other strain components are zero, the elastic energy density is then given by

$$U=3(C_{11}+2C_{12})e^2/2 \tag{3.1.7}$$

U can also be expressed in terms of the adiabatic bulk modulus B defined by $\delta P\equiv B(\delta V/V)$, where V is the crystal volume, δV is its change, and δP is the corresponding pressure change. The differential work done in an adiabatic process is $\delta P\,d\delta V$. Integrating this yields $UV=B(\delta V/V)^2/2=9Be^2/2$, because the dilatation is $\delta V/V=3e$ in the present case. This establishes the relationship

$$B=(C_{11}+2C_{12})/3 \tag{3.1.8}$$

(ii) The next case to consider is a tetragonal shear strain e which changes a displacement according to

$$(x,y,z) \rightarrow (x+ex,\ y-ey,\ z) \tag{3.1.9}$$

Thus, the only nonzero strain components are $e_1 = -e_2 = e$. Then U becomes

$$U = (C_{11} - C_{12})e^2 \tag{3.1.10}$$

(iii) To calculate C_{44} we consider a shear strain e which changes a displacement according to

$$(x,y,z) \rightarrow (x+ey/2,\ y+ex/2,\ z) \tag{3.1.11}$$

This strain contains e_6 as the only nonzero component. Because for cubic symmetry C_{44} equals C_{66}, the energy density is simply

$$U = C_{44}e^2/2 \tag{3.1.12}$$

Although the macroscopic crystal distortion of the Bravais lattice caused by this strain is described by Eq. (3.1.11), microscopically there is a relative displacement $\mathbf{u} = (0,0,u)$, the so-called Kleinman (1962) internal displacement, between two successive atomic planes perpendicular to the z-axis. In other words, the relative displacements between the atoms on the same fcc sublattice are governed by Eq. (3.1.11), but there is an additional induced relative displacement $\mathbf{u}$ between the two sublattices. The directions of the displacements of atoms in a tetrahedral cell are shown in Fig. 3.1. In calculations an arbitrary infinitesimal pair of e and u can be used to obtain the coefficients in the following quadratic expansion of the strain energy density:

$$U = \Phi u^2/2 + Deu + C_{44}^0 e^2/2 \tag{3.1.13}$$

Both Φ and D are constants, and C_{44}^0 would be the shear stiffness coefficient if the internal displacement were not allowed. The force function Φ is related to the transverse optical phonon frequency ω at Γ(the center of Brillouin zone) by $\Phi = \mu\omega^2$, with μ being the reduced mass. Kleinman (1962) defined an internal displacement parameter ζ which is related to the equilibrium value of u by $u = \zeta ae/4$ for fixed e, where a is the lattice constant. Taking the first derivative of U with respect to u in Eq. (3.1.13) and setting it equal to zero, we find that the ζ value is

$$\zeta = -4D/a\Phi \tag{3.1.14}$$

The sought-after C_{44} expression is

$$C_{44} = C_{44}^0 - \zeta^2 a^2 \Phi/16 \tag{3.1.15}$$

Thus the internal displacement is an essential part of C_{44}.

This completes the procedure for calculating the elastic constants for the diamond and zinc blende semiconductors. These procedures will be used in several energy model calculations.

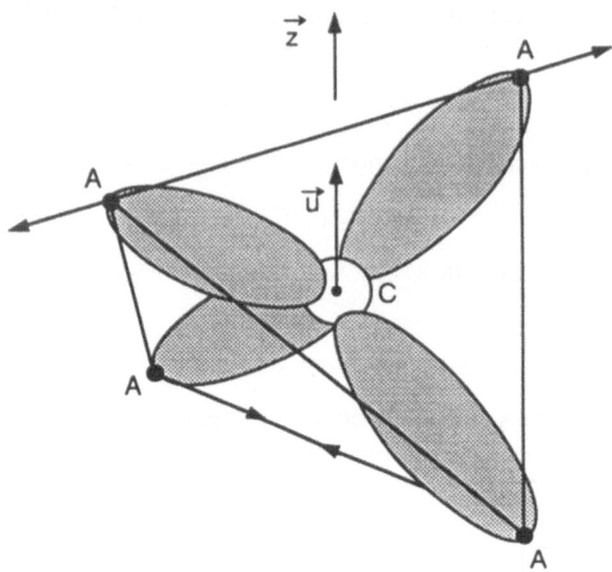

FIGURE 3.1. Distortion of a tetrahedral cell in a zinc blende semiconductor under a shear strain described in Eq. (3.1.11). **u** is the internal displacement between the anion and cation sublattices.

3.2. *AB INITIO* CALCULATIONS

As mentioned in Section 2.1, the dynamic matrix elements $A^{\alpha\beta}_{nm}$ in Eq. (2.1.2) are directly related to the elastic constants. For example, for a Bravais lattice they are related by (Ashcroft and Mermin, 1976)

$$c_{\alpha\beta\mu\nu} = -\frac{1}{2}\sum_{L} L_\alpha A^{\beta\mu}(\mathbf{L})L_\nu \qquad (3.2.1)$$

where the sum is over all the lattice vectors **L**.

Self-consistent density functional theory provides a means for an *ab initio* calculation of the elastic constants. This is accomplished following the recipe for calculating the total energy described in Section 2.3. First evaluate E_g as a function of ionic positions, find the equilibrium configuration, then impose a strain and determine the strain energy following the prescription of Section 3.1 to deduce the elastic constants. The problem then becomes strictly computational. However, the elastic constant calculation is still a challenging task even for a pure semiconductor crystal and is even more so for alloys. Because the strain energy is many orders of magnitude smaller than the total energy, very precise computations are required if one hopes to obtain meaningful elastic constants. So far at least two band structure methods have proven to be reliable for all three elastic constants: the plane-wave method using pseudopotentials (PP-PW) (Nielsen and Martin, 1983, 1985a) and the full-potential linearized muffin-tin-orbital method (FP-LMTO) (Methfessel *et al.*, 1989).

Even if the total energies for different distortions can be calculated accurately, there is still the problem of searching for the equilibrium (minimum energy) atomic positions in a distorted crystal, and in the numerical determination of elastic constants from energy

TABLE 3.1. Comparison between Calculated and Experimental Lattice Constant a, Elastic Constants B, C_{11}–C_{12}, and C_{44}, Kleinmann Internal Distortion Parameter ζ, and the TO Optical Phonon ω in Wave Numbers 1/cm[a]

	Experiment	FP-LMTO[b]	PP-PW[c]	TB
Si				
a	5.431	5.41	5.45	5.431
B	9.923	9.9	9.3	9.923
C_{11}–C_{12}	10.274	10.2	9.89	10.274
C_{44}	8.036	8.3	8.5	8.013
C^0_{44}			11.1	11.30
ζ	0.54[d]	0.51	0.53	0.51
ω	523.00	518.00	521.00	572.00
Ge				
a	5.65		5.59	5.65
B	7.653		7.2	7.653
C_{11}–C_{12}	8.189		8.5	8.189
C_{44}	6.816		6.3	6.189
C^0_{44}			7.7	9.46
ζ			0.44	0.49
ω	303.00		302.00	342.00
GaAs				
a	5.642		5.55	5.642
B	7.69		7.3	7.69
C_{11}–C_{12}	6.63		7.0	6.63
C_{44}	6.04		6.2	5.791
C^0_{44}			7.5	7.83
ζ			0.48	0.50
ω	273.00		268.00	292.00

[a]Also listed are C^0_{44} corresponding to the value without the internal distortion. The FP-LMTO and PP-PW are the *ab initio* theories, and TB is the tight-binding method discussed in the text. All elastic constants are in units of 10^{11} dyne/cm^2. Experimental values are those listed in Table 3.2.

[b]Methfessel *et al.* (1989)

[c]Nielsen and Martin (1985a and b)

[d]Cousins *et al.* (1987)

differences. If the strain energy can be calculated directly without taking the difference between two large energies, or better if the derivatives can be calculated directly, not only can the computational time be shortened but the numerical errors will also be reduced. The quantum mechanical theory of forces and stresses of Nielsen and Martin (1985b) and the closely related direct calculation of elastic constants from linear response theory of Boroni *et al.* (1987) represent the status of efforts in this direction (see also Fig. 3.4). The former has been carried out for all three elastic constants of Si, Ge, and GaAs (Nielsen and Martin, 1985a), while the latter has only been done for the bulk modulus for Si, both based on the PP-PW method. Table 3.1 and Fig. 3.4 compare theoretical calculations and experimental results.

From this comparison, it is fair to say that DFT-LDA is a reliable *ab initio* theory for the elastic constants of crystalline semiconductors. Table 3.1 also lists the results from the empirical tight-binding (TB) calculation based on the model described in Section 2.7. Note that the C_{44}, ζ, and transverse optical (TO) phonon frequency ω values from TB are theory's prediction, but the other quantities are adapted to fit the TB parameters.

3.3. VALENCE-FORCE-FIELD MODEL

While the DFT for the elastic constants requires complicated and accurate *ab initio* calculations which for semiconductors have been obtained only recently, a useful phenomenological microscopic model of elastic constants of all semiconductors has been available for some time. This is the valence-force-field model (VFF). This topic has been reviewed and well analyzed in a paper by Martin (1970). The essence of this model is described next.

3.3.1. Diamond Structure

The most convenient form of VFF is the Keating model (1966), which connects the elastic energy of a semiconductor to crystal deformations by the following relation:

$$\Delta E = \frac{3\alpha}{8d^2}\sum_i [\Delta(\mathbf{r}_i \cdot \mathbf{r}_i)]^2 + \frac{3\beta}{8d^2}\sum_{i>j}\sum [\Delta(\mathbf{r}_i \cdot \mathbf{r}_j)]^2 \tag{3.3.1}$$

where the bond index i runs over all the bonds, but the i and j only sum over those pairs of bonds that are connected to a common atom. In Eq. (3.3.1) d is the equilibrium bond length and $[\Delta(\mathbf{r}_i \cdot \mathbf{r}_j)]$ is the change in the dot product of the two bond vectors that start at the common atom, point along the bond directions, and end at the first-neighbor atoms. Therefore, α is the force constant for bond-length distortions, and β is the force constant for bond-angle distortions.

Following the calculational procedures described in Section 3.1, the U in the VFF under a uniform expansion can be shown to be $U = 2(3\alpha + \beta)\, e^2d^2/\Omega$, where $\Omega = a^3/4$ is the equilibrium volume per unit cell. Thus we find

$$B = (C_{11} + 2C_{12})/3 = (\alpha + \beta/3)/a \tag{3.3.2}$$

For the shear strain described by Eq. (3.1.9), Eq. (3.3.1) yields $U = 4\beta e^2/a$. Thus, according to Eq. (3.1.10),

$$C_{11} - C_{12} = 4\beta/a \tag{3.3.3}$$

Given the shear strain described by Eq. (3.1.11) and with an internal displacement $\mathbf{u} = (0,0,u)$, Eq. (3.3.1) yields the expression

$$U = [\alpha(e-\eta)^2 + \beta(e+\eta)^2]/(8a) \tag{3.3.4}$$

where $u = \eta a/4$. A comparison between Eqs. (3.1.13) and (3.3.4) shows that $\Phi = 16(\alpha+\beta)/a^3$, $D = -4(\alpha-\beta)/a^2$, and $C_{44}^0 = (\alpha+\beta)/a$. Using these results in Eq. (3.1.14), we find that the Kleinman internal displacement parameter in the present model is given by

$$\zeta = (\alpha-\beta)/(\alpha+\beta) = 2C_{12}/(C_{11} + C_{12}) \tag{3.3.5}$$

Equation (3.1.15) then produces

$$C_{44} = 2\alpha\beta/[(\alpha+\beta)a] \tag{3.3.6}$$

TABLE 3.2. Experimental Elastic Constants for Some Cubic Semiconductors and the Parameters of Eq. (3.3.10) Taken from Martin (1970) along with the Force Constants α and β Obtained from Eqs. (3.3.11) and (3.3.12) and the Identity Relations I_K, I_M, and I_{BOM} given by Eqs. (3.3.7), (3.3.17), and (3.5.11), Respectively

	C_{11}	C_{12}	C_{44}	s	α	β	I_K	I_M	I_{BOM}
		(10^{11} dyn/cm^2)			(10^3 dyn/cm)				
C[a]	107.640	12.520	57.740	0.0	129.100	84.573	1.00	1.00	1.02
Si[a]	16.772	6.498	8.036	0.0	49.247	13.951	1.00	1.00	1.13
Ge[a]	13.112	4.923	6.816	0.0	39.438	11.583	1.08	1.08	1.05
AlSb[a]	8.769	4.341	4.076	1.684	33.768	6.653	1.11	1.05	1.08
GaP[a]	14.390	6.520	7.143	3.815	46.965	10.448	1.12	1.05	1.05
GaAs[a]	12.110	5.480	6.040	2.827	40.895	9.159	1.12	1.06	1.05
GaSb[a]	9.089	4.143	4.440	1.569	33.123	7.412	1.10	1.06	1.06
InP[a]	10.220	5.760	4.600	3.766	41.095	6.250	1.20	1.07	1.03
InAs[a]	8.329	4.526	3.959	2.820	33.744	5.531	1.22	1.11	1.00
INSb[a]	6.918	3.788	3.132	1.372	29.909	4.951	1.17	1.11	1.05
ZnS[b]	9.420	5.680	4.360	6.788	37.026	4.571	1.33	1.07	0.95
ZnS[b]	10.790	7.220	4.120	6.788	45.126	4.341	1.28	1.02	1.01
ZnS[b]	9.810	6.270	4.483	6.788	39.947	4.300	1.42	1.13	0.90
ZnS[b]	10.460	6.530	4.630	6.788	41.880	4.828	1.33	1.08	0.95
ZnSe[a]	8.95	5.39	3.984	4.368	34.432	4.716	1.28	1.09	0.98
ZnSe[b]	8.59	5.06	4.06	4.368	34.519	4.673	1.32	1.13	0.95
ZnSe[b]	8.720	5.240	3.920	4.368	35.469	4.603	1.29	1.10	0.98
ZnTe[b]	7.130	4.070	3.120	2.566	29.976	4.452	1.18	1.06	1.05
ZnTe[b]	7.220	4.090	3.080	2.566	30.204	4.558	1.14	1.03	1.08
CdTe[a]	5.33	3.65	2.04	3.105	27.058	2.455	1.34	1.07	0.98
CdTe[b]	6.150	4.300	1.960	3.105	31.546	2.731	1.16	0.94	1.13
HgTe[a]	5.971	4.154	2.259	2.381	30.300	2.542	1.37	1.16	0.96
HgTe[b]	5.63	3.66	2.11	2.381	26.919	2.542	1.37	1.15	0.95

[a]Data quoted from "Landolt-Bornstein Numerical Data and Functional Relationships in Science and Technology," New Series, Vols. 17 and 22.

[b]Listed in the review by Mittra and Massa (1982).

Because the VFF energy depends on only two parameters, the three elastic constants given cannot be independent, but are related to each other by an identity, referred to as the Keating identity,

$$I_K = 2C_{44}(C_{11} + C_{12})/[(C_{11} - C_{12})(C_{11} + 3C_{12})] = 1 \tag{3.3.7}$$

There is nothing fundamental about this identity; it is simply a consequence of a model that has only two parameters, when the symmetry of the solid allows three. The accuracy of the identity is one measure of accuracy of the model.

3.3.2. *Zinc Blende Structure and Coulomb Force*

The Keating identity holds very well for systems with the diamond structure but not so well for the zinc blende compounds (see Table 3.2). One obvious difference between the two structures is the presence of Coulomb interactions arising from charge shifts between the cation and anion sublattices in zinc blende semiconductors. Martin incorporated Blackman's (1959) treatment of the Coulomb forces in the Keating VFF in the following manner: (a) The Coulomb energy was treated as a screened Madelung energy E_M. For example, in a uniformly expanded crystal with a bond length r, the Coulomb energy was taken to be $E_M = -N\alpha_M Z^{*2}e^2/\varepsilon r$, where N is the total number of unit cells, $\alpha_M = 1.6381$ is the Madelung constant, and Z^{*2}/ε is the effective charge defined by the optic-mode splitting (Harrison, 1983b):

$$S = Z^{*2}/\varepsilon = \mu(\omega_l^2 - \omega_t^2)/4\pi e^2 \tag{3.3.8}$$

In Eq. (3.3.8), ω_l and ω_t are respectively the longitudinal and transverse phonon frequencies in the long-wavelength limit. (b) To counterbalance the Coulomb forces, a repulsive force term was added and assumed to contribute to the bond-stretching energy in the form

$$\Delta E_R = \sum_i \frac{\alpha_M Z^{*2}e^2\, \Delta r_i}{4\varepsilon d^2} \tag{3.3.9}$$

With the above two contributions added, the total strain energy is $\Delta E_T = \Delta E + \Delta E_M + \Delta E_R$, where ΔE is the VFF contribution in Eq. (3.3.1) and ΔE_M is the change in the Madelung energy. The energies ΔE_T are expanded in a power series, and only terms up to the second power in the strain are retained. The ΔE_M contributions arising from fixed values of the charge shift $S = Z^{*2}/\varepsilon$ on the atomic sites under different strains were worked out by Blackman (1959). Using these results and defining

$$s \equiv (e^2/d^4)S = e^2 Z^{*2}/d^4\varepsilon \tag{3.3.10}$$

Martin obtained the following modified expressions for the elastic constants:

$$C_{11} + 2C_{12} = (3\alpha + \beta)/a - 0.355s \tag{3.3.11}$$

$$C_{11} - C_{12} = 4\beta/a + 0.053s \tag{3.3.12}$$

$$\zeta = [(\alpha - \beta)/a - 0.294s]/C_M \tag{3.3.13}$$

$$C_{44} = (\alpha + \beta)a - 0.136s - C_M\zeta^2 \tag{3.3.14}$$

where C_M is defined as

$$C_M \equiv (\alpha + \beta)/a - 0.266s \tag{3.3.15}$$

The above equations can be combined to yield

$$\zeta = (2C_{12} - C')/(C_{11} + C_{12} - C') \tag{3.3.16}$$

where $C' = 0.314s$. Since the extra parameter s is fixed by the optic modes and the bond length, the above results combine to lead to a new identity, called the Martin identity:

$$I_M = \frac{2C_{44}(C_{11} + C_{12} - C')}{(C_{11} - C_{12})(C_{11} + 3C_{12} - 2C') + 0.831C'(C_{11} + C_{12})} \tag{3.3.17}$$

Table 3.2 lists a set of experimental values of the elastic constants and the s values for a number of diamond and zinc blende semiconductors. These values are used to compute the force constants α and β and the identity expressions I_k and I_M given in Eqs. (3.3.6) and (3.3.17), respectively. The reason that several sets of data are quoted for some of the systems is to show the uncertainties in the experiments. These results clearly show that the inclusion of the Coulomb energies improves the identity relation. The deviations of I_M from unity are 15% or less. Also listed are the values for another identity relation I_{BOM} from Eq. (3.5.11) based on the bond-orbital model to be derived in Section 3.5.

Martin further studied trends as functions of the bond lengths d and the ionicity f_p of Van Vechten and Phillips (1969). He found that α scales roughly as $1/d^3$; i.e.,

$$\alpha d^3/e^2 = \text{constant} \tag{3.3.18}$$

where e is the electron charge. He also found the ratio between the bond-angle and bond-stretching forces tends to decrease as f_p increases and scales roughly as

$$\beta/\alpha \sim 1 - f_p \tag{3.3.19}$$

He further observed that if S of Eq. (3.3.8) is set equal to f_p and if the α and β values are extrapolated using Eqs. (3.3.18) and (3.3.19) from those fitted to the average values of B and C_{11}–C_{12} for Si and Ge, then all the elastic constants can be predicted from Eqs. (3.3.11) through (3.3.15) to an accuracy of 10%.

It is interesting to compare Eq. (3.3.11) using the results of Eqs. (3.3.18) and (3.3.19) with Cohen's (1985) empirical formula for the bulk modulus:

$$B = (1971 - 22\lambda)/d^{3.5} \tag{3.3.20}$$

where B is in GPa, d in Å, and $\lambda = 0$, 1, and 2, respectively, for the group IV, III–V, and II–VI semiconductors. Both Martin's and Cohen's formulas give B values accurate to better than 10% for all materials tabulated in Table 3.2. However, the B in Eq. (3.3.20) scales as $1/d^{3.5}$, while in VFF it scales as $1/d^4$. The reason that both theories produce accurate bulk moduli while having different bond-length scaling rules is that the bond lengths of semiconductors all fall within a small range of values.

3.4. "EXACT" TIGHT-BINDING CALCULATION

Within the tight-binding prescription in Section 2.4, once the values of the TB parameters and their dependence on the bond length are known, the Hamiltonian at each **k** inside the Brillouin zone (BZ) can be evaluated and the summation over **k** carried out to obtain the band structure energy. When the repulsive energy is added to the band structure energy, the total energy of any specified geometry can be easily calculated. All the elastic constants, associated internal displacements, and transverse optical phonon frequencies are readily calculable following the procedure of Section 3.1. The only important point to note is the k-sum (summation over **k** vectors inside the Brillouin zone). Without strain, the k-sum can be calculated accurately using 10 special k-points (Chadi and Cohen, 1973) in the irreducible wedge of the BZ. Under strain, the crystal symmetry changes, so these special k-points need to be extended to other nonequivalent wedges. However, since the sum of the valence-band energies as a function of **k** is a rather smooth function of **k**, a uniform sampling over the whole BZ converges quickly. A $5 \times 5 \times 5$ grid is sufficiently accurate for all the calculations needed. To avoid the numerical inaccuracy inherent in direct energy subtractions, one can also calculate the second derivatives "exactly" using a perturbation expansion.

This can be accomplished by expansion of the k-dependent Hamiltonian H in powers of the infinitesimal strain parameter e retained to second order

$$H(k) = H_0 + H_1 e + \tfrac{1}{2} H_2 e^2 \tag{3.4.1}$$

where H_0 is the strain-free Hamiltonian, and H_1 and H_2 are respectively the first and second derivatives with respect to e evaluated at $e = 0$. The band energy contribution to the strain coefficient then comes from the second derivative of E_{bs} with respect to e, denoted by

$$\frac{\partial^2 E_{bs}}{\partial e^2} = \sum_{v}\sum_{k} \langle \mathbf{vk}|H_2|\mathbf{vk}\rangle + 2\sum_{v}\sum_{c}\sum_{k} \frac{|\langle \mathbf{vk}|H_1|c\mathbf{k}\rangle|^2}{\varepsilon_v(\mathbf{k}) - \varepsilon_c(\mathbf{k})} \tag{3.4.2}$$

where $\varepsilon_c(\mathbf{k})$ and $|c\mathbf{k}\rangle$ are respectively the eigenenergies and eigenvectors of H_0 for the conduction bands. Similarly $|\mathbf{vk}\rangle$ stands for the valence bands. Note that the intravalence-band contributions in the second-order perturbation sum cancel exactly so they are not needed in Eq. (3.4.2). To evaluate these matrix elements one needs to have the first and the second strain derivatives of the two-center interactions and the direction cosines ξ_i. For the strain parameters e defined in Section 3.1 and for the two-center interactions V that scale as $1/d^n$, the following results are useful:

(1) For the bulk modulus the direction cosines do not change, and we have $\partial V/\partial e = -nV$ and $\partial^2 V/\partial e^2 = n(n+1)V$

(2) For C_{11}–C_{12}, with e specified in Eq. (3.1.9) we get $\partial\xi_i/\partial e = \xi_i(\delta_{i1} - \delta_{i2})$, $\partial^2\xi_i/\partial e^2 = \xi_i[3(\delta_{i1} + \delta_{i2}) - 4]/3$, $\partial V/\partial e^2 = -4nV/3$, where δ_{ij} is the Kronecker delta: $\delta_{ij} = 0$ for $i \neq j$ and $\delta_{ij} = 1$ for $i \neq j$.

(3) For C_{44}, with the strain e given in Eq. (3.1.11) and an internal displacement u, we find $\partial V/\partial e = -n\xi_i\xi_2 V$, $\partial V/\partial u = n\xi_3 V/d$, $\partial\xi_i/\partial e = (\delta_{i1}\xi_2 + \delta_{i2}\xi_1)/2 - \xi_i\xi_1\xi_2$, $\partial\xi_i/\partial u = -\delta_{i3}/d + \xi_i\xi_3/d$, $\partial^2 V/\partial e^2 = n(n-2)V/9$, $\partial^2 V/\partial u^2 = n(n-1)V/3d^2$, $\partial^2\xi_i/\partial e^2 = -\xi_i(\delta_{i1} + \delta_{i2})/12$, $\partial^2\xi_i/\partial u^2 = -2\partial_{i3}\xi_3/d^2$, $\partial^2 V/\partial e\partial u = -n(n+2)\xi_1\xi_2\xi_3/d$, and $\partial^2\xi_i/\partial e\partial u = (\delta_{i1}\xi_2\xi_3 + \delta_{i2}\xi_1\xi_3 + 2\delta_{i3}\xi_1\xi_2 - 6\xi_i\xi_1\xi_2\xi_3)/2d$. Also note that three second derivatives of the band structure energy are needed, namely $\partial^2 E_{bs}/\partial e^2$, $\partial^2 E_{bs}/\partial e\partial u$, and $\partial^2 E_{bs}/\partial u^2$ for the evaluation of C_{44}^0, D, and Φ of Eq. (3.1.13) respectively.

3.5. ANALYTICAL EXPRESSIONS IN THE BOND-ORBITAL MODEL

Since the bond energy E_b is given analytically in Eqs. (2.5.10) and (2.5.11), the bulk modulus can be obtained analytically by taking the second derivatives of E_b with respect to the bond length:

$$B = 2\sqrt{3}\, d_0 \frac{\partial^2 E_b}{\partial d^2} \tag{3.5.1}$$

For group IV semiconductors it takes the simple form (see Eq. (2.5.10))

$$B = -2V_2(1 - 9\alpha_m^2/16)/(\sqrt{3}\, d^3) \tag{3.5.2}$$

For a polar semiconductor, B reads (see Eq. (2.5.11))

$$B = -\frac{2V_2}{\sqrt{3}\, d^3}\left[\alpha_c^2 - \frac{\frac{9}{8}\alpha_c^3(5\alpha_c^2 - 4)(V_{1c}^2 + V_{1A}^2)}{V_2^2 + V_3^2}\right] \tag{3.5.3}$$

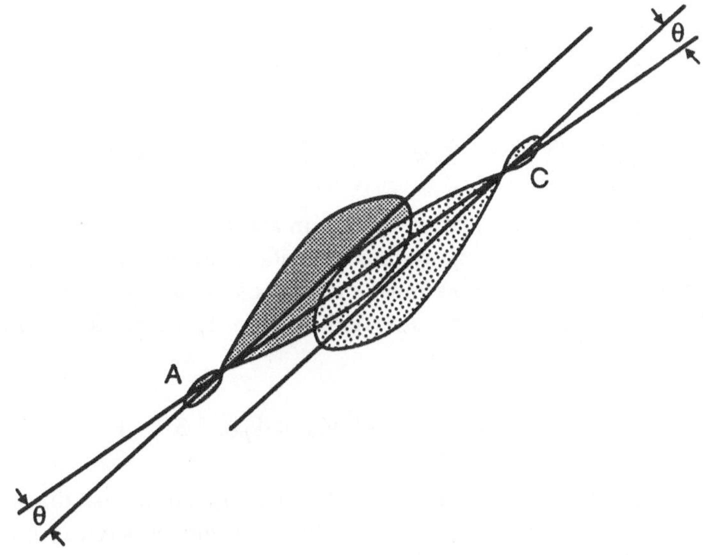

FIGURE 3.2. Misalignment of hybrid orbitals caused by a shear distortion described in Eq. (3.1.9).

The metallicity α_m and covalency α_c were defined in Section 2.5. These expressions show that the bulk modulus varies as $1/d^5$ in the pure covalent case and as $1/d^9$ in the extreme ionic limit $V_3 >> V_2$. Note that this result is different from the $1/d^{3.5}$ dependence in Cohen's (1985) formula and the $1/d^4$ scale factor in the VFF.

Shear strains cause a semiconductor to distort away from perfect tetrahedral symmetry. To deal with the shear elastic coefficient, the BOM has to be modified. A simple approximation is the rigid hybrid model (Harrison, 1983b; van Schilfgaarde and Sher, 1987) in which the hybrid orbitals of each atom are assumed to remain oriented in their original tetrahedral directions despite the lattice distortion. Then, unlike the arrangement in Fig. 2.5, the hybrids of two nearest-neighbor atoms making up the bonding and antibonding states no longer are directed toward each other. There is a misalignment angle θ between each hybrid (as shown in Fig. 3.2) and the line connecting the two atoms. Now the covalent energy V_2 is given by

$$V_2(\theta) = \tfrac{1}{4}[V_{ss\sigma} - 2\sqrt{3}\cos\theta\, V_{sp\sigma} - 3\cos^2\theta\, V_{pp\sigma} + 3(1-\cos^2\theta)V_{pp\pi}] \qquad (3.5.4)$$

The change δV_2 caused by an infinitesimal angular misalignment $\delta\theta$ is then given to lowest order by

$$\delta V_2 = \tfrac{1}{4}(\sqrt{3}\, V_{sp\sigma} + 3V_{pp\sigma} - 3V_{pp\pi})(\delta\theta)^2 \qquad (3.5.5)$$

Under the strain e described in Eq. (3.1.9) for C_{11}–C_{12}, the lowest order in the bond-length change is e^2, while the bond angle change is linear in e with $(\delta\theta)^2 = 2e^2/3$. If one assumes that the metallization coupling is only through the metallic energies V_{1C} and V_{1A}, as has been assumed so far, then the change of the crystal energy is just the change of the band structure energy due to δV_2. Then according to Eq. (3.1.10),

$$C_{11} - C_{12} = \frac{\sqrt{3}}{2d^3} \frac{\partial E_b}{\partial V_2} \frac{\delta V_2}{e^2}$$

$$= \frac{\sqrt{3}}{4d^3} \alpha_c (\sqrt{3}\, V_{sp\sigma} + 3V_{pp\sigma} - 3V_{pp\pi}) \left[1 + \frac{(\frac{3}{4} - \frac{9}{8}\alpha_c^2)(V_{1C}^{x2} + V_{1A}^{x2})}{V_2^2 + V_3^2} \right] \tag{3.5.6}$$

Harrison (1983b) pointed out, however, that in addition to V_1, other interactions such as V_1^x shown in Fig. 2.4 produce important contributions to the shear elastic constant. By arguing that these other contributions must cancel those associated with the change δV_2 arising from the metallization energy in a rigid rotation, Harrison deduced the following expression:

$$C_{11} - C_{12} = \frac{\sqrt{3}}{4d^3} \alpha_c^3 (\sqrt{3}\, V_{sp\sigma} + V_{pp\sigma} - 3V_{pp\pi}) \tag{3.5.7}$$

Under the strain e for C_{44} given in Eq. (3.1.11) and with an internal displacement given by $u = \eta d/\sqrt{3}$ as described in Section 3.1, the bond misalignment angles for the four bonds have the same magnitude with $(\delta\theta)^2 = 2(\eta + e/2)^2/9$. The four bond lengths also change, with the change for one pair given by $\delta r_1 = \delta r_2 = \delta + \varepsilon^2$ and by $\delta r_3 = \delta r_4 = -\delta + \varepsilon^2$ for the other pair, where $\delta = (e - \eta)d/3$ and $\varepsilon = (\eta + e/2)^2 d/9$. If again one assumes that the metallization is only through V_1, then the strain energy density can be shown to be given by

$$U = \frac{9B\delta^2}{2d^2} + 3(C_{11} - C_{12})(\delta\theta)^2$$

$$= \frac{B(e-\eta)^2}{2} + \frac{(C_{11} - C_{12})(\eta + e/2)^2}{3} \tag{3.5.8}$$

For a given strain e, U can be minimized with respect to η, which yields the Kleinman displacement parameter $\zeta = \eta/e$, with ζ given by

$$\zeta = (B - C/3)/(B + 2C/3) \tag{3.5.9}$$

where $C \equiv C_{11} - C_{12}$. Finally from $U = C_{44}\, e^2/2$ the following relationship is established:

$$9/C_{44} = 6/C + 4/B \tag{3.5.10}$$

or

$$I_{\mathrm{BOM}} = 9BC/[C_{44}(6B + 4C)] = 1 \tag{3.5.11}$$

If one includes the effect of V_1^x, the energy density will involve an additional term which couples δr and $\delta\theta$. Then the analysis is no longer simple.

I_{BOM} in Eq. (3.5.11), the Keating identity in Eq. (3.3.7), and Martin's identity in Eq. (3.3.17) all involve combinations of the three supposedly independent elastic constants C_{11}, C_{12}, and C_{44} of a crystal with cubic symmetry. However, the Keating and Martin identities are derived from the VFF model, which is based on a two-parameter (α and β) force model. Thus, there is no fundamental information revealed in the fact that the three parameters C_{11}, C_{12}, and C_{44} are interrelated. However, Eq. (3.5.10) is different. It was derived from an admittedly approximate tight-binding Hamiltonian, but not in a formalism that ensures a relation between B, C, and C_{44}. In fact Eq. (3.5.9) constitutes a proof that a two-parameter elastic theory like the VFF model has a chance to be correct to a fair level of approximation.

The above explicit formulas for the elastic constants in BOM are not much more complicated than the VFF model. However, they, as we have just noted, relate macroscopic forces to intrinsic atomic interactions. It is interesting to note that the simple identity relation Eq. (3.5.10) holds very well. As can be seen in Table 3.2 this result is certainly better than the I_K of Eq. (3.3.7) and is very competitive with Martin's identity I_M of Eq. (3.3.17), which requires the inclusion of his treatment of Coulomb forces. Because the long-range Coulomb energy is not included explicitly in the present derivation, its contribution to the elastic constants is probably small.

While the elastic identity relation derived from the BOM holds well, the numerical values of individual terms, i.e., Eqs. (3.5.2) and (3.5.3) for B, Eq. (3.5.7) for $C_{11}–C_{12}$, and (3.5.9) for C_{44}, are not accurate. These values are shown in Table 3.3 along with the results calculated for the same Hamiltonian based on the description of Section 3.4. The corresponding experimental values can be found in Table 3.4. All values from BOM are substantially smaller than the experimental results, but the relative strengths among the three components for a given system and among different systems hold well. This shows that the BOM as applied to elastic properties is more suited to comparative studies than to predicting individual numbers. Considering the simplicity of these BOM expressions, the agreement with the full-band-structure calculations is commendable. Note that these calculations are based on Harrison's universal TB parameters. Quantitative improvement for both BOM and full-band-structure TB calculations can be obtained by using individual sets of parameters for each material, as will be discussed in the next section.

TABLE 3.3. Comparison of Tight-Binding Theory Using Full Band Structures (BS) and Bond-Orbital Model for Bond Bulk Moduli B and Shear Coefficients $C = C_{11}–C_{12}$ and C_{44}[a]

	B		C		C_{44}	
	BOM	BS	BOM	BS	BOM	BS
C	49.31	47.39	67.15	67.10	44.45	51.78
Si	4.65	4.18	7.19	8.07	5.10	5.29
Ge	2.89	1.91	4.72	6.58	3.69	3.00
AlP	3.81	4.21	4.45	4.83	3.45	4.10
AlAs	3.12	3.66	3.85	4.07	2.91	3.50
AlSb	1.99	2.47	2.75	3.12	2.11	2.54
GaP	3.63	4.40	4.48	5.15	3.69	4.34
GaAs	2.83	3.76	3.75	4.30	3.02	3.66
GaSb	1.76	2.45	2.63	3.40	2.17	2.65
InP	2.35	2.91	2.69	2.93	2.21	2.63
InAs	1.93	2.67	2.38	2.52	1.90	2.32
InSb	1.28	1.97	1.81	2.17	1.48	1.87
ZnS	3.46	3.74	3.53	3.62	2.71	3.30
ZnSe	2.67	3.06	2.75	2.67	2.08	2.53
ZnTe	1.82	2.16	1.88	1.87	1.50	1.78
CdTe	1.24	1.47	1.22	1.14	0.96	1.13
HgTe	1.23	1.72	1.33	1.31	1.11	1.30

[a]All energies are in eV and elastic constants in 10^{11} dynes/cm^2.

TABLE 3.4. Experimental Values of Bond Length d, Bond Energy E_{bond}, Bulk Modulus B, and Shear Coefficient $C = C_{11}-C_{12}$ Used to Determine the Parameters in Tables 3.5 through 3.7[a]

	d	E_{bond}	B	C	C_{44}	ω
C	1.540	-3.68	44.227	95.120	57.740	1332
Si	2.532	-2.32	9.923	10.274	8.036	520
Ge	2.450	-1.94	7.653	8.189	6.816	301
AlP	2.367	-2.13	8.600	6.900	6.150	440
AlAs	2.451	-1.89	7.727	7.160	5.520	361
AlSb	2.656	-1.76	5.817	4.428	4.076	366
GaP	2.360	-1.78	9.143	7.870	7.143	367
GaAs	2.448	-1.63	7.690	6.630	6.040	269
GaSb	2.640	-1.48	5.792	4.946	4.440	231
InP	2.541	-1.74	7.247	4.460	4.600	304
InAs	2.622	-1.55	5.794	3.803	3.959	219
InSb	2.805	-1.40	4.831	3.130	3.132	185
ZnS	2.342	-1.59	7.637	3.990	4.558	279
ZnSe	2.454	-1.29	6.457	3.560	3.984	213
ZnTe	2.637	-1.20	5.090	3.060	3.120	177
CdTe	2.806	-1.10	4.210	1.680	2.040	141
HgTe	2.798	-0.81	4.759	1.817	2.259	116

[a]Also listed are the experimental values of C_{44} and the TO optical phonon mode ω at Γ to be compared with the calculations. All the elastic constants are in units of 10^{11} dynes/cm^2, d in Å, E_{bond} in eV, and ω_{TO} is given in terms of wave numbers in 1/cm. All bond lengths are deduced from the lattice constants quoted by Zallen (1982). The values of E_{bond} are taken from Harrison (1980, Table 7.3) except for AlSb, and ZnTe, which are deduced from the Phillips (1973, Tables 8.2, 8.3) and CdTe and HgTe which can be found in Chen *et al.* (1983). The elastic constants are taken from Table 3.1, and the phonon frequencies are taken from values compiled in "Landolt-Bornstein Numerical Data and Functional Relationships in Science and Technology," New Series, edited by K.-H. Hellwidge, Vols. 17 and 22.

3.6. QUANTITATIVE TIGHT-BINDING MODEL

The foregoing comparisons show that the BOM and the BS calculations predict similar results for the elastic constants and their predictions are only qualitatively correct. To be useful for predicting specific material properties, the theory has to be more quantitative. A simple way to improve the quantitative accuracy of the model is to use the scaling rules for the interactions $V_{\alpha\alpha'}$ in Eq. (2.7.2) and the repulsive energy u in Eq. (2.7.3). There are four parameters for each system, namely the scaling parameter f in Eq. (2.7.4), the powers n and m, and the value u_0. These parameters can be determined by requiring that the model produce the correct experimental values for E_{bond}, d_0, $C_{11}-C_{12}$, and B. Since $C_{11}-C_{12}$ is only governed by $V^0_{\alpha\alpha'}$ in both the BOM and the band calculation, it alone determines the scaling factor f. Then the bond energy E_{bond} can be used to determine u_0. The requirement that the first derivative of E_T be zero at d_0 then determines the ratio of the powers n/m, which couples with the equation for the bulk modulus to yield the values for n and m. One can then use these sets of parameters to calculate other quantities that are not employed in the fitting, e.g., C_{44}, the internal displacement parameter ζ, and the optical phonon frequencies ω at the zone center, to check the validity of the model. If acceptable, then the model can be extended to more complicated systems, such as alloys and superlattices having local environments similar to those of the bulk crystals.

Table 3.4 lists the values of bond lengths, bond energies E_{bond}, the elastic coefficients B, $C \equiv C_{11}-C_{12}$, and C_{44}, and the zone-center TO phonon frequencies ω for a selected group of systems to be studied for the remainder of this section. These are experimental values except for the elastic constants of AlP and AlAs, which were extrapolated. Since BOM and the full-band-structure calculations produce different results for a given TB Hamiltonian, as shown in Table 3.3, in a quantitative application it is desirable to treat the two calculations separately.

3.6.1. *Full-Band-Structure Calculation*

Table 3.5 shows the results for f, n, m, and u_0 obtained from the foregoing fitting procedure using full-band-structure calculations, and the corresponding values of C_{44}, ζ, and ω calculated as a consistency check. The scaling factor f ranges from 1 to 1.4 and tends to decrease with an increase in polarity. In the power dependence of $V_{\alpha\alpha'} \propto (d_0/d)^n$, n ranges from 2.8 to 4.3, which is larger than the $n = 2$ used in Harrison's universal TB parameters. For the repulsive pair energy $u = u_0(d_0/d)^m$, the power m ranges from 3.8 to 6.8. The ratio m/n is then 1.3 to 1.9, which is smaller than the $m/n = 2$ used by Harrison. The calculated values of C_{44} for most systems agree with experiment to 10% or better, except for diamond and ZnS. Note that the experimental data for ZnS are rather disparate. The calculated TO optical phonon modes at Γ in 1/cm units for most group IV and III–V systems also agree with experiments to 10% or better. The discrepancies for the II–VI systems are larger (about 15%). Reliable results for ζ from both experiments and first-principles calculations are available for only a limited number of systems. The calculated ζ in the TB model agrees very well with those results, as was shown in Table 3.1. The present TB calculation also produces the experimental cohesive energies, bond lengths, bulk moduli, and shear coefficients C_{11}–C_{12}, because these quantities are used to fit the parameters.

It is useful to know how predictions are influenced by these parameters and the fitting procedure. Table 3.6 shows results based on the Chadi (1978) procedure in which the TB

TABLE 3.5. Results for Parameters f, n, m, and u_0 Obtained from Fitting of the Bond Energy, Bond Length, Bulk Modulus, and Shear Coefficient C_{11}–C_{12} of Table 3.4 Using the Full-Band Structure Calculation[a]

	f	n	m	u_0	C_{44}	ζ	ω
C	1.390	2.480	3.767	21.924	48.393	0.121	1459
Si	1.326	3.040	5.001	6.938	8.013	0.511	572
Ge	1.388	3.204	5.278	6.415	6.841	0.487	342
AlP	1.294	3.530	5.598	6.435	5.827	0.516	447
AlAs	1.464	3.524	5.430	7.089	5.598	0.459	384
AlSb	1.337	3.268	5.668	4.838	3.944	0.564	354
GaP	1.395	3.705	5.683	7.285	6.857	0.501	382
GaAs	1.397	3.633	5.716	6.530	5.791	0.500	292
GaSb	1.431	3.471	5.717	5.519	4.515	0.536	256
InP	1.323	4.240	6.633	5.603	4.260	0.584	304
InAs	1.300	3.997	6.427	4.962	3.564	0.552	220
InSb	1.353	3.773	6.399	4.350	3.092	0.602	200
ZnS	1.062	2.208	5.996	4.225	3.727	0.632	325
ZnSe	1.134	3.420	5.994	4.260	3.164	0.576	233
ZnTe	1.284	3.306	5.828	4.285	2.813	0.590	205
CdTe	1.171	3.656	6.761	3.092	1.701	0.694	156
HgTe	1.173	3.760	7.074	3.080	2.040	0.716	152

[a]Also listed are the calculated C_{44}, internal displacement parameter ζ, and the TO optical phonon mode ω at Γ. All the elastic constants are in units of 10^{11} dynes/cm^2, u_0 is in eV, and ω are given in terms of wave numbers in 1/cm.

TABLE 3.6. Comparison Between the Two Different Sets of TB Parameters Described in the Text, the Resultant Coefficients u_0, u_1, and u_2, of u in Eq. (4.32), and the Predicted Elastic Constants, Kleinman Internal Displacement Parameters ζ, and Phonon Frequency ω from Chadi's Fitting Scheme

Si	ε_s^A	ε_p^A	ε_s^C	ε_p^C	$V_{ss\sigma}$	$V_{sp\sigma}$	$V_{pp\sigma}$	$V_{pp\pi}$	
Chadi	0.0	7.20	0.0	7.20	−2.03	2.55	4.55	−1.09	
Present	0.0	6.88	0.0	6.88	−2.41	2.59	4.05	−1.15	
Si	u_0	u_1	u_2	C	C_{44}	C_{44}^0	ζ	ω	
Chadi	7.29	−9.98	23.90	10.66	7.89	11.38	0.49	620	
Present	6.93	−9.70	23.42	10.27	7.83	11.39	0.51	592	
GaAs	ε_s^A	ε_p^A	ε_s^C	ε_p^C	$V_{ss\sigma}$	$V_{sp\sigma}^{AC}$	$V_{sp\sigma}^{CA}$	$V_{pp\sigma}$	$V_{pp\pi}$
Chadi	0.0	9.64	5.12	11.56	−1.70	2.40	1.90	3.44	−0.89
Present	0.0	10.09	6.79	14.12	−2.34	2.52	2.52	3.94	−1.12
GaAs	u_0	u_1	u_2	C	C_{44}	C_{44}^0	ζ	ω	
Chadi	5.12	−7.12	18.22	6.36	5.60	8.77	0.54	339	
Present	6.53	-839	19.90	6.63	5.70	8.53	0.54	322	

matrix elements $V_{\alpha\alpha'}$ are scaled as $1/d^2$, and the repulsive pair energy is taken to be $u = u_0 + u_1(d - d_0) + u_2(d - d_0)^2$. The parameter u_0 is set to produce the correct bond energy, u_1 is determined by requiring the correct equilibrium bond length, and u_2 is fixed by the bulk modulus. Two sets of TB parameters are tabulated for each system, one of them is the set used by Chadi (1978, 1979, and 1984), and the other is the set obtained by multiplying Harrison's $V_{\alpha\alpha'}$ by the scaling factor f listed in Table 3.5. For convenient comparison, the zero of the term values is set equal to the s energy of the anion. Despite considerable differences in these two sets of TB parameters, the results of the predictions from both sets are very similar and also very similar to those predicted from the other procedure giving the results in Table 3.5. The only noticeable difference between the predictions in Table 3.6 and Table 3.5 is that the present procedure produces larger phonon frequencies and slightly smaller C_{44} values. The fact that two theories with such disparate parameters fit experiments equally well is a clear warning about where they can be trusted. As long as the atom arrangements are close to those for which the parameters were fitted, these theories are trustworthy. However, in situations where there are large distortions, e.g., relaxed surfaces, dislocation cores, or charged impurities, quantitative predictions must be viewed with caution.

3.6.2. *Quantitative Extended Bond-Orbital Model*

Since it is pointed out by Harrison (1983a,b) that the inclusion of interactions such as V_1^π are essential in the calculation of elastic constants in the BOM, a systematic calculation should include all the first-neighbor interatomic TB parameters in the matrix elements $\langle b|H|a'\rangle$ that enter in the metallization terms in Eq. (2.5.7). Doing so extends the $|a'\rangle$ set to include those belonging to the second-neighbor bonds. This is the extended BOM. The calculation is still simple.

Table 3.7 shows the parameters obtained in the extended BOM. As expected, the parameters are not too different from those fitted in the full-band-structure calculation. In

TABLE 3.7. Results for the Parameters f, n, m, and u_0 Obtained from the Fitting of the Bond Energy, Bond Length, Bulk Modulus, and Shear Coefficient C_{11}–C_{12} of Table 3.4 Using the Extended Bond-Orbital Model Described in the Text[a]

	f	n	m	u_0	C_{44}	ζ	ω
C	1.336	2.901	3.872	20.814	47.691	0.135	1531
Si	1.262	3.251	5.346	6.484	7.472	0.580	562
Ge	1.302	3.886	6.264	5.942	6.171	0.658	333
AlP	1.193	3.443	5.577	5.72	4.936	0.541	452
AlAs	1.369	3.559	5.507	6.498	4.812	0.490	387
AlSb	1.249	3.235	5.669	4.402	3.248	0.619	343
GaP	1.276	3.550	5.575	6.422	5.738	0.542	373
GaAs	1.290	3.445	5.511	5.855	4.724	0.549	281
GaSb	1.318	3.415	5.704	4.923	3.670	0.621	239
InP	1.210	3.812	6.153	4.929	3.398	0.604	304
InAs	1.199	3.456	5.724	4.442	2.755	0.579	216
InSb	1.245	3.411	5.943	3.871	2.367	0.657	188
ZnS	0.968	3.062	5.848	3.618	3.120	0.630	343
ZnSe	1.040	3.097	5.671	3.727	2.611	0.580	247
ZnTe	1.178	3.019	5.573	3.731	2.301	0.601	212
CdTe	1.065	3.198	6.368	2.621	1.403	0.690	165
HgTe	1.059	2.937	6.262	2.523	1.563	0.730	154

[a]Also listed are the calculated C_{44}, internal displacement parameter ζ, and the TO optical phonon mode at Γ. All the elastic constants are in units of 10^{11} dynes/cm^2, u_0 is in eV, and ω is given in terms of wave numbers in 1/cm.

comparison with Table 3.5 and the experimental values in Table 3.4, the extended BOM does well for ω, produces slightly smaller C_{44} values, and slightly larger ζ values.

In conclusion, the TB method is a reasonable approach to the static elastic properties of semiconductors. If carried out rigorously, the TB parameters in Table 3.5 will provide quantitative results for superlattices, alloys, and possibly unreconstructed surfaces, those in which the local environments are similar to the ones in the bulk. The quantitative predictions of BOM are not as good as the BS calculations but are quite reasonable. The fitted parameters given in Table 3.7 allow approximations to be made using the BOM. This is especially useful for more complicated systems, because computationally the BOM is about two orders less complicated and faster than the band structure calculations.

3.7. ELASTICITY IN ALLOYS

Although semiconductor alloys have been widely used and studied, detailed information about their elastic constants is scarce, both experimentally and theoretically. One reason for lack of rigorous calculation of these systems is that the lack of translational symmetry in disordered alloys and complexity of the ordered alloys causes existing theories to be less accurate. Another reason may be attributed to the fact that most properties of these alloys including their elasticity were long thought to be reasonably well approximated by the concentration weighed average of the properties of the constituents. To the extent this is true, we still need to know their deviations from the straight-line average. Let us focus on the simplest elastic constant, the bulk modulus B, or instead the difference between the alloy bulk modulus B and the concentration weighted average value $\overline{B}$; i.e., $\Delta B = B - \overline{B}$. Is ΔB

TABLE 3.8. Bulk Moduli (in 10^{11} dyne/cm) of Ordered Alloys in Three Crystal Structures Calculated from the Present TB Model and the Percentage Deviations $\Delta B = (B - B_{av})/B_{av}$ from the Average Values B_{av} of the Constituent Compounds

	B_{CH}	B_{CA}	B_{CP}	B_{av}	$\Delta B_{CH} \times 100$	$\Delta B_{CA} \times 100$	$\Delta B_{CP} \times 100$
$AlGAAs_2$	7.695	7.693	7.689	7.7085	−0.181	−0.20	−0.257
$AlGaP_2$	8.858	8.858	8.854	8.8715	−0.150	−0.15	−0.201
$GaInSb_2$	5.226	5.202	5.156	0.3115	−1.607	−2.07	−2.923
$AlInAs_2$	6.705	6.691	6.661	6.7605	−0.828	−1.03	−1.475
$InGaAs_2$	6.610	6.57	6.508	6.6420	−1.961	−2.42	−3.474
$InAlP_2$	7.876	7.860	7.774	7.9235	−0.605	−0.08	−1.882
$GaInP_2$	8.007	8.035	8.878	8.1950	−1.437	−1.95	−3.865
Ga_2AsP	8.328	8.291	8.294	8.4165	−1.046	−1.50	−1.457
Ga_2AsSb	6.314	6.198	6.157	6.7410	−6.342	−8.05	−8.662
Ga_2PSb	6.584	6.297	6.188	7.4675	−11.836	−15.68	−17.135
$HgCdTe_2$	4.4697	4.4721	4.4706	4.4845	−0.33	−0.28	−0.31
$HgZnTe_2$	4.8898	4.8872	4.6323	4.9245	−0.71	−0.76	−5.93
$CdZnTe_2$	4.6105	4.6035	4.3375	4.6500	−0.85	−1.00	−6.72

positive or negative? Do the sign and the magnitude of ΔB depend on the state of order? How do we calculate them in a disordered system?

3.7.1. Ordered Alloys

As indicated in the preceding section, one reason for parametrizing the TB model is to use it to interpolate alloy properties from those of the constituents. Table 3.8 lists the results for the bulk moduli of a number of III–V and II–VI alloys calculated from full-TB-band-structure calculations using the parameters given in Table 3.5 and the structural information described in 1.4. Also listed are the average values and percentage deviations from the mean $\Delta B/\bar{B}$. Note that all ΔB values are negative and most of the magnitudes are small, except for Ga_2AsSb and Ga_2PSb; the latter has the largest difference in the constituent compounds among the alloys listed. Although the magnitudes of ΔB get larger for systems with larger differences in the bond lengths, the dependence does not seem to be a simple function of the bond-length difference. The uniformly negative ΔB values also appeared in the first-principles local density functional calculations for ordered Ga_2AsSb alloys by Ferreira *et al.* (1989), as shown in Table 3.9. Similarly, *a* −2% in $\Delta B/\bar{B}$ was obtained for the SiC/AlN alternating layer superlattice along (100) in a self-consistent density functional theory (SCDFT) calculation (Lambrecht and Segall, 1990).

TABLE 3.9. Calculated Bulk Moduli (in 10^{11} dynes/cm^2) for GaAs, GaSb, and Ga_2AsSb Ordered Alloys and the Corresponding Percentage Deviation from the Concentration Weighted Average[a]

	GaAs	GaSb	Ga_2AsSb			Ga_4As_3Sb		Ga_4AsSb_3	
	zb	zb	CA	CH	CP	LU	FA	LU	FA
B	7.46	5.18	6.10	5.92	5.96	6.52	6.58	5.40	5.51
$\Delta B/B \times 100$			−3.5	−6.3	−5.7	−5.4	−4.5	−6.1	−7.7

[a]zb = zinc blende, CA = CuAuI, CH = chalcopyrite, CP = CuPt, LU = luzonite, and FA = fametinite structures (Ferreira *et al.*, 1989).

The reason for the negative values of ΔB, in a very qualitative argument, is due to the fact that the bulk moduli of semiconductors scale inversely as high powers of the lattice constant, and, at the same time, the alloy lattice constant is approximated well by the mean value—Vegard's law (1921). Therefore, the value of B at the mean lattice constant should lie below the straight-line average. Because the TB results for the bulk moduli should not differ qualitatively from those predicted from the VFF, most of the key physics for the bowing of B is contained in a VFF analysis. The major effects in the VFF can in turn be realized from the following simple analysis.

Consider the structure of a CuAuI or chalcopyrite crystal ABC_2. Focus on a local tetrahedral cluster A_2B_2C with the two A atoms and two B atoms on the vertices of a tetrahedron and the C atom near the center. Let the coordinates of the two A atoms be $(-1,1,-1)d/\sqrt{3}$ and $(-1,-1,1)d/\sqrt{3}$, and the two B atoms be at $(1,1,1)d/\sqrt{3}$ and $(1,-1,-1)d/\sqrt{3}$. Let the force constants be k_A and k_B and the equilibrium bond lengths d_A and d_B for the AC and BC bonds, respectively. To attain equilibrium, the central C atom is displaced by $(\varepsilon, 0, 0)d/\sqrt{3}$. We further define mean values $\bar{d} = (d_A + d_B)/2$ and $\bar{k} = (k_A + k_B)/2$, the percentage differences $\delta_0 = (d_A - d_B)/\bar{d}$ and $\Delta_0 = (k_A - k_B)/\bar{k}$, and $d = \bar{d}(1+\delta)$. Then the AC bond is now stretched by an amount $\bar{d}(\delta + \varepsilon/3 - \delta_0/2)$ from its equilibrium value and similarly by $\bar{d}(\delta - \varepsilon/3 + \delta_0/2)$ for the BC bond. The strain energy for any arbitrary δ and ε is given by $\Delta E = \bar{d}^2[k_A(\delta + \varepsilon/3 - \delta_0/2)^2 + k_B(\delta - \varepsilon/3 + \delta_0/2)^2]$. When ΔE is minimized with respect to δ and ε, one finds $\delta = 0$ and $\varepsilon = 3\delta_0/2$, and the minimum ΔE is zero. If the crystal expands uniformly with δ having a fixed small value, then ε is now $\varepsilon = 3\delta_0/2 - \Delta_0\delta$ and $\Delta E = 2\bar{k}(1 - \Delta_0^2/4)\bar{d}^2\delta^2$. Thus, the effective spring constant is

$$k_{\text{eff}} = \bar{k}(1 - \Delta_0^2/4) \tag{3.7.1}$$

which is smaller than the average value $\bar{k}$. This weakening of the restoring force constant in the alloy is due to the internal displacement, represented by ε in the above model, which provides an extra degree of freedom for relaxation in response to the external stress. The bulk modulus B is proportional to k_{eff}/d, so the alloy bulk modulus minus the mean $\bar{B}$ is then

$$\Delta B = \bar{B}(\delta_0\Delta_0 - \delta_0^2 - \Delta_0^2)/4 \tag{3.7.2}$$

where we recall the definitions $\delta_0 = (d_A - d_B)/\bar{d}$ and $\Delta_0 = (k_A - k_B)/\bar{k}$. Since δ_0 and Δ_0 tend to have different signs, the bond-length difference causes an extra negative contribution to ΔB (the first two terms in Eq. (3.7.2). If both the bond-stretching force constant α and the bond-angle restoring force β in the VFF are included, the equilibrium value of ΔE is no longer zero, but the deviation ΔB can still be shown to be similar to Eq. (3.7.2) and is now

$$\Delta B = \frac{\bar{B}}{4}\left[\delta_0\left(\frac{3\Delta\alpha + \Delta\beta}{3\alpha + \beta}\right) - \delta_0^2 - 3\frac{(\Delta\alpha)^2}{(\alpha + 2\beta)(3\alpha + \beta)}\right] \tag{3.7.3}$$

where $\Delta\alpha = (\alpha_A - \alpha_B)$ and $\alpha = (\alpha_A + \alpha_B)/2$, and similarly for $\Delta\beta$ and β. Equation (3.7.3) reduces to (3.7.2) if β is set equal to zero.

The foregoing descriptions illustrate two of the mechanisms both contributing to negative ΔB values, namely the $1/d^q$ scaling of B, with q ranging from 3.5 to 9, and more degrees of freedom for internal relaxation. However, contributions from charge distribution and other "chemical" changes have not been included in these arguments. Quantitative results should be described by the TB formalism because, in addition to the strain energy,

TABLE 3.10. Bulk Moduli of Ordered Binary Alloys AB of the Diamond Semiconductors A and B from Theories and Experiment

SiC	PP–PW[a]	Experiment		
B(C)	50.23	44.23		
B(Si)	9.53	9.92		
B(SiC)	23.4	22.4		
$\Delta B/\bar{B}$(%)	–21	-17		
SiGe	**PP–PW[a]**	**PP–PW[b]**	**ASA[c]**	**FP–LMTO[c]**
B(Si)	9.53	9.8	8.80	9.58
B(Ge)	7.75	7.7	6.25	7.05
B(SiGe)	8.73	8.7	7.38	8.31
$\Delta B/\bar{B}$(%)	1	0	2	0

[a]Martins and Zunger (1986).
[b]Qteish and Resta (1988).
[c]van Schilfgaarde and Berding (1995).

there is also some "chemical effect" built into TB theory. Although ordered compounds have been found in epitaxial growth, the bulk moduli are difficult to measure, because these alloys are not single bulk crystals and because the ordering is only partial.

Not all the mechanisms considered above apply to ordered compounds of the elemental semiconductors, because internal relaxation under pressure may not be allowed, e.g., if the structure is assumed to be zinc blende. Unfortunately, a simple analysis of the elastic constants of these 4–4 compounds cannot be made readily because tight-binding and VFF parameters are not available for atomic pairs that do not exist in the constituent crystals. However, several first-principles calculations have been made on ordered SiC and SiGe (Martins and Zunger, 1986; Qteish and Resta, 1988; van Schilfgaarde and Berding, 1990). The main results are listed in Table 3.10. All theoretical results were calculated for the zinc blende structure. The PP-PW calculation of Martins and Zunger (1986) for SiC gave a –21% value for $\Delta B/\bar{B}$, which is in reasonable agreement with the experimental value of –17%. This is also consistent with the qualitative argument based on the $1/d^5$ scaling of B. For SiGe, some weak ordering has been found in epitaxial films (Ourmazd and Bean, 1985). The calculated $\Delta B/B$ for zinc blende SiGe is either slightly above or just below zero. These differences are small and fall within the uncertainties of the first-principles theory. A conclusion which can be drawn with some assurance is that the bulk moduli of ordered semiconductors are smaller than the straight-line average.

3.7.2. *Disordered Alloys*

The structural energy needed for calculating the elastic constants of a disordered alloy is an ensemble average of the total energy over a configuration distribution of the alloying atoms including strains. Thus, the study of this subject requires knowledge of configuration energies and the statistical distribution of atom configurations in alloys, subjects treated in the next chapter. Here we consider a simple effective-medium elastic model (Chen *et al.*, 1988) which shows why the bulk modulus of a random alloy is slightly weaker than that in an ordered alloy.

This theory starts by assuming that the alloy has an effective lattice constant and effective modulus. When part of the effective alloy medium is replaced by a specified cluster,

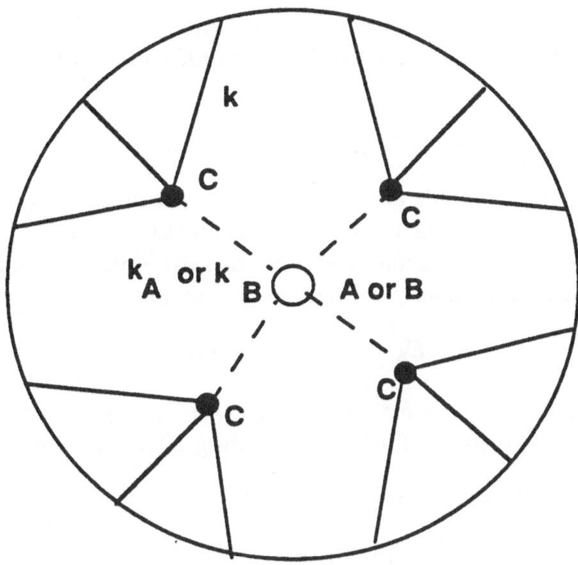

FIGURE 3.3. Schematic diagram of a four-bond cluster embedded in an alloy with an effective spring constant k.

strain energy will be introduced. This strain energy can be taken as the effective energy $\varepsilon(n)$ for that cluster. The index n specifies the size of the cluster, the numbers of A and B atoms, and their arrangement in the cluster. The probability distribution p_n for clusters of type n can be deduced within a statistical theory. The internal strain energy is calculated as $E = M\langle\varepsilon\rangle = M\Sigma p_n\varepsilon(n)$, where M is the ratio of the number of atoms in the alloy to those in a cluster. When the alloy is under an external pressure δP, the effective cluster energy will change by an amount $\delta\varepsilon(n)$, which implies a change of the total energy by an amount $\delta E = M\langle\delta\varepsilon(n)\rangle$. Then the bulk modulus of the alloy can be obtained from $\delta E = \frac{1}{2}(\delta P)^2 V/B$, where V is the alloy volume. The mean-field nature of this approach arises from the fact that the alloy lattice constant and elastic constants are set initially and, in the course of the calculation, are required to become self-consistent. To illustrate this self-consistency procedure, let us consider the following simple spring model for a random pseudobinary alloy $A_{1-x}B_xC$: the cluster corresponds to the four bonds surrounding an "impurity" atom A or B, and the environment of the cluster corresponds to the 12 bonds that connect inwardly to the cluster and outwardly to a rigid lattice of the effective alloy, as shown in Fig. 3.3. Let the spring constants for the pure AC and BC compounds be k_A and k_B respectively, and the effective alloy spring constant be k, with similar notations for the bond lengths d_A, d_B, and d. When an A atom is embedded in the medium, all 16 bonds under consideration will relax, the strain energy can be shown to be

$$\varepsilon(A) = \frac{1}{2}k_1(d - d_A)^2 \tag{3.7.4}$$

where

$$k_1 = 4k_A k/(3k_A + k) \tag{3.7.5}$$

A similar energy $\varepsilon(\mathrm{B})$ is obtained, when a B atom is embedded. The effective bond length d is obtained by minimizing the average cluster energy $E = (1-x)\varepsilon(\mathrm{A}) + x\varepsilon(\mathrm{B})$ with respect to d, which yields

$$d = [(1-x)k_1 d_\mathrm{A} + xk_2 d_\mathrm{B}]/[(1-x)k_1 + xk_2] \tag{3.7.6}$$

When the alloy is compressed, the alloy bond length is reduced to $d(1-e)$, where e is the macroscopic strain corresponding to the external pressure. The pressure-induced strain energy for the 16 bonds in the medium is $8k(de)^2$, or $2k(de)^2$ per four bonds. Embedding an A atom in this compressed medium, we find the total strain energy for the 16 bonds is

$$E_\mathrm{A} = \tfrac{1}{2}k_1(d - d_\mathrm{A} - 4de)^2 \tag{3.7.7}$$

To obtain the extra cluster energy $\delta\varepsilon(\mathrm{A})$ induced by the pressure, we subtract $\varepsilon(\mathrm{A})$ of Eq. (3.7.4) and the background energy for the surrounding 12 bonds from E_A to obtain

$$\delta\varepsilon(\mathrm{A}) = -4k_1(d - d_\mathrm{A})de + 8k_1(de)^2 - 6k(de)^2 \tag{3.7.8}$$

Similarly, the following expression for $\delta\varepsilon(\mathrm{B})$ is obtained when a B atom is embedded:

$$\delta\varepsilon(\mathrm{B}) = -4k_2(d - d_\mathrm{B})de + 8k_2(de)^2 - 6k(de)^2 \tag{3.7.9}$$

Thus, the change of the average cluster energy due to the pressure is given by $\delta E = \langle\delta\varepsilon(n)\rangle = (1-x)\delta\varepsilon(\mathrm{A}) + x\delta\varepsilon(\mathrm{B})$, which, when equated to $2k(de)^2$, leads to a self-consistent equation for the effective spring constant k: $k = (1-x)k_1 + xk_2$. The function k can now be obtained analytically when the expression for k_1 in Eq. (3.7.4) and a similar expression for k_2 are used. The result is

$$k = \langle k\rangle[1 - 3x(1-x)(\delta k/\langle k\rangle)^2] \tag{3.7.10}$$

where $\langle k\rangle = (1-x)k_\mathrm{A} + xk_\mathrm{B}$ is the mean spring constant of the constituents and $\delta k = k_\mathrm{A} - k_\mathrm{B}$ is their difference.

It is interesting to compare this result for a disordered 50–50 alloy; i.e., $k = \bar{k}(1 - \frac{3}{4}\Delta_0^2)$ with the value $\bar{k}(1 - \frac{1}{4}\Delta_0^2)$, in Eq. (3.7.2) for ordered alloys in the CuAuI and the chalcopyrite structures. According to these theories, alloy spring constants are always slightly below the straight-line average, and the bowing is larger for a disordered alloy than for the corresponding ordered compound. Using the effective spring constant in Eq. (3.7.10), the effective average bond length of the alloy becomes

$$d = \langle d\rangle + 4x(1-x)(d_\mathrm{A} - d_\mathrm{B})(k_\mathrm{A} - k_\mathrm{B})k/(3k_\mathrm{A} + k)(3k_\mathrm{B} + k) \tag{3.7.11}$$

which for most alloys lies below the mean value, because the constituent compounds with a large spring constant tend to have a smaller bond length. However, the deviation from $\langle d\rangle$ is extremely small for all III–V pseudobinary alloys.

The available experimental measurements for the elastic constants are rather limited. For $\mathrm{Ga_{1-x}Al_xAs}$, the following linear x dependences were observed (Landolt-Bornstein, 1988): $C_{11} = 11.85 + 0.14x$, $C_{12} = 5.38 + 0.32x$, and $C_{44} = 5.94 - 0.05x$. This lack of detectable bowing is expected, because of the nearly equal bond lengths of the two constituent compounds and corresponding small differences in the elastic constants. The bulk moduli for three II–VI alloy systems were measured in high-pressure x-ray diffraction experiments

TABLE 3.11. Measured Elastic Constants (in 10^{11} dyne/cm^2) of Si–Ge Alloys by Bublik *et al.* (1974)

Alloy	C_{11}	C_{12}	C_{44}
$Si_{0.28}Ge_{0.72}$	16.1 ± 0.8	8.35 ± 0.8	8.55 ± 0.4
$Si_{0.54}Ge_{0.46}$	17.0 ± 0.8		
$Si_{0.64}Ge_{0.36}$	17.1 ± 0.8		

(Quadri *et al.*, 1986). For HgCdTe, the results are similar to those of GaAlAs in that both the bond lengths and the bulk moduli of HgTe and CdTe are so close that the differences between B of the alloys and the pure crystals were beyond experimental resolution. However, a 5% Zn in CdZnTe alloy was found to have a 15% increase in the B value from the pure CdTe value and a 10% Mn in CdMnTe had a 21% decrease. For comparison, the bulk modulus of pure ZnTe is only 20.9% larger than that of CdTe. These significantly large changes in the B values caused by small concentrations cannot be explained by the above considerations.

The qualitative model treated here does not apply to the binary alloys $A_{1-x}B_x$, because in these alloys both the A and B atoms are found on both sublattices, and the local bond-length arrangement is more complicated than the pseudobinary alloys. However, there should be more degrees of relaxation in the disordered binaries than in the ordered compounds. Therefore, we would expect that the bulk modulus of the disordered 50–50 SiGe alloy would have a smaller value than those tabulated in Table 3.10 for the ordered compound. At least based on the arguments presented, we would not expect the B values of the disordered alloy to be significantly larger than the mean values $\bar{B}$. However, the only experimental data available (Bublik *et al.*, 1974 in Table 3.11) find all three elastic constants for the alloys at three different concentrations exceed the values for Si; $\Delta B/\bar{B}$ is as large as 20%, despite the fact that the bond-length difference between Si and Ge is only about 3% and the measured alloy lattice constants only bow slightly below their average. This and the unexplained results for the II–VI alloys point to the need for a more systematic study of the elastic properties of semiconductor alloys, both experimentally and theoretically.

3.8. CONCLUDING REMARKS

As mentioned in Chapter 2, LDA is an accurate *ab initio* theory for elastic constants for semiconductors, as is evident from Table 3.1 for Si, Ge, and GaAs. Figure 3.4 is a more complete compilation of LDA results based on the full-potential LMTO method by van Schilfgaarde and Berding (1995). Besides the standard LDA described in Chapter 2 (columns labeled LD), results from a gradient correction (labeled GC) are also included. The gradient correction gives improvements for the cohesive energies but introduces slightly larger errors in the elastic constants. Figure 3.4 shows the percentage errors of calculated B and C_{44}, except for two materials (InSb and ZnS) with larger errors. LDA elastic constants are generally accurate to within 10% deviation from experimental values. The exceptional cases warrant experimental and theoretical reexamination.

Among the methods discussed in this chapter, the VFF is the simplest because it is a classical model (i.e., without quantum calculations). With only two force constant parameters α and β, VFF provides an accurate description of the elemental semiconductors C, Si, and Ge. When corrected with Martin's Coulomb energies, VFF also produces accurate elastic constants for polar semiconductors.

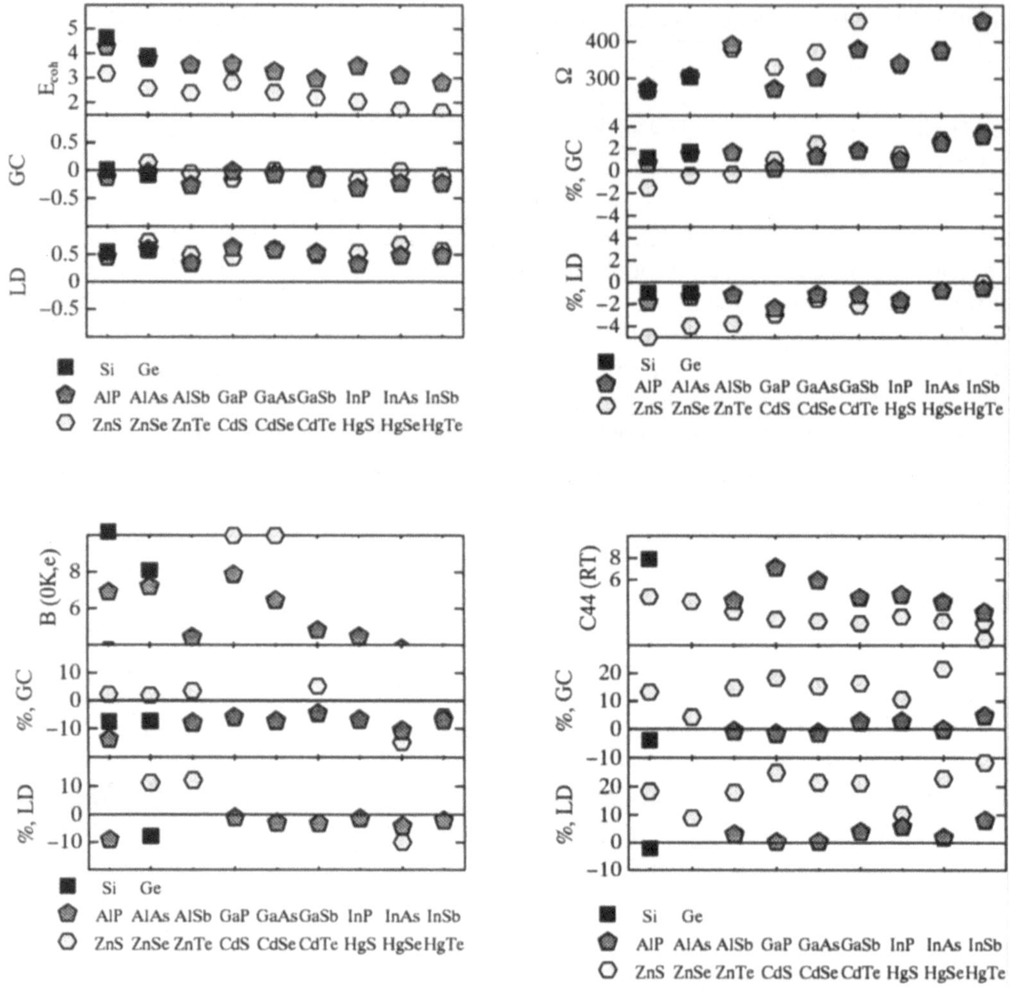

FIGURE 3.4. Full potential, gradient correction (GC), local density (LD) Schrödinger equation self-consistent solutions using the LMTO basis method, for all semiconductors in the zinc blende structure.

The BOM is also simple, although the analytic expressions for elastic constants in BOM are lengthier than those from VFF, because BOM involves quantum calculations of orbital interactions. It is amazing that these long BOM expressions collapse to a very simple relationship, $9/C_{44} = 6/(C_{11} - C_{12}) + 4/B$, which is satisfied by experimental data to 10% for all systems considered. Although Harrison's original universal-based BOM only yields semiquantitative values for the elastic constants, it can be made quantitative by using individualized interactions. Such calculations are useful when detailed information about a particular system is sought.

Finally, the empirical tight-binding model (ETB), which fits the cohesive energy, bond length, bulk modulus, and $C_{11} - C_{12}$, is shown to accurately predict C_{44}, the internal displacement parameters ζ, and the transverse optical phonon frequency ω for the constituent compounds. When extended to ordered alloys, these TB calculations predict that the alloy bulk moduli are slightly softer than the mean of the constituents' values. Available LDA calculations also indicated this trend. A model analysis shows that this trend extends to disordered alloys, although the deviation of B from the mean value is smaller than that for the ordered alloys. However, the few experimental results available have not confirmed this result. On the contrary, some data showed a strengthening of the elastic constants in disordered alloys. Clearly additional experimental and theoretical work is needed in this field. The present TB model, when directly applied to large-scale simulations, is an effective step in this direction.

REFERENCES

Aschcroft, N.W., and N.D. Mermin (1976), *Solid State Physics* (Saunders, Philadelphia), p. 445.

Blackmann, M. (1959), *Phil. Mag.* **3**, 831.

Boroni, S., P. Giannozzi, and A. Testa (1987), *Phys. Rev. Lett.* **58**, 1861.

Bublik, V.T., S.S. Gorelik, A.A. Zaitsev, and A.Y. Polyakov (1974), *Phys. Stat. Sol.* (b), **66**, 427.

Chadi, D.J. (1978), *Phys. Rev. Lett.* **41**, 1062.

Chadi, D.J. (1979), *Phys. Rev. B* **19**, 2074.

Chadi, D.J. (1984), *Phys. Rev. B* **29**, 785.

Chadi, D.J., and M.L. Cohen (1973), *Phys. Rev. B* **8**, 5747.

Chen, A.-B, A. Sher, and W.E. Spicer (1983), *J. Vac. Sci. Technol. A* **1**, 1674.

Chen, A.-B., M.A. Berding, and A. Sher, (1988), *Phys. Rev. B* **37**, 6285.

Cohen, M.L. (1985), *Phys. Rev. B* **32**, 7988.

Cousins, C.S.G., L. Gerwared, J. Staun, B. Selsmark, and B.J. Sheldon (1987), *J. Phys. C* **20**, 29.

Ferreira, L.G., S.-H. Wei, and A. Zunger (1989), *Phys. Rev. B* **40**, 3197.

Harrison, W.A. (1980), *Electronic Structure and the Properties of Solids* (Freeman, San Francisco).

Harrison, W.A. (1983a), *Phys. Rev. B* **27**, 3592.

Harrison, W.A. (1983b), *The Bonding Properties of Semiconductors* (SRI International).

Hirth, J.B., and J. Lothe (1982), *Theory of Dislocations*, 2nd ed. (Wiley, New York).

Keating, P.N. (1966), *Phys. Rev.* **145**, 637.

Kleinman, L. (1962), *Phys. Rev.* **128**, 2614.

Landolt-Bornstein (1988), in *Numerical and Functional Relationship in Science and Technology*, Vol. 22, ed. K.H. Hellemidege (Springer-Verlag, Berlin).

Lambrecht, W.R.L., and B. Segall (1990), private communication.

Love, A.E.H. (1944), *A Treatise on the Mathematical Theory of Elasticity* (Dover, New York).

Martin, R.M. (1970), *Phys. Rev. B* **1**, 4005.

Martins, J.L., and A. Zunger (1986), *Phys. Rev. Lett.* **56**, 1400.

Methfessel, M., C.O. Rodrigues, and O.K. Andersen (1989), *Phys. Rev. B* **40**, 2009.

Mittra, S.S., and N.E. Massa (1982), in *Handbook on Semiconductors*, ed. T.S. Moss (North-Holland, Amsterdam), Vol. 1, Chap. 3.

Nielsen, O.H., and R.M. Martin (1983), *Phys. Rev. Lett.* **50**, 697.

Nielsen, O.H., and R.M. Martin (1985a), *Phys. Rev. Lett. B* **32**, 3792.

Nielsen, O.H., and R.M. Martin (1985b), *Phys. Rev. Lett. B* **32**, 3780.

Ourmazd, A., and J.C. Bean (1985), *Phys. Rev. Lett.* **55**, 765.

Phillips, J.C. (1973), *Bonds and Bands in Semiconductors* (Academic Press, New York and London).

Qteish, A., and R. Resta (1988), *Phys. Rev. B* **37**, 1308.

Quadri, S.B., E.F. Skelton, and A.W. Webb (1986), *J. Vac. Sci. Technol. A* **4**(4), 1971, 1974.

van Schilfgaarde, M., and M.A. Berding (1990) (private communication).

van Schilfgaarde, M., and A. Sher (1987), *Phys. Rev. B* **36**, 4375.

van Schilfgaarde, M., and M.A. Berding (1995), private communication.

Van Vechten, J.A., and J.C. Phillips (1969), *Phys. Rev.* **187**, 1007.

Vegard, L. (1921), *Z. Phys.* **5**, 17.

Zallen, R. (1982), in *Handbook on Semiconductors*, ed. T.S. Moss (North-Holland, Amsterdam), Vol. 1, Chap. 1.

4

Alloy Statistics and Phase Diagrams

This chapter is devoted to the study of phase diagrams of semiconductors and their alloys, and the underlying statistics and free energies. It starts by treating the relationship between phase diagrams and free energies. The concepts of common tangent line and its relationship to quality of chemical potentials and activity coefficients are deduced. Then several approximate but analytical free-energy models frequently encountered in alloy phase diagram studies are introduced. These results provide the necessary background for a comprehensive review of statistical models and thermal data that have worked for semiconductors. The phase diagrams studied here include the liquidus curves of binary melts (such as $Ga_{1-x}As_x$) in equilibrium with the stoichiometric semiconductor compounds (such as GaAs), the miscibility gap of solid solutions, and the liquidus and solidus curves of ternary alloys. To study the detailed statistical properties of pseudobinary alloys, the quasi-chemical approximation (QCA) is generalized from pairs to clusters of arbitrary sizes. A transformation is devised that partitions the alloys' excess energies naturally into two parts: a strain-dominated part that depends only on the concentration, and a smaller "chemical" energy that controls the temperature dependence of the cluster populations. Although this generalized QCA (GQCA) calculation produces free energies and phase diagrams that are similar to those from previous empirical models, the physics is quite different. In previous models, a tendency for phase separation in alloys always implies there is a repulsive energy between the different alloying species. The present theory shows that, while the larger long-range strain energy drives phase separation, in most zinc blende alloys the chemical energies are attractive, so they favor local correlations resembling ordered compounds. These results are shown to be consistent with the conclusions based on LDA and the cluster variational method (CVM). The general mathematical structures of both GQCA and CVM and their limitations in special cases are also discussed.

4.1. MIXING FREE ENERGY, MISCIBILITY GAP, AND ORDER–DISORDER TRANSITIONS

Consider an ideal pseudobinary semiconductor alloy $A_{1-x}B_xC$ in the zinc blende structure in which the alloy atoms A and B randomly occupy their fcc sublattice sites while the C atoms occupy the other fcc sublattice. If the C atoms are treated as spectators, then this pseudobinary alloy behaves like a fcc binary alloy $A_{1-x}B_x$ in the statistical mechanics formalism. However, as evidenced by the EXAFS experiment discussed in Chapter 1, the atomic positions in a real semiconductor alloy distort slightly from the zinc blende sites. Because the bonding is covalent, the C atoms mediate the bond energies and in that sense affect the statistics. It will become clear later that the distribution of A and B atoms in an alloy is never completely random. This has important consequences on many physical properties. For example, a knowledge of structural energies and associated atomic distributions is essential to accurate calculations of phase diagrams.

In this chapter we deal with equilibrium statistics and phase diagrams. The equilibrium state at a fixed temperature T and pressure P is the one with the minimum Gibbs free energy. The equilibrium state may contain only one phase, such as an ordered alloy in a given crystal structure or a disordered solution with a uniform concentration. It may also contain several phases, such as an ordered alloy plus a disordered one or two disordered phases with different concentrations. Gibbs's phase rule (e.g., Landau and Lifshitz, 1986) states that a solution containing n species can have up to a maximum of $n + 2$ phases coexisting in equilibrium. Thus, to determine the equilibrium state at constant temperature and pressure of an alloy requires knowledge of the Gibbs free energies of all possible phases.

Under normal pressures of about 1 atmosphere, the difference between the Gibbs free energy G and the Helmholtz free energy F, $G - F = PV$, is insignificant for a solid or liquid. Thus, it is sufficient to use F for most cases. For a disordered zinc blende pseudobinary alloy $A_{1-x}B_xC$, it is convenient to define a mixing free energy ΔF as a function of x and T at a fixed pressure:

$$\Delta F(x,T) = F(x,T) - (1 - x)F_{AC}(T) - xF_{BC}(T) \quad (4.1.1)$$

where F, F_{AC}, and F_{BC} are, respectively, the Helmholtz free energies for the alloy, pure AC compound, and pure BC compound containing the same number of C atoms. Then ΔF can be written as

$$\Delta F = \Delta E - T\,\Delta S \quad (4.1.2)$$

where ΔE is the mixing energy and ΔS is the mixing entropy defined similarly to ΔF. Again, because the magnitude of $\Delta(PV)$ is small, ΔE and the mixing enthalpy ΔH are used interchangeably.

If $\Delta E > 0$ at $T = 0$, which turns out to be the case for most semiconductor alloys, then ΔF as a function of x at different fixed T will have the schematic shapes of the curves in Fig. 4.1a. The x dependence of ΔF is separated into two regimes by a critical temperature T_c. The curve labeled $T_3 > T_c$ represents a typical high-temperature curve. It tells us that disordered alloys with any concentration fraction x are stable at this temperature. However, the low-temperature curve labeled $T_2 < T_c$ indicates there is a miscibility gap for x within the interval $x_1 < x < x_2$. Here x_1 and x_2 are the points at which the common tangent line (the light horizontal line in Fig. 4.1a) touches the ΔF curve. This curve means that the alloy is thermally stable against decomposition for $x \leq x_1$ and $x \geq x_2$ but not inside the interval $x_1 < x < x_2$. For x inside the gap, the alloy tends to decompose into two alloys with concentrations x_1 and x_2 with proportions P_1 and P_2 governed by the "lever rule," $P_1 = (x_2 - x)/(x_2 - x_1)$ and $P_2 = (x - x_1)/(x_2 - x_1)$, because this decomposition minimizes the free energy.

There are two other special concentration values, x_1' and x_2', called the spinodal points. These occur at inflection points of ΔF at each $T < T_c$; i.e., $\partial^2\Delta F/\partial x^2 = 0$ at these x values.

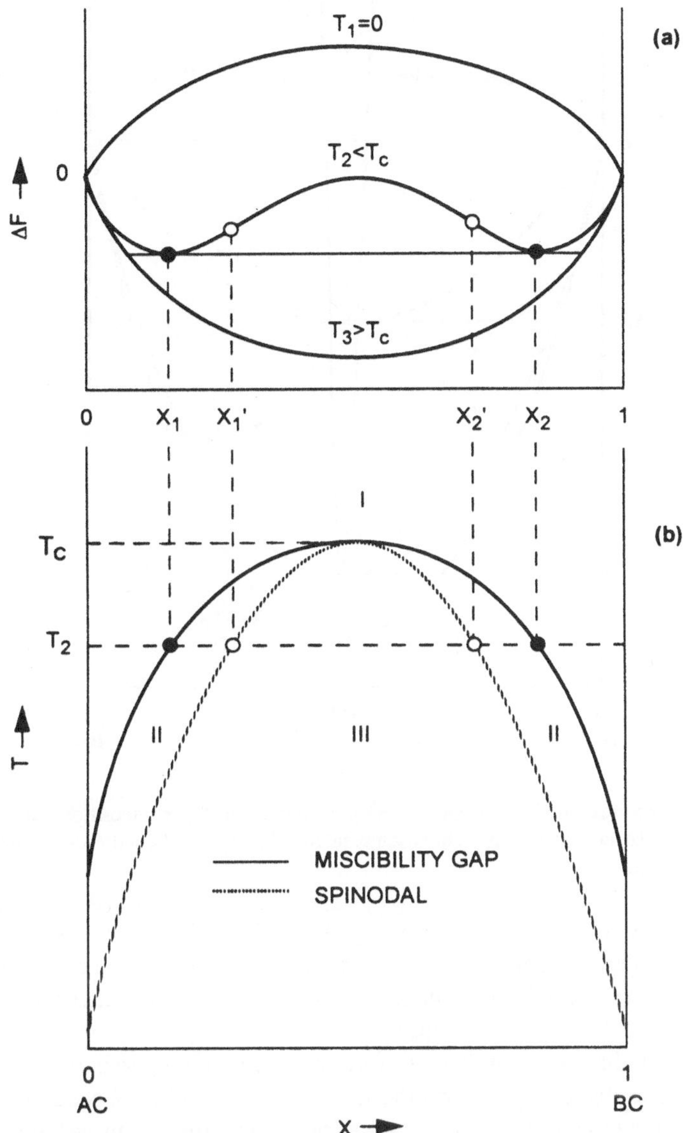

FIGURE 4.1. (a) Schematic of mixing free energies ΔF as a function of composition x of a pseudobinary alloy $A_{1-x}B_xC$ with a positive mixing enthalpy at three temperatures: $T_1 = 0$, $T_2 < T_c$ and $T_3 > T_c$. (b) Schematic miscibility gap and spinodal curves. Regions I, II, and III are stable, metastable, and unstable, respectively.

These values separate x space into unstable and metastable regions. For $x_1 < x < x_1'$ or $x_2' < x < x_2$, the alloy is metastable against local decomposition because the ΔF value for any x in these regions is lower than the lever-rule average value of ΔF with any two compositions in the neighborhood of x. This increase in ΔF acts as a temporary energy barrier against alloy decomposition into its final equilibrium concentrations x_1 and x_2. A uniform alloy with $x_1' < x < x_2'$ is inherently unstable because there is no such decomposition barrier. We can construct the binodal and spinodal curves by continuously varying the temperatures and

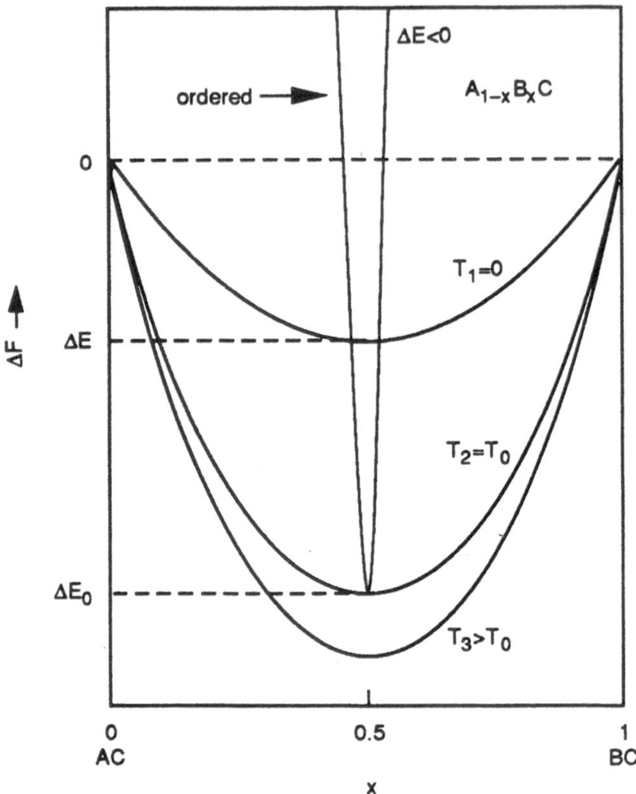

FIGURE 4.2. Schematic picture of mixing free energies ΔF as a function of alloy composition x of a pseudobinary alloy $A_{1-x}B_xC$ with a negative mixing enthalpy at three temperatures: $T_1 = 0$, $T_2 = T_c$, and $T_3 > T_c$. The lighter curve is for an ABC_2 ordered phase.

tracing the values of these gaps, as shown in Fig. 4.1b. These gap values become closer to each other as the temperature increases until a critical temperature T_c is reached, where all the values x_1, x_2, x_1' and x_2' merge into one value. Above T_c, and before melting, the disordered solid phase is stable for all alloy concentrations.

If $\Delta E < 0$ at $T = 0$, there is a tendency to form a long-range ordered alloy. Whether the alloy is ordered or disordered depends on the temperature. Figure 4.2 schematically compares the mixing free energy of the disordered alloy against that of an ordered compound ABC_2. With a negative mixing enthalpy, the energy of the ordered system ΔE_0 is lower than the disordered alloy (i.e., $\Delta E_0 < \Delta E$ at $T = 0$), so that the ordered state is the equilibrium state. As temperature increases, ΔF of the disordered alloy becomes more negative, mainly because the entropy term in Eq. (4.1.2) decreases roughly linearly in T. The system remains in the ordered phase until a transition temperature T_0 is reached when ΔF is equal to ΔE_0. The disordered alloy then becomes the stable state for temperatures greater than T_0.

4.2. ANALYTICAL MODELS

Since pseudobinary alloys $A_{1-x}B_xC$ statistically are similar to binary alloys $A_{1-x}B_x$, the binary results will be used whenever they do not cause any confusion. The statistical theory for binary solid solutions is mathematically equivalent to the three-dimensional Ising model in magnetism. Even for the simplest case with nearest-pair interactions, an analytical solution of this model remains one of the most challenging problems in theoretical physics. However, systematic computational methods are available, as will be discussed later. This section briefly reviews several analytical models, which are useful for illustrating the basic concepts and for semiempirical phase diagram evaluation. This review will focus on disordered alloys with positive mixing energies, because this is the most relevant case for most bulk semiconductors. However, results related to ordered cases will also be discussed.

4.2.1. *Ideal-Solution Model*

The ideal-solution model is equivalent to the case where the mixing energy vanishes, i.e., $\Delta E = 0$ in Eq. (4.1.2). Thus, in this approximation, the mixing free energy ΔF results totally from the additional entropy arising from a random arrangement of N_A A atoms and N_B B atoms on their N lattice sites, with $N = N_A + N_B$ and $x = N_B/N$. The number of different arrangements is

$$\Phi_0 = \frac{N!}{N_A!N_B!} \tag{4.2.1}$$

The corresponding mixing entropy is $\Delta S = k \ln \Phi_0$. Using the Stirling approximation for large N, $\ln N! \cong N \ln N - N$, we obtain the mixing entropy for the random alloy:

$$\Delta S = -Nk[(1-x)\ln(1-x) + x \ln x] \tag{4.2.2}$$

This model always yields negative values of $\Delta F = -T\,\Delta S$ for all x and T. As a function of x, ΔF in Eq. (4.2.2) is always concave upward, so no miscibility gap exists at any T.

4.2.2. *Zeroth Approximation*

When ΔE is not zero but the mixing entropy is still set equal to the random alloy result, Guggenheim (1952) called this model the "strict-regular solution" model, or the zeroth approximation. If, following shapes often experimentally observed in pseudobinary alloys, ΔE is assumed to have an x dependence given by

$$\Delta E = N\mathrm{x}(1-x)\omega \quad \text{or} \quad = \nu \mathrm{x}(1-x)\Omega \tag{4.2.3}$$

then miscibility gaps exist at low temperatures for positive molar interaction parameters Ω. At this point Ω is being treated as an empirical fitting parameter, but later in the chapter it will be connected to various energetic mechanisms. In Eq. (4.2.3), ν is the molar number, $\nu = N/N_0$, and N_0 is Avogadro's number.

Equation (4.2.3), as we now demonstrate, is the expression for the mixing energy if pairwise interaction energies and a random distribution are assumed. Let z be the coordination number. Note that z is the number of nearest-neighbor atoms surrounding a site in the binary $A_{1-x}B_x$ case, but it represents the number of the second-neighbor atoms in a pseudobinary alloy $A_xB_{1-x}C$. For example, the value of z for a zinc blende alloy is 12. Then the total number of pairs involving the A and B atoms is $M = zN/2$. Let M_{AA}, M_{AB}, and M_{BB} be the number of AA, AB, and BB pairs respectively, and let ε_{AA}, ε_{AB}, and ε_{BB} be the corresponding pair interaction energies. We further define the fractions $r_{ij} = M_{ij}/M$. Not all these pair fractions are independent. Let $r = r_{AB}$. Then $r_{AA} = 1 - x - 0.5r$ and $r_{BB} = x - 0.5r$. The mixing energy is governed only by the AB pair fraction r

$$\Delta E = M_{AA}\varepsilon_{AA} + M_{AB}\varepsilon_{AB} + M_{BB}\varepsilon_{BB} - (1-x)M\varepsilon_{AA} - xM\varepsilon_{BB} = Mr\varepsilon \tag{4.2.4}$$

where ε is the excess energy per pair and is defined as

$$\varepsilon = \varepsilon_{AB} - (\varepsilon_{AA} + \varepsilon_{BB})/2 \tag{4.2.5}$$

For a random alloy the pair probability is given by

$$r = 2(1-x)x \tag{4.2.6}$$

so the mixing energy has the form of Eq. (4.2.3), with the interaction parameter given by

$$\Omega = N_0 z\varepsilon \tag{4.2.7}$$

This strict-regular model is the simplest one that contains some aspect of reality, because it relates a positive mixing enthalpy to a miscibility gap and critical temperature T_c. For example, T_c can be obtained explicitly by setting $\partial^2\Delta F/\partial x^2$ equal to zero at $x = 1/2$,* where $\Delta F = \Delta E - T_c\Delta S$ with E in the form of Eq. (4.2.3) and ΔS having the random alloy expression in Eq. (4.2.2). The result is

$$T_c = \frac{\Omega}{2R} = \frac{\omega}{2k} \tag{4.2.8}$$

where $R = N_0 k$ is the universal gas constant, and $\omega = z\varepsilon$.

4.2.3. *First Approximation—The Quasi-Chemical Approximation*

In the zeroth approximation, an energy term is added, but the distribution of A and B atoms is still constrained to be random, a condition not consistent with the nonzero excess pair energy ε. Intuitively one expects that the pair probability r for the AB pair should be smaller than the random value in Eq. (4.2.6) if ε is positive, and vice versa. The first approximation in Guggenheim's notation (1952) corrects this flaw in an approximate way. The correction comes from the mixing entropy. We will introduce a counting procedure that is intuitively more appealing than the one Guggenheim actually used. For a random alloy the number of ways of configuring the A and B atoms, Φ_0, is given by Eq. (4.2.1). For a specified set of M_{AA}, M_{AB}, and M_{BB}, the number of complexions Φ (i.e., ways to arrange A and B atoms) should be just Φ_0 times the probability of having the specified set of numbers of pairs (Sher *et al.*, 1987):

$$\Phi = \frac{N!}{N_A!\,N_B!}\left[\frac{M!}{M_{AA}!\,M_{AB}!\,M_{BB}!}y^{2M_{AA}}(2xy)^{M_{AB}}x^{2M_{BB}}\right] \tag{4.2.9}$$

In Eq. (4.2.9), $y = 1 - x$ is the fractional concentration of A atoms. The term outside the brackets is the total distinguishable number of ways to arrange N_A and N_B atoms on N sites,

*ΔF has its second derivative vanish at $x = 1/2$ in this case because of the assumed symmetric forms of Eqs. (4.2.2) and (4.2.3). In more general cases that extend beyond pair interactions, as we shall see, the second derivative can vanish at different x values.

and the term inside the brackets is the fraction of these ways (i.e., the probability) M_{AA}, M_{AB}, and M_{BB} pairs will be found when the concentration of B atoms is x. This probability is the number of ways of arranging the M_{AA}, M_{AB}, and M_{BB} pairs on M pair sites times the probability of having these pairs in the system, i.e., $(y^2)^{M_{AA}}(2xy)^{M_{AB}}(x^2)^{M_{BB}}$. This expression still neglects the correlations between pairs.

Equation (4.2.9) should be compared with Guggenheim's (1952) combinatorial formula

$$\Phi_G = \frac{N!}{N_A!\,N_B!}\,\frac{(Mx^2)!(Mxy)!(Mxy)!(My^2)!}{M_{AA}!\,M_{AB}!\,M_{BA}!\,M_{BB}!} \tag{4.2.10}$$

Note that the leading terms of $\ln \Phi$ and $\ln \Phi_G$ in the Stirling approximation are the same, so these two expressions give the same mixing entropy. However, Eq. (4.2.9) is an easier form to extend beyond the pair approximation.

The mixing free energy with Φ given by Eq. (4.2.9) and excess energy ε by Eq. (4.2.5) in terms of the pair probability $r \equiv r_{AB} = M_{AB}/M$ and with the constraints $2r_{AA} + r = 2y$ and $2r_{BB} + r = 2x$ becomes

$$\begin{aligned}\Delta F &= Mr\varepsilon - kT \ln \Phi_0 \\ &\quad + MkT[(x - r/2)\ln(x - r/2) + (y - r/2)\ln(y - r/2) \\ &\quad + r\ln(r/2) - 2(x \ln x + y \ln y)]\end{aligned} \tag{4.2.11}$$

The value of r for an equilibrium distribution is the one that minimizes ΔF. Taking the partial derivative $\partial \Delta F/\partial r$ and setting it equal to zero leads to a quadratic algebraic equation for r. The proper solution is

$$r = \frac{4xy}{1 + \{1 + 4xy[\exp(2\varepsilon/kT) - 1]\}^{1/2}} \tag{4.2.12}$$

This expression reduces to several correct limits. As ε approaches 0, r approaches $2xy$, which is the random limit given in Eq. (4.2.6). For large $\varepsilon >> kT$, r decreases exponentially, $r \sim e^{-\varepsilon/kT}$, as expected. If ε is negative, then $r > 2xy$, and for large enough $|\varepsilon| >> kT$ the system is eventually driven into compound formation; i.e., $r \to 2x$ for $x < 1/2$, or $r \to 2y$ for $x > 1/2$. However, when ε is positive and at low T, the present approximation leads to a negative entropy (see Appendix B).

The AB pair probability r in Eq. (4.2.12) can also be obtained by considering the following chemical equilibrium:

$$\mathrm{AA} + \mathrm{BB} \leftrightarrow 2\mathrm{AB} \tag{4.2.13}$$

According to the law of mass action, the equilibrium pair fractions obey the relation

$$\frac{r_{AB}^2}{r_{AA}r_{BB}} = \frac{Z_{AB}^2}{Z_{AA}Z_{BB}} = \frac{(2e^{-\varepsilon_{AB}\beta})^2}{e^{-\varepsilon_{AA}\beta}e^{-\varepsilon_{BB}\beta}} = \frac{4}{e^{2\varepsilon\beta}} \tag{4.2.14}$$

where $\beta = 1/kT$ and the Z_{ij}'s are the partition functions for the indicated pairs. Equation (4.2.14) reduces to

$$r^2 e^{2\varepsilon/kT} = r^2 - 2r + 4xy \tag{4.2.15}$$

The proper solution for r from this equation is given in Eq. (4.2.12). It is in this connection that the first approximation is often referred to as the quasi-chemical approximation (QCA).

The QCA represents an improvement over the zeroth approximation because in it the pair distribution of A and B atoms depends on the pair energies. The level of improvement can be appreciated by comparison with the exact analytical solutions in two dimensions. The binary alloy statistical problem with pair interactions is equivalent to the magnetic Ising model. For $\varepsilon > 0$, the alloy problem can be directly translated into the ferromagnetic Ising model. The only difference is that while the alloy concentration x is specified, x in the magnetic problem is further varied to minimize ΔF. Thus, above T_c and with pair interactions the minimum of ΔF always occurs at $x = 1/2$, which in the magnetic case means no net spin or magnetic moment. However, for temperatures below T_c, ΔF has two minima occurring at x values x_1 and x_2 corresponding to the miscibility gaps indicated in Fig. 4.1a. For magnetism, this situation corresponds to having a net magnetic moment. Thus, the alloy spinodal critical temperature T_c is equivalent to the magnetic phase transition critical temperature.

The value of T_c in the pair model can be written in terms of a parameter λ as

$$T_c = \frac{\Omega}{\lambda R} \tag{4.2.16}$$

where $\Omega = N_0 z\varepsilon$ is the interaction parameter defined in Eq. (4.2.5), and λ is a numerical factor that depends on the model used. For the zeroth approximation we already found $\lambda = 2$ (see Eq. (4.2.8)). The zeroth approximation is equivalent to the Bragg–Williams (1935) approximation to the Ising model. For QCA,

$$\lambda = z \ln\left(\frac{z}{z-2}\right) \tag{4.2.17}$$

This result can be obtained by starting from the expressions for ΔF in Eq. (4.2.11) and r in Eq. (4.2.12), taking the second derivative $\partial^2 \Delta F/\partial x^2$, and setting it equal to zero at $x = 1/2$. In doing so, the conditions $\partial \Delta F/\partial r = 0$ and $\partial r/\partial x = 0$ at $x = 1/2$ are used. Note that QCA for alloys is equivalent to the Bethe–Peierls approximation (Bethe, 1935; Huang, 1987) to the Ising model.

Table 4.1 compares the λ values that enter the T_c formula, Eq. (4.2.17), for a square lattice from different approximations against the exact solution (Onsager, 1944) and for a fcc lattice against the best numerical value available (de Fontaine, 1979). The results obtained from the cluster variation method (CVM), discussed in Section 4.10, are also listed for later use. Progressively more sophisticated statistical theories result in larger λ values and, hence, smaller critical temperatures. The QCA predicts only 79% of the exact λ value in the square

TABLE 4.1. Values of λ for the Transition Temperature $T_c = \Omega_v/\lambda R$ from Different Calculations

Alloy, Ising model	Zeroth approx. Bragg–Williams	QCA, Bethe–Peirls	CVM[a]	Exact or best result
Square lattice	2	2.773	3.332 (square)	3.5255
fcc	2	2.188	2.438	2.4501

[a]CVM indicates the results from the cluster variation method to be discussed in Section 4.10.
After de Fontaine (1979, Table V).

lattice case and gives 89% of the best computer-simulated value for the fcc lattice. Although these statistical models have been used with reasonable success to describe the semiconductor alloy phase diagrams, as discussed in the next three sections, the pair energies do not adequately describe the energetics. Improved models will be considered in later sections.

4.3. PHASE DIAGRAM: COMMON TANGENT LINE AND ACTIVITY COEFFICIENT

So far our calculation of free energies has been confined to solid solutions. To find the phase diagrams one also needs to know the free energies of liquids (or melts). Very often liquids are treated as dense lattice gases, as solids with a high concentration of vacancies, or simply as a disordered system with some effective coordination number. The latter treatment has been used with some success in the calculation of phase diagrams of compound semiconductors and alloys. This and the next sections introduce the basic ideas governing phase diagrams, including (1) the liquidus curve between a binary liquid $A_{1-x}B_x$ and a stoichiometric compound AB, e.g., between a $Ga_{1-x}As_x$ melt and stoichiometric solid GaAs; (2) the liquidus–solidus curves describing the equilibrium between a ternary liquid and a solid pseudobinary alloy; and (3) the general ternary phase diagrams. The basic ideas and most results are in Vieland (1963), Stringfellow and Green (1969), Kikuchi (1981, 1982a,b), Brebrick *et al.* (1983), and Casey and Panish (1978).

As mentioned in Section 4.1 the equilibrium state at a given P and T is the state with minimum Gibbs free energy. To construct a phase diagram, we calculate the free-energy curves as a function of alloy concentrations for all the relevant phases and then find the combination of phases that minimizes the free energy. For example, consider the liquidus–solidus curves between a pseudobinary liquid and the solid solutions $A_{1-x}B_xC$ schematically shown in Fig. 4.3a. The corresponding free-energy curves for both phases as a function of x for a given temperature T_1 are drawn in Fig. 4.3b. Note to the common tangent line that touches the two free-energy curves at x_1 and x_2. This picture illustrates the equilibrium-state behavior as a function of the average alloy concentration x at this temperature. For $x < x_1$ the liquid phase is the stable phase, and for $x > x_2$ the solid solution is the stable phase. For an average concentration lying between x_1 and x_2 the stable state is a mixture of a liquid phase with concentration x_1 and a solid phase of concentration x_2 with proportions governed by the lever rule. The two free-energy curves at this temperature T_1 give two phase boundary points (denoted by L and S) for the liquidus–solidus curves shown in Fig. 4.3a. The full liquidus–solidus curves are generated from a set of such free-energy curves derived by continuously varying the temperature. This procedure for minimizing the free energy using the common tangent line is equivalent to matching the chemical potentials, as will be demonstrated.

The Gibbs free energy G, or in practice the Helmholtz free energy F, for a given phase can be written as a sum of the products of the number of particles N_i and the chemical potentials μ_i of all the atom and molecular species i in that phase:

$$F = G = \sum_i N_i \mu_i \tag{4.3.1}$$

Conversely, the chemical potential can be calculated from

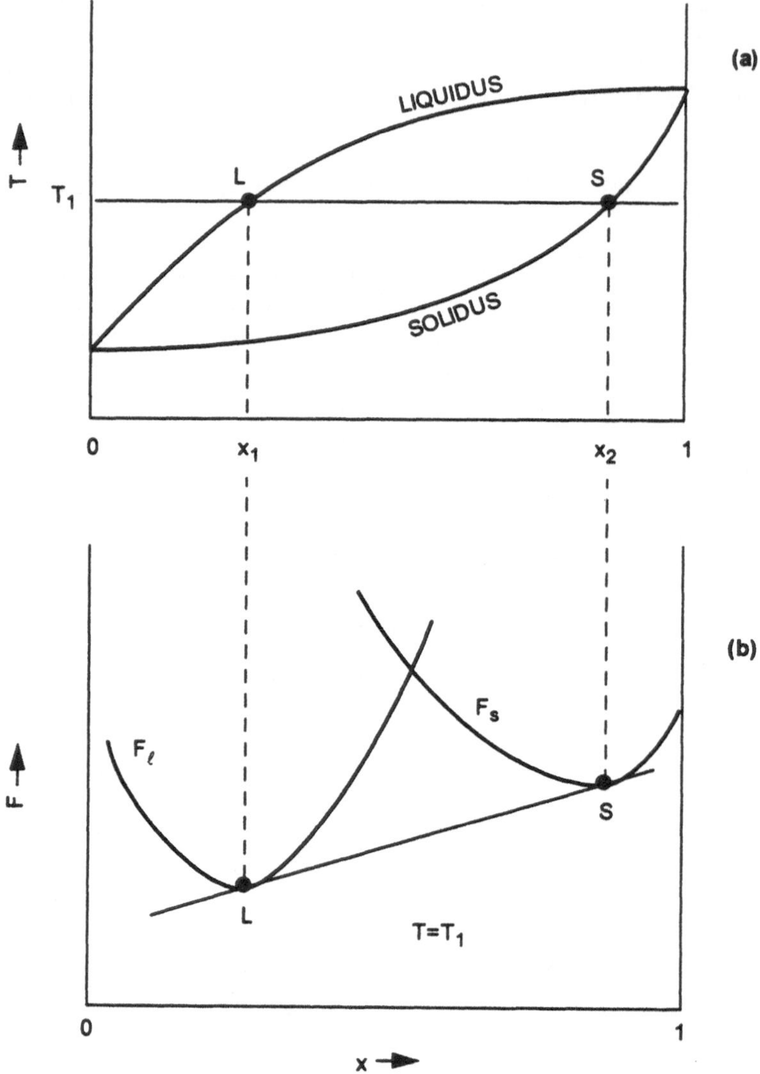

FIGURE 4.3. (a) Schematic liquidus and solidus curves and (b) the corresponding free energies of the liquid F_l and solid F_s at T_1 as a function of alloy composition x.

$$\mu_i = \left(\frac{\partial F}{\partial N_i}\right)_{(N_j,\, i \neq j)} \tag{4.3.2}$$

Now consider the equilibrium between two phases, denoted by α and β, of a binary alloy $A_{1-x}B_x$ at a given temperature T, such as depicted in Fig. 4.3b. Let the common tangent line touch the two free-energy curves F_α and F_β at x_α and x_β respectively. Denote the chemical potential of an A atom in the α phase, $\mu_A^\alpha(x_\alpha, T)$, as μ_A^α, and assign similar meanings to μ_B^α, μ_A^β, and μ_B^β . Algebraically, the common tangent line requires

$$\frac{\partial F_\alpha(x_\alpha, T)}{\partial x} = \frac{\partial F_\beta(x_\beta, T)}{\partial x} \tag{4.3.3}$$

and

$$F_\beta(x_\beta, T) = F_\alpha(x_\alpha, T) + (x_\beta - x_\alpha)\frac{\partial F_\alpha(x_\alpha, T)}{\partial x} \tag{4.3.4}$$

In terms of the chemical potentials, we have

$$F_\alpha(x_\alpha, T) = N[x_\alpha \mu_B^\alpha + (1 - x_\alpha)\mu_A^\alpha] \tag{4.3.5}$$

$$F_\beta(x_\beta, T) = N[x_\beta \mu_B^\beta + (1 - x_\beta)\mu_A^\beta] \tag{4.3.6}$$

$$\frac{\partial F_\alpha(x_\alpha, T)}{\partial x} = N(\mu_B^\alpha - \mu_A^\alpha) \tag{4.3.7}$$

$$\frac{\partial F_\beta(x_\beta, T)}{\partial x} = N(\mu_B^\beta - \mu_A^\beta) \tag{4.3.8}$$

Using these expressions in Eqs. (4.3.3) and (4.3.4) leads to the usual conditions for chemical equilibrium: $\mu_A^\alpha = \mu_A^\beta$ and $\mu_B^\alpha = \mu_B^\beta$.

Very often the activity coefficients γ rather than the chemical potentials μ are used in phase diagram evaluations. The activity coefficient γ_i is related to the chemical potentials by

$$\mu_i - \mu_i^0 = kT \ln(x_i \gamma_i) \tag{4.3.9}$$

where μ_i^0 is the chemical potential for the pure phase of the ith species. For an ideal solution—i.e., a random mixture which is always found in the limit $T \to \infty$—all the activity coefficients become unity: $\gamma_i = 1$. In the zeroth approximation for a binary $A_{1-x}B_x$ (or a pseudobinary $A_{1-x}B_xC$), the activity coefficients can be shown to be given by

$$\gamma_A = e^{x^2(\Omega/RT)} \tag{4.3.10}$$

$$\gamma_B = e^{(1-x)^2(\Omega/RT)} \tag{4.3.11}$$

In QCA, using Eqs. (4.2.11) and (4.2.12), one finds

$$\gamma_A = \left(\frac{1 - 2x + f}{(1 - x)(1 + f)}\right)^{z/2} \tag{4.3.12}$$

$$\gamma_B = \left(\frac{2x - 1 + f}{x(1 + f)}\right)^{z/2} \tag{4.3.13}$$

where f is related to the excess energy ε by

$$f = [1 + 4xy(e^{2\varepsilon\beta} - 1)]^{1/2} \tag{4.3.14}$$

with $\beta = 1/kT$. These results will be used in the next two sections.

4.4. VIELAND'S METHOD AND BINARY LIQUIDUS

The Vieland (1963) method, which establishes the relation between chemical potentials of a stoichiometric solid AB compound and those of the supercooled liquid solution $A_{0.5}B_{0.5}$ through the entropy of fusion and heat capacity differences between the two phases, remains the principal way of calculating phase diagrams between semiconductor solids and liquids.

Consider the equilibrium between a liquid solution $A_{1-x}B_x$ and a solid compound AB. A schematic liquidus curve is shown in Fig. 4.4a, while the corresponding free-energy

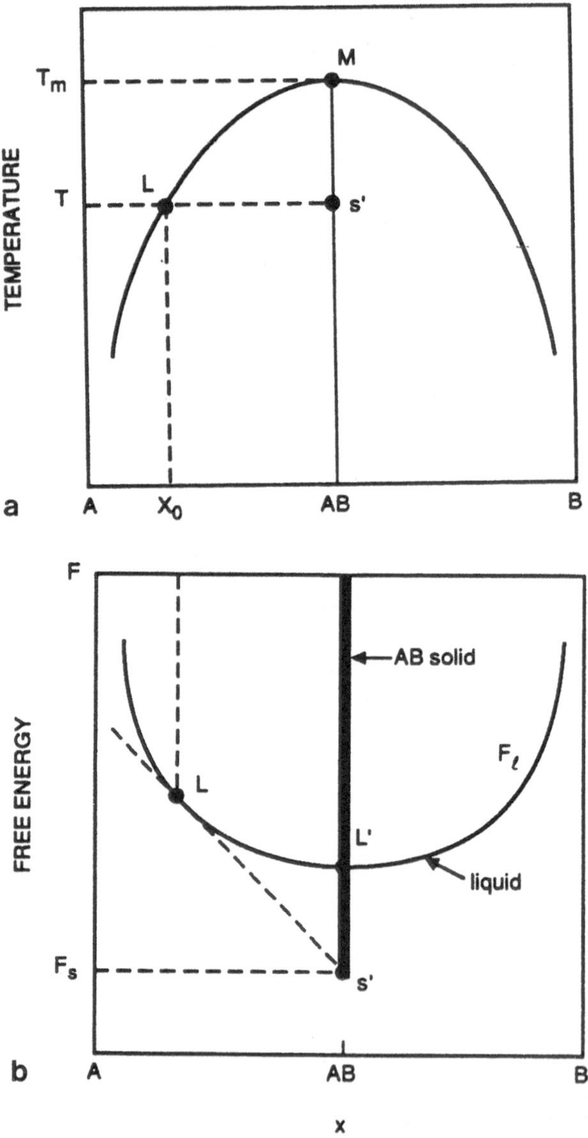

FIGURE 4.4. (a) Schematic liquidus curve for a compound semiconductor, (b) free-energy curve F_l as a function of composition x of a binary liquid at a given temperature T and the corresponding value F_s for the solid compound at the same temperature. Points L and s' coexist.

diagram at a given temperature T is given in Fig. 4.4b. In equilibrium, the chemical potential μ_{AB} per AB pair (unit cell) in the AB compound must be equal to the sum of the chemical potentials in the liquid solution,

$$\mu_{AB} = \mu_A^l(x_0) + \mu_B^l(x_0) \tag{4.4.1}$$

where x_0 is the point at which the common tangent line touches the liquid free-energy curve shown in Fig. 4.4b. The free energy of the solid compound with N unit cells at this temperature is, of course, $F_s = N\mu_{AB}$, and F_s can be related to the free energy of a supercooled 50–50 liquid solution by following a sequence of quasi-equilibrium steps. First, heat 1 mole of an AB crystal from T to its melting temperature T_m, i.e., going from the point marked s' to M in Fig. 4.4a. The change in the free energy in this step is

$$F_s(T_m) - F_s(T) = -\int_T^{T_m} S_s(T')\, dT' \tag{4.4.2}$$

The subscript s denotes the solid phase. The entropy S_s is related to the heat capacity C_s by

$$S_s(T) = S_s(T_m) - \int_T^{T_m} \frac{C_s(T')}{T'}\, dT' \tag{4.4.3}$$

Combining Eqs. (4.4.2) and (4.4.3) yields,

$$F_s(T_m) - F_s(T) = -S_s(T_m)(T_m - T) + \int_T^{T_m} \int_{T'}^{T_m} C_s(T'')\frac{1}{T''}\, dT''\, dT' \tag{4.4.4}$$

The second step is to melt the crystal at T_m, which causes no change in the free energy. This situation corresponds to Fig. 4.4b at a different temperature T_m, with the liquid free-energy curve F_l lowered to pass through the solid free-energy minimum labeled s'.

Finally the liquid is cooled back from T_m to T (a supercooled liquid with its free energy given by the value at L' shown in Fig. 4.4b), which changes the free energy by

$$F_l(T) - F_l(T_m) = -S_l(T_m)(T - T_m) - \int_T^{T_m} \int_{T'}^{T_m} C_l(T'')\frac{1}{T''}\, dT''\, dT' \tag{4.4.5}$$

The subscript l indicates the liquid. The total change is

$$\begin{aligned} F_l(T) - F_s(T) &= [S_l(T_m) - S_s(T_m)](T_m - T) \\ &\quad - \int_T^{T_m} \int_{T'}^{T_m} [C_l(T'') - C_s(T'')]\frac{1}{T''}\, dT''\, dT' \end{aligned} \tag{4.4.6}$$

In terms of chemical potentials, Eq. (4.4.6) can be written for 1 mole of AB compound as

$$\begin{aligned} N_0[\mu_A(0.5) + \mu_B(0.5) - \mu_{AB}] &\equiv N_0\, \Delta\mu_{ls}^{AB} \\ &= \Delta S_m(T_m - T) - \Delta C[T_m - T - T\ln(T_m/T)] \end{aligned} \tag{4.4.7}$$

where $\Delta C = C_l(T) - C_s(T)$ is assumed to be independent of T, and the entropy of fusion is defined as $\Delta S_m = S_l(T_m) - S_s(T_m)$ for 1 mole of AB compound. Using Eq. (4.4.1) for μ_{AB} and

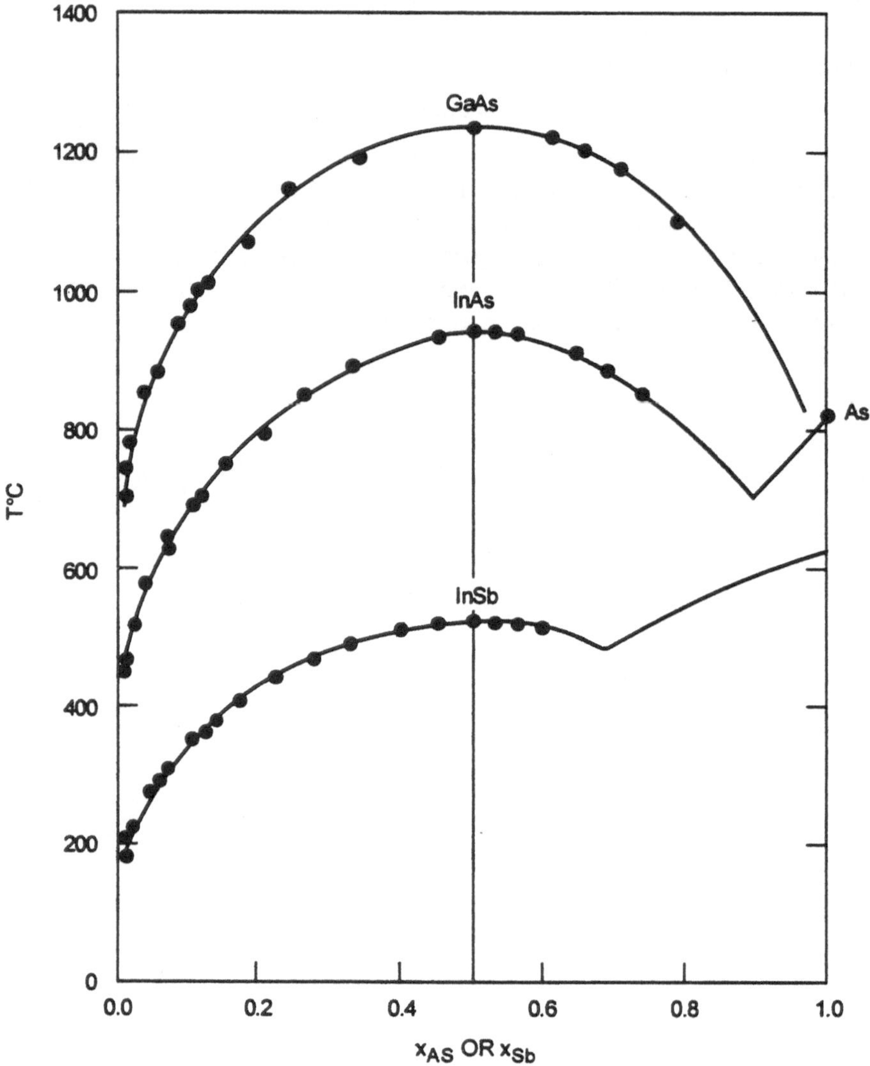

FIGURE 4.5. Liquidus curves of III–V semiconductors. Solid curves are theory, and circles are experiments (Kikuchi, 1981).

the relation between the chemical potentials and activity coefficients in Eq. (4.3.9), we can write Eq. (4.4.7) as

$$\ln\left[\frac{1}{4x_0(1-x_0)}\right]+\ln\left[\frac{\gamma_A(0.5)\gamma_B(0.5)}{\gamma_A(x_0)\gamma_B(x_0)}\right]=\frac{\Delta S_m}{RT}(T_m-T)-\frac{\Delta C}{RT}\left[T_m-T-T\ln\left(\frac{T_m}{T}\right)\right] \quad (4.4.8)$$

Equation (4.4.8) is Vieland's (1963) formula. It is a simple transcendental equation for T as a function of x_0—the liquidus curve. When the liquid solution is in equilibrium with the pure solid B phase, the liquidus is governed by a similar equation:

TABLE 4.2. Parameters Used in the Calculation of Liquidus Curves in Fig. 4.5 and Phase Diagrams in Fig. 4.10

Melting temperature T_m (K)		Entropy of fusion ΔS_m (eu)		Mixing enthalpy parameter Ω (cal/mole)	
GaAs	1511	GaAs	14.7	Ga–As	–4380
InAs	1215	InAs	14.7	In–As	–6070
InSb	803	InSb	13.3	In–Sb	–3980
As	1090	As	3.67	In–Ga	1066
Sb	903	Sb	5.27	As–Sb	610
				GaAs–InAs	2800
				InAs–InSb	2900

After Stringfellow and Green (1969).

$$\ln\left[\frac{1}{x_0\gamma_B(x_0)}\right]=\frac{\Delta S_m^B}{RT}(T_m^B-T)-\frac{\Delta C^B}{RT}\left[T_m^B-T-T\ln\left(\frac{T_m^B}{T}\right)\right] \tag{4.4.9}$$

To actually calculate the liquidus curve using Eqs. (4.4.8) and (4.4.9), one needs to know the activity coefficients γ_A and γ_B, which, of course, depend on the statistical model and the energetics. If QCA is used, these activity coefficients are given explicitly by Eqs. (4.3.12) and (4.3.13). Then the remaining quantities needed are the mixing energy parameter $\Omega = N_0 z\varepsilon$ for the binary liquid, the fusion entropy ΔS_m, and the heat capacity difference ΔC. In most applications ΔC has been taken to be zero, which is a good approximation at high temperature. This approach works reasonably well for the III–V binary compounds. Examples are shown in Fig. 4.5, where calculated liquidus curves (Stringfellow and Green, 1969; Kikuchi, 1981) for GaAs, InAs, and InSb are compared with experiment. The parameters used in the calculations are listed in Table 4.2. Note that the entropy unit eu used for ΔS_m is eu = cal/(K-mole). The symmetrical liquidus curves indicate that the QCA model, with coordination number z = 4, is a reasonable representation of these binary melts. This approach, however, does not work well for the II–VI systems. As shown in Fig. 4.6, the skewed and sharp nature of the experimental curves indicates that these liquid solutions are more complex than the simple picture represented by QCA for the binary solutions. A simple extension of QCA to include the molecule species AB in addition to A and B in the liquid solutions is found to work well, as indicated by comparison with experiment in Figs. 4.6a,b. This extended QCA model is called the associated solution model. Kikuchi (1982a,b) found that repulsive interactions between the molecular and atomic species in the solution are responsible for the skewed nature of the liquidus curves.

4.5. TERNARY PHASE DIAGRAMS

The methods of Section 4.4 are next extended to the phase diagrams for ternary systems. The phase boundary curves are the solidus–liquidus curves that describe the equilibrium between the pseudobinary solid solutions $A_{1-x}B_xC$ and the liquid ternary solution denoted by $A_aB_bC_c$, where a, b, and c are fractional concentrations satisfying normalization $a + b + c = 1$. A schematic three-dimensional phase diagram is shown in Fig. 4.7; it is a plot of the temperature T versus concentrations. The liquidus is the upper surface and the solid pseudobinary phase is the plane lying between the two lines labeled AC and BC and under

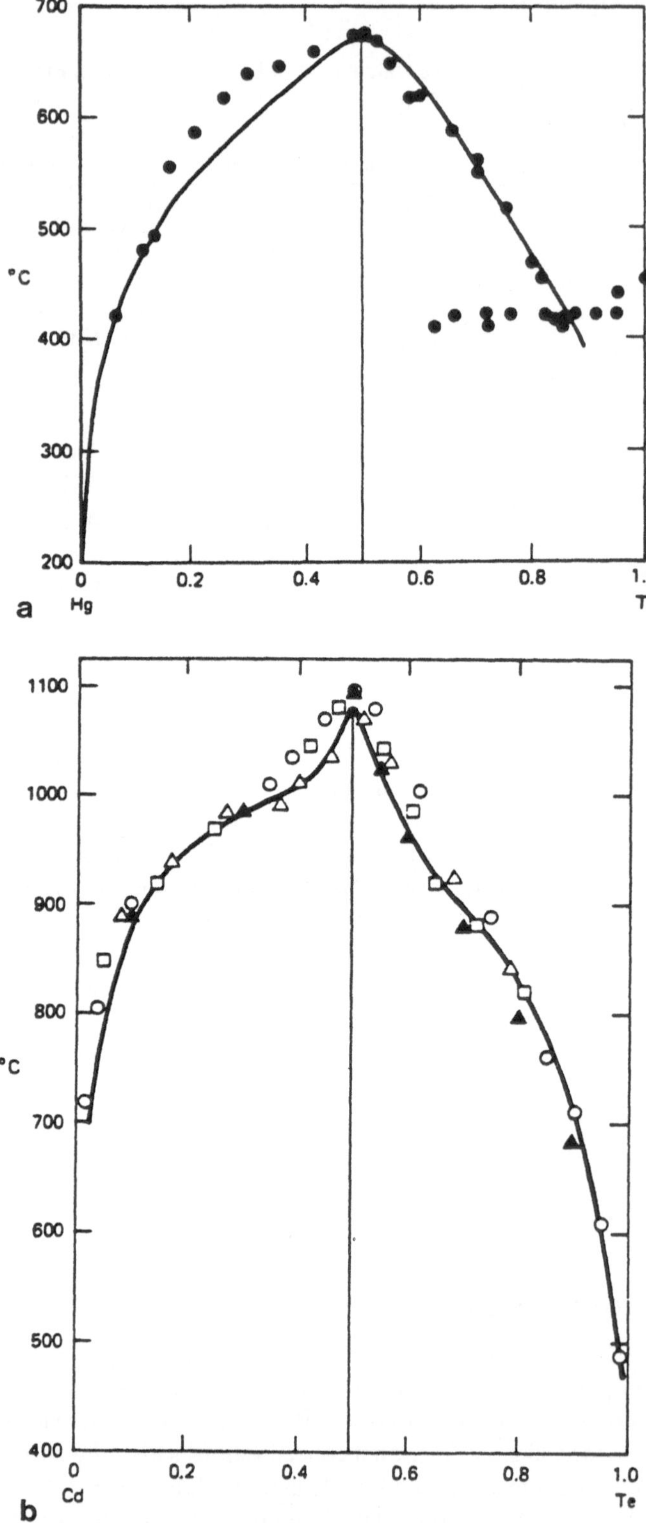

FIGURE 4.6. Binary liquidus. The curve is the theory, and the points are experiments for (a) HgTe and (b) CdTe (Kikuchi, 1982b).

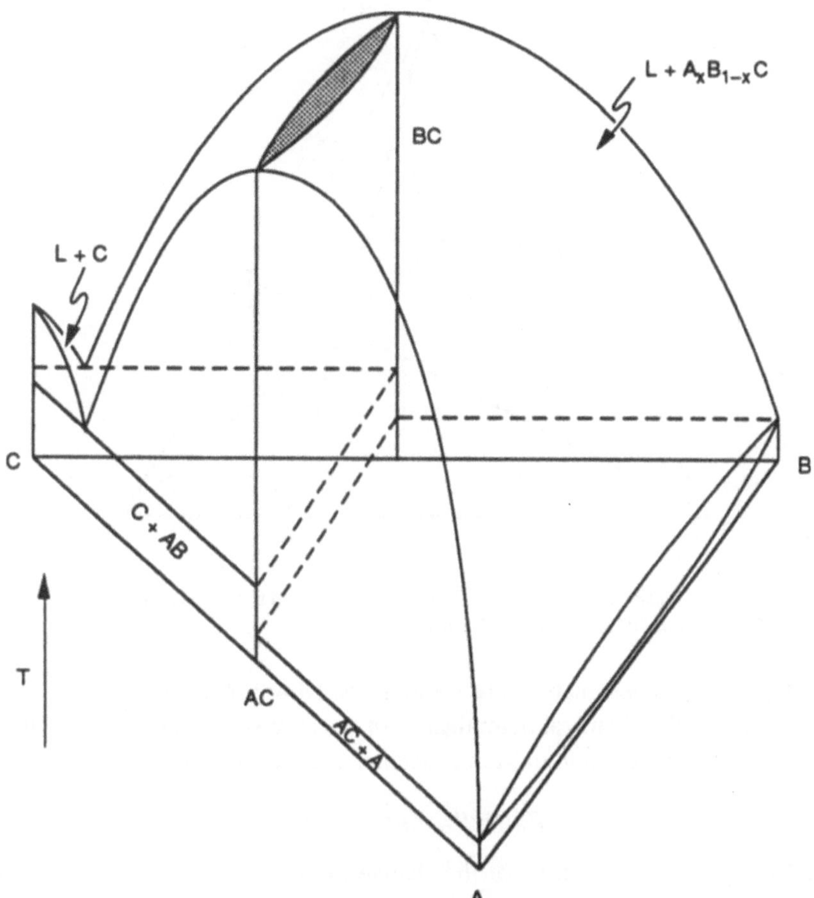

FIGURE 4.7. Schematic A–B–C ternary phase diagram (Stringfellow and Green, 1969). The shaded region is that between the pseudobinary liquidus and solidus curves. (See Fig. 4.10 for an example.)

the solidus curve. A given set of a, b, and c values correspond to a point D inside the base triangle ABC, as shown in Fig. 4.8. The values of a, b, and c are related to the geometric line ratios by $a = DA'/AA'$, $b = DB'/BB'$, and $c = DC'/CC'$. The solidus–liquidus boundaries form a family of curves relating the solid concentration x and the temperature T to the two independent liquid concentrations a and b (note $c = 1 - a - b$). These relations can be obtained by matching the chemical potentials μ described below.

Let us start with μ_{AC} and μ_{BC} for the AC and BC components in the solid solution $A_{1-x}B_xC$. They are related to those in the pure AC and BC compounds through the activity coefficients γ by

$$\mu_{AC}(x,T) = \mu_{AC}^0(T) + kT\ln[(1-x)\gamma_{AC}] \tag{4.5.1}$$

and

$$\mu_{BC}(x,T) = \mu_{BC}^0(T) + kT\ln(x\gamma_{BC}) \tag{4.5.2}$$

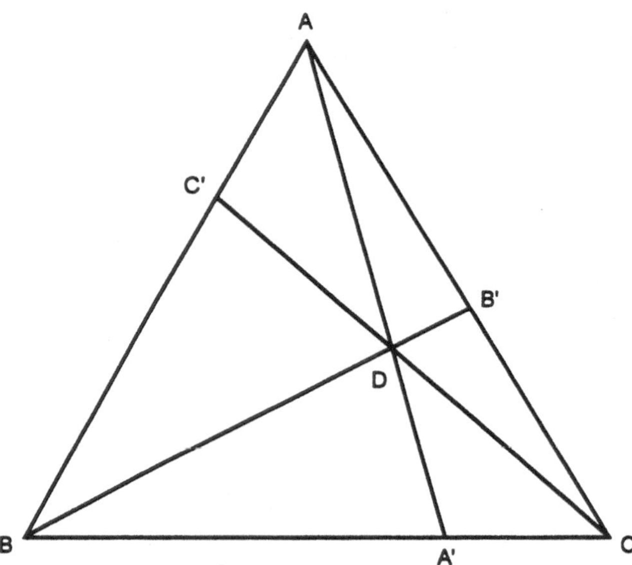

FIGURE 4.8. Any point D inside the triangle ABC defines a unique set of fractional concentrations in ternary alloy $A_aB_bC_c$ by $a = DA'/AA'$, $b = DB'/BB'$, and $c = DC'/CC'$.

where the superscript 0 designates pure compounds. Now Vieland's relation in Eq. (4.4.7) can be used to relate the chemical potentials μ^0 of the pure solid compounds to those of the supercooled 50–50 binary liquid mixtures, denoted μ^{sc}. For example, μ^0_{AC} is given by

$$\mu^0_{AC} = \mu^{sc}_{A} + \mu^{sc}_{C} - \Delta\mu^{AC}_{ls} \tag{4.5.3}$$

Note that $\Delta\mu$ is expressed in terms of the change in the heat capacity ΔC and entropy of fusion ΔS_m between the 50–50 binary liquid and the solid compound AC given by the right side of Eq. (4.4.7). Since the ternary liquid is in equilibrium with the solid pseudobinary, μ_{AC} in the solid is equal to the sum of the chemical potentials for the atomic species in the former by

$$\mu_{AC}(x,T) = \mu^l_A(a,b;T) + \mu^l_C(a,b;T) \tag{4.5.4}$$

A similar expression holds for μ_{BC}. The μ^l for the atomic species in the ternary liquid can be further expressed in terms of those of the pure elementary liquids and the activity coefficients by

$$\mu^l_A = \mu^0_A + kT\ln(a\gamma_A) \tag{4.5.5}$$

Combining Eqs. (4.5.3)–(4.5.5) with Eq. (4.5.1), one obtains the equilibrium condition

$$\ln[(1-x)\gamma_{AC}] = \ln\left(\frac{4ac\gamma_A\gamma_C}{\gamma^{sc}_A\gamma^{sc}_C}\right) + \frac{\Delta\mu^{AC}_{ls}}{kT} \tag{4.5.6}$$

Similarly, Eq. (4.5.2) leads to

$$\ln(x\gamma_{BC}) = \ln\left(\frac{4bc\gamma_B\gamma_C}{\gamma^{sc}_B\gamma^{sc}_C}\right) + \frac{\Delta\mu^{BC}_{ls}}{kT} \tag{4.5.7}$$

Note that the γ^{sc}'s in Eq. (4.5.6) are those in the AC liquid and in Eq. (4.5.7) are the ones for the BC liquid.

To calculate the phase diagram from Eqs. (4.5.6) and (4.5.7), one needs to know the x and T dependence of the activity coefficients γ_{AC} and γ_{BC} for the pseudobinary solid solutions, the T dependence of the activity coefficients γ^{sc} of the supercooled 50–50 binary liquids, and the T and concentration (a, b, and c) dependences of γ_A,γ_B, andγ_C in the ternary liquid solutions $A_aB_bC_c$. In addition, ΔC and ΔS_m are needed for the calculation of $\Delta\mu$. If QCA is used, the activity coefficients for the binary liquid and pseudobinary solid solutions take the forms of Eqs. (4.3.12) and (4.3.13). For the ternary liquid solution, QCA has to be extended to three components. There are now six different kinds of pairs to be considered, namely, AA, BB, CC, AB, BC, and AC. We label these pairs by j, where j runs from 1 to 6, and denote the fractions of each pair as x_j. If independent pair energies $\varepsilon_{\alpha\beta}$ are assumed, there are only three nonvanishing energy parameters, Δ_{AB}, Δ_{AC}, and Δ_{BC}, while $\Delta_{AA} = \Delta_{BB} = \Delta_{CC} = 0$, where we have defined $\Delta_{\alpha\beta} = \varepsilon_{\alpha\beta} - (\varepsilon_{\alpha\alpha} + \varepsilon_{\beta\beta})/2$. Then it is straightforward to write the mixing energy as

$$\Delta E = \sum_{j=1}^{6} M x_j \Delta_j \tag{4.5.8}$$

where $M = Nz/2$, with z being an effective coordination number, taken to be 4. The mixing entropy is $\Delta S = k \ln \Phi$, with

$$\Phi = \frac{N!}{N_A! N_B! N_C!} \frac{M!}{\prod_{j=1}^{6} M_j!} \prod_{j=1}^{6} (x_j^0)^{M_j} \tag{4.5.9}$$

where $M_j = x_j M$ and x_j^0 are the *a priori* probabilities; e.g., $x_{AA}^0 = a^2$ and $x_{AB}^0 = 2ab$, etc. This Φ is a simple extension of Eq. (4.2.9). The mixing free energy then reads

$$\Delta F = \sum_{j=1}^{6} M\left[x_j\Delta_j + kTx_j \ln\left(\frac{x_j}{x_j^0}\right)\right] + NkT(a \ln a + b \ln b + c \ln c) \tag{4.5.10}$$

Note that not all the x_j are independent. They are constrained to have the right numbers of A, B, and C atoms for a given set of fractional concentrations a, b, and c:

$$2x_{AA} + x_{AB} + x_{AC} = 2a \tag{4.5.11}$$

$$2x_{BB} + x_{AB} + x_{BC} = 2b \tag{4.5.12}$$

$$2x_{CC} + x_{CB} + x_{AC} = 2c \tag{4.5.13}$$

Minimization of ΔF with respect to the x_j under these imposed constraints leads to the following coupled equations:

$$(2a - x_{AB} - x_{AC})(2b - x_{AB} - x_{BC}) = x_{AB}^2 \exp(2\Omega_{AB}/zRT) \tag{4.5.14}$$

$$(2a - x_{AB} - x_{AC})(2c - x_{AC} - x_{BC}) = x_{AC}^2 \exp(2\Omega_{AC}/zRT) \tag{4.5.15}$$

$$(2b - x_{AB} - x_{BC})(2c - x_{AC} - x_{BC}) = x_{BC}^2 \exp(2\Omega_{AB}/zRT) \tag{4.5.16}$$

where the mixing energy parameters are defined as $\Omega_{\alpha\beta} = N_0 z \Delta_{\alpha\beta}$, with the effective coordination number z taken to be 4. After Eqs. (4.5.11) through (4.5.16) are solved for x_j, they are used in Eq. (4.5.10) to obtain the functional dependence of ΔF on the concentrations a, b, and c and on the temperature T. Then the activity coefficients can be calculated. For example, γ_A is calculated from

$$kT\ln(a\gamma_A) = \left(\frac{\partial \Delta F}{\partial N_A}\right)_{N_B, N_C} \tag{4.5.17}$$

Thus, the parameters needed to complete the specification of a ternary phase diagram in QCA are ΔC and ΔS_m for both AC and BC compounds, the mixing energy parameters Ω_{AB}, Ω_{AC}, and Ω_{BC} for the ternary liquid and $\Omega_{AB(C)}$ for the pseudobinary solid solution. It is a good approximation to set $\Delta C = 0$. There are four independent variables, say, a, b, x, and T, and two equations, Eqs. (4.5.6) and (4.5.7), with which to work. One can fix T and solve for a (and hence $c = 1 - a - b$) and x as functions of b. The plot of the contour of a, b, and c for the same T in the Gibbs triangle ABC is an isotherm. If the solidus concentration x is fixed, the contour of a, b, and c in the Gibbs phase plot is the isosolidus concentration line. Figure 4.9 shows both sets of these plots for the In–Ga–As system reported by Stringfellow

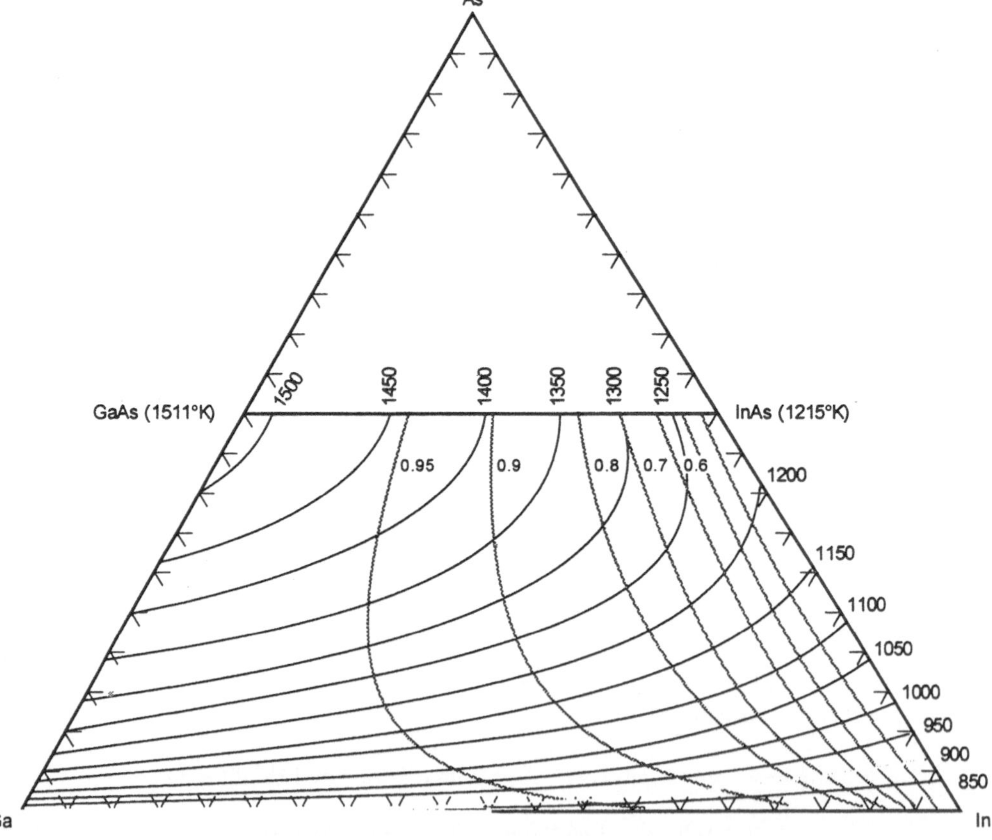

FIGURE 4.9. Isotherms and isosolidus concentration lines for phase equilibrium between pseudobinaries $In_{1-x}Ga_xAs$ and the corresponding ternary liquid (Stringfellow and Green, 1969).

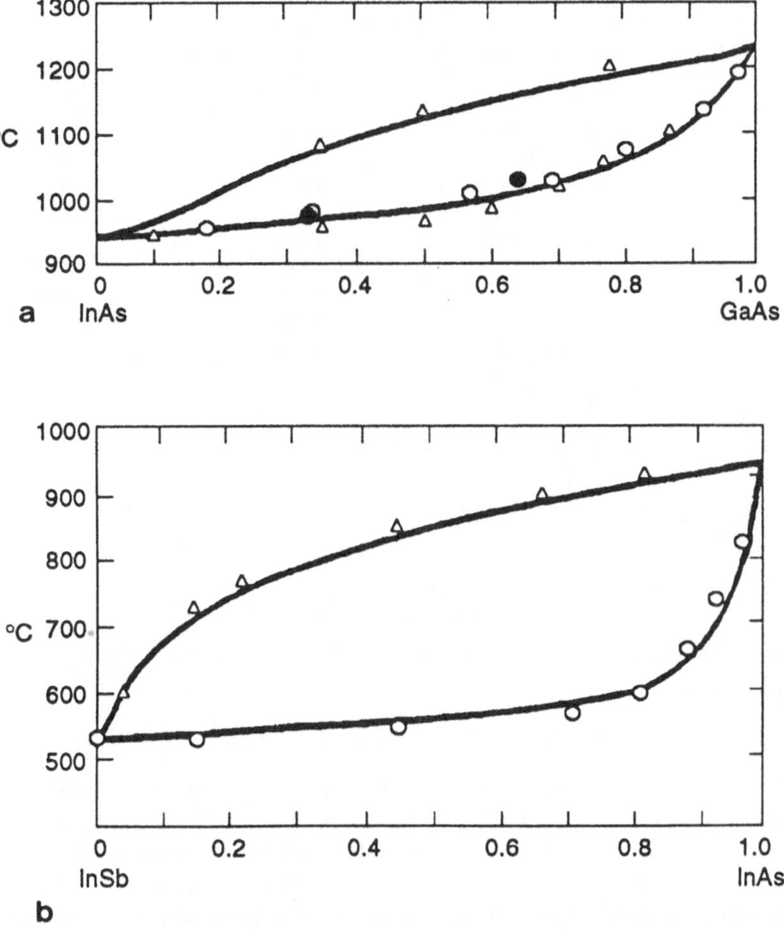

FIGURE 4.10. Liquidus and solidus curves for $In_{1-x}Ga_xAs$ and $InSb_{1-x}As_x$. Solid curves are calculations and the circles are the experiments. (After Kikuchi 1981.)

and Green (1969), using the parameters given in Table 4.2. We can fix $c = 0.5$ in the ternary and solve for x and T as functions of a. Then we obtain the liquidus and the solidus curves along the $c = 1/2$ line, as shown in Fig. 4.10, also reported by Stringfellow and Green and further verified by Kikuchi (1981). The agreement between theory and experiment is remarkably good.

4.6. PHASE DIAGRAM DATA AND SIMPLE MIXING ENTHALPY MODELS

Before moving into more rigorous statistical theory and more systematic energetic calculations, it is useful to summarize the semiempirical phase diagram data and simple statistical and energetic models that have been used in practical calculations.

From a survey of phase diagram calculations of ternary and quaternary semiconductors, Stringfellow (1974) preferred the following forms for mixing free energies. For the III–V liquid solutions, the preferred form (per mole) is

TABLE 4.3. Values of $\Omega_L = \Omega_0 - \Omega_1 T$ in Eq. (4.6.1) for the Liquid Mixing Enthalpy Parameter Ω_L in cal/mole and Other Parameters in the Liquid Phase

System	Ω_L (cal/mole)	ΔS_m (eu)	T_m (°C)
AlP	2800 – 4.80 T	15.0	2530
AlAs	600 – 12 T	15.6	1770
AlSb	1230 – 10 T	14.74	1065
GaP	28000 – 4.8 T	16.8	1465
GaAs	5160 – 9.16 T	16.64	1238
GaSb	4700 – 6 T	15.8	710
InP	4500 – 4 T	14.0	1070
InAs	3860 – 10 T	14.52	942
InSb	3400 – 12 T	14.32	525
Al–Ga	104		
Al–In	1060		
Ga–In	1060		
P–As	1500		
As–Sb	750		

After Stringfellow (1974).

$$\Delta F_l = x(1-x)(\Omega_0 - \Omega_1 T) + RT(x \ln x + y \ln y) \tag{4.6.1}$$

For the III–III and V–V liquids and the III–V pseudobinary solids, the strict regular solution model is his preference:

$$\Delta F = x(1-x)\Omega + RT(x \ln x + y \ln y) \tag{4.6.2}$$

Table 4.3 lists the values of $\Omega_L = \Omega_0 - \Omega_1 T$ and other phase diagram data, including the entropy of fusion ΔS_m, the melting temperatures T_m for the III–V binary systems, and the mixing energy parameters for the III–III and V–V interactions in the liquid solutions. In Table 4.4, the experimental values of Ω based on Eq. (4.6.2) are compared with the mixing enthalpies calculated from several theoretical models to be discussed presently. One has to be cautious in comparing the experimental Ω with the calculated values, because the former may include additional contributions from entropy terms not contained in the second term of Eq. (4.6.2).

One simple model for estimating the mixing energy is the so-called delta lattice parameter model (DLP) suggested by Stringfellow (1974). He observed that the experimental Ω values for the III–V pseudobinary alloys $A_{1-x}B_xC$ correlated strongly with the difference in the lattice constant $\Delta a = a_{BC} = a_{AC}$, as shown in Fig. 4.11 in a log–log plot generated by Stringfellow (1974). The best straight-line fit gives a slope of 2.45, yielding the fitted function

$$\Omega_{\text{fit}} = 1.174 \times 10^6 (\Delta a / a)^{2.45} \tag{4.6.3}$$

where $a = (a_{AC} + a_{BC})/2$. This result prompted Stringfellow to correlate with Phillips and Van Vechten's (1970) dielectric model, which relates the alloy formation energy to the energy-gap parameter E_h (see Eq. (2.6.4)), which in turn is roughly proportional to $a^{-2.5}$. By simply assuming that the formation energy has the form

$$E = -\kappa / a^{2.5} \tag{4.6.4}$$

for every semiconductor, then the mixing enthalpy is given by

TABLE 4.4. Mixing Energy Parameters Ω in kcal/mole[a]

System	DLP[b]	FM[c]	IMP[d]	MZ[e]	CS[f]	GQCA[g]	WF[h]	Exp
(Ga,Al)P			0.00		–0.05	0.01		
(GaAl)As	0.02	0.03	0.00	0.02	–0.07	0.03	0.30	0.0[i]
(Ga,Al)Sb	0.02	0.03	0.02	0.02	–0.15	0.12		0.0[i]
(Ga,In)P	3.63	2.94	3.00	4.56	2.54	2.85	3.07	3.25[j], 3.40[l], 3.50[i]
(Ga,In)As	2.81	2.42	2.36	2.49	1.60	2.19	2.35	1.65[j], 2.00[k], 3.00[i]
(Ga,In)Sb	1.85	1.83	1.77	2.53	0.81	1.72		1.48[j], 1.90[i]
(In,Al)P			2.77		2.55	2.66		
(In,Al)As	2.81	2.37	2.32	3.60	2.17	2.26	2.50[i]	
(In,Al)Sb	1.46	1.45	1.49	2.06	1.36	1.46		0.60[i]
(Cd,Zn)Te	1.97	1.63	1.73	2.12	1.24	1.59	2.29	1.34[n]
(Hg,Cd)Te			0.00		–0.07	0.03	0.38	0.72[o], 1.40[n]
(Hg,Zn)Te	1.81	1.48	1.56	1.91	1.50	1.55	1.88	3.00[n]
Al(P,As)			0.65		0.76	0.67		
Ga(P,As)	0.98	0.66	0.70	1.15	0.94	0.74	0.91	0.40[i], 1.00[m], 1.26[p]
In(P,As)	0.58	0.52	0.52	0.72	0.57	0.52		0.40[i]
Al(As,Sb)			3.38		4.09	3.45		
Ga(As,Sb)	3.35	2.76	2.81	4.38	3.67	2.92		4.00[l], 4.27[p], 4.50[i]
In(As,Sb)	2.29	2.17	2.23	2.89	2.52	2.20		2.25[i], 2.90[k]
Al(P,Sb)			6.99		8.32	6.94		
Ga(P,Sb)			6.36		8.66	6.57		
In(P,Sb)			5.08		5.76	4.97		
Zn(S,Se)			0.98		0.90	0.95		
Zn(Se,Te)	3.11	2.12	2.23	2.91	2.26	2.16		1.55
Zn(S,Te)			6.45		6.20	6.12		

[a]For models in which the mixing enthalpy ΔH is a function of both x and T, Ω is defined as $\Delta H = x(1-x)\Omega$ for a random alloy $T = \infty$ at $x = 0.5$.

[b]Stringfellow (1974), WDLP in Eq. (4.6.5).

[c]Fedders and Muller (1984), Ω_{scale} in Eq. (4.6.7).

[d]Chen and Sher (1985), Ω_{imp} in Eq. (4.6.13)

[e]Martins and Zunger (1984), VFF model.

[f]Chen and Sher (1985), VFF plus BOM perturbation.

[g]Sher *et al.* (1987), the 16-bond model in Section 4.9.

[h]Wei *et al.* (1990), LDA calculation described in Section 4.11.

[i]Panish and Ilgemes (1972).

[j]Foster (1972).

[k]Antypas (1970b).

[l]Foster and Woods (1971).

[m]Antypas (1970a).

[n]Laugies (1973).

[o]Su *et al.* (1985), value taken at 1000 K.

[p]Stringfellow (1969, 1973, 1974).

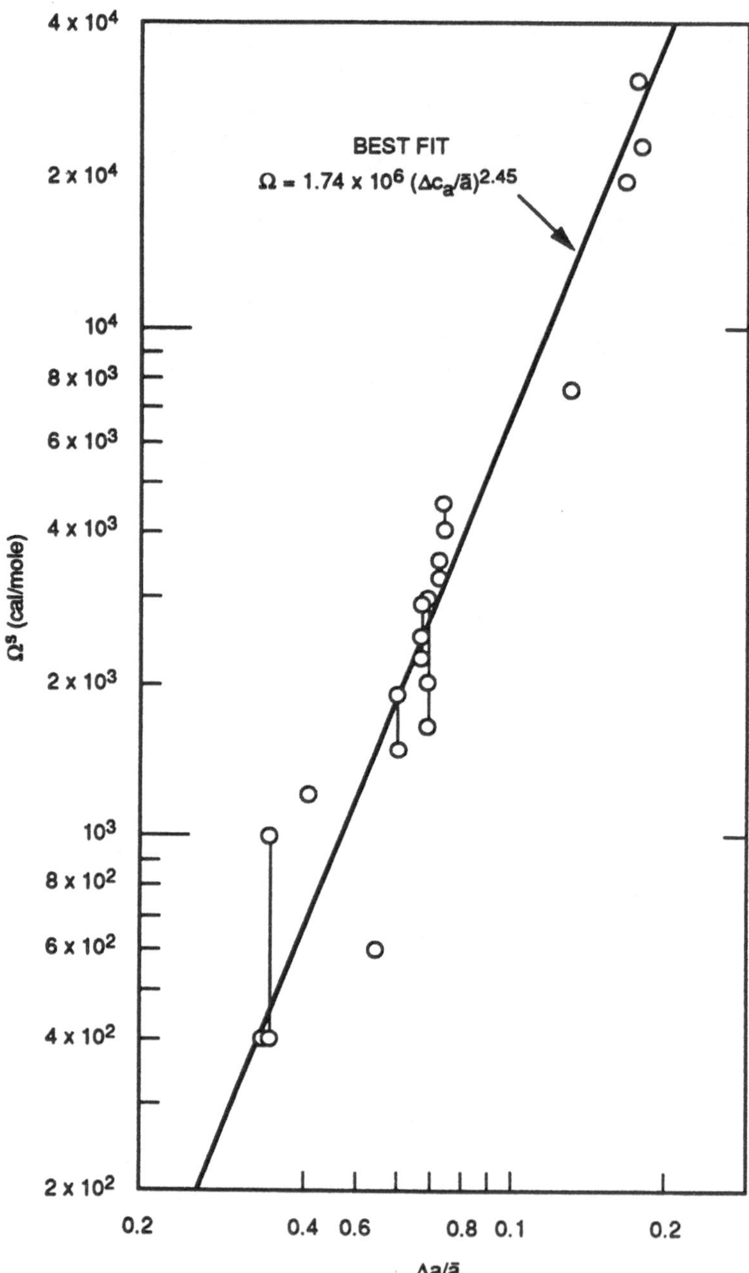

FIGURE 4.11. Log–log plot of experimental values of Ω versus percentage lattice constant difference. The solid line represents the best straight-line fit (Stringfellow, 1974).

$$\begin{aligned}\Delta H &= E(\text{alloy}) - xE(\text{BC}) - (1-x)E(\text{AC}) \\ &= -\kappa[a_{\text{alloy}}^{-2.5} - xa_{\text{BC}}^{-2.5} - (1-x)a_{\text{AC}}^{-2.5}] \\ &= 4.375\kappa x(1-x)(\Delta a)^2/a^{4.5} = x(1-x)\Omega_{\text{DLP}} \end{aligned} \tag{4.6.5}$$

In Eq. (4.6.5) the alloy lattice constant a_{alloy} is assumed to be the concentration weighted average (Vegard's law), and only terms to second order in Δa in the Taylor expansion have been kept. If κ is treated as an adjustable parameter, the best fit to the III–V alloy data yields $\kappa = 1.15$ for Ω in kcal/mol and Δa and a in Å. This model gives very good results, as indicated in Table 4.4.

Arguing from a different point of view, Fedders and Muller (1984) suggested that the Δa^2 dependence in Eq. (4.6.5) might originate from strain energy. If the alloying atoms A and B are held rigidly on their fcc sublattice sites and if the alloy lattice constant is taken to be the concentration weighted average (Vegard's law), then the strain energy in the alloy can be shown to be $\Delta E = x(1-x)\Omega_{\text{FM}}$, with mixing energy parameter

$$\Omega_{\text{FM}} = \tfrac{9}{2}BV(\Delta a/a)^2 \tag{4.6.6}$$

where B is the average bulk modulus and V is the average molar volume. Because B in semiconductors varies as a^{-n}, with n ranging from 4 to 9 (see Chapter 3), Ω_{FM} is seen to behave like $(\Delta a)^2/a^{n-1}$. Thus, the power-law dependencies on Δa and a in Ω_{DLP} and Ω_{FM} are similar. However, Ω_{FM} is about four times larger than the experimental values. In fact a scaled-down strain contribution,

$$\Omega_{\text{scale}} = 0.226\Omega_{\text{FM}} \tag{4.6.7}$$

yields results comparable to those of Ω_{DLP}. The reduction of Ω_{FM} is mainly due to bond-length relaxation neglected in the above derivation.

If strain energy is the main contribution to the mixing enthalpy, then the valence-force-field model (VFF) given in Eq. (3.3.1) should provide a means to formulate a systematic calculation. One can replace an A atom in the AC crystal by an impurity B atom and calculate the substitution energy ΔE(B in AC) and *vice versa* for ΔE(A in BC). Then the mixing enthalpy parameter is given by (Chen and Sher, 1985) $\Omega_{\text{imp}} = N_0\omega_{\text{imp}}$ with

$$\omega_{\text{imp}} = [\Delta E(\text{A in BC}) + \Delta E(\text{B in AC})]/2 \tag{4.6.8}$$

To calculate these energies precisely in VFF, one has to allow the atoms in the first several shells about the impurity to relax. The results of such calculation (Martins and Zunger, 1984; Chen and Sher, 1985) give good results for mixing energy and bond lengths between the impurity and the first-shell host atoms. However, it was found (Chen and Sher, 1985) that by neglecting the bond-angle terms in the VFF ($\beta = 0$) and at the same time truncating the range of the atomic relaxation at the second shell, a simple spring model similar to that shown in Fig. 3.3 yields nearly the same results as the extended VFF for both the impurity bond lengths and alloy mixing energy. It is instructive to examine this simple spring model closely. Let the 16 bonds before the substitution be the host BC bonds with VFF force constant α and bond length d. When the central B atom is removed and an A atom is substituted in its place, the bond lengths d_1 of the first shell and d_2 of the second shell are different from d. Assuming $\beta = 0$ and the second-shell atoms are held firmly in place, the strain energy is

$$\Delta E(\text{A in BC}) = 4\left[\tfrac{3}{2}\alpha_I(d_1 - d)^2\right] + 12\left[\tfrac{3}{2}\alpha(d_2 - d)^2\right] \tag{4.6.9}$$

where the subscript I denotes the impurity bond in its own crystal (i.e., the AC bond in the pure AC crystal). Let $\delta = (d_1 - d)/d$ and $\delta_0 = (d_I - d)/d$. Then in the relaxed configuration $(d_2 - d)/d = -\delta/3$ and Eq. (4.6.9) becomes

$$\Delta E(\text{A in BC}) = 4\left[\tfrac{3}{2}\alpha_I(\delta - \delta_0)^2 + \tfrac{1}{2}\alpha\delta^2\right]d^2 \tag{4.6.10}$$

Minimizing ΔE with respect to δ yields the following value of δ at the equilibrium position:

$$\delta = \frac{\delta_0}{1 + \frac{1}{3}(\alpha/\alpha_I)} \tag{4.6.11}$$

If both α and α_I are nearly equal, then

$$\delta \cong \frac{3}{4}\delta_0 \tag{4.6.12}$$

and ΔE (A in BC) $= 3\alpha\delta_0^2 d^2/2$. Similar results can be obtained for ΔE(B in an AC host). Equation (4.6.9) then yields the following estimate for the mixing energy parameter from this impurity model:

$$\omega_{\text{imp}} = 3\alpha\delta_0^2 d^2/2 \tag{4.6.13}$$

Note that this model, Ω_{imp} is precisely one quarter of Ω_{FM}. Thus, this simple relaxed lattice model not only explains why the strain energy is reduced by 1/4 but also gives a very good prediction of the value $\Gamma = 0.75$ for the lattice relaxation parameter [compare Eq. (4.6.12) with the definition of Γ in Eq. (1.3.2) and Table 1.3] found in pseudobinary alloys from the EXAFS experiments. While this short-range strain model with β set to zero produces accurate results, keep in mind that it is not the way nature actually behaves. The approximation works only because effects from long-range relaxation nearly cancel the lowest-order modifications caused by the β terms. If one asks questions that involve shears, or for more detailed structural information, e.g., the second-neighbor distances, then effects due to β and long-range relaxation must be included.

4.7. GENERALIZED QUASI-CHEMICAL THEORY

There are at least two reasons to extend the theory beyond the first approximation. First, effects of neglected statistical correlations should be included. Second, and perhaps more important, the microscopic energies that govern the statistics may not be well represented by simple pair energies. In this section, QCA is generalized to deal with clusters of any size.

Let each cluster contain n atoms on the sublattice occupied by the A and B atoms, and let $J+1$ be the number of different configurations in which the A and B atoms can be arranged on a cluster with a distinct energy ε_j, for $j = 0, 1, \ldots, J$. For example, in the case of a tetrahedral cluster in which ε_j differs only when the number of B atoms on the cluster changes, the index j will equal the number of B atoms on the cluster and therefore takes on values $j = 0, 1, 2, 3,$ and 4. However, for a square with two A atoms and two B atoms on the corners, the energy for two atoms on the same side may be different from that when they are on a diagonal, so J may be larger than n. Obviously, the degeneracies of arrangements with the same ε_j tend to decrease if $J > n$ since the total number of configurations is fixed; i.e., for a four-site case, it is 2^4. Let M be the total number of clusters in the alloy, M_j the number of j-type clusters so that $M = \Sigma M_j$, and $x_j = M_j/M$ the fraction of j-type clusters. Let us further define n_j^{A} (n_j^{B}) to be the number of A (B) atoms in a j-type cluster, and the notation n_j will stand for n_j^{B}. Since the concentration x is fixed, the set of fractions x_j are constrained to satisfy

$$nx = \sum_{j=0}^{J} n_j x_j \tag{4.7.1}$$

The mixing energy for a specified set $\{M_j\}$ is then

$$\Delta E = \sum_{j=0}^{J} M_j \varepsilon_j - M[(1-x)\varepsilon_{\mathrm{A}}^0(n) + x\varepsilon_{\mathrm{B}}^0(n)] \tag{4.7.2}$$

where $\varepsilon_{\mathrm{A}}^0(n)$ and $\varepsilon_{\mathrm{B}}^0(n)$ are the energies of an n-atom cluster in the pure AC and pure BC crystals, respectively, as indicated by the superscript 0. We note that the cluster energy $\varepsilon_0 = \varepsilon_{\mathrm{A}}(n)$, which is the energy in the alloy for a cluster containing all A atoms, in general differs from $\varepsilon_{\mathrm{A}}^0(n)$, because of the influence of strain and different chemical bonding. Let $\varepsilon_{\mathrm{A}} = \varepsilon_0 - \varepsilon_{\mathrm{A}}^0(n)$ and similarly for $\varepsilon_{\mathrm{B}} = \varepsilon_J - \varepsilon_{\mathrm{B}}^0(n)$. Then Eq. (4.7.2) can be written as

$$\Delta E = M[(1-x)\varepsilon_{\mathrm{A}} + x\varepsilon_{\mathrm{B}}] + M\Sigma_j x_j \Delta_j \tag{4.7.3}$$

where the reduced excess energies Δ_j are defined as

$$\Delta_j = \varepsilon_j - \frac{n-n_j}{n}\varepsilon_0 - \frac{n_j}{n}\varepsilon_J \tag{4.7.4}$$

Note that by definition Δ_0 and Δ_J are zero.

All the temperature dependence arising from the statistical mechanics is, as we shall see, contained in the x_j factors. Therefore, the first term of ΔE in Eq. (4.7.3) does not influence the temperature dependence arising from averaged statistical quantities for a given x. The mixing entropy can be calculated from $k \ln \Phi$, with Φ given by (an extension of Eq. (4.2.9))

$$\Phi = \frac{N!}{N_{\mathrm{A}}!N_{\mathrm{B}}!} \frac{M!}{\prod_j M_j!} \prod_j (x_j^0)^{M_j} \tag{4.7.5}$$

where x_j^0 is the *a priori* probability, i.e., the random-alloy value

$$x_j^0 = g_j(1-x)^{(n-n_j)} x^{n_j} \tag{4.7.6}$$

and g_j is the degeneracy of clusters with energy ε_j. The resulting mixing entropy takes a simple form

$$\Delta S = -Nk(x \ln x + y \ln y) - Mk \sum_j x_j \ln\left(\frac{x_j}{x_j^0}\right) \tag{4.7.7}$$

Equations (4.7.2) and (4.7.7) thus combine to give an explicit expression for the mixing free energy $\Delta F = \Delta E - T\Delta S$ in terms of x_j.

To find the equilibrium cluster probability distribution $\{\bar{x}_j\}$, one takes the partial derivatives $\partial\Delta F/\partial x_j$ and sets them to zero. Although there are $J + 1$ unknown x_j's, there are two independent constraints, one given by Eq. (4.7.1) and one that says the total probability is unity:

$$\sum_j x_j = 1 \tag{4.7.8}$$

This leaves us with $J - 1$ nonlinear coupled equations to be solved for the thermally averaged $\{\bar{x}_j\}$. While the set $\{\bar{x}_j\}$ are the equilibrium values of the more general set $\{x_j\}$, we leave the bar off since it causes no confusion. A simple approach to the numerical problem is to use the Lagrange multiplier formalism in the constrained variational calculation for the $\{x_j\}$ set; i.e.,

$$\frac{\partial}{\partial x_j}\left\{\frac{\Delta F}{M} - \lambda_1\left[\left(\sum_j x_j\right) - 1\right] - \lambda_2\left[\left(\sum_j n_j x_j\right) - nx\right]\right\} = 0 \tag{4.7.9}$$

This relates x_j to the Lagrange multipliers λ_1 and λ_2

$$x_j = x_j^0 \exp[(\lambda_1 + \lambda_2 n_j - \Delta_j)/kT] \tag{4.7.10}$$

The normalization condition (4.7.8) is then used to eliminate λ_1 and obtain the result

$$x_j = \frac{x_j^0 \exp[(\lambda_2 n_j - \Delta_j)/kT]}{\sum_j x_j^0 \exp[(\lambda_2 n_j - \Delta_j)/kT]} \tag{4.7.11}$$

The algebra is simplified by defining a new variable $\eta = xe^{\lambda_2\beta}/y$. Then

$$x_j = g_j \eta^{n_j} e^{-\beta\Delta_j}/Z \tag{4.7.12}$$

where $\beta = 1/kT$, and Z is a cluster grand partition function

$$Z = \sum_j g_j\, \eta^{n_j} e^{-\beta\Delta_j} \tag{4.7.13}$$

In general, the thermal average of any dynamic variable D is

$$\overline{D} = \sum_j D_j x_j \tag{4.7.14}$$

Then the expression for x_j in Eq. (4.7.12) is used in the other constraint equation, Eq. (4.7.1), to arrive at the following nth-order polynomial equation for the unknown parameter η:

$$\bar{n}_j = \sum_j n_j x_j = \sum_j \frac{n_j g_j \eta^{n_j} e^{-\beta\Delta_j}}{Z} = nx \tag{4.7.15}$$

We refer to the above argument as the generalized QCA (GQCA), while QCA is reserved for the pair interaction case.

To illustrate the usefulness of the equations developed in this section, let us consider the case in which the cluster energies ε_j depend only on the number of A and B atoms in the cluster. For a cluster containing n atoms there are a total of $n + 1$ distinct energies of the cluster type $A_{(n-j)}B_{(j)}$, with j ranging from 0 to n and with $n_j = j$. Then the cluster degeneracy is simply

$$g_j = \frac{n!}{j!\,(n-j)!} \tag{4.7.16}$$

In terms of η and Z the mixing free energy is given by

$$\Delta F = Nl(y\varepsilon_A + x\varepsilon_B) + NkT[(1-nl)(x \ln x + y \ln y) - l \ln Z + nlx \ln \eta] \tag{4.7.17}$$

where we have defined $l \equiv M/N$. Notice that a term from ΔS has exactly canceled the temperature-dependent term $\Sigma_j x_j\Delta_j$ in the enthalpy.

Equation (4.7.17) provides a useful form for treating the analytical behavior of GQCA, but this discussion will be put off to Section 4.10 after we have introduced the mechanisms determining the energies ε_j and their numerical values. Before leaving this section we note that the cluster distribution function in Eq. (4.7.12) could have been written directly had we used a grand canonical ensemble for the clusters from the outset. Then η is the absolute activity for the B atoms in this ensemble. The whole derivation can also be done using a steepest descents argument for the partition function (Sher *et al.*, 1987). While these methods are more elegant, the essential physics is the same as the GQCA derived in the Lagrange multiplier formalism. These calculations can be summarized in four steps: (1) calculate the cluster energies ε_j and from them the reduced excess energies Δj defined in Eq. (4.7.4); (2) solve the polynomial equation (4.7.14) to obtain η; (3) use η and the energies to obtain the cluster distribution from Eq. (4.7.12) and use this distribution to calculate all the statistical averaged properties for the alloy; and (4) in particular, use the set $\{x_j\}$ in Eq. (4.7.3) to obtain the mixing enthalpy ΔE, and in Eq. (4.7.7) to obtain ΔS, which then enables phase diagram calculations and thermodynamic studies.

Further discussion of GQCA will be made in Appendices 4A through 4C. Appendix 4A presents several analytical formulas needed for the calculation of the critical temperature T_c and the miscibility gap. The behavior of T_c is studied in Appendix 4B. These special cases are also examined to illustrate the similarity and difference between the simple models in Section 4.2 and the results for semiconductors to be considered in the next few sections. In Appendix C, the low-temperature limit of GQCA is examined, and situations in which a flaw may arise are pinpointed.

4.8. INTERNAL STRAIN AND CLUSTER ENERGIES

Strictly speaking, QCA and GQCA imply short-ranged interaction energies. However, the energy models in Section 4.6, which are consistent with the experiments, indicate that strain is the most important contribution to mixing energy in most pseudo-binary alloys. The exceptions are the near-lattice-constant-matched alloys (e.g., GaAlAs) and HgCdTe. Since the strain is long ranged and its energy is shared among bonds throughout the crystal and is mutually interactive in nature, it can never be separated into isolated contributions. To write the mixing enthalpy as a sum of cluster energies, these cluster energies must contain contributions extending beyond the cluster size. Such an effect has already been included in the energy models used in Section 4.6. This section is designed to provide a basis for treating the internal strain energy in alloys (Chen *et al.*, 1988) and obtaining its contribution to cluster energies for insertion into the GQCA calculation.

Consider an alloy divided into M nonoverlapping clusters. In general, alloy strain energy can be written as the sum of the strain energies of these clusters and the interaction among the clusters. Focus on a particular cluster denoted n, detach this cluster from the alloy, and allow both the cluster and the remaining medium to relax. When the cluster is reconnected to the medium, there will be a change in the energy. This energy change, denoted h_{nm}, is defined as the interaction energy between the cluster and its surrounding medium labeled by m. The total strain energy of the alloy can be written as

$$E = \sum_{n} (\varepsilon_n + \tfrac{1}{2}h_{nm}) \tag{4.8.1}$$

Here ε_n is a residual strain energy for the nth cluster when it is detached from the alloy and is in its equilibrium configuration. The sum in Eq. (4.8.1) is over all clusters in the alloy. The factor 1/2 is inserted to eliminate double counting. In an alloy, Eq. (4.8.1) can be replaced by an ensemble average

$$E = M \sum_{n} \sum_{m} (\varepsilon_n + \tfrac{1}{2}h_{nm}) p_n P_{nm} \tag{4.8.2}$$

where p_n is the probability that a cluster will be of the n type (e.g., specified by the number of A and B atoms and their arrangements) and P_{nm} is a conditional probability that the surrounding environment is in state m when the cluster is in state n. Note that while n in Eq. (4.8.1) denotes the location of the cluster, n in Eq. (4.8.2) also designates the state of this cluster. From Eq. (4.8.2), an effective cluster energy $\varepsilon(n)$ can be defined as

$$\varepsilon(n) = \sum_{m} (\varepsilon_n + \tfrac{1}{2}h_{nm}) P_{nm} \tag{4.8.3}$$

These energies can be expressed explicitly within the following elastic continuum model.

Consider a spherical cluster embedded in an elastic medium. The interaction energy can be written as

$$h_{nm} = \tfrac{1}{2}\mu_n(a_n - r)^2 + \tfrac{1}{2}\nu_m(b_m - r)^2 \tag{4.8.4}$$

where the equilibrium radius r is to be determined by minimizing h_{nm}. The lengths a_n and b_m are the "natural" radii, and μ_n and ν_m are the force constants of the cluster and the medium, respectively. For a macroscopic semiconductor cluster of radius R, $\mu = 12\pi BR$, and a good approximation for ν is $\nu = 3.2\pi\ (C_{11} - C_{12} + 3C_{44})$ (Chen and Sher, 1985). The value of r corresponding to the minimum value of h_{nm} is given by

$$r_{nm} = (\mu_n a_n + \nu_m b_m)/(\mu_n + \nu_m) \tag{4.8.5}$$

and the actual interaction energy becomes

$$h_{nm} = \tfrac{1}{2}\mu_n \nu_m (a_n - b_m)^2/(\mu_n + \nu_m) \tag{4.8.6}$$

Next cast the average energy into a form in which a_n and b_m are decoupled. To do this, an effective radius r_e is introduced such that

$$\sum_n \sum_m \frac{p_n P_{nm} \mu_n \nu_m (a_n - b_m)^2}{\mu_n + \nu_m}$$
$$= \sum_n \sum_m \frac{p_n P_{nm} \mu_n \nu_m [(a_n - r_e)^2 + (b_m - r_e)^2]}{\mu_n + \nu_m} \tag{4.8.7}$$

which imposes the following condition on r_e:

$$\sum_n \sum_m \frac{p_n P_{nm} \mu_n \nu_m (a_n - r_e)^2 (b_m - r_e)^2}{\mu_n + \nu_m} = 0 \tag{4.8.8}$$

A very quick and important result for the internal strain energy can be obtained from these two equations by neglecting the correlation between p_n and P_{nm} and approximating all the force constants by the concentration weighted average values μ and ν. This result is

$$r_e = \langle a_n \rangle = \langle b_m \rangle, \langle (a_n - r_e)^2 \rangle = \langle (b_n - r_e)^2 \rangle = \langle \Delta r^2 \rangle, \text{ and}$$
$$E = M[\varepsilon + \tfrac{1}{2} \mu\nu \langle \Delta r^2 \rangle / (\mu + \nu)] \tag{4.8.9}$$

where ε is the average value $\langle \varepsilon_n \rangle$.

Equation (4.8.9) shows that there are three essential features contributing to the internal strain energy in this cluster theory: the mean square fluctuation of the cluster radii $\langle \Delta r^2 \rangle$, the effective force constant which is $\mu\nu/(\mu + \nu)$, and the mean strain energy ε of isolated clusters.

This result further suggests a way to use the effective medium to obtain cluster energies in an alloy. For example, if one assumes that the effective medium has an effective elastic force constant ν and radius r_e, then replacing the medium by a cluster n introduces a strain energy having the same expression as Eq. (4.8.6) with ν_m replaced by ν and b_m by r_e. If all the strain energy is assigned to this cluster, we obtain an effective cluster energy:

$$\varepsilon_c(n) = \varepsilon_n + \tfrac{1}{2} \mu_n \nu (a_n - r_e)^2 / (\mu_n + \nu) \tag{4.8.10}$$

We can see that the alloy mean energy $E = M \Sigma p_n \varepsilon_c(n)$ reduces to the result in Eq. (4.8.9) when μ_n is taken to be the mean value μ. The effective medium nature enters if we take r_e to be determined by the condition that the average energy E be a minimum; i.e.,

$$\sum_n \frac{p_n \mu_n \nu (a_n - r_e)}{\mu_n + \nu} = 0 \tag{4.8.11}$$

Application of this self-consistent approach to the elastic constants has already been considered in Section 3.7.

Let us now apply the general ideas behind Eqs. (4.8.10) and (4.8.11) to the cluster energies, and GQCA to a systematic discussion of the statistical properties of pseudobinary alloys $A_{1-x}B_xC$. Because of the special atomic arrangements of these alloys, the bonds are the natural entities to use for energy counting while it is more convenient to count the atoms for statistical arrangements. All the bonds in these alloys can be clearly separated into AC and BC bonds. A convenient way to generate the cluster is to assign the four AC bonds surrounding an A atom to that atom, and similarly assign the four BC bonds to the central B atom. In this way the smallest cluster is a one-atom four-bond cluster. Following the ideas behind Eq. (4.8.10) we start with an effective medium and replace the center atom by an A atom, the four surrounding bonds by AC bonds, and calculate the interaction energy to obtain the one-atom cluster energy ε_A. In a similar way we can obtain ε_B by replacing the cluster with a B atom and its four BC bonds. Note that the mixing energy ΔE in Eq. (4.7.3) is not zero, because ε_A and ε_B in an effective medium are different from their respective values ε_{A0} and ε_{B0} in their respective pure constituent crystals. If the effective spring model of Fig. 3.3 is used, then $\mu = 6\alpha_A$, $\nu = 2\alpha$, and we have (following Eqs. (4.8.10) and (4.7.4)) the excess energy of an A cluster given by

$$\Delta_A = \frac{3\alpha_A\alpha}{3\alpha_A + \alpha}(d^0_{AC} - d)^2 = \frac{3\alpha_A\alpha}{3\alpha_A + \alpha}(1 - x)^2\delta_0^2 d^2 \tag{4.8.12}$$

where δ_0 is $(d^0_{AC} - d^0_{BC})/d$ with d being the virtual crystal approximation (VCA) bond length. A similar expression can be obtained for Δ_B. When d and all the α's in Δ_A and Δ_B are replaced by their mean values, the resulting ΔE is only modified by a few percent from those based on the actual expressions in Eq. (4.8.12). The resulting mixing enthalpy is given by the simple expression $\Delta E = x(1-x)\alpha 3\delta_0^2 d^2/2$. This is the same result as in Eq. (4.6.13) for the mixing enthalpy parameter. Since each cluster contains only one alloying atom and none of the clusters share the same alloying atom, the number of possible ways to arrange the clusters is exactly Φ_0 in Eq. (4.2.1), so that the total mixing free energy in this model is precisely Stringfellow's preferred form given in Eq. (4.6.2). However, now the origin of the mixing enthalpy is the long-range strain energy, which is different from the nearest-neighbor pair energy assumed in the zeroth approximation (see Section 4.2.2 and Appendix 4B).

The next simplest cluster set for the pseudobinary alloy contains 4 alloy atoms and 16 bonds, depicted in Fig. 4.11. This is the smallest cluster size that can provide a meaningful description of the local statistical correlations among the alloying atoms. All the bonds associated with each alloy atom are contained within the cluster which alleviates the counting problem. Since the results of this 16-bond cluster have important implications to many properties, the results warrant a detailed discussion, which will be done in the next section.

4.9. SIXTEEN-BOND MICROCLUSTERS

As mentioned earlier, the 16-bond clusters shown in Fig. 4.12 are the smallest nonoverlapping clusters permitting examination of local statistical correlations of alloy atoms in a pseudobinary alloy $A_{1-x}B_xC$. Each cluster contains four alloy atoms A and B on the vertices of a tetrahedron. The 16 bonds in a cluster include the 4 bonds connecting the 4 alloy atoms to the central C atom and the 12 bonds connecting the 4 alloy atoms to 12 C atoms on the periphery. In the absence of external stresses, there are five kinds of clusters distinguished by distinct cluster energies. These clusters can be labeled according to the numbers of A and B atoms in the cluster, namely A_4B_0, A_3B_1, A_2B_2, A_1B_3, and A_0B_4, to be labeled by $j = 0$ to J with $J = n = 4$, following the notation in Section 4.7. Each cluster now occupies one cube in the crystal. Thus, the ratio of the number of clusters M to the number of unit cells (or C atoms) N is $M/N = 1/4$. Since each cluster shares no alloy atoms with any other cluster, the

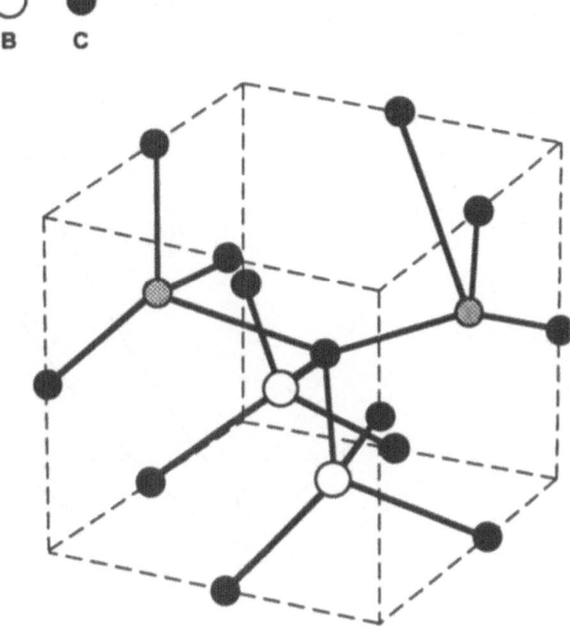

FIGURE 4.12. Schematic picture of a 4-atom, 16-bond cluster in a pseudobinary alloy. This figure shows an A_2B_2 cluster in a cubic cage.

total number of alloy configurations Φ in Eq. (4.7.5) for a given set of numbers of clusters $\{M_j\}$ is exact. Any error lies in approximating the Hamiltonian as a sum of independent cluster contributions.

The effective cluster energies can be calculated in a manner similar to what was done for the one-atom clusters in Section 4.8. To circumvent the complicated atomic relaxation that can occur in the medium outside a cluster, the effective spring model results deduced in the single-impurity case can be extended to the present case. To be more explicit, we start with a VCA medium and replace 16 bonds in this medium by a cluster. The atoms in the cluster including the 12 peripheral C atoms are allowed to relax, but the atoms in the medium are fixed at their VCA sites. To compensate the imposed rigidity, the bond-angle restoring force β in VFF is taken to be zero for all interactions involving those bonds connecting the peripheral C atoms to the medium. With this simplification the strain energies and atomic positions can be treated without further approximations in VFF through a minimization of the strain energy. These strain energies, according to Eq. (4.8.10), can be assigned as the strain contribution to the cluster energy.

Contributions to the cluster energies from modifications of the chemical bonds in the alloy can be estimated using the bond-orbital model (BOM) of Section 2.5 by calculating the changes in the metallization energy of each bond, i.e., $\Delta\varepsilon_b$ in Eq. (2.5.7), from its pure-crystal value (Chen and Sher, 1985). The contributions to $\Delta\varepsilon_b$ due to antibonding states $|a'\rangle$ coming from the same cluster can be treated exactly, while those from the other clusters are approximated as statistically averaged values.

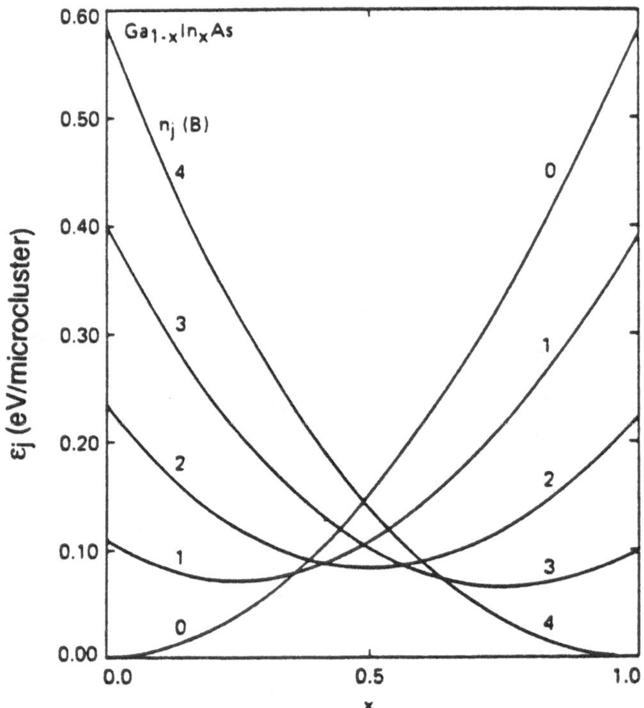

FIGURE 4.13. Cluster excess energies of the 16-bond clusters described in the text in $Ga_{1-x}In_xAs$ for clusters with different numbers of In atoms (Sher *et al.*, 1987).

Figure 4.13 shows the calculated cluster excess energies as a function of x in $Ga_{1-x}In_xAs$. For a small x value, the ε_j energies are larger for larger j, because a larger j means there are more InAs bonds in the cluster, and therefore it is harder to fit it into a medium with a lattice constant close to that of a GaAs crystal. Similarly, at $x = 1/2$, ε_2 for the A_2B_2 cluster is the smallest energy because this cluster has the best match to the 50–50 VCA alloy lattice. Clearly, these energies are dominated by strain. However, it is difficult to judge the local correlation directly from these energies. To do this it is better to work with the reduced cluster energies Δj defined in Eq. (4.7.4) because it is these energies that drive the cluster populations, given by Eq. (4.7.12).

Figure 4.14 shows the reduced cluster energies Δj for $Ga_{1-x}In_xAs$. By definition Δ_0 and Δ_4 are zero. The other Δj differ drastically from the ε_j set, not only in magnitude but also in their x dependence. These reduced cluster energies are all negative and have a weak linear dependence on x. The three lines are nearly parallel to each other, indicating they have the same x dependence. Δ_1 and Δ_3 are nearly equal and are less negative than Δ_2. The negative values of the Δ_j qualitatively imply that A and B atoms do not like to segregate. These general features of cluster energies are found to be common in all III–V and II–IV alloys with appreciable (few percent) lattice mismatches. For the lattice-matched alloys such as $Ga_{1-x}Al_xAs$ and $Hg_{1-x}Cd_xTe$, all the energies are small and the effects caused by $\{\Delta j\}$ are negligible.

One way to measure the local correlation is to look at the average values of the cluster population $x_j = M_j/M$ against those in a random alloy x_j^0 given in Eq. (4.7.6) with g_j given

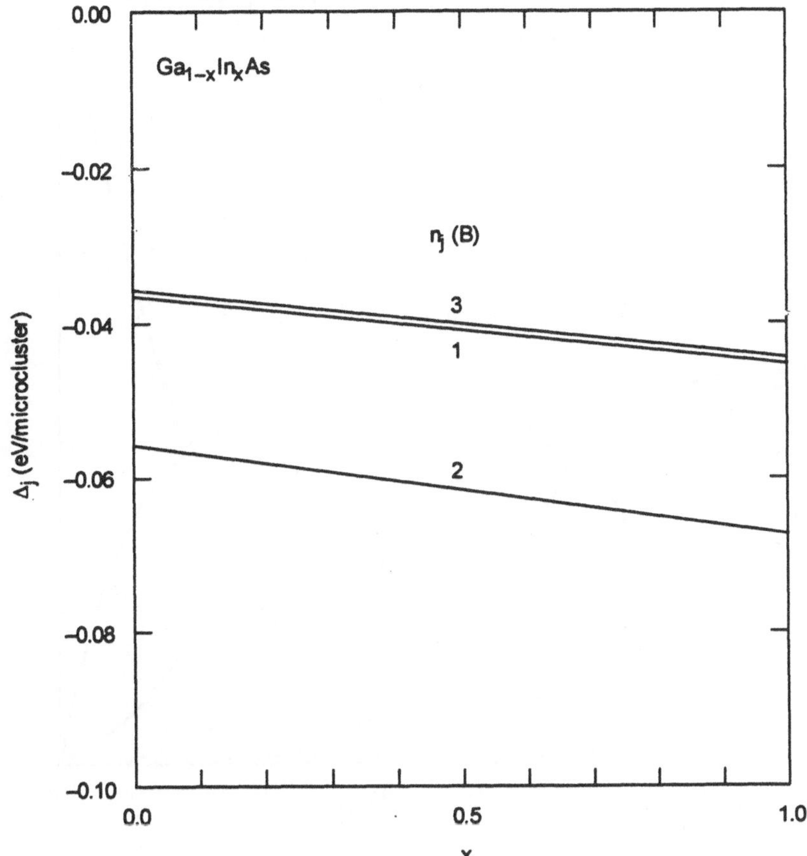

FIGURE 4.14. The corresponding reduced cluster energies Δ_j for $Ga_{1-x}In_xAs$ (Sher *et al.*, 1987).

in Eq. (4.7.15). The average cluster populations can be calculated from Eq. (4.7.12) after η is determined by solving Eq. (4.7.14). For a random alloy the cluster distribution is given by the Bernoulli distribution $x_j^0 = \binom{4}{j}x^j(1-x)^{n-j}$, as shown in Fig. 4.15. Figures 4.16a,b show plots of the deviations from randomness, $x_j - x_j^0$ for different clusters as a function of alloy concentration x at equilibration temperatures of 600 K and 1500 K, respectively. These two sets of curves are similar in shape but different in magnitude. As expected from the fact that $\Delta_1, \Delta_2, \Delta_3 < 0$ and $\Delta_0 = \Delta_4 = 0$, the $j = 1, 2, 3$ cluster populations, where they are large, are seen to be enhanced by the Δ_j, while the $j = 0$ and $j = 4$ populations are reduced. These cluster populations are seen to be sensitive to temperature. For example, at $x = 0.50$, the deviation decreases from 25% to 5.5% as T increases from 600 to 1500 K. For a given sample, the actual temperature to be used to calculate the cluster population depends on the "last" temperature T_f at which the alloy atoms cease to diffuse. If T_f is larger than the sample temperature T, then the alloy is in a metastable state, and T_f should be used instead of T.

After the cluster distribution is calculated, the alloy mixing enthalpy is determined from Eq. (4.7.3). Figure 4.17 shows ΔE as a function of x for $Ga_{1-x}In_xAs$ at four temperatures: $T = 300$, 600, 1000, and 1500 K. The ΔE is seen to be dominated by the T-independent part

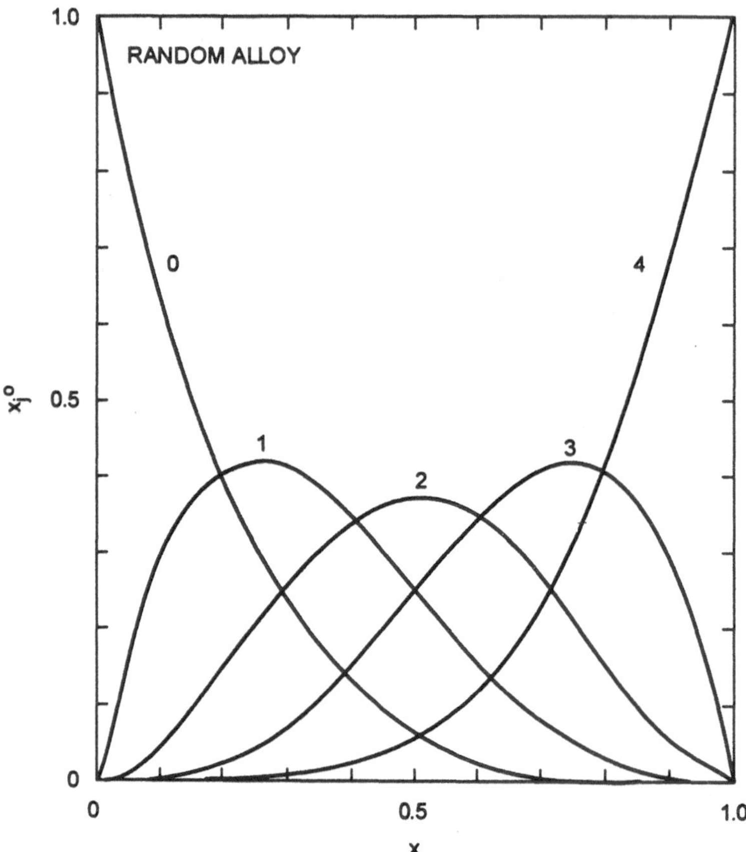

FIGURE 4.15. Four-atom cluster distribution for a random alloy (Sher *et al.*, 1987).

given by the first two terms on the right side of Eq. (4.7.3), which can be attributed to the strain energy. The T-dependence is seen to be very weak, because the Δj's are small. ΔE increases with T because that promotes increases of population of clusters with higher energies. The curves in Fig. 4.17 indicate that the approximation $\Delta E = x(1-x)\Omega$ with a constant positive Ω, such as that used in Section 4.6, should not introduce much error in the free-energy calculation. However, a positive value of Ω does not necessarily imply that A and B atoms repel each other microscopically. This is evident from Fig. 4.16, which shows enhanced population for the A_2B_2 clusters relative to those of a random alloy.

The total mixing free energy, ΔF, can now be calculated from $\Delta E - T\,\Delta S$ with ΔS given by Eq. (4.7.7). The results for $Ga_{1-x}In_xAs$ and $GaAs_{1-x}Sb_x$ are shown in Fig. 4.18. These curves show that these alloys have miscibility gaps at low temperature. These results are similar to those based on pair-potential QCA. Therefore, it is not surprising that replacement of the pair-QCA ΔF for pseudobinary solid solutions by the present GQCA results produce nearly the same phase diagrams as those in Fig. 4.10 (Patrick *et al.*, 1987). This also shows that the phase diagrams are in the class of phenomena that are rather insensitive to details of

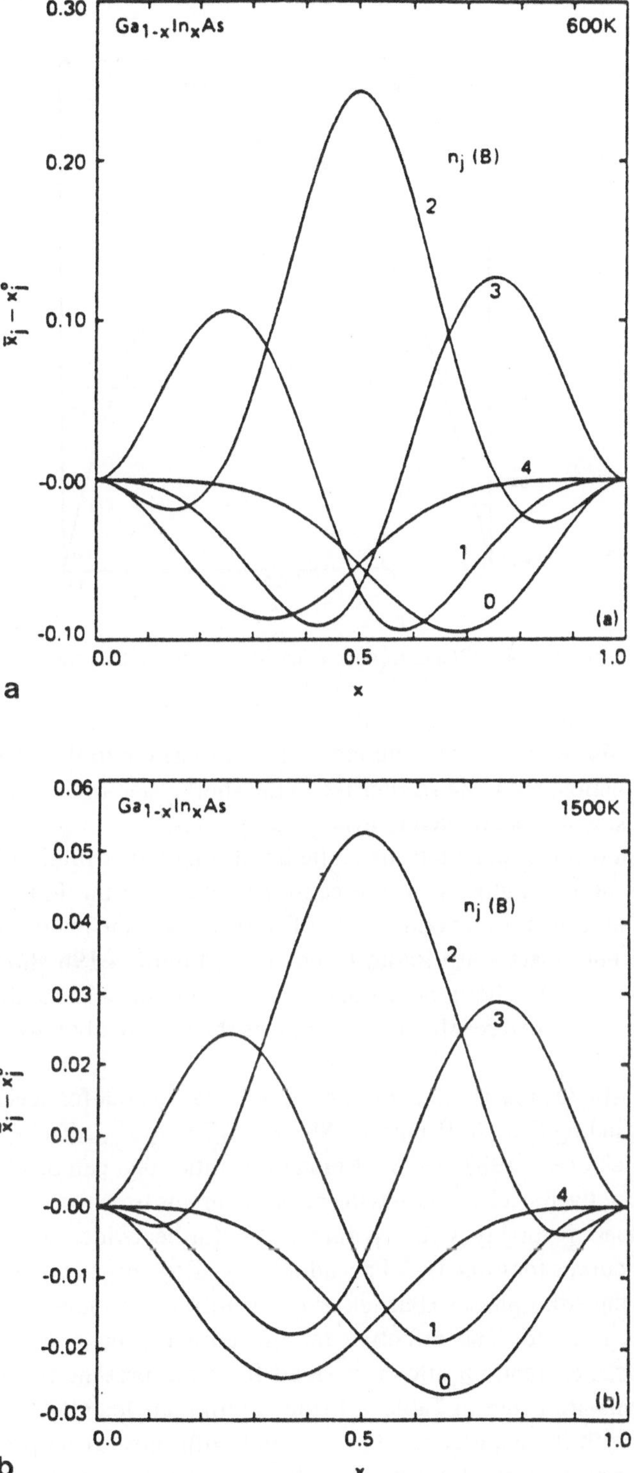

FIGURE 4.16. Deviations of cluster populations in $Ga_{1-x}In_xAs$ from random distribution at T = (a) 600 K and (b) 1500 K. For reference the random populations $\{x_j\}$ are plotted against x in Fig. 4.14 (Sher *et al.*, 1987).

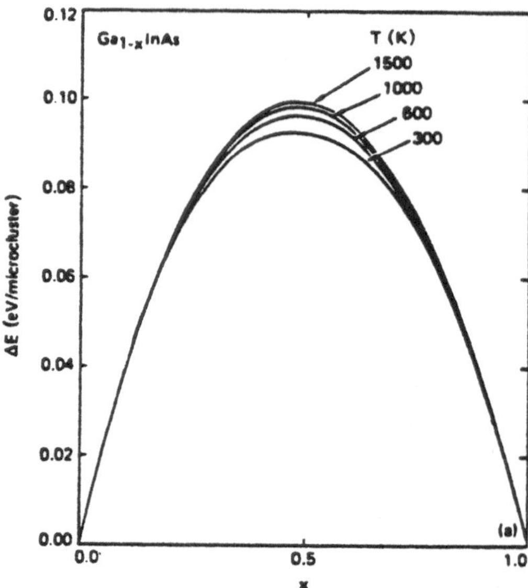

FIGURE 4.17. Mixing enthalpies of $Ga_{1-x}In_xAs$ as a function of x at four temperatures (Sher *et al.*, 1987).

the short-range-order statistical distributions. Effects sensitive to their local environment, e.g., vacancy formation, are more responsive to the short-range-order state than those such as phase diagrams, which are related to average properties.

A by-product of these calculations is the bond-length distribution in alloys. Figure 4.19a shows the bond lengths from the central C atoms to the four alloy atoms as a function of x for all five clusters considered. The spreads in the lengths of the same kind of bond in different clusters are found to be small. Figure 4.19b shows their average values and the rms widths about the mean of Ga—As and In—As bonds over the cluster populations. The calculated result clearly supports the bimodal bond-length distribution revealed by EXAFS.

Besides $Ga_{1-x}In_xAs$, the above studies have been carried out for several other pseudobinary alloys, including $Ga_{1-x}In_xP$, $GaAs_{1-x}Sb_x$, $Hg_{1-x}Cd_xTe$, $Hg_{1-x}Zn_xTe$, $Cd_{1-x}Zn_xTe$, and $ZnSe_{1-x}Te_x$ (Patrick *et al.*, 1987, 1988). Among quantities that can be checked accurately against experiment, the bond lengths and the phase diagrams from the 16-bond cluster, QCA calculations compare favorably with experiments. As further evidence, Fig. 4.20 shows the liquidus–solidus curves for three II–VI pseudobinary alloys based on the 16-bond GQCA calculations for the solid phases (Patrick *et al.*, 1988). Agreement between predictions and experiments is good. The effective mixing enthalpy parameters for completely disordered alloys, i.e., random alloys, derived from the present 16-bond model for a number of systems are listed in Table 4.4 (the column labeled GQCA). These are seen to be in accord with the experimental values and with those from previous estimates. However, the attractive interactions between the alloy atoms, and the associated changes in the cluster populations found from the calculations, have not been clearly demonstrated experimentally.

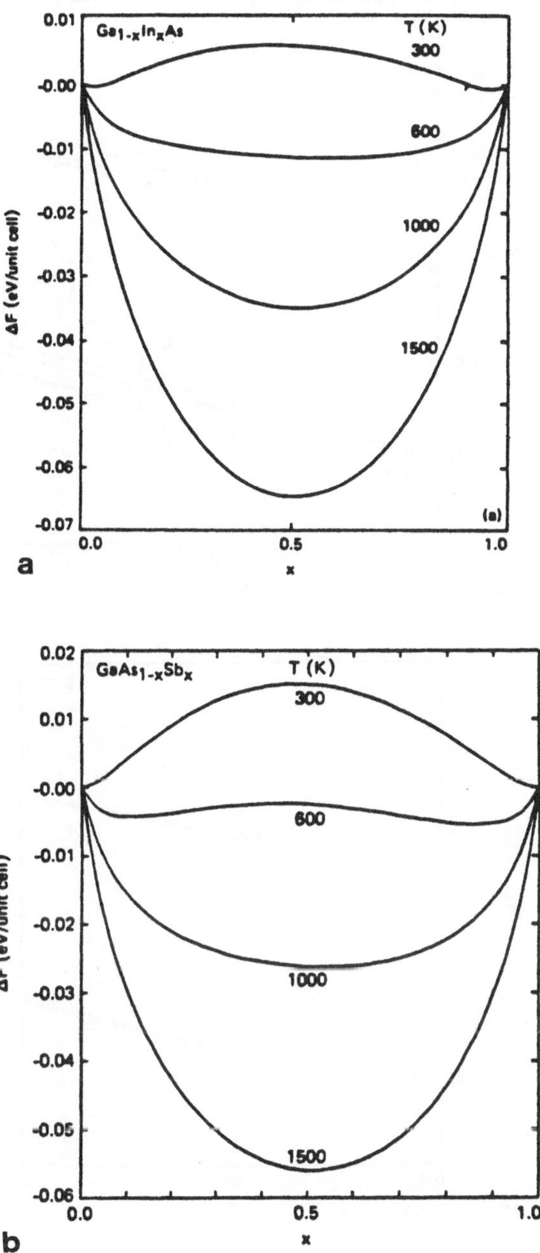

FIGURE 4.18. Mixing free energies of (a) $Ga_{1-x}In_xAs$ and (b) $GaAs_{1-x}Sb_x$ as a function of x at four temperatures (Sher *et al.*, 1987).

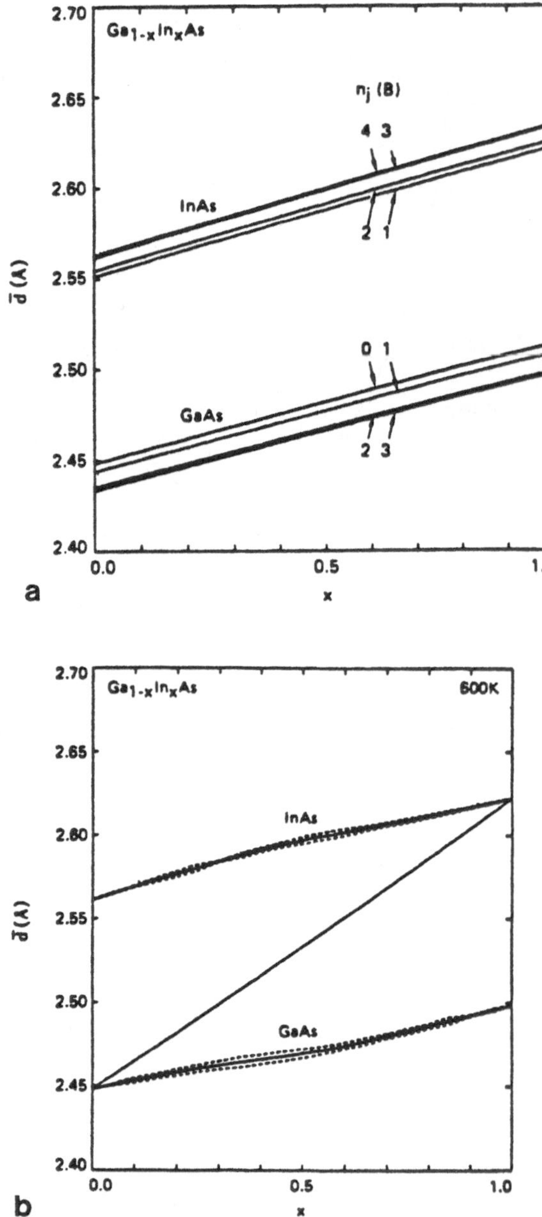

FIGURE 4.19. (a) The Ga–As and In–As bond lengths in different 16-bond clusters in $Ga_{1-x}In_xAs$ as a function of x. (b) The average bond lengths at $T = 600$ K. The figure also shows the rms widths of the distribution (Sher *et al.*, 1987).

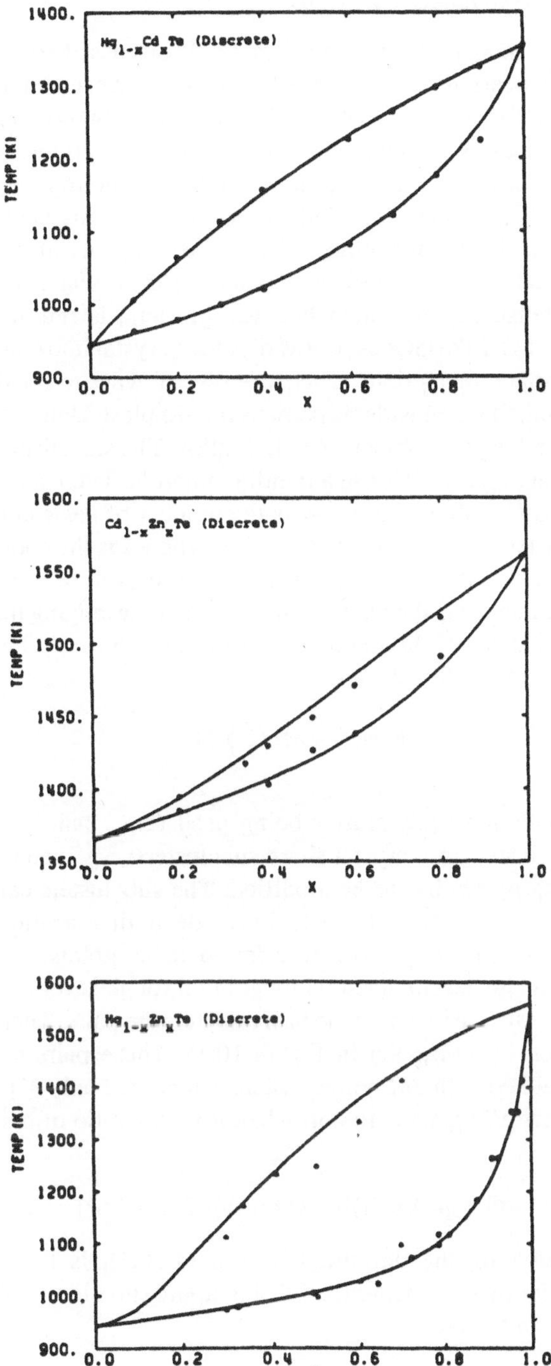

FIGURE 4.20. Calculated liquidus–solidus curves for three II–VI pseudobinary alloys using the 16-bond GQCA results for the solid phase and their comparison with experiments (Patrick *et al.*, 1988).

4.10. CLUSTER VARIATIONAL METHOD

The mixing energy models and statistical methods discussed so far have provided an adequate description of local correlation and phase diagrams in pseudobinary semiconductor alloys. All the results in these models were obtained without relying on heavy computations. However, modern computers do enable sophisticated calculations, and more rigorous calculations should be considered. This section will discuss an improved statistical theory, and the next section will review recent efforts to combine this statistical method with first-principles LDA calculations for semiconductor pseudobinary alloys.

A statistical method first proposed by Kikuchi (1951), which improves GQCA by incorporating some statistical correlation between clusters, is referred to as the cluster variational method (CVM). Like GQCA, CVM divides a crystal into clusters that are chosen to contain all the important interactions. There are several ways to obtain the CVM cluster configurations. The method we consider is perhaps the simplest. Unless specified, we ignore the spectator C atoms in the pseudobinary $A_{1-x}B_xC$ alloy. The statistical counting is then the same as that in the binary $A_{1-x}B_x$. We use a familiar notation. There are N lattice sites equal to the number of atoms $N = N_A + N_B$, and M is the number of basic clusters in the crystal. For example, if the clusters are pairs, then $M = zN/2$, where z is the coordination number. If the basic cluster is a square, then $M = N$ in a square lattice, but $M = 3N$ in a simple cubic lattice. The quantity of interest is the number of ways, Φ, of arranging the A and B atoms on the lattice for a specified set of the numbers of different types of clusters $\{M_j\}$. If these clusters were all independent, then the answer would be

$$g(M,s) = M!/\prod_j M_j! \tag{4.10.1}$$

where the index s labels the basis cluster being used (e.g., pair, triangle, square, etc.). However, these basis clusters in a crystal share subclusters; e.g., squares may share pairs. Therefore, the counting $g(M,s)$ has to be modified. The subclusters can also share further subclusters, and further corrections have to be made until reaching the final smallest subclusters—the single-atom clusters to be referred to as points. The ways systematic corrections are made can be best understood by going through several examples.

Consider the simplest case in which the basis clusters are pairs. Then $M = zN/2$, and the uncorrected Φ is given by $g(M,\text{pair})$ in Eq. (4.10.1). These pairs share points. The M independent pairs would contain $2M$ points, whereas there are only N points in the crystal. A logical way to correct $g(M,\text{pair})$ is to normalize it with a ratio of point arrangements as follows:

$$\Phi = g(M,\text{pair})[g(N,\text{point})/g(2M,\text{point})] \tag{4.10.2}$$

where $g(N,\text{point})$, following the definition of Eq. (4.10.1), is the number of random configurations Φ_0 in Eq. (4.2.1). Using the Stirling approximation $\ln N! = N \ln N - N$, the mixing entropy $\Delta S = k \ln \Phi$ becomes

$$\Delta S = -Nk\sum c_j \ln c_j - Mk\left[\sum h_j y_j \ln y_j - 2\sum c_j \ln c_j\right] \tag{4.10.3}$$

where, for later convenience, c_1 and c_2 stand for fractions x and $1 - x$, respectively, $y_j = M_j/M$ is the pair probability of cluster type j, and h_j is the degeneracy of cluster type j.

Note that M_j is the number of clusters of the type identified by specifying the arrangement of A and B atoms in the cluster. It does not include other types with the same cluster energy. Thus, h_jy_j is equivalent to the cluster population y_j in QCA. With this understanding, the CVM is exactly the same as QCA in the pair approximation.

Next consider square clusters on a square lattice. Let M now be the total number of squares in the crystal. Then $M = N$. The number of pairs in the crystal is actually $L = 2N$. If all the M squares were independent (e.g., separated from one another), then the total number of pairs would be $4M$ and the total number of atoms (points) would also be $4M$. The first correction to $g(M,\text{square})$ comes from the pair normalization factor

$$f_{\text{pair}} = \frac{g(L,\text{pair})}{g(4M,\text{pair})} \tag{4.10.4}$$

However, both the denominator and numerator of this factor have to be further corrected with point normalizations in a form similar to that inside the brackets of Eq. (4.10.2). Explicitly, the corrected Φ takes the form

$$\Phi = g(M,\text{square})\left[\frac{g(L,\text{pair})}{g(4M,\text{pair})}\right]\left[\frac{g(8M,\text{point})}{g(4M,\text{point})}\right]\left[\frac{g(N,\text{point})}{g(2L,\text{point})}\right] \tag{4.10.5}$$

In this equation the second factor is the pair normalization, the third factor is the point normalization to correct the denominator of the second factor, and the last factor is to correct the numerator of the second factor. The associated mixing entropy becomes

$$\begin{aligned}\Delta S &= -k[(8M - 4M - 2L + N)\textstyle\sum c_j \ln c_j + (L - 4M)\sum h_jy_j \ln y_j + M\sum g_jz_j \ln z_j]\\ &= -kN(\textstyle\sum c_j \ln c_j - 2\sum h_jy_j \ln y_j + \sum g_jz_j \ln z_j)\end{aligned} \tag{4.10.6}$$

where y_j and z_j are, respectively, the pair and square probabilities, and h_j and g_j are their respective degeneracies.

Now it becomes straightforward to write the expression for ΔS for square clusters in a (three-dimensional) simple cubic lattice, for which $M = 3N$, $L = 3N$, and the expression for Φ in Eq. (4.10.5) is still valid, so

$$\Delta S = -kN(7\textstyle\sum c_j \ln c_j - 9\sum h_jy_j \ln y_j + 3\sum g_jz_j \ln z_j) \tag{4.10.7}$$

As a final example, consider the tetrahedron clusters in a fcc lattice. There are a total of $M = 2N$ tetrahedral clusters and $L = 6N$ pairs in the crystal, and there are six sides and four atoms in a tetrahedron. Equation (4.10.5) has to be modified slightly for the present case to read

$$\Phi = g(M,\text{tetrahedron})\left[\frac{g(L,\text{pair})}{g(6M,\text{pair})}\right]\left[\frac{g(12M,\text{point})}{g(4M,\text{point})}\right]\left[\frac{g(N,\text{point})}{g(2L,\text{point})}\right] \tag{4.10.8}$$

The entropy is then given by

$$\Delta S = -kN(5\textstyle\sum c_j \ln c_j - 6\sum h_jy_j \ln y_j + 2\sum g_jz_j \ln z_j) \tag{4.10.9}$$

It is useful to observe a rule that governs the coefficients in front of the summation signs inside the brackets for all CVM expressions for ΔS: If we multiply each coefficient by the number of atoms in the cluster that term represents and sum these products, the result is 1. This rule is useful for checking if the configurations are properly normalized.

Up to now we did not have to distinguish between binary and pseudobinary alloys. However, for the tetrahedron case there are two kinds of tetrahedra in pseudobinaries: one contains a central C atom, and the other is empty. If the energies are assumed to be associated only with bonds, we can ignore those tetrahedra without a central C atom. Then $M = N$ should be used in Eq. (4.10.8). The first set of brackets is dropped and $g(12M, \text{point})/g(2L, \text{point})$ is 1, so we have

$$\Delta S = -kN(-3\sum c_j \ln c_j + \sum g_j z_j \ln z_j) \tag{4.10.10}$$

Note that the pair contribution disappears from this expression. This is understandable because the clusters treated this way share only points. It is important to point out that Eq. (4.10.10) is exactly the same as the GQCA result in Eq. (4.7.7) for the case of tetrahedra with the identification of x_j with $g_j z_j$. This example demonstrates that alloy statistics for pseudobinaries are not necessarily identical to those for binaries.

The equilibrium values of cluster probabilities such as $\{y_j\}$ and $\{z_j\}$ are determined by minimization of the mixing free energies in a manner similar to GQCA. Likewise, not all cluster probabilities are independent. Since there are more variables to be determined and more constraint conditions to be satisfied in CVM than in GQCA. Kikuchi (1981) worked with the grand potential $\phi = F - \Sigma \mu_j N_j$ and designed an iterative procedure—the "natural iteration method"—to attack the problem. Impressive results have been obtained from CVM with small clusters. For example, Table 4.1 shows that CVM predicts critical temperatures very well for the square lattice using square clusters and for the fcc lattice using tetrahedron clusters. There are also examples in which CVM obtains correct phase diagrams, whereas the Bragg–Williams approximation fails quantitatively (Kikuchi, 1974). This method has recently been incorporated in the phase diagram studies of semiconductor alloys using energies derived from *ab initio* LDA calculations. The results will be discussed in the next section.

4.11. *AB INITIO* CALCULATIONS

Because self-consistent LDA has been successful predicting the structural properties of ordered semiconductors, it is desirable to attempt to apply this *ab initio* theory to alloys. There are several challenges in such an undertaking. First, the mixing enthalpies in semiconductor alloys are typically about 1 kcal per mole or 1/23 eV per atom, but the total energies [i.e., each term on the right side of Eq. (4.7.2)] are many orders larger. Very precise and accurate numerical calculations are required to get meaningful results. However, if carried out precisely, LDA has been demonstrated to be able to predict changes in energies correctly, despite its larger error in predicting total energies. The next, and more challenging difficulty, is alloy disorder. To compute the total energy of an alloy quantum mechanically, one needs to solve the Schrödinger equation for a Hamiltonian which does not have lattice translational symmetry, which is so indispensable in the traditional band structure theory. Finally, to use a statistical theory such as CVM, the excess energy has to be decomposed into short-ranged

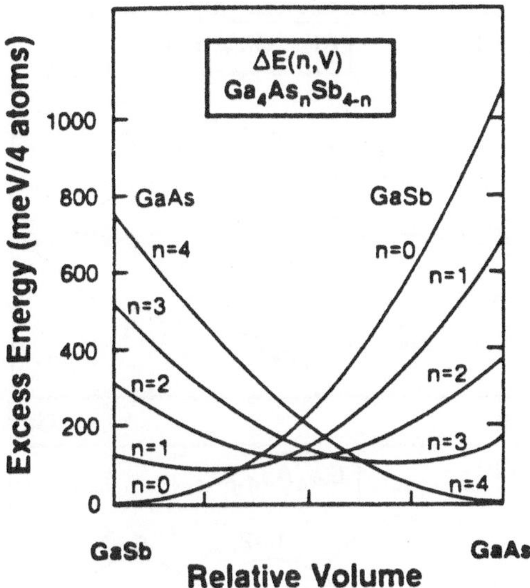

FIGURE 4.21. Volume-dependent excess energies for $Ga_4As_nSb_{4-n}$ clusters obtained by Ferreira *et al.* (1989).

multisite correlation energies, including only the single-site, pair, etc., up to a manageable cluster size containing only a handful of sites. To circumvent these difficulties, Connolly and Williams (1983), working on metal alloys, proposed that these multisite correlation energies be deduced from ordered systems that are composed of the same atoms. This scheme allows a direct application of the first-principles theory in the calculation of the energy parameters. This theory has been extended to semiconductor alloys, including considerable refinements, by Wei *et al.* (1990) and Ferreira *et al.* (1989) (to be referred to as WF), and respectable results have been obtained for the phase diagrams and alloy equilibrium properties. Their main results are summarized below.

The cluster excess energies in WF are calculated as a function of relative cell volume. Figure 4.21 shows an example for $Ga_4As_nSb_{4-n}$ clusters. These energy curves are similar to the x-dependent excess energies shown in Fig. 4.13, because it is a good approximation to scale the alloy cell volume linearly with x. The overall implications of these excess energies are also very much in accord with what we learned from Section 4.9. The magnitudes of mixing enthalpies are generally dominated by terms driven by lattice-constant mismatches. For the lattice-matched alloys, the excess energies are very small, their effects fall within calculational uncertainties and therefore may be neglected. For lattice-mismatched systems, ΔE is positive and, as we saw in Section 4.9, is dominated by a temperature-independent strain contribution. The reduced excess energies that govern the atomic statistical correlation are small and negative. This implies anticlustering, as shown in Fig. 4.22 for the deviation of cluster populations from random-alloy values. These curves are qualitatively the same as those obtained in Section 4.9. The mixing enthalpy parameters Ω for random alloys (at $T = \infty$) were calculated and are listed in Table 4.4, where they can be compared with previous estimates. The miscibility gaps and spinodals are also calculated and are shown in Fig. 4.23. These curves display considerable asymmetry about $x = 1/2$, a behavior also found from the experimental data for $GaSb_{1-x}As_x$. The calculated critical temperatures, indicated as T_{MG},

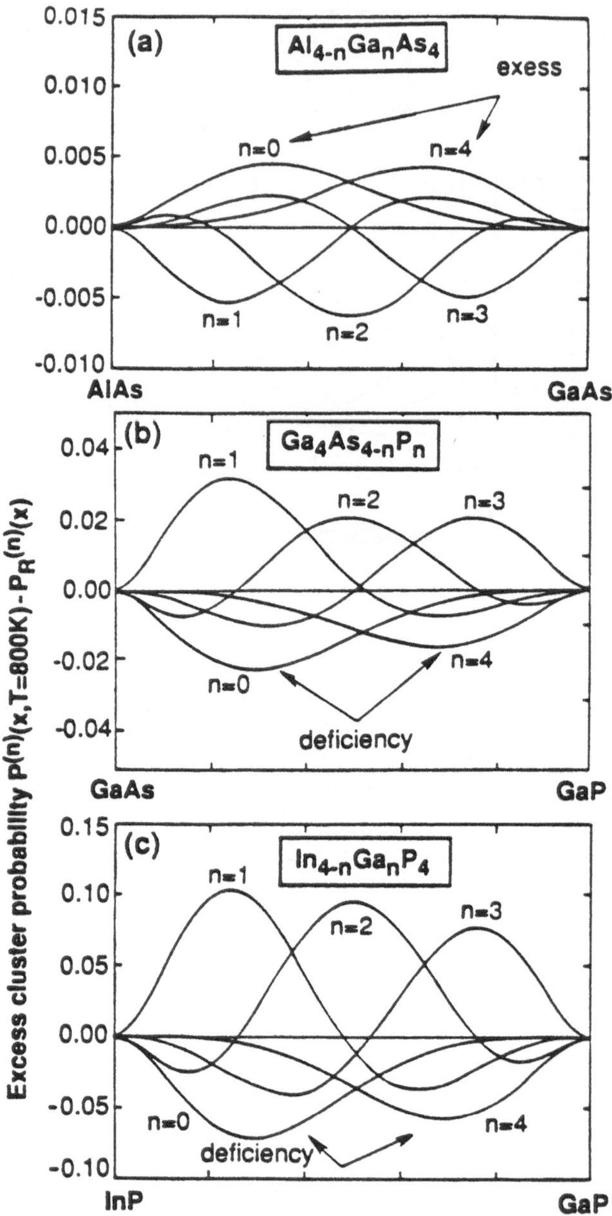

FIGURE 4.22. Deviations of cluster populations from random distributions for five III–V pseudobinary alloys calculated by Wei *et al.* (1990).

and the concentration x_{MG}, where the bimodal curves peak, are tabulated in Table 4.5. For systems for which experimental data are available, the agreement between theory and experiments is good. Finally, the predications of the bimodal bond-length distributions are similar to those predicated from the 16-bond cluster model in Section 4.9. However, to

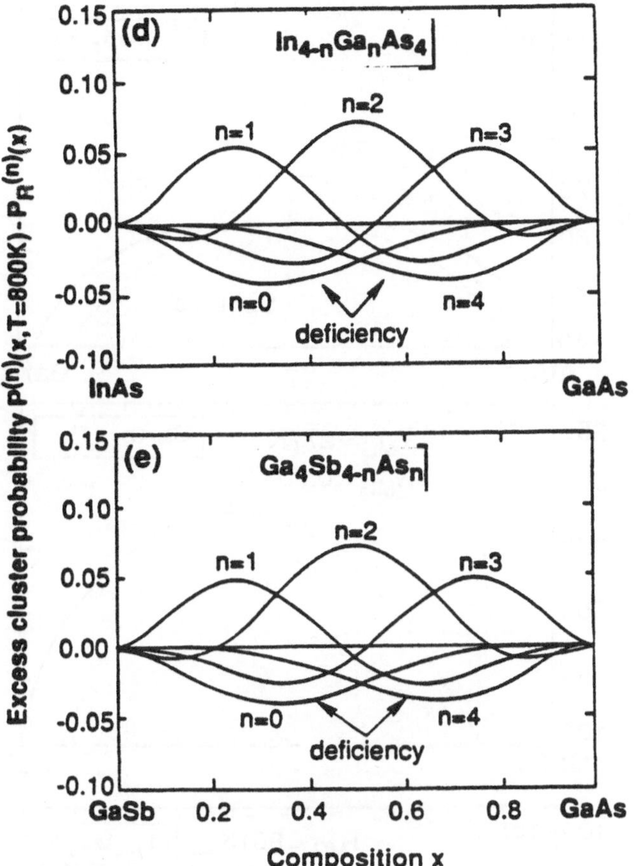

FIGURE 4.22. (continued)

TABLE 4.5. Calculated Critical Temperatures T_{MG} for Binodal and Spinodal Curves and Corresponding Compositions x_{MG} Compared to Experiments

	Theory		Experiment	
System	T_{MG} (K)	x_{MG}	T_{MG} (K)	x_{MG}
$Al_{1-x}Ga_xAs$	64	0.49		
$In_{1-x}Ga_xP$	961	0.676	933	0.62
$In_{1-x}Ga_xAs$	630	0.77		
$GaAs_{1-x}P_x$	277	0.603		
$GaSb_{1-x}As_x$	1080	0.595		
$Hg_{1-x}Cd_xTe$	84	0.60		
$Hg_{1-x}Zn_xTe$	455	0.56		
$Cd_{1-x}Zn_xTe$	605	0.623		

After Wei *et al.* (1990).

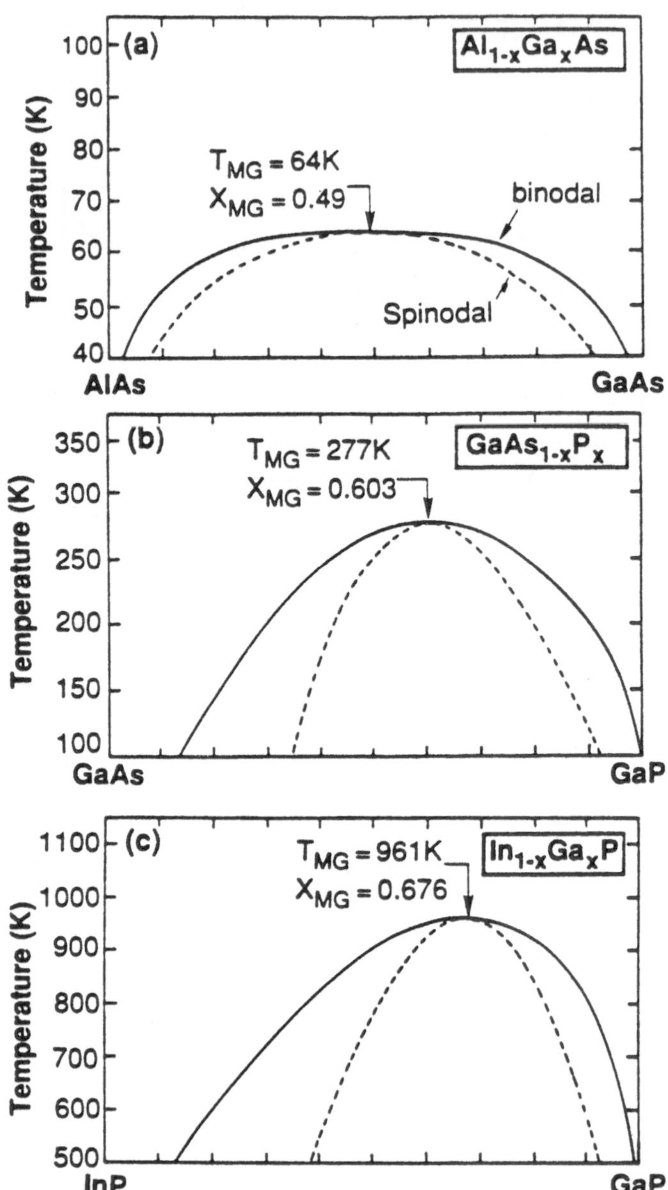

FIGURE 4.23. Miscibility gap and spinodal curves for five III–V pseudobinary alloys calculated by Wei *et al.* (1990).

produce the kind of accuracy in the energies required for the phase diagram calculation, very detailed and careful computational efforts are needed. The results presented here represent the state of the art in the LDA calculations.

However, the Connolly–Williams approach is only a kind of interpolation between *ab initio* energies of the ordered systems and alloys. The multisite correlation energies deduced from this approach represent interpolation parameters for the total energies. They depend on

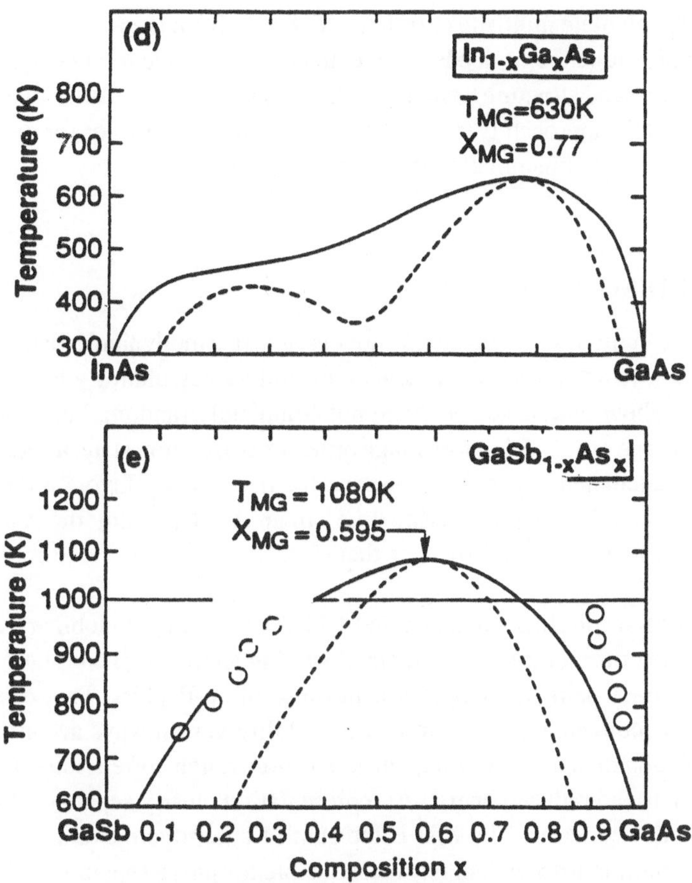

FIGURE 4.23. (continued)

the ordered systems chosen in the parametrization. The volume-dependent cluster energies are unlikely to be sufficient for calculating alloy properties sensitive to the cluster shapes, e.g., shear moduli and lattice vibrations. These are fine points that require improvements.

A different way to approach the *ab initio* calculation is to attack the disordered problem directly. If the fluctuations of the alloy potentials from their virtual crystal approximation (VCA) is small, then the next leading correction to VCA can be obtained from perturbation theory. This should work for most semiconductor alloys except for systems with large potential fluctuations such as $Hg_{1-x}Cd_xTe$ (Spicer *et al.*, 1982). A more general but more difficult approach is to extend an effective medium alloy theory, the coherent potential approximation (CPA) (to be discussed in Chapter 5), to clusters and to achieve triple self-consistency: consistency between the cluster distribution and the Hamiltonian, between the Hamiltonian and the electron density, and between the self-energy operator Σ in the cluster CPA theory and the potential fluctuations. Currently this theory has been carried out only for metal alloys, and only within single-site KKR-CPA (Gyorffy and Stocks, 1979) with a random distribution. There is also a theory developed along with CPA, the so-called generalized perturbation method (GPM) (Ducastelle and Gautier, 1976), which is intended to systematically calculate multisite correlation energies from the CPA results. However, to use GPM in semiconductors, it needs to be extended to include total energy contributions from other than the band structure term in Eq. (2.2.13) and to allow lattice distortions. Major work is needed if this approach is to achieve the same degree of rigor for disordered alloys as self-consistent density functional theory now enjoys dealing with crystalline semiconductors.

4.12. CONCLUDING REMARKS

This chapter summarizes statistical models and thermodynamic data for disordered semiconductor alloys and provides a practical method for calculating alloy phase diagrams. The calculations show that these alloys are not completely random, but they exhibit some degree of short-range order. This short-range order, or correlation state, affects various alloy properties. The calculated cluster distributions for most alloys favors a tendency toward "compound formation" or "anticlustering"; i.e., in an $A_{1-x}B_xC$ alloy the A atoms prefer to be next to B rather than A atoms. However, there is insufficient data yet to confirm or dispute these results.

The theoretical results in Sections 4.9 and 4.11 show that in pseudobinary semiconductor alloys $A_{1-x}B_xC$, the miscibility gaps and anticlustering short-range correlations can coexist where A–A and B–B pairs are suppressed in favor of A–B pairs. This behavior is quite different from the conventional picture that miscibility gaps always accompany A–A and B–B pair population enhancement (i.e., clustered short-range correlation). The transformation, Eq. (4.7.4), to the reduced energy Δj representation clearly separates the enthalpy ΔE in Eq. (4.7.3) into two terms. The first is the sum of a large temperature-independent but strain- and concentration-dependent term responsible for phase separation and a small term containing Δj which dictates the temperature dependence and short-range correlation. The concentration dependencies of the cluster energies arise from approximations to the long-range strain energies that allow them to be associated with localized n-atom clusters. Sections 4.8 and 4.9 provide a method for obtaining these effective cluster energies.

Once the cluster energies are obtained, either through the combination of TB and elastic models or from LDA using the Connolly–Williams method, the GQCA or CVM can be employed to obtain the solid solution's free energy and phase diagram. However, the calculation of the liquid states' free energies, at the moment, is still totally empirical. Additional effort is needed in this area. For solid solutions, the TB model in the preceding chapter should be accurate enough for a direct Monte Carlo calculation of the free energies and other statistical quantities without going through the intermediate step of calculating the cluster energies and the approximate GQCA or CVM entropies. However, the present TB method may not work for liquids, because while the model parameters are fitted to tetrahedral structure the liquid configurations contain a distribution of coordination numbers. To extend TB to liquids, the model has to incorporate more general structures and distortions. Molecular dynamic and Monte Carlo simulation using LDA and the order-N algorithm mentioned at the end of Chapter 2 is a direct method to attack this problem.

REFERENCES

Antypas, G.A. (1970a), *J. Electrochem. Soc.* **117**, 700.
Antypas, G.A. (1970b), *J. Electrochem. Soc.* **117**, 1393.
Brebrick, R.F., C.-H. Su, and P.-K Liao (1983), in *Semiconductors and Semimetals* (Academic Press), p. 171.
Casey, Jr., H.C. and M.B. Panish (1978), *Heterostructure Lasers*, Part B, "Materials and Operation Characteristics" (Academic Press, New York).
Chen, A.-B., and A. Sher (1985), *Phys. Rev. B* **32**, 3695.
Chen, A.-B., A. Sher, and M.A. Berding (1988), *Phys. Rev. B* **37**, 6285.
Connolly, W.D., and A.R. Williams (1983), *Phys. Rev. B* **27**, 5169.
de Fontaine, D. (1979), *Solid State Phys.* **34**, 73.
Fedders, P.A., and M.W. Muller, (1984), *J. Phys. Chem. Solids* **45**, 685.
Ferreira, L.G., S.-H. Wei, and A. Zunger, (1989), *Phys. Rev. B* **40**, 3197.
Foster, L.M. (1972), Electrochem. Soc. Extended Abs. Housing Meeting, P147.
Foster, L.M., and J.F. Woods (1971), *J. Electrochem. Soc.* **118**, 1175.
Guggenheim, A. (1952), *Mixtures* (Clarendon, Oxford).
Gyorffy, B.L., and G.M. Stocks (1979), in *Electrons in Disordered Metals and Metallic Surfaces*, NATO Advanced Study Institute, Vol. 42, eds. P. Phariseau, B.L. Gyorffy, and L. Schine (Plenum, New York).
Kikuchi, R. (1951), *Phys. Rev.* **81**, 988.
Kikuchi, R. (1974), *J. Chem. Phys.* **60**, 1071.
Kikuchi, R. (1981), *Physica* **103B**, 41.
Kikuchi, R. (1982a), *Calphad* **6**, 1.
Kikuchi, R. (1982b), *J. Vac. Sci. Technol.* **21** (1), 129.
Landau, L.D., and E.M. Lifshitz (1986), *Statistical Physics*, 3rd ed. Part 1, trans. J.B. Skes and M.J. Kearsley (Pergamon Press).
Laugies, A. (1973), *Rev. Phys. Appl.* **8**, 259.
Martins, J.L., and A. Zunger (1984), *Phys. Rev. B* **30**, 6217.
Onsager, L. (1944), *Phys. Rev. 65*, 117.
Panish, M.B., and Ilgemes (1972), *Prog. Solid State Chem.* **7**, 39.
Patrick, R., A.-B. Chen, and A. Sher (1987), *Phys. Rev. B* **36**, 6385.
Patrick, R.A., A.-B. Chen, A. Sher, and M.A. Berding (1988), *J. Vac. Sci. Technol.* **A6**(4), 2643.
Phillips, J.C., and J.A. Van Vechten (1970), *Phys. Rev. B* **2**, 2147.
Sher, A., M. van Schilfgaarde, A.-B. Chen, and W. Chen (1987), *Phys. Rev. B* **36**, 4279.
Spicer, W.E., J.A. Silberman, J. Morgan, I. Lindau, J.A. Wilson, A.-B. Chen, and A. Sher (1982), *Phys. Rev. Lett.* **49**, 948.
Stringfellow, G.B. (1974), *J. Cryst. Growth* **27**, 21.
Stringfellow, G.B., and P.E. Green (1969), *J.Phys. Chem. Solids.* **30**, 1779.
Stringfellow, G.B. (1973), *J. Phys. Chem. Solids* **34**, 1748.
Su, C.H., P.K. Liao, and R.F. Brebrick (1985), *J. Electrochem. Soc.* **132**, 942.
Vieland, L.J. (1963), *Acta. Mat.* **11**, 137.
Wei, S.-H., L.G. Ferreira, and A. Zunger, (1990), *Phys. Rev. B* **41**, 8240.

APPENDIX 4A. ANALYTICAL FORMULAS OF GQCA

To gain a deeper understanding of GQCA here we examine some analytical expressions. Calculations of critical temperatures and phase diagrams involve derivatives of ΔF with respect to x, which in turn require the derivatives of η and Z defined in Section 4.7. Taking the derivative of $\Sigma x_j = 1$ and using the expression for x_j in Eq. (4.7.12), we find

$$nx\frac{\partial \ln \eta}{\partial x} - \frac{\partial \ln Z}{\partial x} - \frac{1}{kT}\left(\overline{\frac{\partial \Delta_j}{\partial x}}\right) = 0 \tag{4.A.1}$$

Taking the x derivative of Eq. (4.7.15) gives a different relation:

$$n(\overline{n_j^2})\frac{\partial \ln \eta}{\partial x} - nx\frac{\partial \ln Z}{\partial x} - \frac{1}{kT}\left(\overline{n_j\frac{\partial \Delta_j}{\partial x}}\right) = 0 \tag{4.A.2}$$

Equations (4.A.1) and (4.A.2) can be solved to yield the expression

$$\frac{\partial \ln \eta}{\partial x} = n + \frac{1}{kT}\left[n_j\left(\overline{\frac{\partial \Delta_j}{\partial x}}\right) - nx\left(\overline{\frac{\partial \Delta_j}{\partial x}}\right)\right](\overline{n_j^2} - x^2n^2)^{-1} \tag{4.A.3}$$

A similar expression can be obtained for $\partial \ln Z/\partial x$ from Eqs. (4.A.1) and (4.A.3).

Next take $\partial/\partial x$ and $\partial^2/\partial x^2$ of ΔF in Eq. (4.7.17) and use the above equations to simplify the results:

$$\frac{1}{N}\frac{\partial \Delta F}{\partial x} = l\frac{\partial}{\partial x}[y\varepsilon_A(x) + x\varepsilon_B(x)] + l\left(\overline{\frac{\partial \Delta_j}{\partial x}}\right) + kT\left[(1-nl)\ln\left(\frac{x}{y}\right) + ln \ln \eta\right] \tag{4.A.4}$$

$$\frac{1}{N}\frac{\partial^2 \Delta F}{\partial x^2} = -2\omega_{\text{eff}}(x, T) + kT\left[\frac{1-nl}{xy} + \frac{ln^2}{\overline{n_j^2} - x^2n^2}\right] \tag{4.A.5}$$

where $\omega_{\text{eff}}(x,T)$ is defined as

$$\omega_{\text{eff}}(x, T) = -\frac{l}{2}\left\{\frac{\partial^2}{\partial x^2}[y\varepsilon_A(x) + x\varepsilon_B(x)] + \frac{\partial}{\partial x}\left(\overline{\frac{\partial \Delta_j}{\partial x}}\right) + n\left[\left(\overline{n_j\frac{\partial \Delta_j}{\partial x}}\right) - nx\left(\overline{\frac{\partial \Delta_j}{\partial x}}\right)\right](\overline{n_j^2} - x^2n^2)^{-1}\right\} \tag{4.A.6}$$

Note that ω_{eff} vanishes if none of the energies ε_A, ε_B, or $\{\Delta j\}$ depend on x. The critical temperature T_c and the critical concentration x_c are determined from the conditions that both $\partial\Delta F/\partial x$ and $\partial^2\Delta F/\partial x^2$ vanish. Then T_c satisfies the following equation, assuming that x_c is known:

$$kT_c = \frac{2\omega_{\text{eff}}}{(1-nl)/x_cy_c + ln^2/(\overline{n_j^2} - x_c^2n^2)} \tag{4.A.7}$$

This equation will be used to examine various mechanisms affecting T_c in Appendix B.

APPENDIX 4B. CRITICAL TEMPERATURE IN GQCA

Let us critically examine Eq. (4.A.7), make connections to various conclusions drawn in this chapter, and identify various physical mechanisms driving the critical temperature. The predicted behavior is most easily discerned by examining several special instructive cases.

One interesting case to consider is that all Δj are zero and the strain terms take the forms $\varepsilon_A(x) = \varepsilon_A^0x^2$, $\varepsilon_B(x) = \varepsilon_B^0y^2$, and $\varepsilon_A^0 = \varepsilon_B^0 = \varepsilon^0$. These energy expressions are fair approximations to the results shown in Fig. 4.13. This case corresponds to the zeroth approximation for ΔF (Section 4.2.2), which has a random-alloy mixing entropy, and a mixing enthalpy given by $\Delta E = Nxy\omega$, with $\omega = l\varepsilon^0$. Then T_c is given by

$$kT_c = l\varepsilon^0/2 \tag{4.B.1}$$

It can be checked that Eqs. (4.B.1) and (4.2.8) have the same form. However, this new result is for general clusters, and the physics is quite different from that in the pair model considered in Section 4.2. Here the alloy mixing enthalpy parameter $\omega = l\varepsilon^0$ originates from long-range strain energy while ω in the pair model comes from the reduced pair energy, $\omega = z\Delta_1$, where Δ_1 is called ε in Eq. (4.2.5).

Next consider the case that has nonzero Δj in addition to the strain energy just discussed, but for simplicity assume Δj consists only of pair potentials summed over the cluster constituents. Then ΔF is a symmetric function of x about $x = 1/2$ for a given T. The critical temperature T_c occurs at $x_c = 1/2$. At $x = 1/2$, the solution for η is $\eta = 1$. This result can be obtained by inspecting Eq. (4.7.15) and knowing that interchanging the A and B atoms does not change the number of AB pairs in a cluster so $\Delta j = \Delta_{n-j}$ and also $g_j = g_{n-j}$. Using $x_c = 0.5$ and $\eta = 1$, Eq. (4.A.7) becomes

$$kT_c = \frac{l\varepsilon^0}{2(1 - nl) + 2ln^2/(4\overline{n_j^2} - n^2)} \tag{4.B.2}$$

where

$$\overline{n_j^2} = \frac{\Sigma n_j^2 g_j \exp(-\Delta_j/kT_c)}{\Sigma g_j \exp(-\Delta_j/kT_c)} \tag{4.B.3}$$

Note that if isolated clusters are used, $nl = 1$. Then higher-order terms in ω_{eff} must be retained to have a nonzero T_c if ε^0 is zero because $4\overline{n_j^2} > n^2$. However, a vanishing ε^0 does not happen in general, because if clusters do not share atoms the intercluster interactions have to be included in a manner described by Eq. (4.8.3), which gives rise to concentration-dependent cluster energies.

It is instructive to study the behavior of T_c given by Eq. (4.B.2) for pair clusters with $x = \frac{1}{2}$, because the equation is greatly simplified and the interplay of ε and l can be examined closely. With $l = z/2$ (z is the coordination number) and $n = 2$, T_c now obeys the explicit equation

$$T_c/T_0 = 2/(2 - z + ze^{-\Delta_l/kT_c}) \tag{4.B.4}$$

where T_0 is the critical temperature for the case where Δ_1 is zero; i.e., $kT_0 = l\varepsilon^0/2$ (Eq. (4.B.1)). Since Δ_1 is much smaller than ε^0 in semiconductors, as numerical results in Section 4.9 reveal, it is a good approximation to keep only the lowest-order correction due to Δ_1. We then have

$$T_c = T_0(1 + 2\Delta_l/\varepsilon^0) \tag{4.B.5}$$

So the correction to the critical temperature $T_c - T_0$ scales linearly with Δ_1. This linear dependence holds reasonably well until the magnitude of Δ_1 is comparable to ε^0.

It is also interesting to see that Eq. (4.10.5) reduces readily to the first-approximation result of Section 4.2 in the limit as ε^0 goes to zero (i.e., $\varepsilon^0 << \Delta_1$):

$$kT_c = \Delta_1/\ln\left(\frac{z}{z-2}\right) \tag{4.B.6}$$

This case may arise in lattice-matched alloys (e.g., $Ga_{1-x}Al_xAs$ or $Hg_{1-x}Cd_xTe$) where strain energies are small. However, in these cases T_c is so small that it is unmeasurable.

Finally, we should be cautious not to push the approximation in GQCA into the low-temperature regime. If the clusters are chosen such that $nl > 1$, the mixing entropy can become negative in GQCA. Examples are shown in Appendix C. Note that this also occurs in the CVM, because it is shown in Section 4.10 that in the pair cluster case QCA and CMV are identical.

Thus, the final conclusion is that T_c is dominated by strains in all lattice-mismatched semiconductor alloys and has only small contributions from the excess energies responsible for their correlation state. In lattice-matched semiconductor alloys where the strain terms are small, T_c is usually so small that it introduces no observable effects.

APPENDIX 4C. GQCA AT LOW TEMPERATURE

We begin by referring to Eqs. (4.7.5), (4.7.6), and (4.7.7), for Φ, x^0, and ΔS, respectively. The entropy ΔS can be written as

$$\frac{\Delta S(x,T)}{N} = (1-nl)s_R + \frac{l\bar{\Delta}_j}{T} + lk(\ln Z - nx\ln\eta) \tag{4.C.1}$$

where

$$s_R \equiv -k[x\ln x + (1-x)\ln(1-x)]$$

is the entropy per particle for a random alloy of concentration x.

Let us now examine Eq. (4.C.1) in the pair case with a positive reduced pair energy Δ_1 where, at $x = 1/2$, $s_R(1/2) = k\ln 2$, it becomes

$$\frac{\Delta S(1/2,T)}{N} = (1-\tfrac{z}{2})k\ln 2 + \frac{z}{2T}\frac{\Delta_1}{(1+e^{\Delta_1/kT})} + \frac{z}{2}k\ln(1+e^{-\Delta_1/kT})$$

$$= \begin{cases} k\ln 2 & \text{for } \dfrac{\Delta_1}{kT} \ll 1 \\ (1-\tfrac{z}{2})\ln 2 & \text{for } \dfrac{\Delta_1}{kT} \gg 1 \end{cases} \tag{4.C.2}$$

Obviously since $z > 4$ for all three-dimensional cases of interest, the low-temperature limit, where populations of some clusters are strongly enhanced and others suppressed, constitutes a case where the entropy becomes negative and therefore nonphysical.

The origin of the problem can be traced back to Eq. (4.7.5). Return now to general n-atom clusters. For the case with all Δj's positive semidefinite, the low T limit is $M_0 = yM$, $M_J = xM$, and $M_j = 0$ for $j \neq 0,J$. Then Φ becomes

$$\Phi = \frac{N!}{N_A!N_B!}\frac{M!}{(yM)!(xM)!}y^{nyM}x^{nxM} \tag{4.C.3}$$

Using Stirling's approximation in the form $N! \cong (N/e)^N$, this equation becomes

$$\Phi = [(x^x y^y)^{(nl-l-1)}]^N \tag{4.C.4}$$

In the special case where no substituted atom contributes to more than one cluster, $nl = 1$ and Eq. (4.B.4) takes the proper value $\Phi = M!/\,[(xM)!(yM)!]$ corresponding to the number of ways in which two kinds of clusters can be arranged. When $nl - l - 1 > 0$, Φ is less than unity and the excess entropy becomes nonphysical. Both the one-atom 4-bond microclusters of Section 4.8 and the four-atom 16-bond cluster of Section 4.9 correspond to arrangements where $nl = 1$, so no nonphysical negative ΔS is encountered. However, in the pair interaction case $nl = z$ and the nonphysical result is present.

5

Band Structure Theory

Semiconductor electronics provides an excellent demonstration of the close connection between modern engineering and quantum physics. It was an understanding of the electronic structure of semiconductors in the 1940s that led to the invention of the first transistor (Bardeen and Brattain, 1948; Brattain and Bardeen, 1948; Shockley and Pearson, 1948)—the backbone of modern computers. Since then, quantum mechanics has been an integral part of the progress of modern electronics technology. As devices become smaller, reaching submicron dimensions where electrons and holes traverse the active region of devices without experiencing a collision (the ballistic transport regime), quantum effects become even more important. Band structure studies deal with the energy levels and wave functions of electrons in materials and their relations to material properties. This chapter will begin by introducing the basic concepts of energy bands (Section 5.1). We will then describe two of the simplest band structure methods used for crystalline semiconductors—the tight-binding method (Section 5.2) and the plane-wave method (Section 5.3). The important band structure results for pure semiconductors are summarized in Section 5.4. The difficulties associated with the aperiodic potentials in an alloy and their effects on band gaps are discussed in Section 5.5. The remaining sections are devoted to the treatment of disordered alloys using the Green function methods, including the coherent potential approximation (CPA) (Soven, 1967; Velicky *et al.*, 1968; Kirpatrick *et al.*, 1970) and the perturbation method (PT). Both CPA and PT will be formulated for semiconductor alloys in this chapter and will be used to treat effects on band-edge states. These formalisms will be applied to detailed calculations of band structures of semiconductor alloys in Chapter 7.

Sections 5.1 to 5.4 are intended to remind the reader of the essential features of energy bands in periodic structures that will be used later for alloys. Many concepts, such as the Bloch theorem and reciprocal lattice, are simply used without explanation. If readers encounter unfamiliar concepts that remain unclear after Section 5.4 is completed, we recommend they consult an elementary solid-state physics text (e.g., Kittel, 1986) before continuing.

5.1. FORMATION OF ENERGY BANDS

The electronic energy levels and the associated wave functions in a solid are governed by the Schrödinger equation with a many-body Hamiltonian H:

$$H\,\Psi = E\,\Psi \tag{5.1.1}$$

where H includes all the terms in Eq. (2.2.1). There are about 10^{23} electrons in a typical solid. These electrons interact with the atomic nuclei and with each other via Coulomb forces. They also have to obey the Pauli (1925) exclusion principle. A rigorous solution to this many-body problem is not currently possible. Nearly all band structures calculated are based on a one-electron approximation, in which the many-body interactions are incorporated into an effective one-electron potential V. The one-electron Schrödinger equation

$$(p^2/2m + V)\psi_\alpha = E_\alpha \psi_\alpha \tag{5.1.2}$$

is then solved to obtain the energy levels E_α and the associated wave functions ψ_α, where α denotes the quantum numbers specifying an eigenstate.

These ε_α and ψ_α may have different ranges of validity for different assumptions about the effective potential. For example, within the context of the local density functional approximation (LDA) for the ground state energy discussed in Chapter 2, these band energies and wave functions may be sufficient to accurately predict the charge densities, relative total energies, and other equilibrium properties, but they do not accurately predict excitation spectra in semiconductors. In particular, band gaps calculated from LDA are too small. However, it is commonly assumed that an effective potential can be constructed to describe the excitation spectra. This assumption can be justified in the many-body theory (Hedin and Lundquist, 1971; Hybertsen and Louie, 1987) if, in the jargon of this theory, the quasi-particle self-energy is treated as part of the potential function. Examples of empirical band structures with conduction bands agreeing with experiments, at least near the band edges, will be given in Section 5.4 and Chapter 7.

It is instructive to examine the qualitative nature of band formation as atoms come together to become a solid. In Fig. 5.1a, the solid curve represents the potential for an isolated atom, and the energy level E_a denotes the highest occupied level. The free-atom wave function ψ_a associated with this energy level, being confined by an atomic potential well, is only appreciable in the vicinity of the atom. As atoms move close to each other in a solid, the potential between atoms decreases, due mainly to superposition of atomic potentials, as depicted by the dashed curves. A wave function in the solid that satisfies the Schrödinger equation at the free-atom energy E_a is no longer confined to the vicinity of a single atom. As will be seen, the energy eigenstates in a crystal will have a periodic probability density, given by $|\psi_\alpha(\mathbf{r})|^2$, as a function of $\mathbf{r}$.

The energy levels also rearrange themselves as atoms are collected into a solid. Starting with about 10^{23} free atoms, each atom has a state with the same energy E_a—a level with 10^{23}-fold degeneracy. As atoms come close to each other, these degenerate states interact with each other. They are shifted up and down to form a dense set of levels in a finite energy range around E_a. This is an energy band. Figure 5.1b schematically illustrates this band formation. In general, there is more than one atomic valence level participating in the formation of bands in a solid. For the semiconductors and alloys that we will emphasize, the most important parts of the bands are derived primarily from the outermost partially filled atomic s and p levels.

A simple but useful quantity to describe the energy bands is the electronic density of states (DOS), defined as $\mathcal{D}(E) = \Sigma_\alpha\, \delta(E - E_\alpha)$, where $\delta(E - E_\alpha)$ is a Dirac delta function, and $\mathcal{D}(E)$ is the number of energy levels per unit energy at E. Therefore, $\mathcal{D}(E)\, dE$ is the number

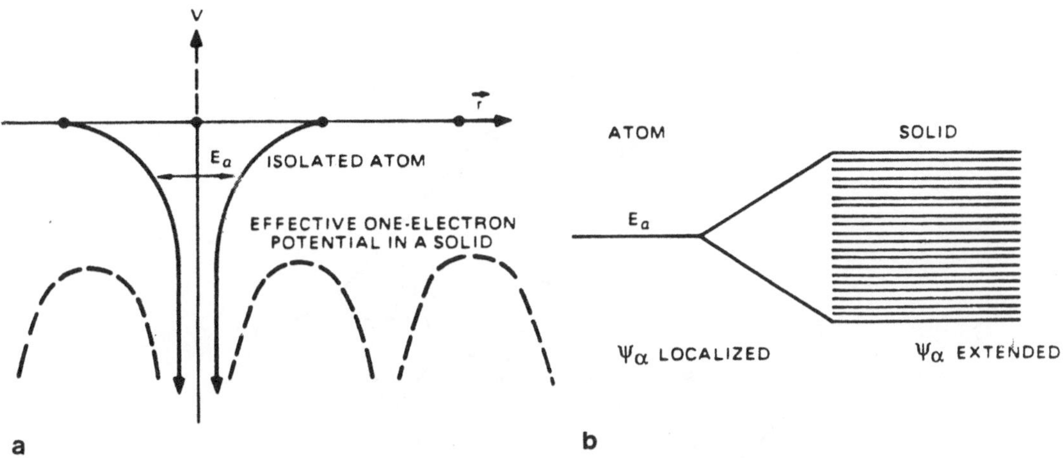

FIGURE 5.1. Free atom and solid electronic states: (a) potential energy diagram; (b) electronic states.

of states with their energies in a small interval between E and $E + dE$. Figure 5.2 displays the DOS per unit cell for GaAs calculated by Cohen and Chelikowsky (1989) in the energy range of interest. At zero temperature, electrons occupy the lowest possible levels. For a pure semiconductor at zero temperature, all the states below E_V, which is set to zero in Fig. 5.2, are occupied, while all the levels above this energy are empty. These occupied bands are called the valence bands, and the unoccupied bands are the conduction bands. A gap exists between the top of the valence bands E_V and the bottom of the conduction bands E_C (at 1.52 eV in Fig. 5.2) in which there are no energy levels. This is the well-known forbidden energy gap, or simply the band gap $E_g = E_C - E_V$. This general feature of DOS is a characteristic of all semiconductors and their alloys.

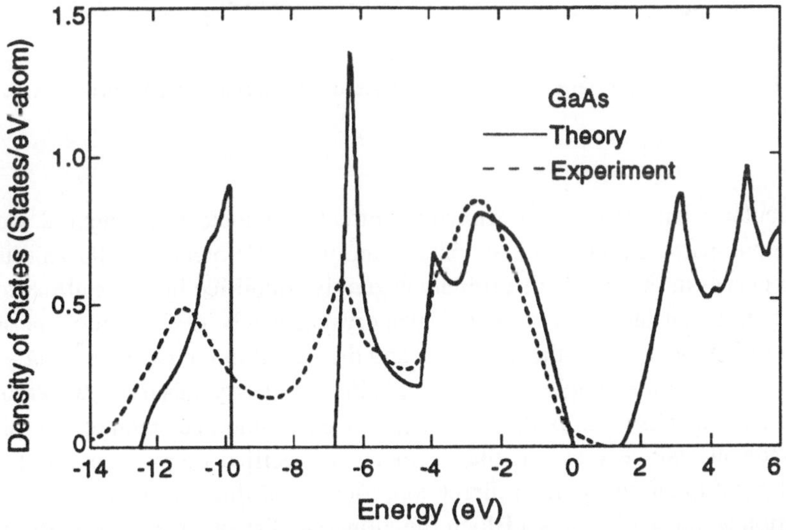

FIGURE 5.2. Calculated electronic density of states for GaAs compared to x-ray photoemission spectroscopy measurements (Ley *et al.*, 1973). (From Cohen and Chelikowsky, 1989.)

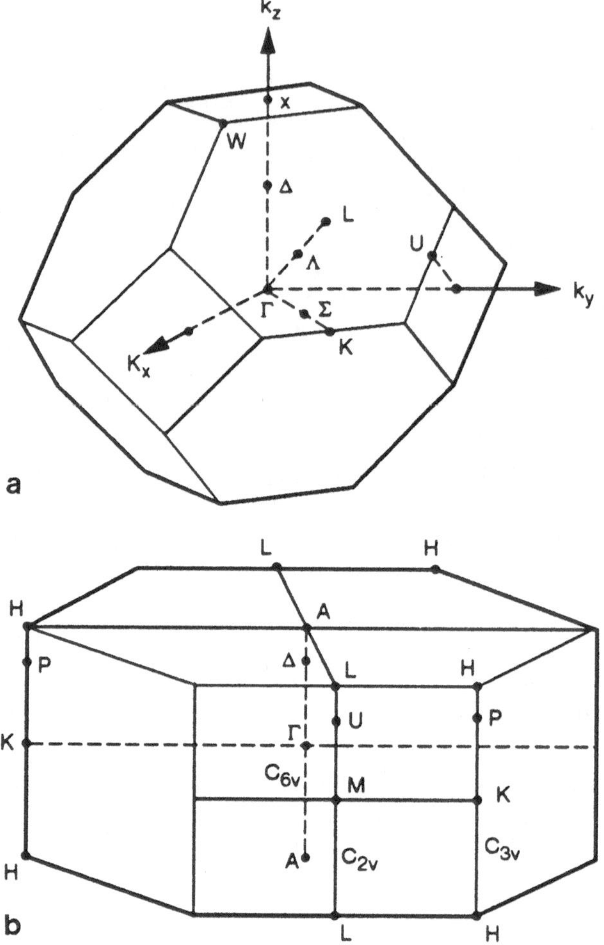

FIGURE 5.3. Brillouin zone for (a) the zinc blende crystal and (b) the Wurtzite crystal.

To calculate a band structure, even with an effective one-electron potential, we still must deal with a one-particle potential that contains around 10^{23} potential wells, as illustrated by the dashed curves in Fig. 5.1. This problem is greatly simplified in a crystalline solid which possesses lattice translational symmetry. Such crystals can be divided into a periodic array of N smallest identical cells—the unit cells, and the Hamiltonian $H(\mathbf{r}) = p^2/2m + v(\mathbf{r})$ is the same in every cell. For example, diamond and zinc blende crystals have two atoms per unit cell, while wurtzite has four. Mathematically, lattice translational symmetry implies $H(\mathbf{r}) = H(\mathbf{r} + \mathbf{L})$ for any lattice vector $\mathbf{L}$ that connects a specified point in one unit cell to the equivalent point in another cell. A direct consequence of this symmetry is that the energy eigenfunction ψ in Eq. (5.1.2) is a Bloch function, one that satisfies the condition

$$\psi_{\mathbf{k}}(\mathbf{r}+\mathbf{L}) = e^{i\mathbf{k}\cdot\mathbf{L}}\psi_{\mathbf{k}}(\mathbf{r}) \qquad (5.1.3)$$

where the quantum number is now the crystal wave vector **k**. For a crystal containing N unit cells, there are also N nonequivalent **k** vectors represented by N points uniformly distributed in the wave vector space within a region called the first Brillouin zone (BZ). Both diamond and zinc blende semiconductors have a fcc Bravais lattice, while the wurtzite semiconductors have an hcp lattice. Figure 5.3 shows the first Brillouin zones of these two lattices which are Fourier transforms of the crystal structures with boundaries selected so each band can accommodate two electrons (one with spin up and the other with spin down) within the zone.

Using the Bloch theorem, the band structure calculation for a crystal amounts to choosing a wave vector **k** and finding the energies $E_{\mathbf{k}}$ and corresponding wave functions ψ that satisfy both the Schrödinger equation and the Bloch condition Eq. (5.1.3). The potential and wave function required in such calculations only need to be specified in one of those N unit cells. Many methods have been developed to solve this problem (Callaway, 1974, Chapter 4). These methods can start with very different basis sets and arrive at the same results. For example, the LCAO (linear combination of atomic orbitals) method (Wang and Klein, 1981) that produced the band structures shown in Fig. 2.3 uses the localized atomic-like orbitals as the basis set. On the other hand, a plane-wave basis is often used in pseudopotential calculations (Cohen and Chelikowsky, 1989). These two methods not only provide simple analytical models of the formation of bands but also allow detailed calculations, so they will be discussed further.

5.2. LCAO AND THE EMPIRICAL TIGHT-BINDING METHOD

In LCAO, which is also referred to as the tight-binding method (Callaway, 1964), one starts with a set of local orbitals $\phi_\alpha(\mathbf{r}-\mathbf{R}_n)$ centered about the atomic sites $\mathbf{R}_n$ as the basis functions in a linear variational calculation of the eigenenergy and eigenvalues of the Schrödinger equation, where α denotes the type of orbitals (e.g., s or p orbitals). In the Dirac notation $\phi_\alpha(\mathbf{r}-\mathbf{R}_n)$ is written as $|n\alpha\rangle$. Expanding the eigenfunction in this basis,

$$|\psi\rangle = \sum_n \sum_\alpha C_{n\alpha}|\alpha n\rangle \tag{5.2.1}$$

the Schrödinger equation (5.1.2) can be cast into a matrix form

$$\sum_{n\alpha} H_{n'\alpha',n\alpha}C_{n\alpha} = E\sum_{n\alpha} S_{n'\alpha',n\alpha}C_{n\alpha} \tag{5.2.2}$$

where H is the Hamiltonian matrix and S is the overlap matrix. Both H and S are $M \times M$ matrices, with M being the product of the total number of atoms N_T (N_T equals the number of atoms in a unit cell times the number of cells N) and the number of orbitals per atom J (i.e., $M = N_T J$). For a solid containing $N_T = 10^{23}$ atoms, a direct diagonalization of this huge matrix is impossible. However, in a crystal the Bloch theorem simplifies this problem.

If there is only one atom per unit cell, the atomic position vectors $\mathbf{R}_n$ are the lattice translation vectors **L**. We can choose a wave vector **k** and demand that the eigenfunction be a Bloch function obeying Eq. (5.1.3). This can be achieved by first constructing the Bloch basis from a linear combination of these atomic orbitals:

$$|\mathbf{k}\alpha\rangle = \frac{1}{\sqrt{N}}\sum_{\mathbf{L}} e^{i\mathbf{k}\cdot\mathbf{L}}|\mathbf{L}\alpha\rangle \tag{5.2.3}$$

where N is again the number of unit cells and $\mathbf{L}$ sums over all the lattice vectors. Then expand $|\psi\rangle$ in this basis:

$$|\psi_{\mathbf{k}}\rangle = \sum_{\alpha} a_{\alpha}|\mathbf{k}\alpha\rangle \tag{5.2.4}$$

Then Eq. (5.2.2) reduces to

$$\sum_{a} H_{\alpha',\alpha}(\mathbf{k})a_{\alpha}(\mathbf{k}) = E\sum_{\alpha} S_{\alpha',\alpha}(\mathbf{k})a_{\alpha}(\mathbf{k}) \tag{5.2.5}$$

where the $\mathbf{k}$-dependent Hamiltonian matrix elements are related to those in the local basis by

$$H_{\alpha',\alpha}(\mathbf{k}) = \sum_{\mathbf{L}} e^{i\mathbf{k}\cdot\mathbf{L}} H_{0\alpha',\mathbf{L}\alpha} \tag{5.2.6}$$

A similar relation holds for the overlap matrix elements $S_{\alpha',\alpha}(\mathbf{k})$. Note that Eq. (5.2.5) in this one-atom-per-unit-cell case is a $J\times J$ matrix equation rather than the $N_{\mathrm{T}}J \times N_{\mathrm{T}}J$ original LCAO equation—a tremendous reduction.

The LCAO calculation for a given potential is thus straightforward. The most difficult part in this calculation is the evaluation of the matrix elements, especially the three-center integrals that include two wave functions and one potential, each centered on a different site. A popular LCAO scheme is based on Gaussian orbitals, in which each atom is assigned orbitals of the type $\phi_{lm}(r) = r^l \exp(-\lambda r^2)Y_{lm}$, with several λ values and several angular-momentum components indexed by l and m. The band structures shown in Fig. 2.3 are the results of an LDA calculation using a Gaussian basis. Many other forms of local orbitals have been used in connection with local density functional theory. For example, the LMTO (linearized muffin-tin orbitals) (Andersen *et al.*, 1985; van Schilfgaarde and Methfessel, 1991), which facilitates fast and accurate computations, uses orbitals that are solutions to the radial Schrödinger equation for full potentials at special energies.

In the semiempirical TB calculation, very often the local orbitals are assumed to be orthonormal; i.e., $\langle\mathbf{L}\alpha|\mathbf{L}'\alpha'\rangle = \delta_{\mathbf{L}\mathbf{L}'}\delta_{\alpha\alpha'}$. Then the Bloch basis functions of Eq. (5.2.3) are also orthonormal, $\langle\mathbf{k}\alpha|\mathbf{k}'\alpha'\rangle = \delta_{\mathbf{k}\mathbf{k}'}\delta_{\alpha\alpha'}$. This means that the overlap matrix S is an identity matrix, and the secular equation becomes the usual eigenvalue problem for a Hermitian matrix. The transformation from a set of nonorthogonal basis to the orthogonal basis in a crystal can always be made such that the local symmetry of the orbitals is still preserved (Chen and Sher, 1982). If the matrix elements between these orthonormal local orbitals are short-ranged, the band structure will be completely determined by a handful of these matrix elements, which may be fitted using experimental data. This is the basis for empirical tight-binding (ETB) calculations.

It is instructive to work out a simple ETB model: one-orbital per atom in a one-dimensional crystal with only the on-site and nearest-neighbor (NN) interaction being nonzero. Since we have only one orbital per atom, it is sufficient to specify the orbitals by the site index n, which also indexes the lattice vector through $L(n) = na$, where a is the lattice constant. This Hamiltonian matrix between two local orbitals at sites n and m then takes the form

$$H_{nm} = \begin{cases} \varepsilon_0, & \text{if } n = m \\ -\gamma, & \text{if } n = m \pm 1 \\ 0, & \text{otherwise} \end{cases} \tag{5.2.7}$$

Since there is only one orbital per atom, there is only one Bloch basis for each crystal wave vector **k**. The allowed wave vectors lying within the first Brillouin zone are given by $k = 2s\pi/Na$, with the integer s taking values within the range $-N/2 < s \leq N/2$. Then there is only one energy per k given by

$$E(k) = \langle k|H|k\rangle = \sum_n H_{0n} \exp(ikna) = \varepsilon_0 - 2\gamma \cos(ka) \tag{5.2.8}$$

This result shows that the original N-fold degenerate level ε_0 (of the N separated atoms) now becomes a band that contains N states, each specified by a wave vector **k**. These closely spaced energy levels lie in an energy range between $\varepsilon_0 - 2\gamma$ and $\varepsilon_0 + 2\gamma$. Therefore, this band has a bandwidth of 4γ centered at the original energy level ε_0 (see the lower band represented by the dashed line in Fig. 5.4).

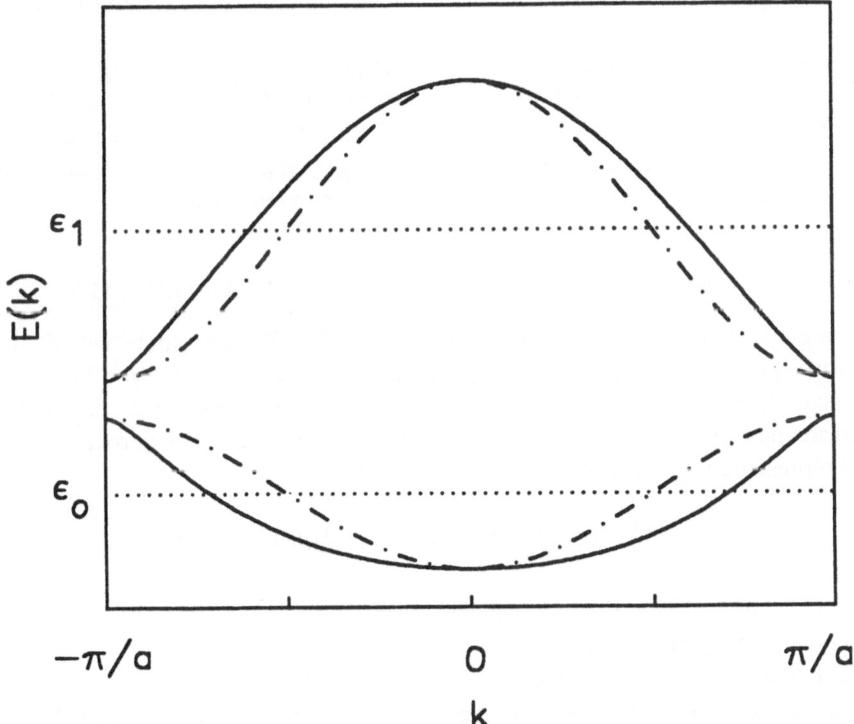

FIGURE 5.4. The two-band model of Eq. (5.2.9) (dashed lines) and of Eq. (5.2.10) (solid lines).

If there are two orbitals per atom with the atomic energy levels ε_0 and ε_1 and the first-neighbor interactions $-\gamma_0$ and γ_1 among the same kinds of orbitals, these two levels will evolve into two bands of the form

$$E_0(k) = \varepsilon_0 - 2\gamma_0 \cos(ka)$$
$$E_1(k) = \varepsilon_1 + 2\gamma_1 \cos(ka) \tag{5.2.9}$$

The different signs in the interactions, for example, may reflect the different symmetry of the s and p orbitals. Now if the energy separation $\varepsilon_1 - \varepsilon_0$ is larger than $2(\gamma_0 + \gamma_1)$, there is an energy gap $E_g = (\varepsilon_1 - \varepsilon_0) - 2(\gamma_0 + \gamma_1)$ occurring at $k = \pi/a$. If there are also interactions between the two different kinds of orbitals, say $\langle n,\alpha |H| n \pm 1, \beta\rangle = \pm\lambda$, then the two bands will repel each other at certain k points in the zone. These energies can be obtained analytically and are given by

$$E_{\pm}(k) = \frac{E_0(k) + E_1(k)}{2} \pm \frac{1}{2}\sqrt{[E_1(k) - E_0(k)]^2 + 16\lambda^2 \sin^2(ka)} \tag{5.2.10}$$

In Fig. 5.4, $E_0(k)$ and $E_1(k)$ are shown as dashed lines, and $E_+(k)$ and $E_-(k)$ are solid lines. This two-band model illustrates the origin of the band gap.

In a realistic ETB description of semiconductors, the minimum basis set required contains four orbitals per atom, i.e., one s- and three p states. Since there are two atoms per unit cell in a diamond and zinc blende semiconductor, the Hamiltonian matrix for a given $\mathbf{k}$ is 8×8. When the spin-orbit interaction is included, the order of the matrix doubles. In the total energy calculation considered in Chapters 2 and 3, the spin-orbit interaction was not included, because it has very little effect on the total energy. The first-neighbor interaction model seems to work reasonably for the structural energies, because only the valence bands, those occupied, are needed for the total energy calculation. However, to simulate the conduction bands, one has to include second-neighbor and even longer-range interactions. Examples of these applications will be discussed further later in this chapter and in Chapter 7.

5.3. PLANE-WAVE METHOD AND EMPIRICAL PSEUDOPOTENTIALS

In an empty lattice in which the potential energy V is flat and can be set equal to zero, the energy eigenfunctions are those of the kinetic energy operator $K = p^2/2m$. For a given wave vector $\mathbf{k}$ inside the first Brillouin zone, the eigenfunctions that satisfy the Bloch theorem are the crystal plane waves denoted by $|\mathbf{k} + \mathbf{G}\rangle$, where $\mathbf{G}$ is a reciprocal lattice vector. In the $\mathbf{r}$-representation, these plane waves take the form

$$\phi_{\mathbf{k}+\mathbf{G}}(\mathbf{r}) = \langle \mathbf{r} | \mathbf{k} + \mathbf{G}\rangle = \frac{1}{\sqrt{\Omega}} \exp[i(\mathbf{k} + \mathbf{G}) \cdot \mathbf{r}] \tag{5.3.1}$$

where Ω is the crystal volume. The associated kinetic energy is $T(\mathbf{k} + \mathbf{G}) = \hbar^2(\mathbf{k} + \mathbf{G})^2/2m$. Each $\mathbf{G}$ designates a band inside the BZ.

These plane waves are orthonormal; $\langle \mathbf{k}' + \mathbf{G}' | \mathbf{k} + \mathbf{G}\rangle = \delta_{\mathbf{k}'\mathbf{k}}\delta_{\mathbf{G}'\mathbf{G}}$. They form a complete basis set. The energy eigenfunctions in a crystal for a given potential V can then be expanded in plane waves. The energy eigenfunction for a chosen wave vector $\mathbf{k}$ only involves those plane waves belonging to the same $\mathbf{k}$:

$$|\psi_{\mathbf{k}}\rangle = \sum_{G} b(\mathbf{k}+\mathbf{G})|\mathbf{k}+\mathbf{G}\rangle \tag{5.3.2}$$

Then the Schrödinger equation (5.1.2) reduces to a matrix eigenvalue problem

$$\sum_{\mathbf{G}} \{[T(\mathbf{k}+\mathbf{G}) - E]\delta_{G'G} + \langle \mathbf{k}+\mathbf{G}'|V|\mathbf{k}+\mathbf{G}\rangle\} b(\mathbf{k}+\mathbf{G}) = 0 \tag{5.3.3}$$

For weak potentials this expansion converges rapidly, and the calculation is very efficient. Even with moderate potentials, the plane-wave method is still useful because the basis set is unbiased and the matrix elements can be easily calculated.

Thus, a key to the simplicity found within the plane-wave method is weak potentials. Inevitably, this method is used in connection with so-called pseudopotentials V_{ps} (Harrison, 1966; Cohen and Chelikowsky, 1989). Pseudopotentials are constructed to work with pseudo wave functions ψ_{ps}, which resemble real wave functions ψ outside the atomic core region but are smooth in the core region where the real wave functions usually wiggle about nodes. Pseudopotentials are constructed such that if ψ satisfies the Schrödinger equation for a full potential V, then ψ_{ps} satisfies the same equation with V replaced by V_{ps}. The way V_{ps} is deduced from V can be found in a number of books, for example Harrison (1966). The advantage of pseudopotentials is that they allow the neglect of the deep atomic core states and deal only with those valence orbitals responsible for the chemical bonding. One important development in *ab initio* pseudopotential theory is the so-called norm-conserved pseudopotential for LDA calculations (Hamann *et al.*, 1979; Bachelet *et al.*, 1982). These pseudopotentials are not weak, and an extensive number of plane waves is needed. However, these pseudopotentials are still much weaker than the real potentials, and the structural properties calculated from these potentials are as accurate as those based on full potentials. All the potentials V to be used in this section are assumed to be pseudopotentials.

For a local crystal potential, i.e.,

$$\langle \mathbf{r}|V|\mathbf{r}'\rangle = V(\mathbf{r})\delta(\mathbf{r}-\mathbf{r}')$$

the potential matrix elements in the plane-wave basis depend only on the difference between the reciprocal lattice wave vectors,

$$\langle \mathbf{k}+\mathbf{g}'|V|\mathbf{k}+\mathbf{g}\rangle = \mathcal{V}(\mathbf{g}'-\mathbf{g}) \tag{5.3.5}$$

where $\mathcal{V}(\mathbf{g})$ is the Fourier transform of $V(\mathbf{r})$:

$$\mathcal{V}(\mathbf{g}) = \frac{1}{\Omega}\int e^{-i\mathbf{g}\cdot\mathbf{r}}V(\mathbf{r})\, d^3r \tag{5.3.6}$$

If a crystal contains several atoms per unit cell, the total crystal potential can be written as the sum of potentials v_j centered on atoms:

$$V(\mathbf{r}) = \sum_{L_j} v_j(\mathbf{r}-\mathbf{L}-\tau_j) \tag{5.3.7}$$

where $\mathbf{L}$ specifies the unit cells, and j is the index of an atom in a unit cell displaced by τ_j from $\mathbf{L}$. Then $\mathcal{V}(\mathbf{g})$ can be written as the sum of the Fourier transforms of the atomic potentials $\mathcal{v}_j(\mathbf{g})$ as

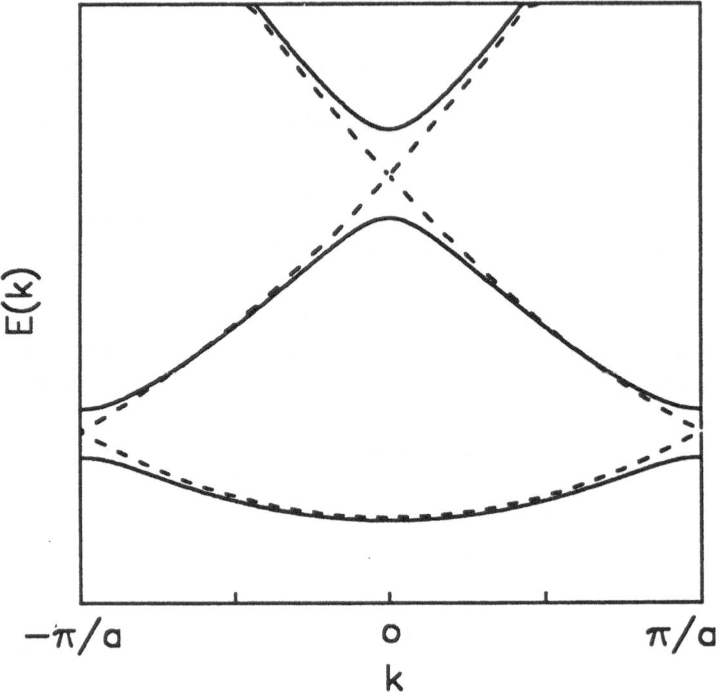

FIGURE 5.5. Free-electron bands (dashed lines) and bands with nonzero crystal potentials (solid curves).

$$\mathcal{V}(\mathbf{g}) = \sum_j e^{i\mathbf{g}\cdot\tau_j} v_j(\mathbf{g}) \tag{5.3.8}$$

If the atomic pseudopotentials are weak and are assumed to be spherically symmetrical, then $\mathcal{V}(\mathbf{g})$ is a function of only the magnitude g of $\mathbf{g}$, and only $v(g)$ values at several g's are needed to reach convergence. Because the whole band structure only depends on these few form factors and because there is usually enough experimental information, such as optical transition energies, these few parameters can be deduced. This is the basic idea behind the empirical pseudopotential method (EPM).

It is instructive to illustrate the way the free-electron bands are affected by the potential and how the gaps are formed in a one-dimensional model. In Fig. 5.5, the dashed lines are the free-electron bands $T(\mathbf{k}+\mathbf{G}) = \hbar^2(\mathbf{k}+\mathbf{G})^2/2m$ as a function of $\mathbf{k}$ inside the BZ. Note that the allowed values of the reciprocal lattice vectors in the one-dimensional crystal are $G = 2n\pi/a$, with n being any integer and a the lattice constant. To see how these energies are affected by a crystal potential, we need only consider the interactions among those bands at the same $\mathbf{k}$; i.e., along vertical lines, because the crystal potential is periodic, it does not connect states of different wave vectors. If the potential is weak, perturbation theory will give a good estimate of the changes it causes. For a nondegenerate state $|\mathbf{k}+\mathbf{G}\rangle$, the first correction to the energy is $\langle \mathbf{k}+\mathbf{G}\,|V|\,\mathbf{k}+\mathbf{G}\rangle = \mathcal{V}(0)$, which is taken to be zero, because a constant diagonal term can always be included in the unperturbed Hamiltonian. The second-order correction is given by

$$\Delta E_2 = \sum_{G'} \frac{|\mathcal{V}(\mathbf{G}' - \mathbf{G})|^2}{T(\mathbf{k}+\mathbf{G}) - T(\mathbf{k}+\mathbf{G}')} \tag{5.3.9}$$

Starting with the lowest energy associated with $k = G = 0$, its energy is shifted downward because all the other states are at higher energies and in second-order perturbation theory states repel. As k increases, Eq. (5.3.9) can be used until k approaches the BZ boundary π/a, where the two lowest unperturbed bands become degenerate. Then degenerate perturbation theory has to be used. At $k = \pi/a$, the two degenerate bands are represented by two reciprocal lattice vectors $G = 0$ and $G' = -2\pi/a$. According to Eq. (5.3.5), the potential that couples these two states is $\mathcal{V}(g)$, with $g = G - G' = 2\pi/a$. The shifts of these twofold degenerate levels are obtained by diagonalizing the following 2×2 potential matrix in the degenerate subspace:

$$\mathrm{V} = \begin{pmatrix} 0 & \mathcal{V}(g) \\ \mathcal{V}^*(g) & 0 \end{pmatrix} \tag{5.3.10}$$

which gives two values $|\mathcal{V}(g)|$ and $-|\mathcal{V}(g)|$ for the energy shifts. This level thus splits into two levels with a separation of $2|\mathcal{V}(g)|$. If there are two electrons per atom, the lowest band is the valence band and the second band is the conduction band, and we have a semiconducting polymer with a gap $E_g = 2|\mathcal{V}(g)|$. The above considerations can be applied to every band, and the results are those indicated by the solid lines in Fig. 5.5. It is interesting to note that the two lowest bands in Figs. 5.4 and 5.5 are similar despite the fact that they are derived from two almost opposite band concepts.

5.4. BAND GAPS AND EFFECTIVE MASSES

While the density of states provide an understanding of the average properties of the electronic states, more detailed features of the band structures are needed to characterize their electronic properties. The main features of band structures are often illustrated by plots of E versus $\mathbf{k}$ along several symmetry directions in the BZ. The band structures shown in Figs. 5.6 through 5.9 represent perhaps the most accurate theoretical calculations to date for semiconductors. They were calculated by Chelikowsky and Cohen (1976), using the EPM. However, in their work, nonlocal pseudopotentials, in addition to the local ones, are used to provide more accurate calculation of the optical properties. The bands in Figs. 5.6 to 5.9 are for the cubic semiconductors. They are plotted from L to Γ, Γ to X, X to U, and K to Γ (see the fcc BZ shown in Fig. 5.2, where these symmetry points are labeled). In units of $2\pi/a$, these $\mathbf{k}$ points are (1/2, 1/2, 1/2) for L, (0, 0, 0) for Γ, (1, 0, 0) for X, (1, 1/4, 1/4) for U, and (3/4, 3/4, 0) for K. Note that U and K are two equivalent points, so the energies at these two $\mathbf{k}$ points are the same. The zero of energy is chosen to be the top of the valence band $E_v = 0$. Note that for some systems, such as GaAs, InP, and CdTe, E_v and the bottom of the conduction band E_c occur at the same $\mathbf{k}$ point, normally at the center of the Brillouin zone Γ. These semiconductors are designated direct-gap semiconductors. Note that all the wurtzite semiconductors shown in Fig. 5.9 are also direct-gap systems. Silicon is an indirect-gap semiconductor; its E_v is still at Γ, but E_c is at a different $\mathbf{k}$ point, approximately located at (0.8,0,0)

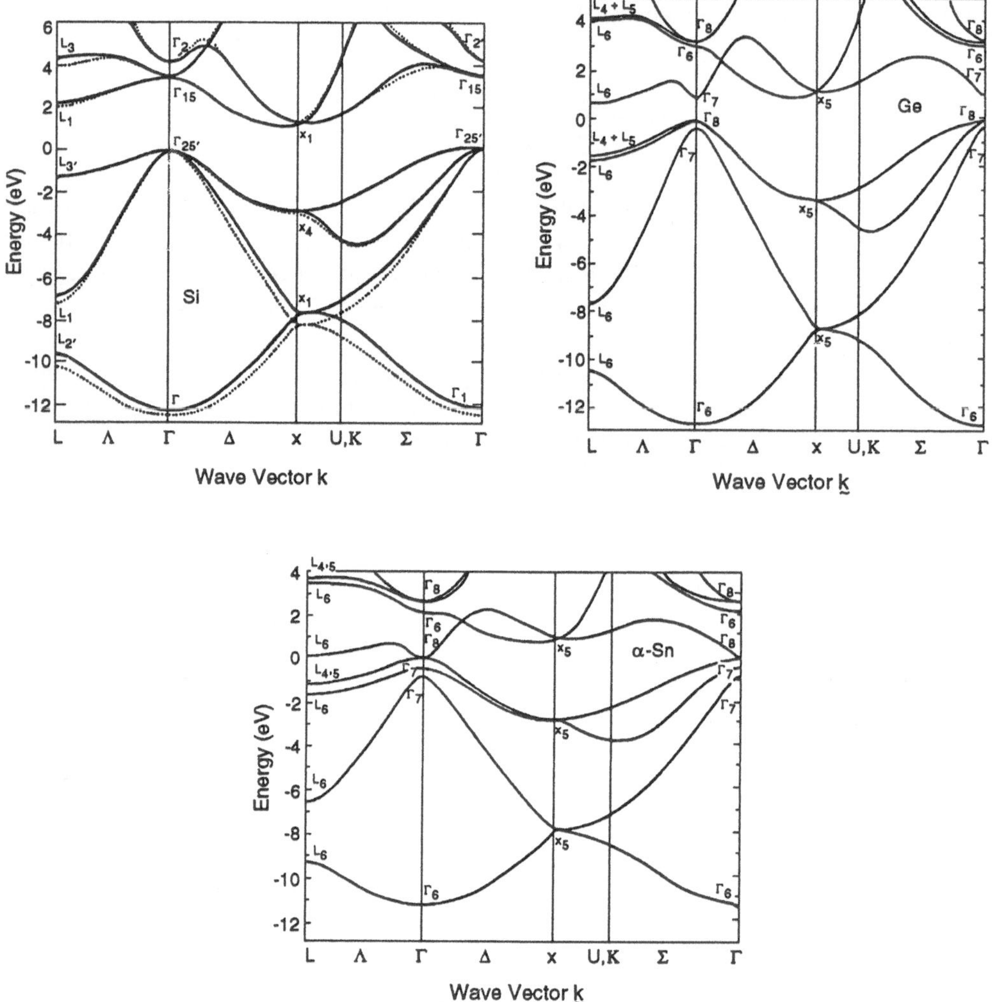

FIGURE 5.6. Band structures of Si, Ge, and α-Sn. (From Cohen and Chelikowsky, 1989.)

and the five other equivalent **k** pcints. GaP and AlAs also have indirect gaps. In general, the direct-gap systems are more effective in optoelectronic applications than are the indirect-gap semiconductors, because in the former electrons can make direct transitions between E_c and E_v by absorption or emission of a photon of energy equal to the band gap. This occurs because photons with an energy equal to the band gap E_g have very little momentum E_g/c. To conserve both energy and momentum, direct optical transitions occur nearly vertically in the band diagram. To excite an electron from E_v to E_c of an indirect-gap semiconductor, phonons must be created or absorbed to conserve both energy and momentum. Because these phonon-assisted processes are much slower than direct transitions, the indirect-gap semiconductors are much less efficient as light-emitting diodes and cannot support lasers. On the other hand, indirect-gap semiconductors are the materials of choice for devices in which long minority carrier recombination lifetimes are beneficial. Table 5.1 contains a collection of

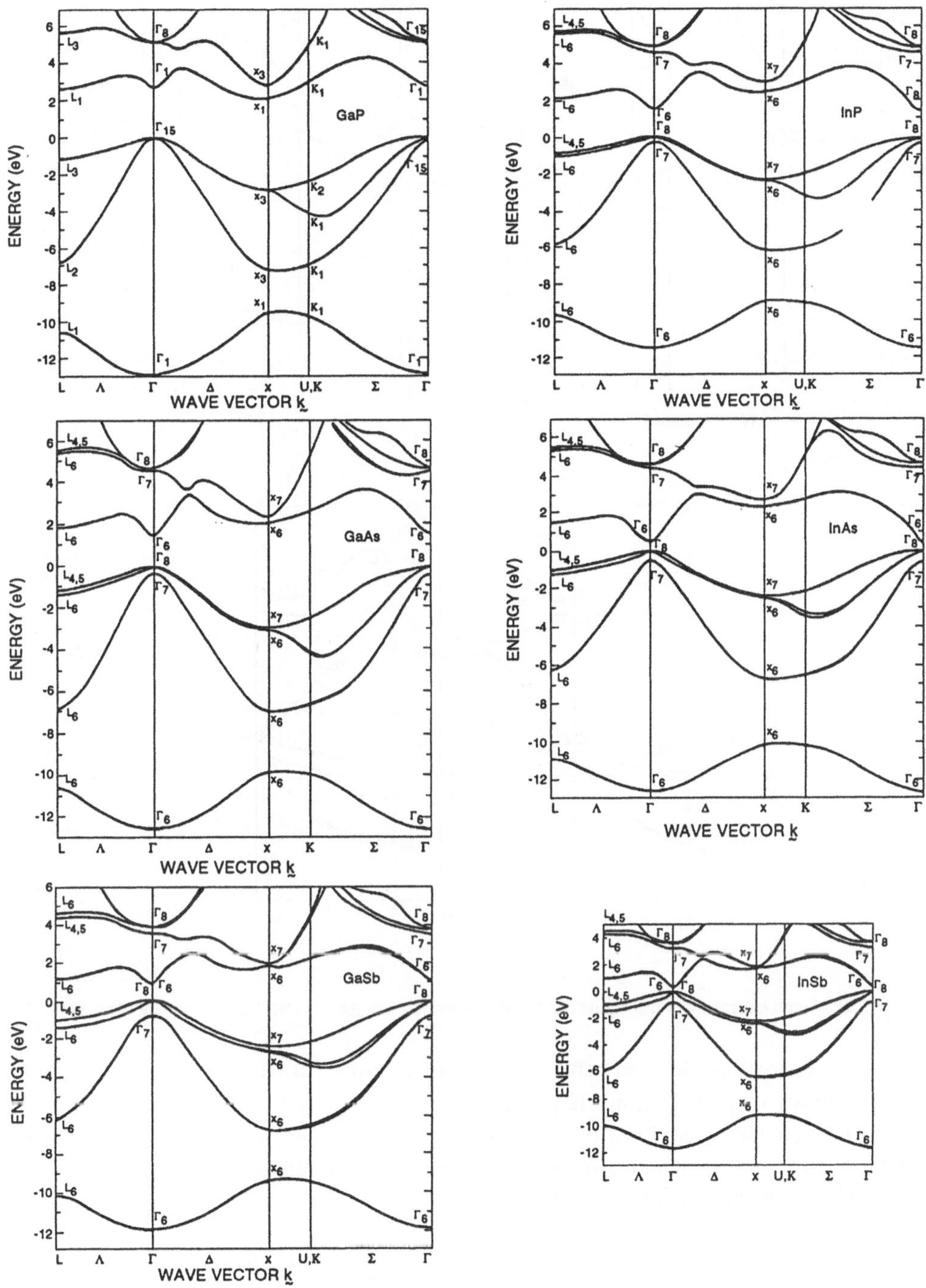

FIGURE 5.7. Band structures of selected III–V zinc blende semiconductors. (From Cohen and Chelikowsky, 1989.)

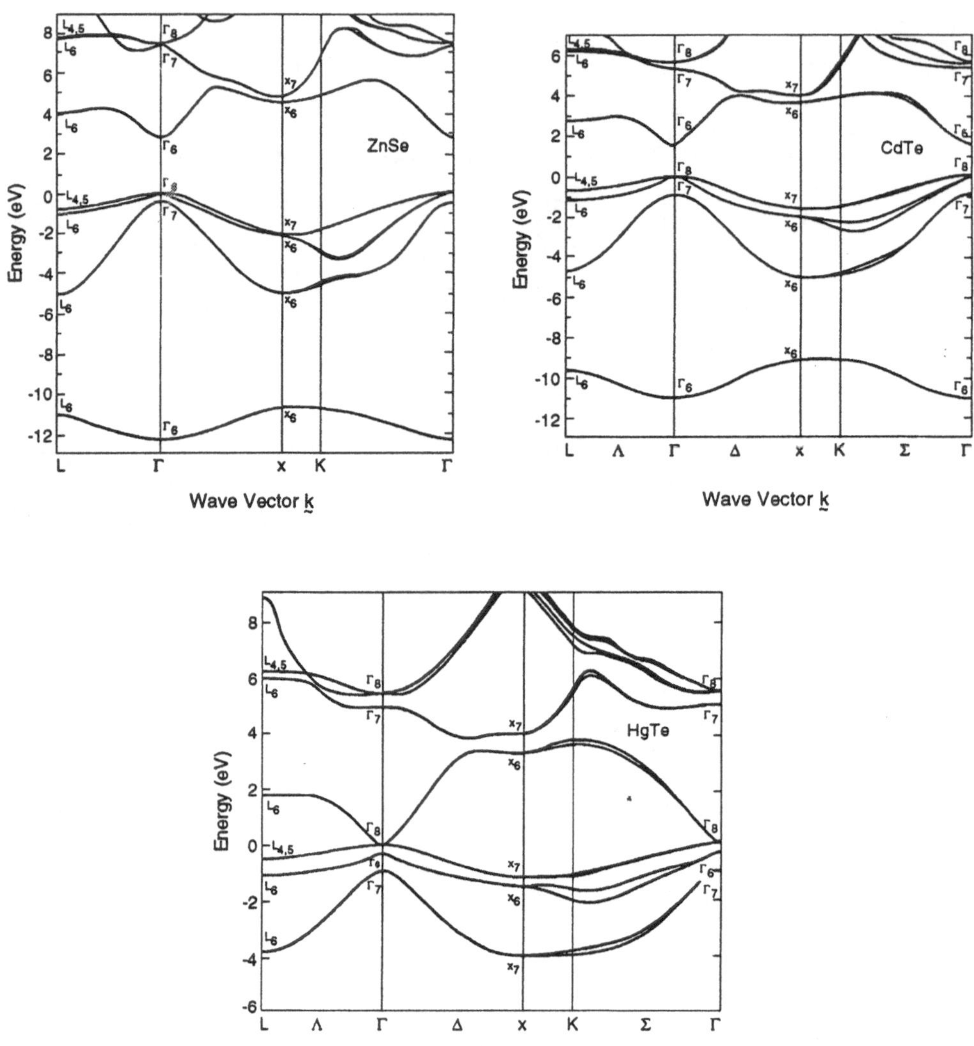

FIGURE 5.8. Band structures of three zinc blende II–VI semiconductors. (From Cohen and Chelikowsky, 1989.)

band-gap information. Detailed band structures of a number of III–V and II–VI compounds and their alloys are presented in Chapter 7.

Besides the band gap, the effective masses at E_v and E_c are also of vital importance in electrical properties. In general, the band structure in the vicinity of a k_0 point, i.e., $\mathbf{k} = \mathbf{k}_0 + \delta\mathbf{k}$, can be expanded in a Taylor series:

$$E(\mathbf{k}) = E(\mathbf{k}_0) + \nabla_k E \cdot \delta\mathbf{k} + \frac{1}{2} \sum_{\alpha} \sum_{\beta} \frac{\partial^2 E}{\partial k_\alpha \partial k_\beta} \delta k_\alpha \delta k_\beta + \cdots \tag{5.4.1}$$

where δk_α is the α-component of $\delta\mathbf{k}$. Up to second order in δk, this expression is usually written as

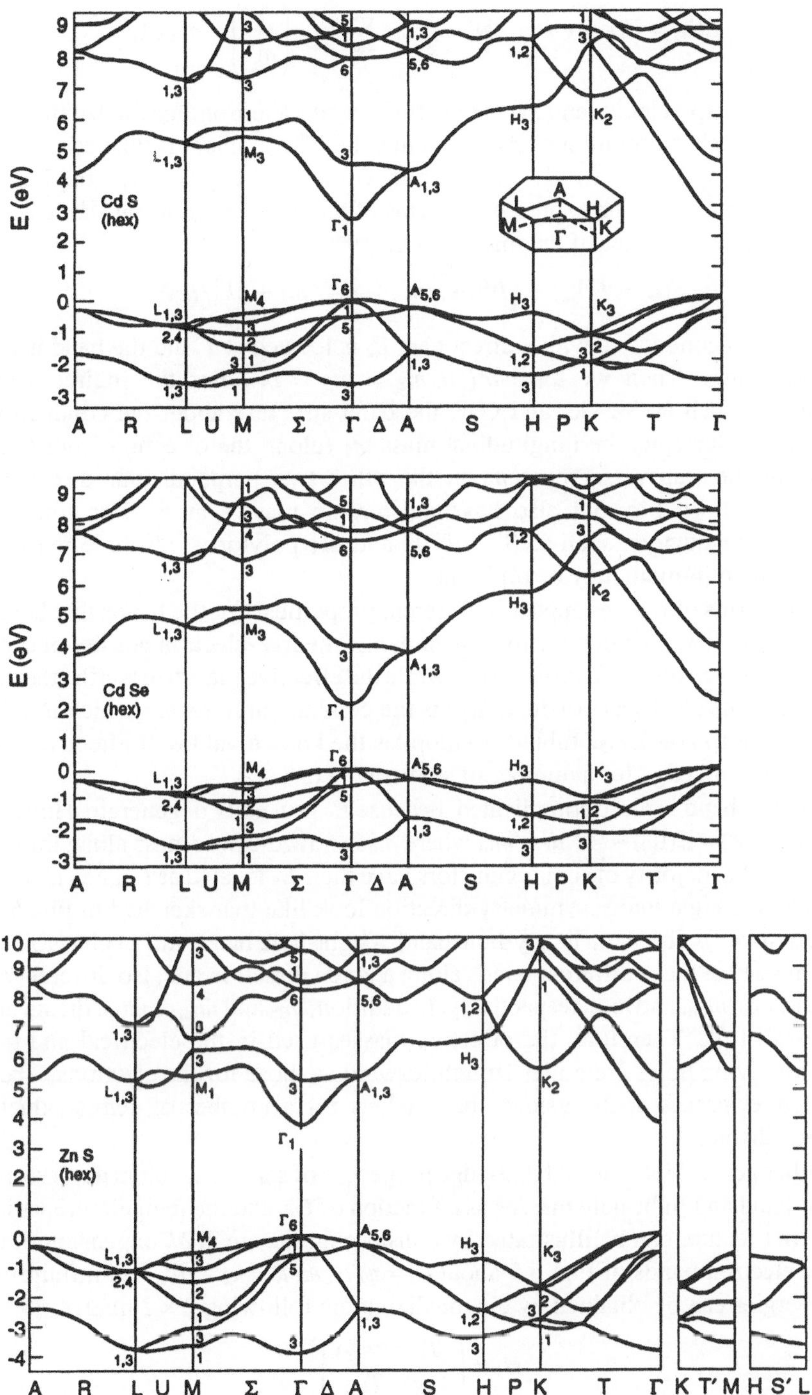

FIGURE 5.9. Band structures of three wurtzite II–VI semiconductors. (From Cohen and Chelikowsky, 1989).

$$E(\mathbf{k}) = E(\mathbf{k}_0) + \hbar \mathbf{v}_k \cdot \delta\mathbf{k} + \frac{1}{2}\hbar^2 \sum_\alpha \sum_\beta \left(\frac{1}{m^*}\right)_{\alpha\beta} \delta k_\alpha \delta k_\beta \tag{5.4.2}$$

where $\mathbf{v}_k$ is the group velocity and m^* is the effective mass tensor. For the band-edge states E_c or E_v, $E(\mathbf{k}_0)$ is either a minimum or maximum at $\mathbf{k}_0$. The gradient is then zero, so the first correction term is quadratic in δk. At the conduction band edge, where the band is not degenerate, Eq. (5.4.2) can be expressed in terms of the three components δk_1, δk_2, and δk_3 along the three principal axes of the mass tensor:

$$E(\mathbf{k}) = E(\mathbf{k}_0) + \tfrac{1}{2}\hbar^2(\delta k_1^2/m_1 + \delta k_2^2/m_2 + \delta k_3^2/m_3) \tag{5.4.3}$$

For the cubic semiconductors with a direct gap, E_c is located at Γ and the band is isotropic about its minimum. Then we have $m_1 = m_2 = m_3 = m^*$. For the indirect-gap cubic semiconductors, such as Si, Ge, and GaP, the band structure about the conduction-band minimum is not isotropic, the longitudinal mass m_l (along the direction from k_0 to Γ) is different from the transverse mass m_t in the other two perpendicular directions. The hexagonal systems in Fig. 5.9 also have anisotropic masses at E_c. For some indirect hexagonal semiconductors, such as SiC of 4H and 6H polytopes, all three masses at the conduction-band minimum may be different.

Different forms of masses appear in different properties. For the bands that behave like Eq. (5.4.3) at small δk, the effective mass to simulate the free-electron density of states near the band edge is given by $m_{ds} = (m_1 m_2 m_3)^{1/3}$, while the effective mass to describe the transport coefficient (e.g, the electrical conductivity) is the conduction mass m_c, which is calculated as $3/m_c = 1/m_1 + 1/m_2 + 1/m_3$. Table 5.1 compiles the known values of effective masses at the conduction-band edge for a number of semiconductors.

The valence band is more complicated, because E_v is usually degenerate. However, one still can write $E(\mathbf{k}) = E(\mathbf{k}_0) - \frac{1}{2}\hbar^2\, dk^2/m_h^*$, where m_h^* is now the hole mass along the direction of $\delta\mathbf{k} = \mathbf{k} - \mathbf{k}_0$. The majority of semiconductors have their E_v located at Γ. The valence-band-edge structures along a major symmetry direction look like that sketched in Fig. 5.10 (see also Figs. 5.6 to 5.9). There is a heavy-hole band, a light-hole band, and a spin-orbit split-off band. The masses of these different bands along a given direction are also different and will be denoted as m_{lh}, m_{hh}, and m_{soh} respectively. In addition, m_{lh} and m_{hh} change drastically with the directions of $\delta\mathbf{k}$. Therefore, the effective masses used in the electrical and transport properties involving holes are not as straightforward as those for the electrons. The values of the effective masses at the valence-band edges for a number of semiconductors are discussed in Chapter 7.

Two other general features of band-edge properties of semiconductors deserve mention: conduction-band and light-hole masses as a function of E_g, and the temperature variation of gaps. The first feature can be illustrated by considering the small δk expansion of the 1-D nearly-free-electron bands of Fig. 5.5 about $k_0 = \pi/a$. At $k = k_0 - \delta k$ with infinitesimal δk, the band energies can be obtained by diagonalizing the following 2×2 matrix:

$$H_k = \begin{pmatrix} T_k & \mathcal{V}(g) \\ \mathcal{V}(g) & T_{k-g} \end{pmatrix} \tag{5.4.4}$$

where $g = 2\pi/a$ and $T_k = \hbar^2 k^2/2m$ is the kinetic energy. At k_0 the two energies are given by $\varepsilon_c = \lambda + |\mathcal{V}(g)|$ and $\varepsilon_v = \lambda - |\mathcal{V}(g)|$ with $\lambda = \hbar^2 k_0^2/2m$, so the band gap is $E_g = 2|\mathcal{V}(g)|$. At $k = k_0 - \delta k$, for infinitesimal δk the two energies are

TABLE 5.1. Energy Band Gaps E_g, Spin-Orbit Splitting $\mathcal{D}$ of the Top of Valence Band, and Conduction-Band Edge Effective Mass Ratios to the Free-Electron Value for Selected Semiconductors[a]

	E_g (eV)		Direct or indirect	m_c		
Structure	Low T	Room T		m_l	m_t	
C (diam.)		5.50	i	1.4	0.36	0.006
Si (diam.)	1.1700	1.1242	i	0.9163	0.1905	0.0441
Ge (diam.)	0.785	0.664 (291 K)	i	1.57	0.0807	0.3
SiC (3C)	2.416 (2 K)		i	0.677	0.247	
SiC (2H)	3.330		i			
SiC (4H)	3.20		i			
SiC (6H)	3.0230	2.86	i	1.5	0.25	
BN (zb)		6.4	i			
BN (w)		5.2	i			
AlN (w)		6.28	d			
GaN (w)	3.503 (1 K)	3.44	d		0.27	
InN (w)	2.11 (78 K)	2.05	d		0.11	
BP (zb)		2.4	i			
AlP (zb)	2.505 (4 K)		i			
GaP (zb)	2.350	2.272	i	4.8	0.252	0.080
InP (zb)	1.4236 (1.6 K)	1.344	d		0.0765	0.108
AlAs (zb)	2.229 (4 K)	2.153	i	1.11	0.19	0.30
GaAs (zb)	1.51914	1.424	d		0.066	0.341
InAs (zb)	0.4180 (4.2 K)	0.354 (295 K)	d		0.0239	0.38
AlSb (zb)	1.686 (27 K)	1.615	i			0.673
GaSb (zb)	0.8113 (2 K)	0.75	d		0.0412	0.765
InSb (zb)	0.2368 (2 K)	0.169	d		0.01359	0.850
ZnO (w)	3.4376 (1.6 K)		d		0.24	
ZnS (zb)	3.78 (19 K)	3.68 (295 K)	d		0.34	
ZnS (w)		3.76	d		0.28	0.086
CdS (w)	2.573 (80 K)	2.485 (293 K)	d		0.23	0.062
CdS (zb)	2.53		d		0.14	
ZnSe (zb)	2.82 (10 K)	2.70 (293 K)	d		0.13	0.40
ZnSe (wb)	2.874 (4.2 K)	2.834	d			
CdSe (w)	1.829 (80 K)	1.751 (293 K)	d		0.12	0.416
HgSe (zb)	−0.205 (80 K)	−0.061	d			
ZnTe (zb)	2.3941 (1.6 K)	2.28 (293 K)	d		0.11	0.396
CdTe (zb)	1.606 (2 K)	1.49	d		0.090	0.91
HgTe (zb)	−0.3025 (4.4 K)	−0.141	d		0.03	0.80

[a]The values for the IV–IV and III–V systems are taken from Madelung (1991) and those for the II–VI compounds are from Landolt and Bornstein (1988). Unless indicated, low T means 0 K and room T means 300 K.

$$\varepsilon_{\pm} = [T_k + T_{k-g} \pm \sqrt{(T_k - T_{k-g})^2 + 4|\mathcal{V}(g)^2|}\,]/2 \tag{5.4.5}$$

Expanding these energies about k_0 and keeping terms up to δk^2, we have

$$\varepsilon_{+} = \varepsilon_{\mathrm{C}} + \frac{\hbar^2 \delta k^2}{2m}\left(1 + \frac{4\lambda}{E_g}\right)$$

$$\varepsilon_{-} = \varepsilon_{\mathrm{V}} + \frac{\hbar^2 \delta k^2}{2m}\left(1 - \frac{4\lambda}{E_g}\right) \tag{5.4.6}$$

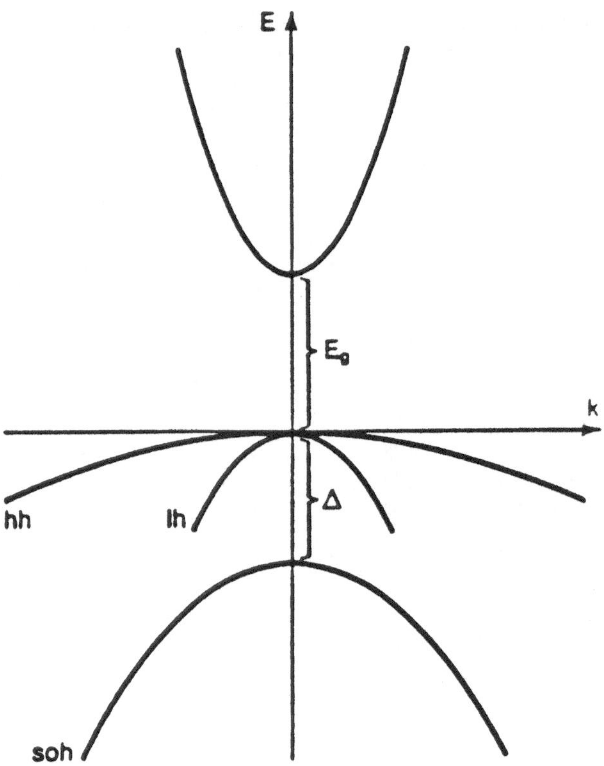

FIGURE 5.10. Schematic band structure for a direct-gap semiconductor (Cohen and Chelikowsky, 1989).

Writing $\varepsilon_+ = \varepsilon_c + \hbar^2\delta k^2/2m_c$ and $\varepsilon_- = \varepsilon_v - \hbar^2\delta k^2/2m_h$, we obtain the following expressions for the effective masses:

$$m/m_c = 1 + 4\lambda/E_g$$

$$m/m_h = -1 + 4\lambda/E_g \tag{5.4.7}$$

If $\lambda >> E_g$, which is true for the present case, then both m_c and m_h are small compared to the free-electron mass and scale linearly with E_g. Expressions similar to Eq. (5.4.7) for a realistic semiconductor can be obtained by "**k** · **p**" perturbation theory (Kane, 1982). In a two-band model for a direct gap, the expressions are:

$$m/m_c = 1 + (2/m)|\langle\psi_c|\mathbf{p}|\psi_v\rangle|^2/E_g$$

$$m/m_{lh} = -1 + (2/m)|\langle\psi_c|\mathbf{p}|\psi_v\rangle|^2/E_g \tag{5.4.8}$$

where $\langle\psi_c|\mathbf{p}|\psi_v\rangle$ is the matrix element of the linear momentum operator between the conduction-band-edge state $|\psi_c\rangle$ and the light-hole state $|\psi_v\rangle$ at the valence-band edge.

Note that the quadratic expansions in Eq. (5.4.6) are only valid for very small δk, such that $\lambda\hbar^2\delta k^2/2m << |\mathcal{V}(g)|^2$. For $\lambda\hbar^2\delta k^2/2m > |\mathcal{V}(g)|^2$, the bands vary nearly linearly with δk.

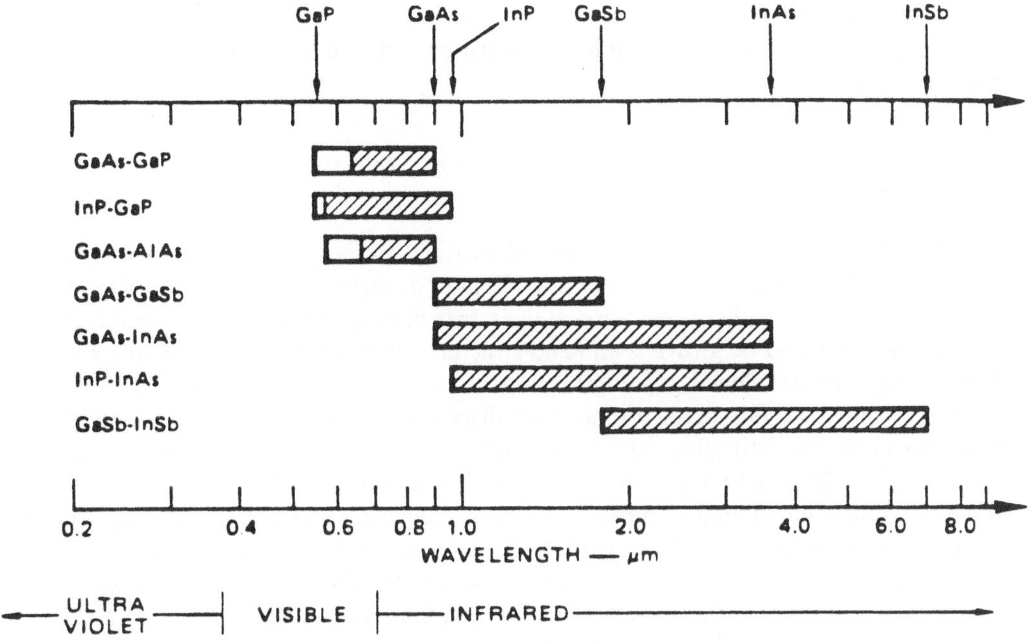

FIGURE 5.11. Room-temperature wavelength range covered by several commonly used III–V ternary alloys.

This situation is often encountered in narrow-gap semiconductor alloys, such as $Hg_{0.8}Cd_{0.2}Te$. This departure from a parabolic band will modify all formulas for transport properties based on the effective-mass approximation.

Finally, according to Table 5.1, band gaps change with the temperature. Most gaps shrink as the temperature increases except for some sufficiently narrow gaps, that widen. This effect is dominated by the electron–phonon interaction (Cohen and Chadi, 1980, and references therein) and can be qualitatively explained by second-order perturbation theory (SOPT) (Krishnamurthy *et al.*, 1988). The scattering of electrons by phonons becomes progressively more important as the temperature increases and more phonons are excited. Through SOPT, the phonon-coupled interactions with the other conduction-band states tend to lower the energy of the conduction-band-edge state, while the valence-band states tend to push it upward. For wide gaps, the conduction-band edge E_c is depressed because the repulsion between E_c and the valence bands is smaller than that between E_c and the nearby conduction bands. The conduction-band states also push down on the valence-band edge while the lower valence-band states push upward. It turns out that even for wider-gap materials (e.g., GaAs), the conduction-band states win and the valence-band energy is depressed (Krishnamurthy *et al.*, 1994). In the wide-gap materials the conduction-band edge is depressed faster than the valence-band edge, so the gap decreases with increasing temperature. However, in a narrow-gap semiconductor such as $Hg_{0.8}Cd_{0.2}Te$, the repulsion between E_c and the heavy hole, having a high density of states, slows the conduction bands, and the net result is an upward shift of E_c. If this upward interaction causes E_c to descend slower than E_v, the gap is widened as the temperature increases. There are also small contributions to $E_g(T)$ arising from lattice dilatation with temperature that must be included to arrive at accurate predictions.

5.5. BAND STRUCTURE OF SEMICONDUCTOR ALLOYS: PROBLEMS AND APPLICATIONS

Alloying is an effective way to control material properties. Here we focus on the variability of band structures. As shown in Chapter 4, many semiconductors form alloys across the whole range of concentration at their growth temperatures. This wide miscibility range allows alloys to be grown with band structures adjusted for specific applications—which is often termed "band-gap engineering." Optoelectronics applications offer a good example to illustrate this point. Figure 5.11 displays the band gaps E_g of several III–V semiconductors and their alloys along with a characteristic photon wavelength λ_g defined by $E_g = hc/\lambda_g$. The gap for each alloy covers the whole range between two end-point values of the constituent compounds. The shaded regions indicate alloys with direct gaps, and the open regions are those with indirect gaps. Note that none of the pure compounds listed have a direct gap large enough ($E_g \geq 1.65$ eV) to have a λ_g in the visible range 750 nm $\geq \lambda_g \geq$ 400 nm. However, the three alloy systems $Ga_{1-x}Al_xAs$, $Ga_{1-x}In_xAs$, and $GaAs_{1-x}P_x$ at some x values do have direct gaps corresponding to visible λ_g values, and have been used for light-emitting diodes (LED) and lasers. Another good example is the II–VI alloy $Hg_{1-x}Cd_xTe$. This alloy has a continuous direct gap ranging from –0.3 eV (for HgTe)* to 1.6 eV (for CdTe), and has become a prime material for long-wavelength infrared ($E_g \leq 0.1$ eV) applications.

All these examples show that semiconductor alloys can provide band gaps that are not accessible by their constituent compounds. Clearly the band gap is the most important feature of the alloy band structure. For most device applications, one needs to know the values of E_g and the effective masses as functions of alloy concentration as well as the concentration at which the direct gap and indirect gap cross, if this occurs. The states near the gaps are also very important, because they are needed for studying high-field and ballistic transport. (This point is discussed further in Chapter 6.) Finally, a good band theory has to cover a wide range of bands well away from the conduction- and valence-band edges, so that a full range of optical and photoelectric properties can be studied.

The most important effect of alloying on electronic structure arises from disorder in the potentials introduced by different constituents. There are two kinds of disorder: compositional disorder caused by a nonperiodic occupation of atomic sites by the alloying atoms; and positional disorder arising from shifts of the atomic positions away from periodic sites. Semiconductor alloys made from lattice-mismatched constituents do have appreciable positional disorder as evidenced by the bimodel bond-length distribution discussed in Chapter 1. Without periodicity in the potential, the crystal wave vectors **k** no longer are good quantum numbers, so the Bloch theorem (Eq. 5.1.3) does not strictly apply. The band structure calculational methods discussed so far no longer work. On the other hand, the materials properties of many semiconductor alloys do not behave too differently from those

*HgTe is a zero-gap semimetal. The negative gap here means that the band-gap states are inverted; i.e., Γ_6 lies below the Γ_8 level (see Fig. 5.8 and Fig. 7.38).

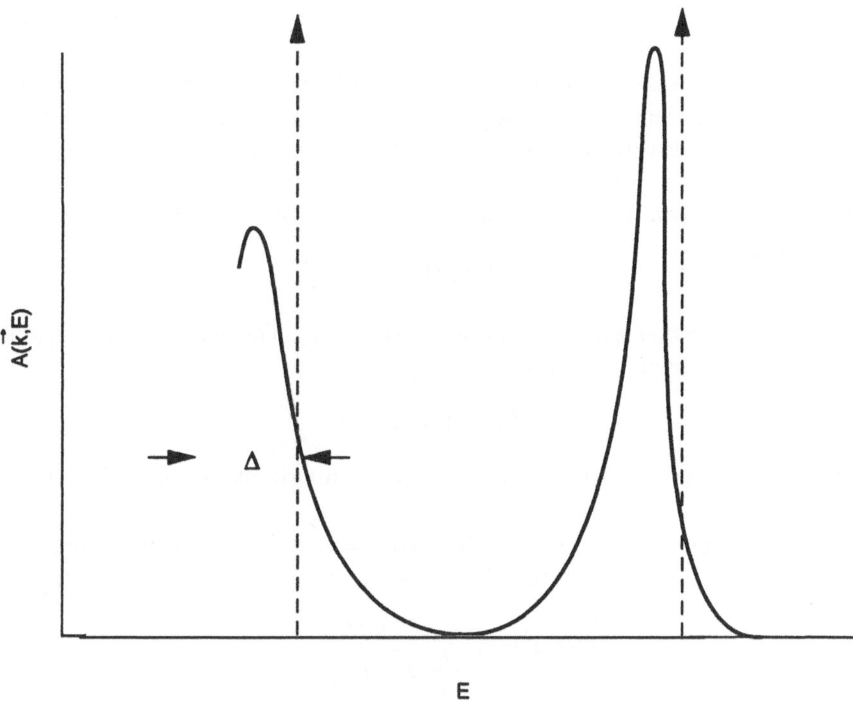

FIGURE 5.12. Schematic energy dependence of the spectral density of states $A(\mathbf{k}, E)$ for a disordered alloy (solid line) and a perfect crystal (dashed line).

of pure semiconductor compounds. This implies that the concepts of band dispersion and effective masses, to some level of approximation, are still valid. These features can be rationalized if we view alloys as effective media and their electronic states as quasi-particle states which can propagate in the media, but with damping. More specifically, a quasi-particle state in an alloy can be represented by its spectral density of states $A(\mathbf{k}, E)$ schematically shown as a function of energy E in Fig. 5.12. The peak position of $A(\mathbf{k}, E)$ can then be identified with the dispersion $\varepsilon(\mathbf{k})$, and the width $\Delta(\mathbf{k})$ may be related to the lifetime $\tau(\mathbf{k})$ of the state by $\tau \approx \hbar/\Delta$. Note that a Bloch energy eigenstate has a zero width, so its spectral density of states is a Dirac delta function $A(\mathbf{k}, E) = \delta[E - \varepsilon(\mathbf{k})]$. As we will show, the shapes of spectral density of states may become quite complicated in an alloy in which the disorder potential has large fluctuations. What is needed next is a theory that will allow an unambiguous definition and a systematic calculation of $A(\mathbf{k}, E)$.

5.6. GREEN FUNCTION AND SPECTRAL DENSITY OF STATES

This section is devoted to determination of relations between the Hamiltonian, or equivalently a Green function, and the electronic density of states, the spectral density, and the self-energy of the effective alloy medium. These relationships are between operators and are obtained without further approximations beyond those used to obtain the one-electron Hamiltonian. While these relations are quite elegant and general, they are difficult to evaluate, even numerically. Approximations enabling their evaluations are the subject of Sections 5.7 and 5.8.

One effective way to obtain the spectral density of states is to use the Green function. Given an effective one-electron potential $V(\mathbf{r})$ for the alloy, the values of V will depend on the alloy configuration governed by the occupation of the atomic sites and the positions of these sites. In a given configuration, the one-electron Hamiltonian is given by $H = p^2/2m + V$. The Green function operator as a function of energy E is defined as

$$G(E + i\circ) \equiv 1/(E + i\circ - H) \tag{5.6.1}$$

where the symbol $\circ$ denotes a positive infinitesimal energy. The density of states, as shown later, is related to the Green function by

$$\mathcal{D}(E) = -\mathrm{Im}\ \mathrm{Tr}\ G/\pi \tag{5.6.2}$$

where the trace Tr means the sum of all the diagonal matrix elements, and Im means the imaginary part.

In terms of the eigenenergies ε_α of H (i.e., $H|\alpha\rangle = \varepsilon_\alpha|\alpha\rangle$), the density of states is given by $\mathcal{D}(E) = \Sigma_\alpha \delta(E - \varepsilon_\alpha)$. The Green function in the eigenenergy basis takes the form

$$G(E + i\circ) = \sum_\alpha \frac{|\alpha\rangle\langle\alpha|}{E + i\circ - \varepsilon_\alpha} \tag{5.6.3}$$

The trace of G in this representation is $\mathrm{Tr}\ G = \sum_\alpha 1/(E + i\circ - \varepsilon_\alpha)$. Since $1/(x + i\circ) = P(1/x) - i\pi\delta(x)$, with P indicating the principal part, we have $-\mathrm{Im}\ \mathrm{Tr}\ G/\pi = \sum_\alpha \delta(E - \varepsilon_\alpha) = \mathcal{D}(E)$, which is what we set out to demonstrate. Although this proof uses the energy eigenfunctions and eigenvalues, the actual calculation of $\mathcal{D}(E)$ does not require diagonalization of the Hamiltonian, which in general is quite difficult. In Eq. (5.6.2), Tr G can be evaluated in any convenient complete basis, because the trace of an operator is invariant under any unitary transformation.

The spectral density of states can now be defined. For a macroscopic alloy system, the measured density of states is the statistically averaged value of $\mathcal{D}$ in Eq. (5.6.2): $\mathcal{D}(E) = -\mathrm{Im}\ \mathrm{Tr}\langle G\rangle/\pi$, where $\langle\ \rangle$ means the average over all alloy configurations. The averaged Green function $\langle G\rangle$ possesses crystal translational symmetry, so its matrix will be block diagonal, where each block belongs to a wave vector $\mathbf{k}$ of the average crystal. Symbolically this is written as $\langle G\rangle = \Sigma G_k$. The state density $\mathcal{D}(E)$ can now be expressed as the sum of the spectral density of states:

$$\mathcal{D}(E) = \sum_{\mathbf{k}} A(\mathbf{k}, E) \tag{5.6.4}$$

where $A(\mathbf{k}, E)$ is related to G_k by

$$A(\mathbf{k}, E) = -\mathrm{Im}\ \mathrm{Tr}\ G_k/\pi \tag{5.6.5}$$

To develop systematic methods for calculating $\langle G\rangle$ and G_k, we first start with the average potential $\langle V\rangle$. Since $\langle V\rangle$ also has crystal translational symmetry, we can solve the eigenvalue problem of the average Hamiltonian $\langle H\rangle = p^2/2m + \langle V\rangle$, using the Bloch theorem:

$$\langle H\rangle|\alpha\,\mathbf{k}\rangle = \varepsilon_\alpha(\mathbf{k})|\alpha\,\mathbf{k}\rangle \tag{5.6.6}$$

If we stop here and take this result as the band structure for the alloy, we are making what is called the virtual crystal approximation (VCA). One of the most severe approximations made by VCA is the neglect of the widths in the spectral density of states caused by disorder. VCA also severely misrepresents the energy spectra in cases where potential differences between the constituent atoms are large.

To find systematic corrections to VCA, let us define a reference Green function associated with $\langle H\rangle$ as

$$g(z) = 1/(z - \langle H\rangle) \tag{5.6.7}$$

where $z = E + i0$. In terms of g and the potential difference $U = V - \langle V\rangle$ the alloy Green function G can be written as $G = g + gUG$ or, when solved, as

$$G = g + gTg \tag{5.6.8}$$

where T is the T matrix associated with the scattering potential U:

$$T = U/(1 - gU) \tag{5.6.9}$$

Keep in mind that G, g, T, U, and H are in general noncommuting operators, so their ordering in products must be maintained. Taking the configurational average of Eq. (5.6.8) leads to

$$\langle G\rangle = g + g\langle T\rangle g \tag{5.6.10}$$

It is useful to define a self-energy operator Σ by

$$\langle G(z)\rangle \equiv 1/(z - \langle H\rangle - \Sigma) \tag{5.6.11}$$

A comparison of Eqs. (5.6.8) and (5.6.11) shows that

$$\Sigma = \langle T\rangle/(1 + g\langle T\rangle) \tag{5.6.12}$$

Similarly the average Green function for each $\mathbf{k}$-block can be written as $G_k = 1/[z - \langle H(\mathbf{k})\rangle - \Sigma(\mathbf{k})]$, with the k-diagonal self-energy given by

$$\Sigma(\mathbf{k}) = \langle T(\mathbf{k})\rangle/[I + g(\mathbf{k})\langle T(\mathbf{k})\rangle] \tag{5.6.13}$$

In these expressions, the operators I (the identity), $H(\mathbf{k})$, $\Sigma(\mathbf{k})$, $T(\mathbf{k})$, and $g(\mathbf{k})$ are matrices for a given wave vector $\mathbf{k}$; and Σ, g, and T are also functions of the energy parameter z. In the basis of eigenvectors of $\langle H\rangle$ in Eq. (5.6.6), the matrix elements of $H(\mathbf{k})$ and $g(\mathbf{k})$ are trivial; e.g., $g_{\alpha\beta}(\mathbf{k}, z) = \delta_{\alpha\beta}/[z - \varepsilon_\alpha(\mathbf{k})]$. All the burden of the calculation of $A(\mathbf{k}, E)$ is on $\langle T(\mathbf{k})\rangle$. The evaluation of $\Sigma(k)$ or $\langle T(\mathbf{k})\rangle$ becomes the central issue of alloy theory.

5.7. PERTURBATION THEORY AND BOWING OF FUNDAMENTAL GAPS

In this section a perturbation theory will be introduced into the above formalism to provide a qualitative understanding of the band-gap variation as a function of alloy concentration x. Because of the importance of band gaps in devices, the gap variation of semiconductor alloys has been the subject of many experimental and theoretical studies. Experimentally, the measured gap E_g in a pseudobinary semiconductor alloy $A_{1-x}B_xC$ as a function of x is generally lower than the concentration-weighed average value

TABLE 5.2. Experimental Bowing Parameters b for Several Ternary Alloys[a]

Alloy	b (eV)
GaInP	0.40–0.88
AlInAs	0.23
GaInAs	0.28–0.56
GaAlAs	0.18–0.47
InAsP	0.23–0.78
GaAsP	0.17–0.21
InAsSb	0.58
GaInSb	0.43
ZnSSe	0.06
ZnSeTe	1.23
ZnSTe	3.00
ZnCdS	0.3
CdSSe	0.54
CdSeTe	0.87
CdSTe	2.0

[a]After Van Vechten and Bergstresser (1970).

$\overline{E}_g = (1-x)E_g(\mathrm{AC}) + xE_g(\mathrm{BC})$ of the constituent compounds. The difference between E_g and $\overline{E}_g$ is found to be reasonably well described by a quadratic function of the concentration:

$$E_g = \overline{E}_g - bx(1-x) \tag{5.7.1}$$

The constant b in Eq. (5.7.1) is called the bowing parameter and in general is positive. Table 5.2 lists the experimental bowing parameters (Van Vechten and Bergstresser, 1970) for a collection of III–V and II–V ternary alloys. The major source of error in determining these bowing parameters is the concentration measurement. However, this table shows unambiguously that the concentration variation of gaps bows below their average values.

Many simple theories have been suggested to predict this result. The main issue centers around the contribution from alloy disorder. In VCA the fluctuations in the potential are neglected, so the nonlinear variation of the gap must arise from either a nonlinear modification of the effective potential or the nonlinear dependence of the energy levels on the potential. Van Vechten and Bergstresser (1970) found no *a priori* way to show that the bowing due to these effects will always lie below the average gaps. They proposed that the bowing parameter contains two terms, $b = b_i + b_e$, where b_i is called the intrinsic contribution (the VCA contribution) and b_e is the extrinsic contribution attributed to alloy disorder. The extrinsic bowing was assigned a simple form $b_e = \delta^2/A$, where δ represents the root-mean-square fluctuation in the potentials and A is an adjustable bandwidth parameter. Objections to the alloy disorder contribution were raised on the grounds (Hill, 1974) that experimental alloying causes very little broadening of the fundamental optical E_0 spectra (the transition between the top of the valence band the lowest conduction-band state at Γ) in modulation-reflective experiments (Williams and Rehn, 1968; Alibert *et al.*, 1972). It will be shown that there is no contradiction between contentions that alloy disorder contributes to the bowing of the gap but does not broaden the E_0 spectra.*

*The E_0 transition refers to the band-gap transition between the Γ_{8v} and Γ_{6c} states in a direct-gap semiconductor.

For a weak potential fluctuation, we can reduce the general theory presented above to the weak limit. Using Eq. (5.6.9), the self-energy operator in Eq. (5.6.12) can be approximated as

$$\Sigma \simeq \langle T\rangle = \langle U + UgU\rangle = \langle UgU\rangle \tag{5.7.2}$$

Note that in the last step of Eq. (5.7.2), we set $U = V - \langle V\rangle$ so $\langle U\rangle = 0$. To illustrate how the factor $x(1-x)$ of Eq. (5.8.1) arises, let us consider the case where the effective potential V can be written as the sum of all site contributions, and the alloy atoms are randomly arranged on their sublattice sites. In other words, this model only contains random compositional disorder. The disorder potential U then takes the form $U = \sum U_n$, where $U_n = V_n - \langle V_n\rangle$ and the summation only runs over the fcc sublattice sites where the alloy atoms A and B reside (the C atoms reside in the other fcc sublattice of the zinc blende structure). The probability of having an A atom on a substitutional site is $1-x$ and that for a B atom is x. V_n takes the value $V_A(\mathbf{r}-\mathbf{R}_n)$ or $V_B(\mathbf{r}-\mathbf{R}_n)$, depending on whether an A or a B atom sits on site n. If an A atom occupies the nth site, then $U_n = V_A - [(1-x)V_A + xV_B] = x(V_A - V_B) \equiv xW_n$, where the potential difference W_n is defined as $V_A(\mathbf{r}-\mathbf{R}_n) - V_B(\mathbf{r}-\mathbf{R}_n)$. Similarly if a B atom is at site n, then $U_n = -(1-x)W_n$. In terms of W_n, the self-energy operator in Eq. (5.7.2) becomes

$$\begin{aligned}\Sigma &= \sum_n \langle U_n g U_n\rangle + \sum_{n\neq m}\sum \langle U_n g U_m\rangle \\ &= \sum_n \langle U_n g U_n\rangle + \sum_{n\neq m}\sum \langle U_n\rangle g\langle U_m\rangle = \sum_n \langle U_n g U_n\rangle \\ &= (1-x)x^2\sum_n W_n g W_n + x(1-x)^2\sum_n W_n g W_n \\ &= x(1-x)\sum_n W_n g W_n \end{aligned} \tag{5.7.3}$$

The second equality in Eq. (5.7.3) is a consequence of g being independent of the configurational average and, for $n \neq m$, the average of the product $\langle U_n g U_m\rangle$ being the product of the averages for an uncorrelated random distribution. The third equality stems from the fact that, by definition, $\langle U_n\rangle$ is 0. The last two equalities are just algebra.

The shift from the VCA energy and the width of the spectral density of states are dominated by the diagonal matrix element of the self-energy operator:

$$\sigma_\alpha(\mathbf{k},E) = \langle\alpha\mathbf{k}|\Sigma|\alpha\mathbf{k}\rangle = x(1-x)\sum_{\beta\mathbf{k}'}\frac{\sum_n |\langle\alpha\mathbf{k}|W_n|\beta_{\mathbf{k}}'\rangle|^2}{E + i0 - \varepsilon_\beta(\mathbf{k}')} \tag{5.7.4}$$

The real part of σ at $E = \varepsilon_\alpha(\mathbf{k})$ corresponds to the potential-fluctuation-induced shift of the energy from the VCA value $\varepsilon_\alpha(k)$:

$$\eta_\alpha(\mathbf{k}) = x(1-x)P\sum_{\beta\mathbf{k}'}\frac{\sum_n |\langle\alpha\mathbf{k}|W_n|\beta_{\mathbf{k}}'\rangle|^2}{\varepsilon_\alpha(\mathbf{k}) - \varepsilon_\beta(\mathbf{k}')} \tag{5.7.5}$$

where P indicates the principal part. At the conduction-band minimum $\varepsilon_\alpha(k) = E_c$, because of the presence of the gap, the major contribution to the summation comes from the states just above E_c. This results in a negative value of η. In contrast, at the top of the valence band $\varepsilon_\alpha(k) = E_v$, the major contribution comes from states just below E_v, so the shift is positive. This combination gives a bowing of the gap below the VCA value and qualitatively explains the $x(1-x)$ dependence.

We note that these arguments may not hold for a narrow-gap alloy such as $Hg_{0.8}Cd_{0.2}Te$ ($E_g = 0.1$ eV), for which the sign of η depends on a detailed cancellation of the sums in Eq. (5.7.5) between the conduction and valence bands. We also note that, besides the disorder effect discussed, the VCA band structure also contributes to the nonlinear concentration variation of the gap. In this connection, semiconductors having larger lattice constants tend to have smaller band gaps. The band gaps can be assumed to scale as some inverse power of the bond length, $E_g \propto 1/d^n$, which, for example, can be seen by comparing the Γ gaps among AlAs, GaAs, and InAs, or among GaP, GaAs, and GaSb. We also learned from Chapter 1 that the Vegard law for the average bond length $d = (1-x)d_{AC} + xd_{BC}$ holds very well in pseudobinary alloys $A_{1-x}B_xC$. These two facts can be used to argue that qualitatively the VCA bands will have a nonlinear gap variation of the form

$$E_g(\text{VCA}) - \overline{E} \simeq -x(1-x)n(n+1)(\Delta d/d_0)^2 E_0/2$$

where $d_0 = (d_{AC} + d_{BC})/2$, $\Delta d = d_{AC} - d_{BC}$, and $E_0 = [E_g(\text{AC}) + E_g(\text{BC})]/2$. This result thus demonstrates that VCA also produces a positive contribution to the bowing parameter b of Eq. (5.7.1). Although a quantitative description of band-gap bowing requires a detailed calculation, many trends follow this qualitative argument.

The width in the spectral density of states for a state with energy $E = \varepsilon_\alpha(\mathbf{k})$ can be related to the imaginary part of σ by $\Delta = -\text{Im}\,\sigma$:

$$\begin{aligned}\Delta_\alpha(\mathbf{k}) &= \pi x(1-x)\sum_n\sum |\langle\alpha\mathbf{k}|W_n|\beta\mathbf{k}'\rangle|^2\delta[E-\varepsilon_\beta(\mathbf{k}')] \\ &\simeq \pi x(1-x)|\langle\alpha\mathbf{k}|W_n|\alpha\mathbf{k}\rangle|^2\mathcal{D}(E)\end{aligned} \qquad (5.7.6)$$

For both states at E_c and E_v, the widths approach zero because the density of states $\mathcal{D}$ is zero at both band edges. This explains why alloy disorder contributes little to the broadening of the E_0 spectra.

5.8. MULTIPLE SCATTERING THEORY AND THE COHERENT POTENTIAL APPROXIMATION

The discussion in the preceding section relies on perturbation theory. This approximation works only when the fluctuations in the alloy potential are small, the so-called weak-scattering limit. When the potential fluctuations are large, an improved theory is needed. In this section the general multiple-scattering theory and an effective-medium theory, the coherent potential approximation (CPA) (Soven, 1976), will be introduced. CPA proves to be useful for both weak- and strong-scattering cases.

Multiple-scattering theory assumes that the scattering potential U in Eq. (5.7.6) is the sum of site contributions $U = \sum U_n$. There is no unique way to chose the U_n's. As long as the sum accurately reproduces the total potential U, reliable answers will be given by the theory. Generally the computational speed is increased if the U_n's can be selected to have short ranges. Associated with each site potential U_n an atomic scattering operator t_n can be defined,

$$t_n = U_n/(1 - gU_n) \tag{5.8.1}$$

The operator t_n accounts for scattering off-site n to all orders in U_n. The total T matrix of Eq. (5.6.6) can be expanded in the following multiple-scattering series:

$$T = \sum_n t_n + \sum_{n \neq m}\sum t_n g t_m + \sum_{n \neq m}\sum_{m \neq j}\sum t_n g t_m g t_j + \cdots \tag{5.8.2}$$

The first term is the sum of single-scattering contributions from all sites. The second term is the sum of double-scattering contributions, and the third term is from triple-scattering contributions, etc. Note that no two successive site indices are the same. For example, in the third term, m must be different from both n and j, but n can be the same as j. Now the configuration average can be performed on Eq. (5.8.2). If all sites are statistically independent, such as for the random alloy, then the average T matrix takes the form

$$\langle T \rangle = \sum_n \langle t_n \rangle + \sum_{n \neq m}\sum \langle t_n \rangle g \langle t_m \rangle = \sum_{n \neq m}\sum_n\sum \langle t_n g \langle t_m \rangle g t_j \rangle + \cdots \tag{5.8.3}$$

If statistical correlations in the triple- and higher-order scattering terms are neglected (i.e., if all the t_n in Eq. (5.8.3) are replaced by $\langle t_n \rangle$, then we are making what is called the single-site approximation (SSA). In this approximation, the series can be summed into a closed form to yield a self-energy of the form

$$\Sigma = \sum_n \Sigma_n \tag{5.8.4}$$

with the site self-energies given by

$$\Sigma_n = \langle t_n \rangle / (1 + g\langle t_n \rangle) \tag{5.8.5}$$

The SSA with the self-energy given explicitly above is also known as the average T-matrix approximation (ATA).

The coherent potential theory is a self-consistent method for generating the average Green function $\langle G \rangle$. The theory starts by assuming $\langle G \rangle$ to have the form of Eq. (5.6.11) with an unknown self-energy operator Σ. When expressed in terms of $\langle G \rangle$, the G for a particular alloy configuration is given by

$$G = \langle G \rangle + \langle G \rangle T \langle G \rangle \tag{5.8.6}$$

where T is similar to Eq. (5.6.9) except that U is replaced by $V - \Sigma$. Now take the average over both sides of Eq. (5.8.6) to obtain $\langle G \rangle = \langle G \rangle + \langle G \rangle \langle T \rangle \langle G \rangle$, from which we deduce $\langle T \rangle = 0$ or, more explicitly,

$$\langle T \rangle = \langle (U - \Sigma)/[1 - \langle G \rangle (U - \Sigma)] \rangle = 0 \tag{5.8.7}$$

Equation (5.8.7) is a self-consistent equation for Σ, because $\langle G \rangle$ is also a function of Σ, as can be seen from Eq. (5.6.11).

Because Eq. (5.8.7) still involves the total alloy scattering potential $U = V - \langle V \rangle$, its formal solution is not easier than the original eigenvalue problem. Further approximations are needed to make the theory tractable. The simplest approximation is the single-site approximation. Expanding the multiple-scattering series with respect to the effective Green function $\bar{G} = 1/(z - \langle H \rangle - \Sigma)$, with Σ given by the site contributions as in Eq. (5.8.3), produces the following expression for $\langle T \rangle$:

$$\langle T \rangle = \sum_n \langle t \rangle_n + \sum_{n \neq m} \sum \langle t \rangle_n \bar{G} \langle t \rangle_m + \sum_{n \neq m} \sum \sum_j \langle t_n \bar{G} \langle t_m \rangle \bar{G} t_j \rangle + \cdots \tag{5.8.8}$$

where the atomic scattering operator has become

$$t_n = (U_n - \Sigma_n)/[1 - \bar{G}(U_n - \Sigma_n)] \tag{5.8.9}$$

Now if $\langle t_n \rangle$ is set equal to zero for every site, the first three terms of (5.8.8) vanish. The nonzero terms begin at the fourth order in the atomic t matrices. If all the statistical correlations are neglected (i.e., in SSA), then $\langle t_n \rangle = 0$ for all n implies $\langle T \rangle = 0$. This is the CPA. The general self-consistent CP equation Eq. (5.8.7) reduces to the following equation in CPA:

$$\langle t_n \rangle = \langle (U_n - \Sigma_n)/[1 - \bar{G}(U_n - \Sigma_n)] \rangle = 0 \tag{5.8.10}$$

In summary, both ATA and CPA are SSA. ATA is the non-self-consistent version in which the site self-energy is calculated from Eq. (5.8.5). CPA is a self-consistent SSA in which Σ_n is determined from the self-consistent equation, Eq. (5.8.10). CPA is the best SSA, because once Σ_n satisfies (5.8.10), there is no further correction to the self-energy within SSA. To demonstrate this point, let us assume that we have obtained some self-energy $\tilde{\Sigma} = \sum_n \tilde{\Sigma}_n$ from a method other than CPA. We can then apply ATA to find corrections to $\tilde{\Sigma}$:

$$\Sigma_n = \tilde{\Sigma}_n + \langle \tilde{t}_n \rangle / (1 + \tilde{G} \langle \tilde{t}_n \rangle) \tag{5.8.11}$$

where $\tilde{G} = 1/(E + i0 - \langle H \rangle - \tilde{\Sigma})$ and $\tilde{t}_n$ is like that of Eq. (5.8.9) with Σ_n replaced by $\tilde{\Sigma}_n$ and $\bar{G}$ by $\tilde{G}$. We can see that if $\tilde{\Sigma}_n$ is the CPA self-energy, then $\langle \tilde{t}_n \rangle = 0$ according to Eq. (5.8.10), so there is no further correction to the self-energy in SSA. The CPA equation (5.8.10) is generally still difficult to solve. In practice some iterative procedure has to be employed. It has been shown (Chen, 1973) that Eq. (5.8.11) can be used as an iteration method, named "iterative ATA (IATA)," in which convergence to CPA is guaranteed starting from the VCA Green function.

5.9. A SINGLE-BAND ALLOY MODEL

Before discussing applications of the preceding theory to real semiconductor alloys, it is useful to consider a simple band model (Velicky *et al.*, 1968; Chen *et al.*, 1972) that provides an opportunity to glean insight into the analytical behavior of the theory. Going through this model will also display the basic steps needed to carry out a CPA calculation. Since the mathematical structure in the CPA calculation for a pseudobinary $A_{1-x}B_xC$ alloy is essentially the same as that for a binary $A_{1-x}B_x$ alloy, for simplicity we shall consider the binary alloy in this section. The alloy Hamiltonian in this model is a single-orbital, or single-band, tight-binding Hamiltonian:

$$H = H_0 + V = \sum_n |n\rangle\varepsilon_n\langle n| + \sum_{n\neq m}\sum |n\rangle h_{nm}\langle m| \tag{5.9.1}$$

The hopping matrix elements h_{nm} between sites n and m are assumed to be independent of alloy configuration, but the diagonal element ε_n takes the values ε_A or ε_B, depending on whether an A atom or a B atom is at site n. Here the local basis is assumed to be orthonormal; i.e., $\langle n|m\rangle = \delta_{nm}$. The eigenvectors and the eigenvalues of $\langle H\rangle$ are given by $\langle H\rangle|\mathbf{k}\rangle = \varepsilon(\mathbf{k})|\mathbf{k}\rangle$, where $|\mathbf{k}\rangle$ is the Bloch function constructed from a linear combination of the local orbitals as described by Eq. (5.2.3), and the band energy $\varepsilon(\mathbf{k})$ takes the form

$$\varepsilon(\mathbf{k}) = \bar{\varepsilon} + \sum_{n\neq 0} h_{0n}e^{i\mathbf{k}\cdot\mathbf{R}_n} \tag{5.9.2}$$

where $\bar{\varepsilon}$ denotes the average value $\bar{\varepsilon} = (1-x)\varepsilon_A + x\varepsilon_B$.

In this model the scattering potential U_n is site diagonal, $U_n = |n\rangle(\varepsilon_n - \bar{\varepsilon})\langle n|$, as is the site CPA self-energy: $\Sigma_n = |n\rangle\sigma\langle n|$, where σ is a complex number whose values are to be obtained from the CPA self-consistent equation (5.8.9). By expanding Eq. (5.8.9) in a power series of $u_n - \Sigma_n$, it is easy to show that only the site-diagonal matrix elements of the effective Green function $\langle n|\bar{G}|n\rangle$ appear in the CPA calculation. It can also be shown that $\langle n|\bar{G}|n\rangle$ is independent of the site index n and takes the form

$$\langle n|\bar{G}|n\rangle = F(z) = \frac{1}{N}\sum_k \frac{1}{z - \varepsilon(\mathbf{k}) - \sigma} = \int \frac{\rho_0(\varepsilon)}{z - \varepsilon - \sigma}\,d\varepsilon \tag{5.9.3}$$

where ρ_0 is the density of states per atom for the average Hamiltonian $\langle H\rangle$ given by

$$\rho_0(\varepsilon) = \frac{1}{N}\sum_k \delta[\varepsilon - \varepsilon(\mathbf{k})] \tag{5.9.4}$$

Thus, for the present random-alloy model, the CPA equation Eq. (5.8.10) becomes a scalar equation for the self-energy function $\alpha(z)$ and can be reduced to the following form for a given energy z:

$$\sigma(z) = -(\varepsilon_A - \bar{\varepsilon} - \sigma)F(z)(\varepsilon_B - \bar{\varepsilon} - \sigma) \tag{5.9.5}$$

The solution to this equation is in general still nontrivial, because $F(z)$ in Eq. (5.9.5) is an integral with the integrand containing the unknown σ.

It is instructive to examine a model density of states that was used extensively to analyze this alloy theory (Velicky *et al.*, 1968; Chen *et al.*, 1972). This model density of states has a semielliptical form

$$\rho^0(\varepsilon) = 2\sqrt{1-\varepsilon^2}/\pi \tag{5.9.6}$$

for $-1 \le \varepsilon \le 1$, and $\rho^0 = 0$ otherwise. Implicit in this model, the energy unit is half the width of the band, and the center of the band is at zero. The VCA density of states for $\bar{\varepsilon} \neq 0$ is $\rho_0(\varepsilon) = \rho^0(\varepsilon - \bar{\varepsilon})$ (compare Eq. (5.9.2) with Eq. (5.9.4)). One advantage of using this model density of states is that the site-diagonal Green function F has an analytic expression. To see this, let us define

$$F^0(z) = \int \frac{\rho^0(\varepsilon)}{z-\varepsilon}\, d\varepsilon \tag{5.9.7}$$

Using the expression in Eq. (5.9.6) for ρ^0, we can integrate to obtain an analytic form for F^0:

$$F^0(z) = 2z - 2\sqrt{z-1}\,\sqrt{z+1} \tag{5.9.8}$$

Then $F(z)$ in Eq. (5.9.3) becomes $F(z) = F^0(z - \bar{\varepsilon} - \sigma)$. Alternatively σ can be expressed in terms of F by

$$\sigma = z - \bar{\varepsilon} - F/4 + 1/F \tag{5.9.9}$$

No generality is lost by setting $\varepsilon_A = -\delta/2$ and $\varepsilon_B = \delta/2$. Then Eq. (5.9.5) can be reduced to a cubic equation for F:

$$F^3 - 8zF^2 + (16z^2 - 4\delta^2 + 4)\,F - 16\,(z - \bar{\varepsilon}) = 0 \tag{5.9.10}$$

When solved at a real energy $z = E$, only one of the three roots of F will satisfy the analytic behavior required of the Green function corresponding to $z = E + i\text{o}$. For example, for the case in which a pair of complex roots exist, the correct root must have a negative imaginary part, because the density of states per atom for the alloy is given by

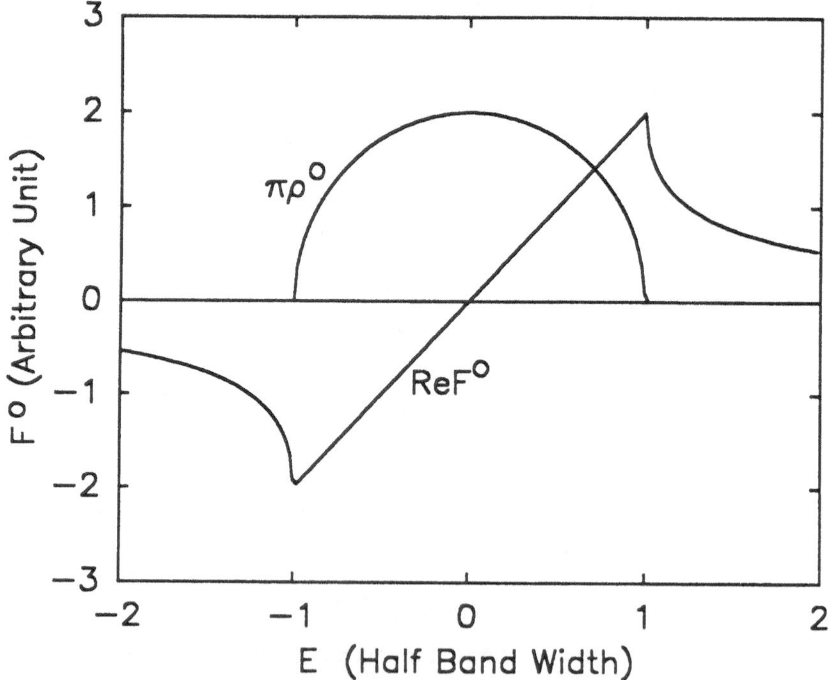

FIGURE 5.13. Green function $F^0 = \text{Re}F^0 - i\pi\rho^0$ of Eq. (5.9.8) for $Z = E + i\text{o}$.

$$\rho(E) = -\text{Im}\, F(E + i0)/\pi \tag{5.9.11}$$

The real part of F can then be analytically extended outside the E region where all the three roots are real. Once F is obtained, σ can be calculated readily from Eq. (5.9.9).

Figure 5.13 shows both the real and imaginary parts of F^0. Figure 5.14 shows the densities of states as a function of energy for several alloys with three scattering strengths δ = 0.25, 0.5, and 0.75, and for several concentrations. Here δ is related to the site energy by $\varepsilon_A = -\delta/2$ and $\varepsilon_B = \delta/2$. For weak scattering ($\delta = 0.25$), the shapes of densities of states for different concentrations do not show appreciable changes. The only detectable change is the change of center of gravity and the edges, because the VCA site-diagonal energy changes according to $\bar{\varepsilon} = (x - 1/2)\delta$. The perturbation theory results in Eqs. (5.7.5) and (5.7.6) should be a good approximation, which for the present case corresponds to $\sigma = \eta - i\Delta$ with η and Δ given by

$$\eta(\text{E}) = x(1 - x)\delta^2 \text{Re}\, F^0(E - \bar{\varepsilon}) \tag{5.9.12}$$

$$\Delta(E) = x(1 - x)\delta^2 \pi \rho^0(E - \bar{\varepsilon}) \tag{5.9.13}$$

In the case with a moderate scattering strength $\delta = 0.5$, the shapes of the DOS can be seen to deviate from the VCA results, and perturbation theory becomes inaccurate. For the strong-scattering case $\delta = 0.75$ and at low concentration, the minority band splits from the host band. The tendency for band splitting is clearly marked in the DOS for all concentrations. These results show that CPA is a theory that can predict the correct qualitative behavior of the electronic states in disordered alloys. It is the starting point of a correct and systematic theory of alloy electronic structures.

5.10. MOLECULAR CPA FOR ZINC BLENDE ALLOYS

The single-band model will now be extended to multiple bands for zinc blende semiconductor alloys. We recall that calculations in Section 4.10 showed that pseudobinary semiconductor alloys $A_{1-x}B_xC$ only deviate slightly from random alloys, in which the substitutional atoms A and B occupy the sites of one fcc sublattice with probabilities $1 - x$ and x respectively. Hence, taking the configuration averages over random site distributions is a good first approach. However, in Chapter 1 it was shown that the nearest-neighbor bond lengths have a bimodal distribution if the constituent compounds have a large bond-length difference. Within the tight-binding Hamiltonian, these effects can be dealt with using the so-called molecular CPA (Hass *et al.*, 1984a; Lempert *et al.*, 1987). Figure 5.15 sketches the local bonding around a substitutional atom. The lobes represent the sp^3 hybrid orbitals (see Eq. (2.5.2)) for the central atom, chosen to be the alloying atom A or B, and its four first-neighbor C atoms. If these eight hybrid orbitals h_1 through h_8 are assigned to a unit cell, now labeled n, and if the fluctuations in the tight-binding Hamiltonian are limited only to the site-diagonal and the first-neighbor matrix elements, then the alloy scattering potential can be written as a sum of independent contributions U_n from different unit cells:

$$U = H = \bar{H} = \sum_n U_n \tag{5.10.1}$$

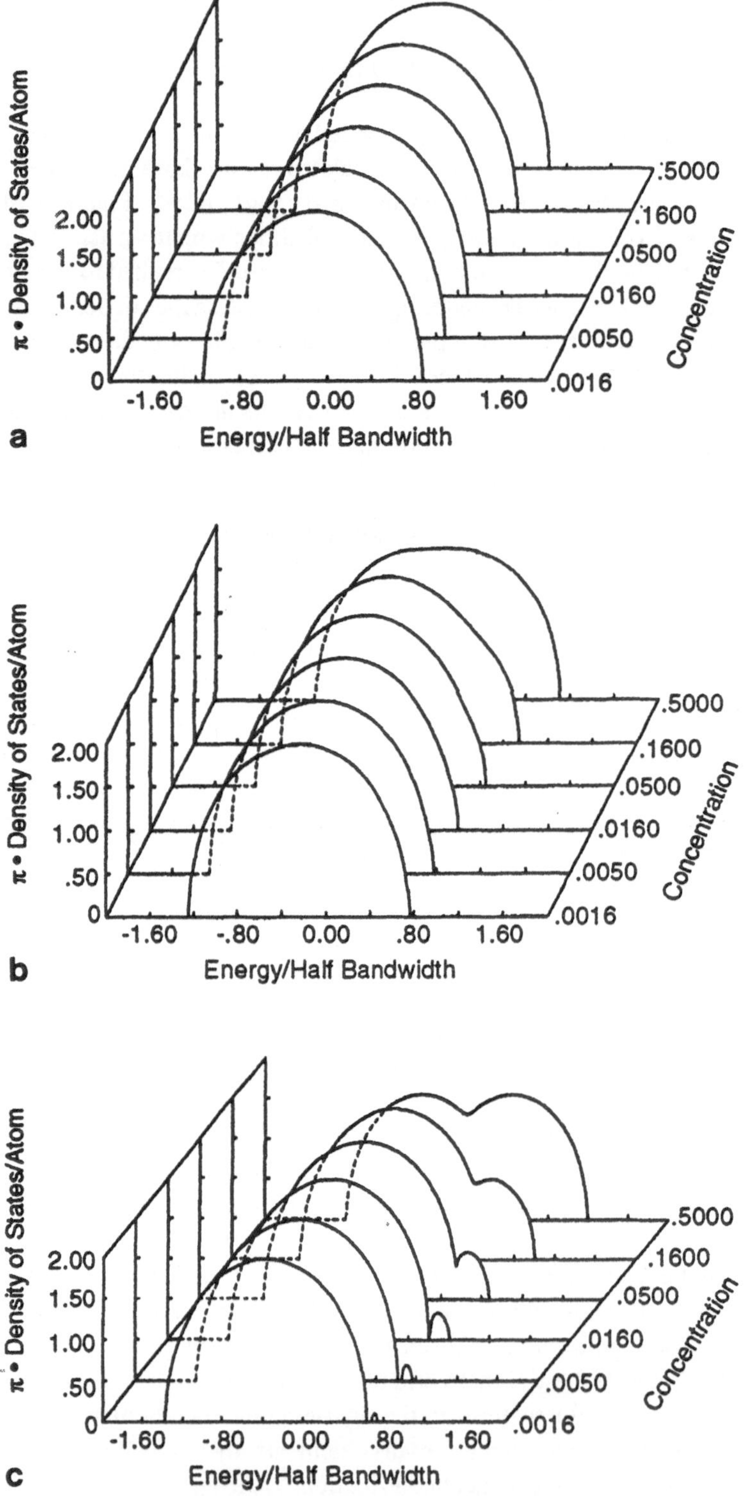

FIGURE 5.14 . Density of states calculated with the model $\rho^{(0)}(E)$ in the coherent potential approximation for a variety of concentrations. (a) $\delta = 0.25$; (b) $\delta = 0.5$; (c) $\delta = 0.75$. (From Velicky *et al.*, 1968.)

$$U_n = \sum \sum_{\alpha\alpha'} |n\alpha\rangle \Delta^n_{\alpha\alpha'} \langle n\alpha'| \tag{5.10.2}$$

Here n is the unit-cell location index and identifies the kind of substitutional atoms (A or B) in that cell; α is the index for the sp^3 hybrids h_α, where α ranges from 1 to 8 as indicated in Fig. 5.15. The 8×8 U_n matrix has the following structure:

$$U_n = \begin{bmatrix} \Delta_0 & \Delta_1 & \Delta_1 & \Delta_1 & \Delta_2 & 0 & 0 & 0 \\ \Delta_1 & \Delta_0 & \Delta_1 & \Delta_1 & 0 & \Delta_2 & 0 & 0 \\ \Delta_1 & \Delta_1 & \Delta_0 & \Delta_1 & 0 & 0 & \Delta_2 & 0 \\ \Delta_1 & \Delta_1 & \Delta_1 & \Delta_0 & 0 & 0 & 0 & \Delta_2 \\ \Delta_2 & 0 & 0 & 0 & 0 & 0 & 0 & 0 \\ 0 & \Delta_2 & 0 & 0 & 0 & 0 & 0 & 0 \\ 0 & 0 & \Delta_2 & 0 & 0 & 0 & 0 & 0 \\ 0 & 0 & 0 & \Delta_2 & 0 & 0 & 0 & 0 \end{bmatrix} \tag{5.10.3}$$

In terms of Harrison's bond-orbital terminology, Δ_0, Δ_1, and Δ_2 are respectively the fluctuations in the hybrid energy ε_h, the metallicity energy V_1, and the covalent energy V_2. If site n is occupied by an A atom, then $\Delta_0 = \varepsilon_h^A - \langle \varepsilon_h \rangle$, where $\langle \varepsilon_h \rangle$ is the concentration weighed average. Similarly, we have $\Delta_1 = V_1^A - \langle V_1 \rangle$ and $\Delta_2 = V_2^A - \langle V_2 \rangle$.

With the scattering potential matrix defined, the IATA method specified by Eq. (5.8.11) then becomes an 8×8 matrix equation, which can be iterated to obtain a converged CPA self-energy. However, the matrix equation can be simplified by symmetry considerations.

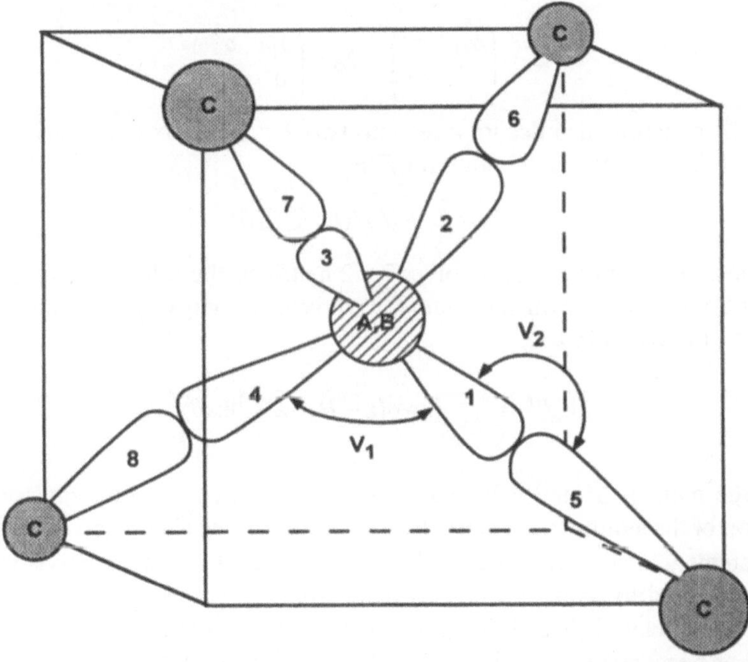

FIGURE 5.15. Sketch of the eight sp^3 hybrids belonging to a unit cell with an alloying atom A or B at the center.

Because of the tetrahedral symmetry in the average Green function, the eight sp^3 hybrid orbitals can be regrouped into states of A_1 and T_2 symmetry. The A_1 state formed by a linear combination of h_1 through h_4 is just the central atom $|s\rangle$ state, and the three T_2 states are simply its $|p_x\rangle$, $|p_y\rangle$, and $|p_z\rangle$ states. The other set of A_1 and T_2 states are formed from h_5 through h_8 in the following linear combinations:

$$|A_1\rangle = (|h_5\rangle + |h_6\rangle + |h_7\rangle + |h_8\rangle)/2$$

$$|T_{2x}\rangle = (|h_5\rangle + |h_6\rangle - |h_7\rangle - |h_8\rangle)/2$$

$$|T_{2y}\rangle = (|h_5\rangle - |h_6\rangle + |h_7\rangle - |h_8\rangle)/2$$

$$|T_{2z}\rangle = (|h_5\rangle - |h_6\rangle - |h_7\rangle + |h_8\rangle)/2 \qquad (5.10.4)$$

In terms of these symmetrized states, the scattering potential matrix becomes a block diagonal matrix of four 2×2 matrices, one for the A_1 symmetry and the other three for the T_2 symmetry. The A_1 block has the form

$$U_s = \begin{pmatrix} \Delta_s & \Delta_2 \\ \Delta_2 & 0 \end{pmatrix} \qquad (5.10.5)$$

where Δ_s is the fluctuation of the s-term value of the central atom, $\Delta_s = \varepsilon_s^{\mathrm{A}} - \langle \varepsilon_s \rangle$. The three T_2 blocks are identical and are similar to U_s, except that Δ_s is replaced by the p-term value fluctuation Δ_p. In these symmetrized local basis, both the site Green function and self-energy matrices are block diagonal in four 2×2 matrices, one for the A_1 symmetry and three for the T_2 symmetry. The 2×2 self-energy matrices for the A_1 and T_2 symmetry, denoted Σ_s and Σ_p, respectively, have the forms

$$\Sigma_s = \begin{pmatrix} \sigma_{s1} & \sigma \\ \sigma & \sigma_{s2} \end{pmatrix}, \quad \Sigma_p = \begin{pmatrix} \sigma_{p1} & \sigma' \\ \sigma' & \sigma_{p2} \end{pmatrix} \qquad (5.10.6)$$

The IATA equation also decomposes into two 2×2 matrix equations for Σ_s and Σ_p, respectively. For example, the iteration for Σ_s is

$$\Sigma_s = \tilde{\Sigma}_s + \langle \tilde{t}_s \rangle / (I + \tilde{g}_s \langle \tilde{t}_s \rangle) \qquad (5.10.7)$$

Although the above iteration only involves 2×2 matrices, the calculation of $\tilde{g}_s$ requires the inversion of 8×8 matrices which contain the whole self-energy operator and a summation over $\mathbf{k}$ in the first Brillouin zone:

$$g_s^{\alpha\beta} = \frac{1}{N} \sum_k \langle \mathbf{k} s \alpha | (z - \overline{H} - \tilde{\Sigma})^{-1} | \mathbf{k} s \beta \rangle \qquad (5.10.8)$$

Here N is the number of unit cells and $|\mathbf{k}\, s\alpha\rangle$ is the Bloch function constructed from a superposition of the symmetrized states $|ns\alpha\rangle$ over the lattice sites as given by Eq. (5.2.3).

For systems which contain heavy elements such as HgTe, the relativistic corrections have to be included to produce a correct band structure description. In the tight-binding model, the scalar relativistic s-shift can be incorporated in the term values. The spin-orbit interaction can be represented by site-diagonal interactions such as those between the spin-up p_x and the spin-down p_y states of the same atom

$$\lambda_j = \langle jp_x\uparrow|H_{so}|jp_y\downarrow\rangle \tag{5.10.9}$$

where j indicates the cation or anion on the site. In other words, each atom only requires a parameter for the spin-orbit coupling. The inclusion of spins doubles the size of the TB Hamiltonian. However, the CPA calculation so far developed is not overly complicated. The 16×16 matrices for the site scattering potential, Green function, and self-energy are again block diagonal into 2×2 matrices: two for the Γ_6 symmetry, two for Γ_7, and four for Γ_8. These symmetry states are the linear combinations of the A_1 and T_2 states with spins. For example, from the s, p_x, p_y, and p_z states of the central atom the following symmetry states can be constructed:

$$
\begin{aligned}
|\Gamma_6^\alpha\rangle &= |s\uparrow\rangle \\
|\Gamma_6^\beta\rangle &= |s\downarrow\rangle \\
|\Gamma_7^\alpha\rangle &= (|p_x\uparrow\rangle - i|p_y\uparrow\rangle)/\sqrt{3} - |p_z\downarrow\rangle/\sqrt{3} \\
|\Gamma_7^\beta\rangle &= (|p_x\downarrow\rangle + i|p_y\downarrow\rangle)/\sqrt{3} + |p_z\uparrow\rangle/\sqrt{3} \\
|\Gamma_8^\alpha\rangle &= (|p_x\uparrow\rangle + i|p_y\uparrow\rangle)/\sqrt{2} \\
|\Gamma_8^\beta\rangle &= (|p_x\downarrow\rangle + i|p_y\downarrow\rangle)/\sqrt{6} - \sqrt{2/3}\,|p_z\uparrow\rangle \\
|\Gamma_8^\gamma\rangle &= (|p_x\uparrow\rangle - i|p_y\uparrow\rangle)/\sqrt{6} + \sqrt{2/3}\,|p_z\downarrow\rangle \\
|\Gamma_8^\mu\rangle &= (|p_x\downarrow\rangle - i|p_y\downarrow\rangle)/\sqrt{2}
\end{aligned}
\tag{5.10.10}
$$

The Greek superscripts label different partners of the same symmetry. The other sets of the Γ_6, Γ_7, and Γ_8 symmetry states can be obtained from the A_1 and T_2 states of Eq. (5.10.4) using the same linear combinations as in Eq. (5.10.10). These 16 basis functions with respect to a given site A will be referred to as the Γ-symmetrized orthogonalized local orbital (Γ-OLO) centered at A.

Since Green functions plays an important role in alloy calculations and their matrix elements for different zinc blende semiconductors have similar energy dependences, it is useful to examine the Green function (GF) matrix elements in these symmetrized local bases for a typical semiconductor. As mentioned, the GF in these site-centered basis are block diagonal into eight 2×2 matrices. Each 2×2 GF matrix g, for a given representation, has the structure

$$g = \begin{pmatrix} g_1 & g_2 \\ g_3 & g_4 \end{pmatrix} \tag{5.10.11}$$

with $g_2 = g_3$. Figures 5.16 through 5.19 show plots of these GF matrix elements for GaAs in the Γ-OLO centered at an As site.* Those in Fig. 5.16 are in the Γ_6 representation, and those in Fig. 5.16 are in Γ_8. The Γ_7 representation is not presented, because the GF matrix elements as functions of energy are similar to those in Γ_8. For comparison, Figs. 5.18 and 5.19 show

* Remember the GF matrix elements are complex functions $g_j(E - i0)$ of energy biased slightly below the real axis in the convention being used here. Also keep in mind when viewing Figs. 5.15 to 5.18 that the density of states is proportional to Im $g_j(E - i0)$.

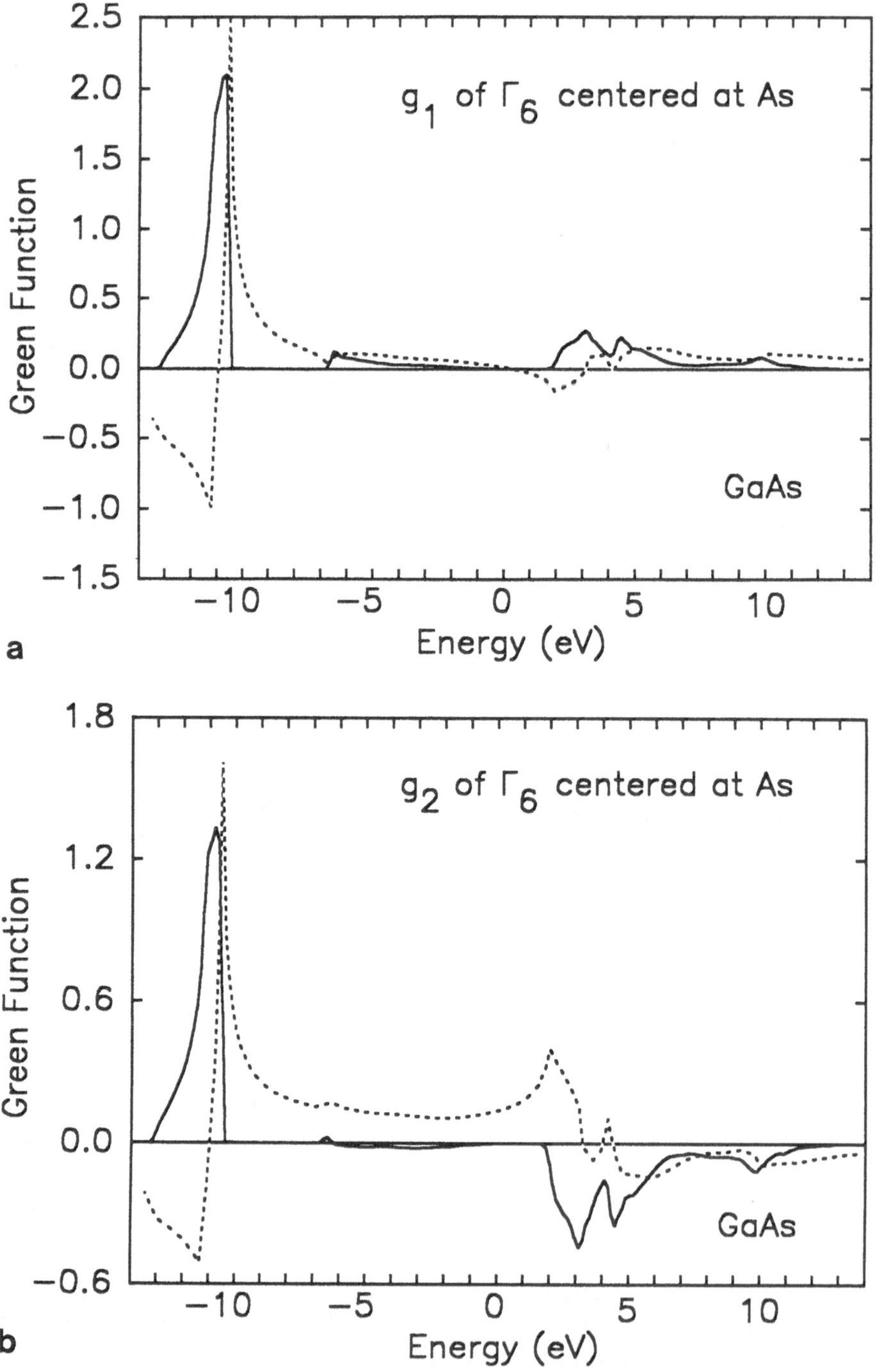

FIGURE 5.16a,b,c. The GF matrix elements defined in Eq. (5.10.11) in the Γ_6 representation centered at an As site. Dashed curves: real part. Solid curves: imaginary part.

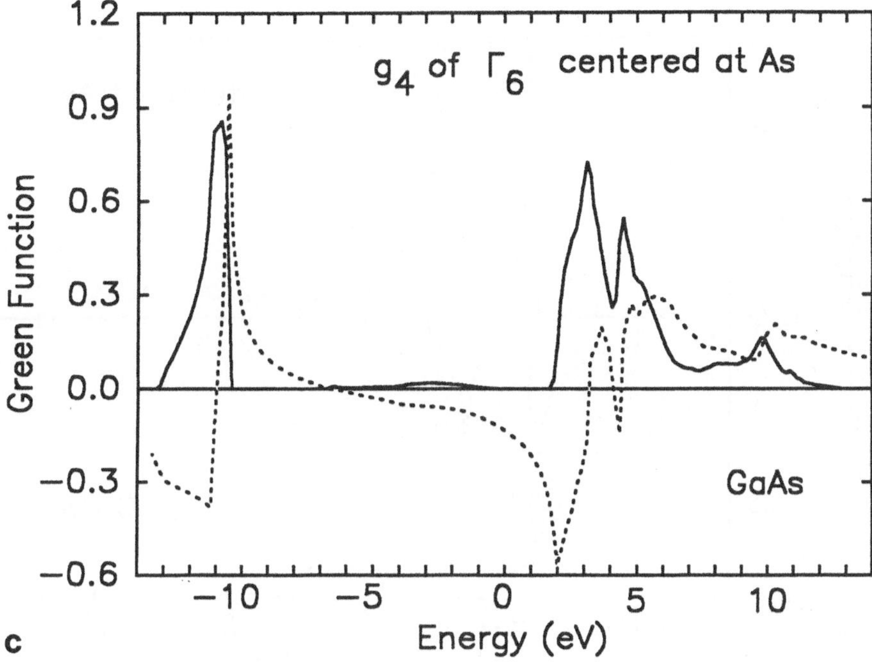

FIGURE 5.16. (continued)

the GF matrix elements in the Γ-OLO basis centered at a Ga site. These GF matrix elements are calculated from a band structure to be discussed in Chapter 7 (Fig. 7.2) using the formula

$$g_{\alpha\beta}(E - i\text{o}) = \frac{1}{N} \sum_{n} \sum_{\mathbf{k}} \frac{C_{\alpha}(n\mathbf{k})C_{\beta}^{*}(n\mathbf{k})}{E - i\text{o} - \varepsilon_{n}(\mathbf{k})} \tag{5.10.12}$$

where $\varepsilon_n(\mathbf{k})$ is a band energy and $C_\alpha(n\mathbf{k})$ is the expansion coefficient of the eigenstate in the Bloch basis constructed from Γ-OLO of type α. In the actual computation, the imaginary part (the solid curves) is calculated first from Eq. (5.10.12) by replacing the factor $1/(E - i\text{o}\, \varepsilon)$ with $\pi\delta(E - \varepsilon)$. The real parts (the dashed curves) are then calculated from the imaginary part using an energy convolution similar to Eq. (5.9.7). These plots are intended to show the general features of g in a zinc blende semiconductor. The sharp structures are associated with Van Hove (1953) singularities which are intrinsic to any density-of-state function in a perfect crystal. The potential and structural disorder in an alloy tend to dampen these sharp structures. However, the qualitative behavior to be discussed below is the same.

Recall that the imaginary part of the Green function is directly related to the density of states (DOS). In the present case, the diagonal matrix elements g_1 and g_4 are related to the component DOS by

$$\rho\alpha(E) = \text{Im}\, g_{\alpha\alpha}(E - i\text{o})/\pi \tag{5.10.13}$$

In the notation of Eq. (5.10.11), the component DOS associated with g_1 and g_4 are designated as ρ_1 and ρ_4, respectively. Specifically, ρ_1 of the Γ_6 representation centered at an As site (see Fig. 5.16a) contains only the pure As s-states, while ρ_1 of Γ_8 centered at As contains only the As p-states. Similarly, ρ_1 of the Γ_6 representation centered at a Ga site in Fig. 5.18a

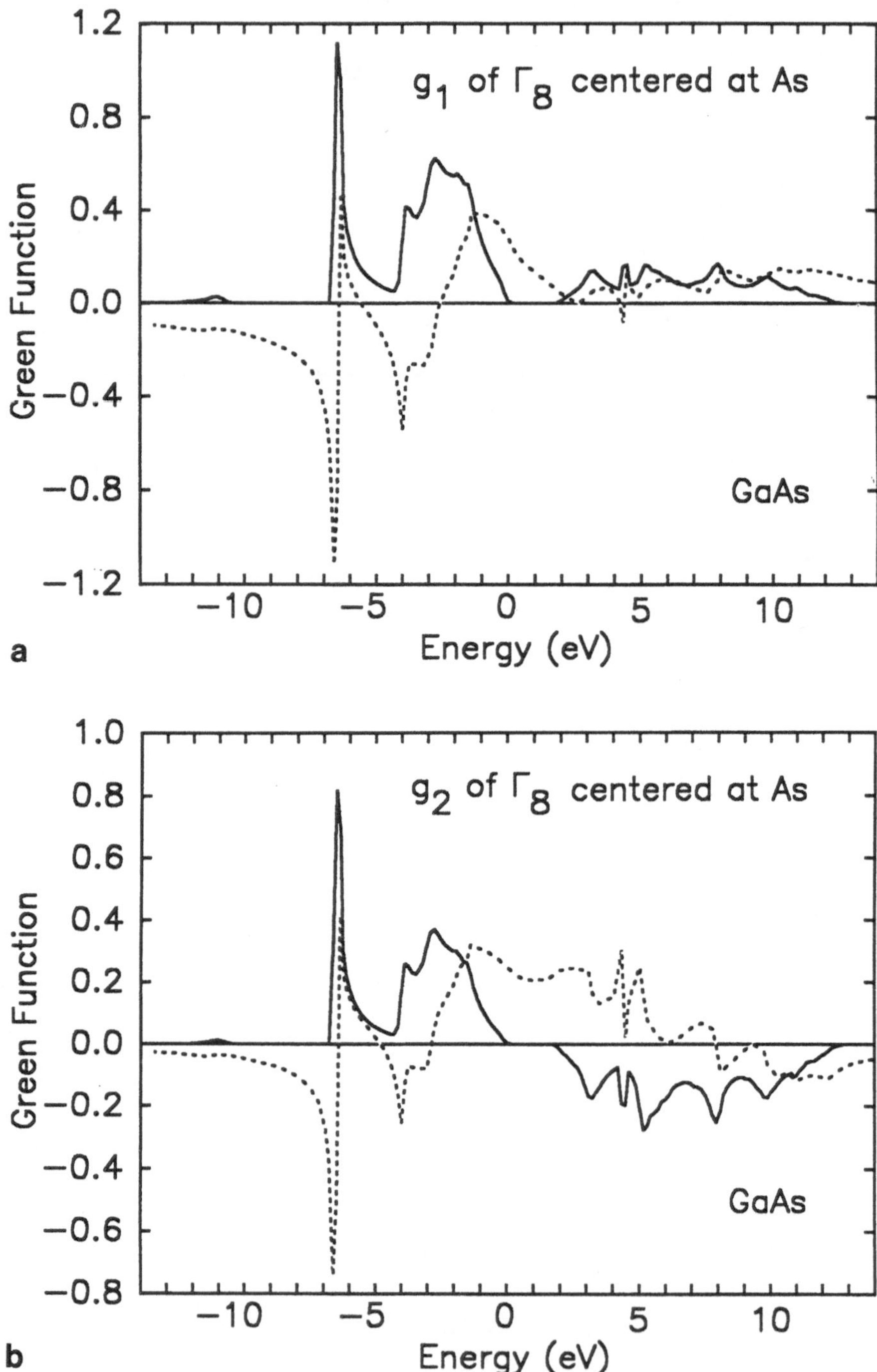

FIGURE 5.17a,b,c. The GF matrix elements defined in Eq. (5.10.11) in the Γ_8 representation centered at an As site. Dashed curves: real part. Solid curves: imaginary part.

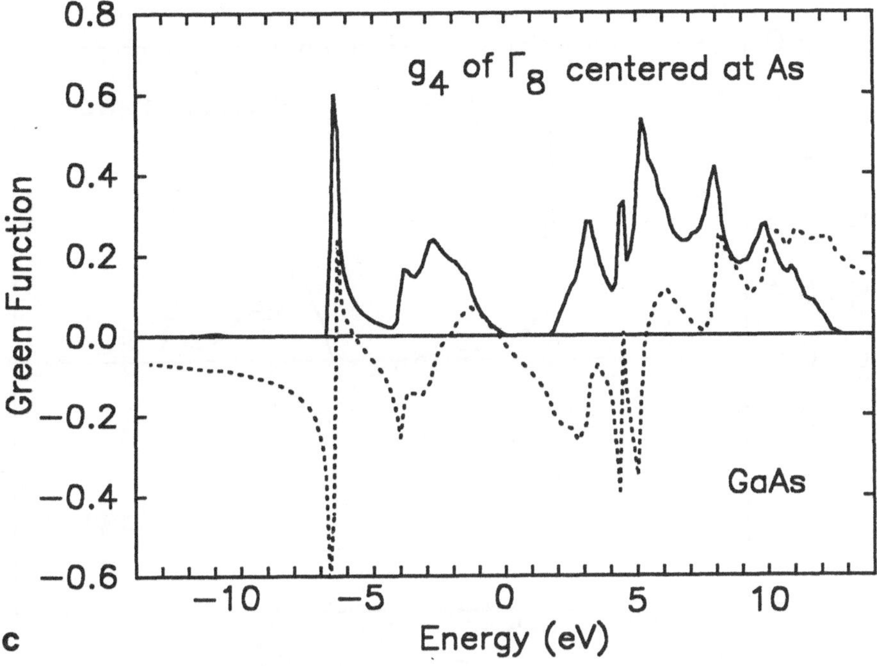

FIGURE 5.17. (continued)

contains only the pure Ga s-states, and ρ_1 of Γ_8 centered at Ga contains only the Ga p-states. The majority of the anion s-states reside in the low-valence bands below −10 eV. The rest are in the conduction bands. There is essentially no contribution from these states to the major valence bands between 0 and −6 eV. In contrast, the cation s-states (Fig. 5.18a) reside more in the conduction bands than in the valence bands. Those in the valence bands form a sharp structure at the bottom of the major valence DOS (at −6.0 eV). While most of the anion (As) p-states (solid curve in Fig. 5.17a) are located in the major valence bands (from −6 to 0 eV), they also make an appreciable contribution to the conduction-band states. The Ga p-states (Fig. 5.19a) are seen to spread over all the bands. All the component DOS $\rho_4 = \mathrm{Im}(g_4)/\pi$ contain both the s- and p-states of the four first-neighbor atoms surrounding the central atom in question. The shape of ρ_4 is similar to that of ρ_1 in a given representation, but the division of states between valence and conduction bands is different. The imaginary parts of the off-diagonal GF matrix elements (i.e., $\mathrm{Im}(g_2)$) also have shapes that can be identified with ρ_1 and ρ_4 in the same representation. The difference is that $\mathrm{Im}(g_2)$ becomes negative in the conduction-band energy range. This change of signs occurs because the product $C_\alpha C_\beta^*$ in Eq. (5.10.12) tends to be positive for a valence-band state (the bonding state) and negative for the conduction-band states (the antibonding states). Finally, the real parts also have singular structures which are directly related to the sharp structures of the imaginary parts. The real part of the Green function is intimately related to the alloy disorder–induced shift of the energy spectra, and these results will be used in the next section to discuss band-gap bowing.

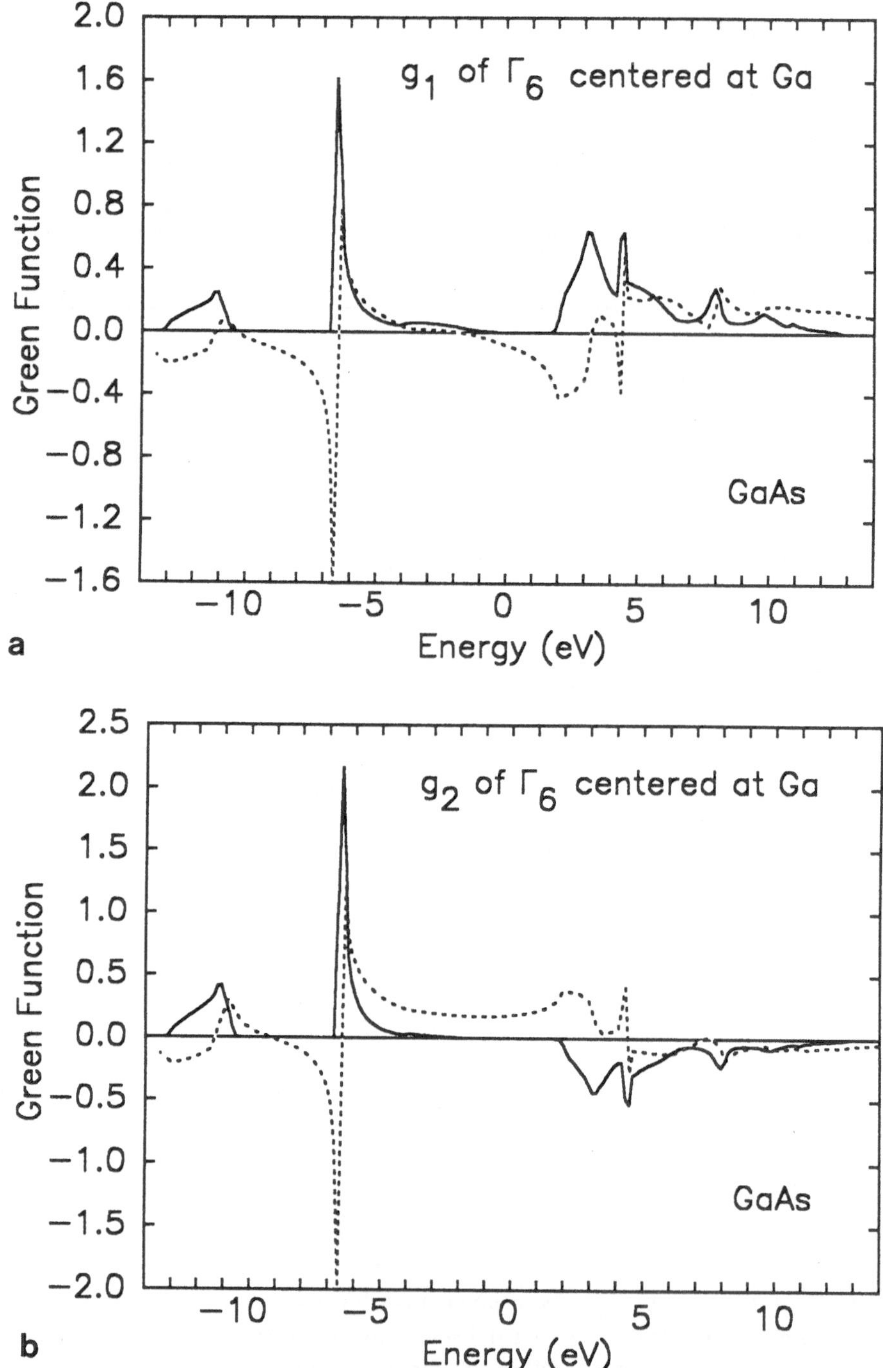

FIGURE 5.18a,b,c. The GF matrix elements defined in Eq. (5.10.11) in the Γ_6 representation centered at a Ga site. Dashed curves: real part. Solid curves: imaginary part.

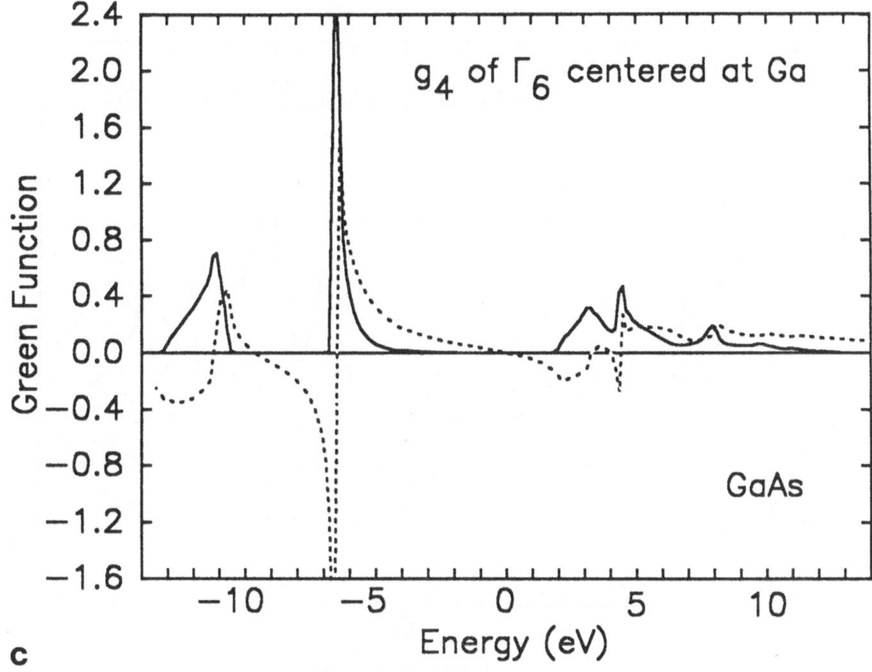

FIGURE 5.18. (continued)

Now we return to molecular CPA. Each of the 2×2 self-energy matrices belonging to each partner of a given representation also has a structure similar to the site Green function g in Eq. (5.10.11). An IATA iteration similar to that in Eq. (5.10.7) can be applied to this case. Once the self-energy is obtained, the spectral density of states

$$A(\mathbf{k}, E) = -\frac{1}{\pi} \sum_{\alpha} \langle \mathbf{k}\alpha | (E + i0 - \overline{H} - \Sigma)^{-1} | \mathbf{k}\alpha \rangle \tag{5.10.14}$$

can be calculated for a given $\mathbf{k}$ to study the energy spectra and alloy scattering lifetime.

A good example of such an application is the calculation for $Hg_{1-x}Cd_xTe$ by Hass *et al.* (1984b), using a second-neighbor tight-binding Hamiltonian, but with the alloy disorder limited to diagonal and first-neighbor matrix elements. Figure 5.20 shows plots from their calculation of the spectral density functions as functions of E for several $\mathbf{k}$ along the (100) and (111) directions for the $Hg_{0.7}Cd_{0.3}Te$ alloy. The VCA band structures are also plotted. The spectral density functions have sharp Lorentzian line shapes for states near the conduction- and valence-band edges. The spectra are broadened more for the states deep inside the valence and conduction bands. States at energies around 4 eV and –5 eV show a strong deviation from a simple line-shape function. These results are caused by a large difference in the s-term values between Cd and Hg atoms due mainly to the large scalar relativistic s-shift in Hg (Spicer *et al.*, 1982).

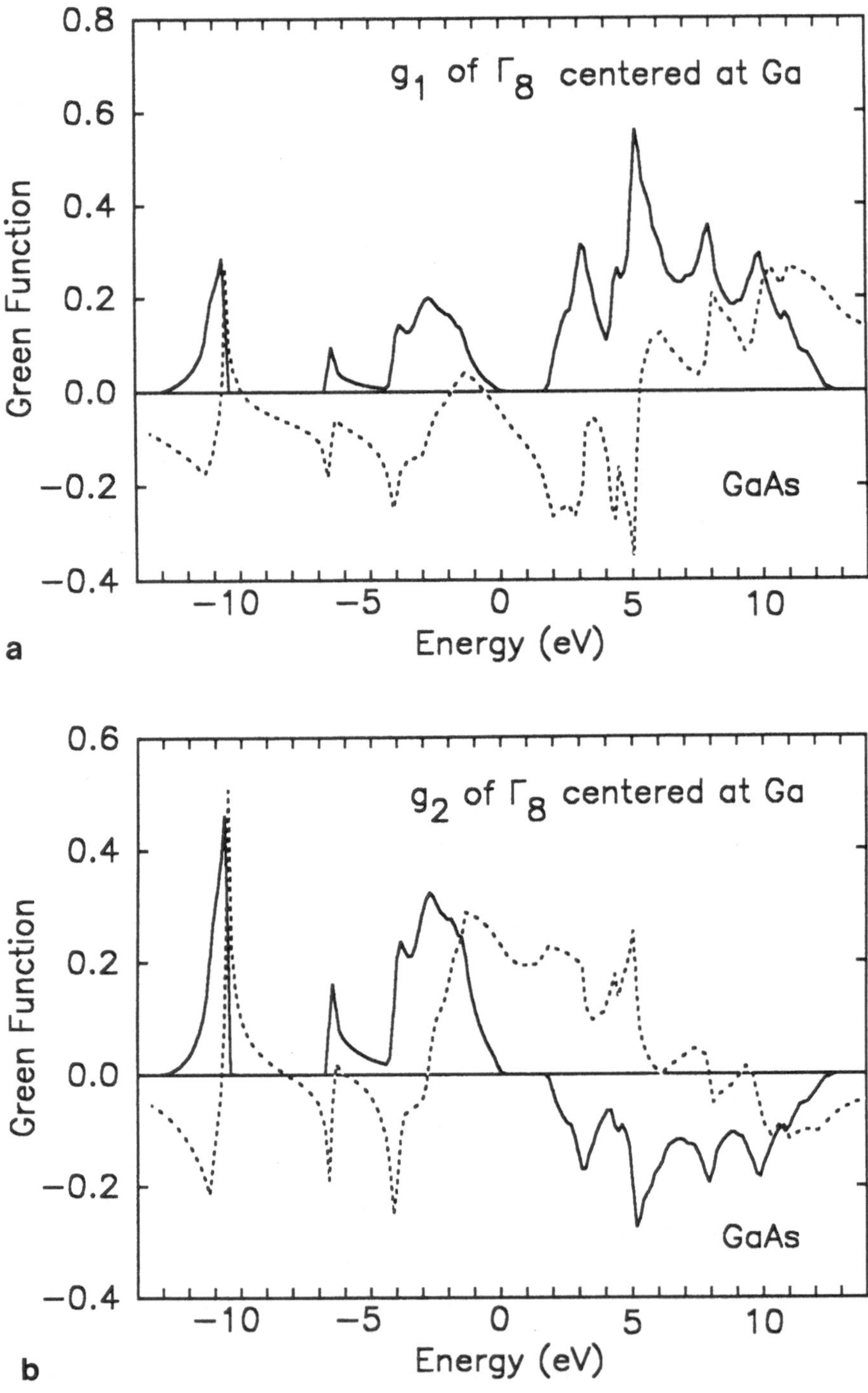

FIGURE 5.19a,b,c. The GF matrix elements defined in Eq. (5.10.11) in the Γ_8 representation centered at a Ga site. Dashed curves: real part. Solid curves: imaginary part.

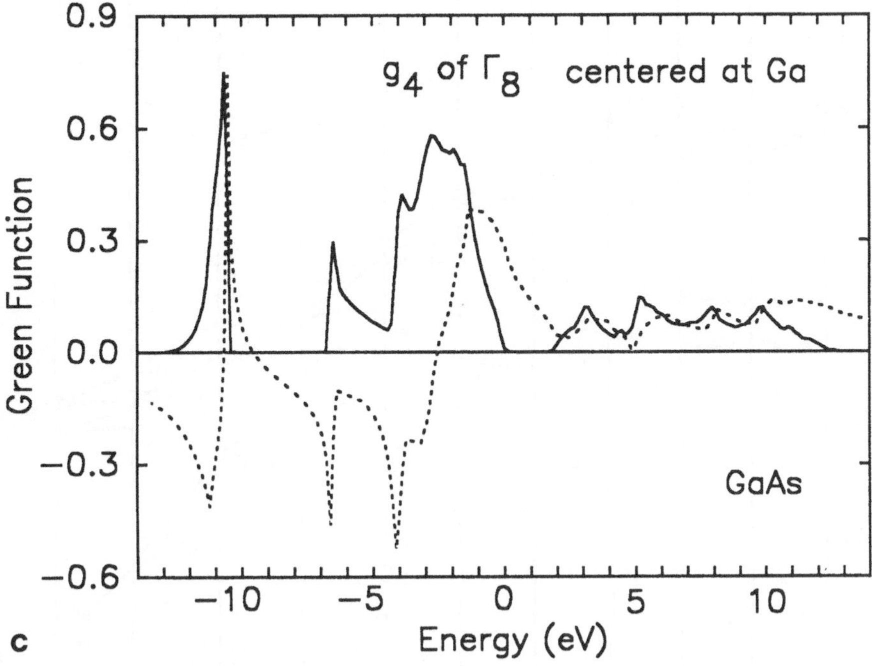

FIGURE 5.19. (continued)

5.11. EFFECTS OF DIAGONAL AND OFF-DIAGONAL DISORDER ON BAND-EDGE PROPERTIES

Molecular CPA theory developed in Section 5.10 will be used in Chapter 7 for a detailed calculation of band structures. The CPA calculations are time consuming, because the Green function, as indicated in Eq. (5.10.8), requires the inversion of a 16×16 matrix at every E and every $\mathbf{k}$ in each iteration toward a converged self-energy in Eq. (5.10.7). On the other hand, the perturbation theory (PT) discussed in Section 5.7 can, when it is valid, significantly reduce the computational time and provide better insight into the effects of alloy disorder. In this section, we shall demonstrate that perturbation theory not only gives a qualitative understanding but also provides accurate values for the energy states near band gaps—the most important states in semiconductors.

Using Eq. (5.7.3), the notations in the preceding section, and perturbation theory, the 2×2 matrices for the site self-energy in the Γ_6, Γ_7, and Γ_8 representations take the form

$$\Sigma = x(1-x)(U_B - U_A)g(U_B - U_A) \tag{5.11.1}$$

where the potential difference has a typical form

$$U_B - U_A = \begin{pmatrix} \delta & \Delta \\ \Delta & 0 \end{pmatrix} \tag{5.11.2}$$

For example, for the Γ_6 representation, $\delta = \varepsilon_s^B - \varepsilon_s^A$ and Δ is the difference $\Delta = V_2^B(\Gamma_6) - V_2^A(\Gamma_6)$ between the V_2 coupling of the $|A_1\rangle$ states given by Eq. (5.10.4) when an A or B atom is located at the central site. If the first-neighbor interaction scale as an inverse

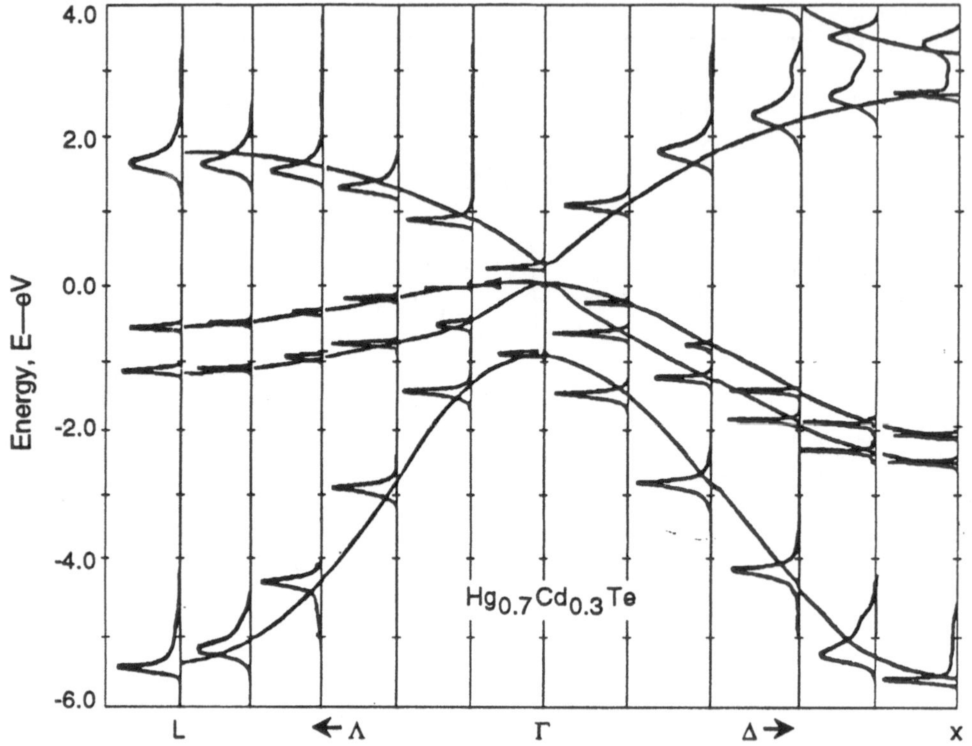

FIGURE 5.20. Band structure shown as smooth solid curves with line shapes at selected K points superimposed.

power of the bond length, then Δ is negative if $d_A > d_B$ and positive if $d_A < d_B$, because the value of $V_2(\Gamma_6)$ in our notation is a negative number. Table 5.2 contains the values of δ and Δ in both the Γ_6 and Γ_8 representations for several III–V and II–VI semiconductor alloys (Krishnamurthy *et al.*, 1988). These parameters are derived from a band model similar to that to be discussed in detail in Chapter 7.

The 2×2 self-energy matrices of Γ_6, Γ_7, and Γ_8 symmetries can be written as

$$\Sigma = \begin{pmatrix} \sigma_1 & \sigma_2 \\ \sigma_3 & \sigma_4 \end{pmatrix} \tag{5.11.3}$$

For cation-substituted alloys with moderate s-term value differences, the most important part of the self-energy is σ_1 in the Γ_6 representation, which will affect the cation $|s\rangle$ dominated E_c. For an anion-substituted alloy with appreciable fluctuations in the p-term values, such as $InP_{1-x}As_x$, in addition to the Γ_6 self-energies, σ_1 in the Γ_8 representation is also important, as it governs the anion p-dominated E_v. If the constituent compounds have an appreciable bond-length difference, then σ_2 and σ_4 are also important. By direct comparison between CPA and perturbation calculations, we can assess the accuracy of the latter.

Figures 5.21a,b compare CPA and PT calculations of the real and imaginary self-energy elements σ_1 and σ_4 of the Γ_6 representation for the cation-substituted alloy $Ga_{0.5}In_{0.5}As$. We note that this alloy has both moderate diagonal and off-diagonal disorder. We see that PT and CPA agree well for energies within 1 eV of the band edges, and the difference between the two calculations starts to grow beyond this range. The agreements between PT and CPA for other components, which are not shown, are comparable. Figures 5.22a,b show the comparison between CPA and PT for the σ_1 and σ_4 elements in the Γ_8 representation for the anion-substituted alloy $InP_{0.5}As_{0.5}$. The agreement once again is excellent over the energy range presented, because the alloy scattering is not as strong as that in $Ga_{0.5}In_{0.5}As$. Although these calculations used the Green functions based on a specific set of band structures (to be discussed in Chapter 7), the results are rather independent of the band models, so long as they have the correct band gaps and bandwidths. Judging from these comparisons and from the scattering parameters in Table 5.3, we can conclude that PT is accurate when applied to the states near the band edges. This result is very important, because these are the states that govern the electrical and transport properties in most semiconductor devices.

Now that PT has been validated for the band-edge states, we can use it to describe two important band-edge properties: band-gap bowing and scattering rates of band-edge states due to alloy disorder. When an alloy has both diagonal and off-diagonal disorder, we might suppose that these two disorder types reinforce each other in their effect on bowing and alloy scattering. However, it was first pointed out by Hass *et al.* (1984a) that the off-diagonal disorder reduces the bowing parameter caused by the diagonal disorder in $Ga_{1-x}In_xAs$. We will show that these two disorder types can be constructive or destructive, depending on the signs of the disorder parameters and on the states considered.

First, let us consider the shift of the conduction-band edge, δE_c, induced by alloy disorder. To calculate δE_c we need to know the nature of the state at the bottom of the conduction band in the virtual crystal approximation:

$$|\Gamma_6^c\rangle = c|c\rangle + a|a\rangle \tag{5.11.4}$$

where $|c\rangle$ and $|a\rangle$ are the $\mathbf{k} = 0$ Bloch basis functions constructed from the Γ_6 local orbitals of the cation and anion, respectively. In PT, the shift is given by $\delta E_c = \mathrm{Re}\langle\Gamma_6^c|\Sigma|\Gamma_6^c\rangle$. The part of Σ entering this calculation is the Γ_6 representation evaluated at the VCA conduction-band edge. Using Eqs. (5.11.1) through (5.11.4), we obtain

TABLE 5.3. Diagonal, δ, and Off-Diagonal, Δ, Scattering Parameters for Several III–V and II–VI Semiconductor Alloys, $A_xB_{1-x}C^a$

	Γ_6 representation		Γ_8 representation	
Alloy	δ	Δ	δ	Δ
		Cation-substituted alloys		
$Ga_xIn_{1-x}As$	0.89	1.05	0.29	0.55
$Ga_xIn_{1-x}Sb$	1.25	1.06	0.12	0.26
$Hg_xCd_{1-x}Te$	1.44	−0.02	0.18	−0.40
$Hg_xZn_{1-x}Te$	1.00	−0.52	0.25	−0.71
		Anion-substituted alloys		
$InP_{1-x}As_x$	0.58	−0.26	−0.47	0.35
$GaAs_xSb_{1-x}$	1.50	1.20	0.07	0.56
$InAs_xSb_{1-x}$	1.26	1.08	−0.09	0.31

[a]We define $\delta = \varepsilon^B - \varepsilon^A$ and $\Delta = V_2^B - V_2^A$, where ε are the diagonal term values and V_2 are the covalent matrix elements. All energies are in eV.

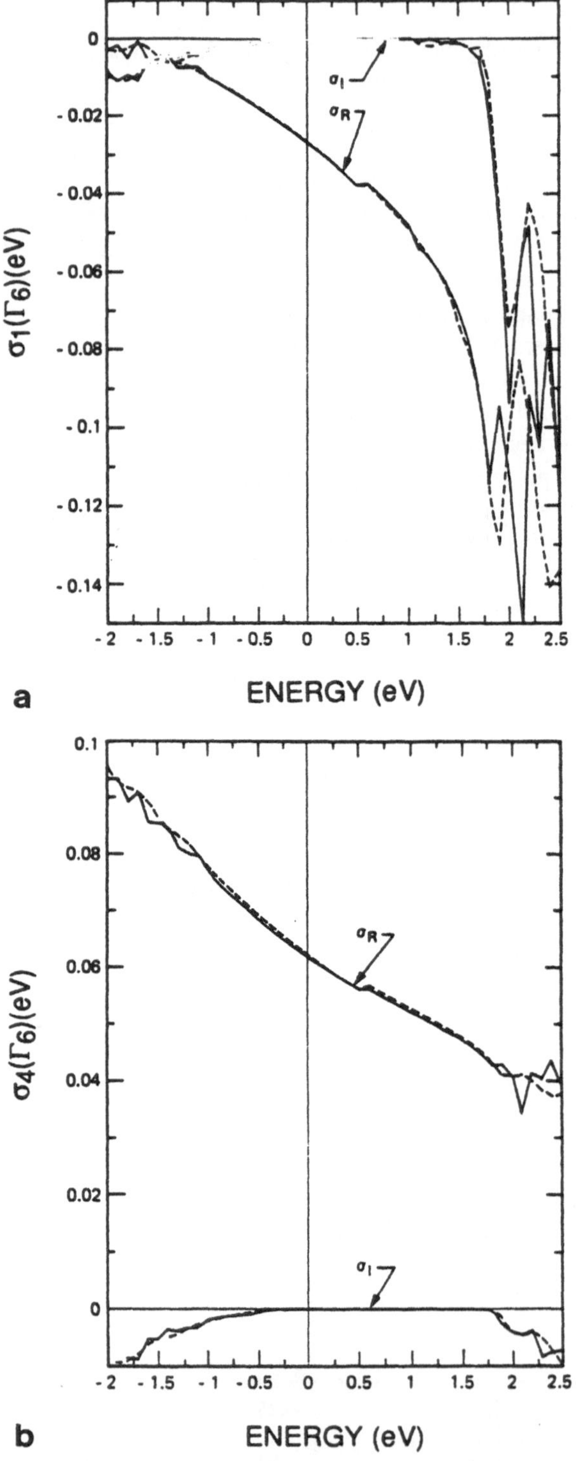

FIGURE 5.21. Comparison between PT (solid curve) and CPA (dashed curve) for $Ga_{0.5}In_{0.5}As$. (a) Diagonal cation self-energy σ_1 and (b) diagonal anion self-energy σ_4 for the Γ_6 representation (Krishnamurthy *et al.*, 1988).

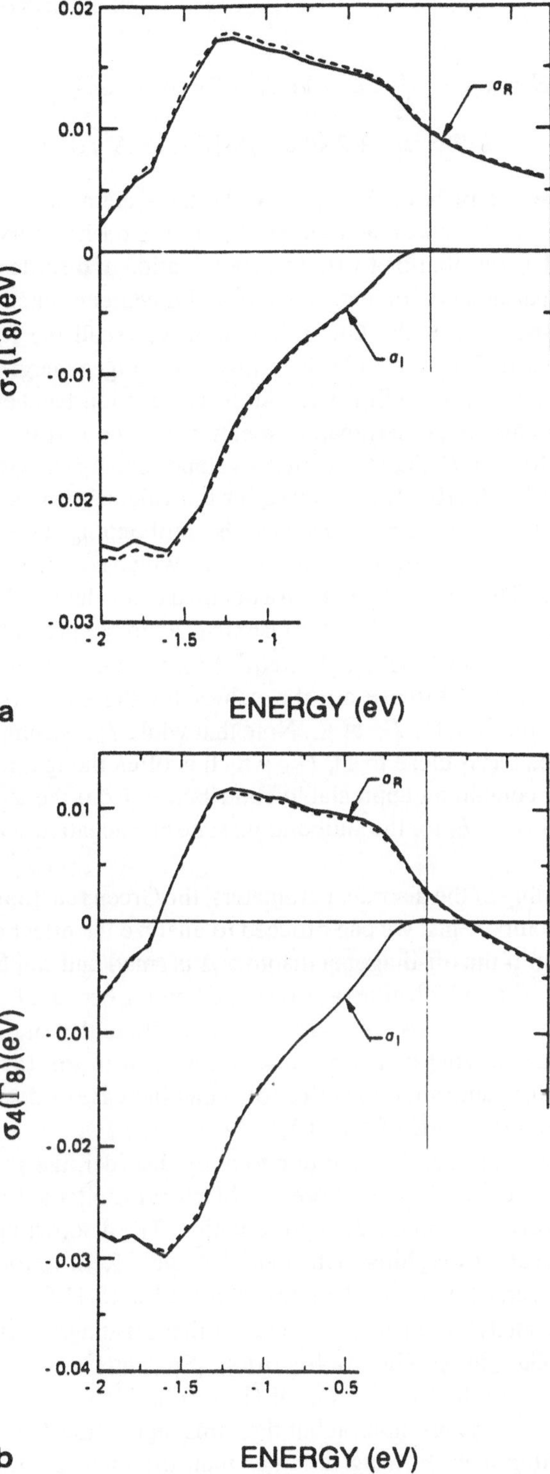

FIGURE 5.22. Comparison between SPT (solid curve) and CPA (dashed curve) for $InAs_{0.5}P_{0.5}$. (a) Diagonal cation self-energy σ_4 and (b) diagonal anion self-energy for the Γ_8 representation (Krishnamurthy *et al.*, 1988).

$$\delta E_c = x(1-x)\{[|c|^2\delta^2 + |a|^2\Delta^2 + 2\mathrm{Re}(a^*c)\,\Delta\delta]f_1$$
$$+ [2|c|^2\,\Delta\delta + 2\mathrm{Re}(a^*c)\Delta^2]f_2 + |c|^2\Delta^2 f_4\} \tag{5.11.5}$$

where f_1, f_2, and f_4 are respectively the real part of the three elements g_1, g_2, and g_4 of the VCA Green function in the Γ_6 representation, similar to the dashed curves shown in Fig. 16a. Note that the disorder contribution to δE_c is a combination of disorder parameters δ and Δ, the wave function expansion coefficients a and c, and the Green function g.

To appreciate the structure of the Green function, we recall the form of the Green function matrix elements in Eq. (5.10.12) in terms of the eigenenergies $\varepsilon_n(\mathbf{k})$ and their eigenvectors and the dashed curves in Fig. 5.18. At the conduction-band edge f_1 is negative, because the conduction band is predominantly a cation s-state, and the conduction-band states are substantially closer to E_c than those in the valence band. f_2 is positive, because the sum in the numerator of Eq. (5.10.12) is positive for the valence-band states (the bonding states) and negative for the conduction-band states (the antibonding states). For f_4 the anion s-states reside mainly at the bottom of the valence bands, which compete with the fewer but closer states just above E_c. This value depends on a detailed cancellation. While f_4 is negative in Fig. 5.18 for GaAs, it is slightly positive for $Ga_{0.5}In_{0.5}As$ in the present model.

Table 5.4 lists the values of $|c|^2$, $|a|^2$, and $2\mathrm{Re}(a^*c)$ for the same alloys as those in Table 5.3. For cation-substituted alloys these are the values for the state at E_c (Γ_{6c}). For the anion-substituted alloys, they are for Γ_{8v} at E_v. Note that while Γ_{8v} is completely dominated by the anion p-states (i.e., $|a|^2$ is close to 1), Γ_{6c}, which is often thought to be derived from the cation s-states, does contain an appreciable contribution from the anion s-states. Also note that $2\mathrm{Re}(a^*c)$ is positive at E_c for the antibonding state and negative at E_v for the bonding state.

With this understanding of the disorder parameters, the Green function matrix elements, and the wave function components, we can proceed to analyze the effect of disorder on the conduction-band bowing. If the off-diagonal disorder Δ is small and can be neglected, then δE_c reduces to $x(1-x)|c|^2\,\delta^2 f_1$, which gives a downward bowing because f_1 is negative. When the off-diagonal disorder Δ is not zero and its sign is opposite to that of δ, the effect of Δ is to increase the downward bowing of E_c, because every term in Eq. (5.11.5) is negative. Finally, if Δ and δ have the same sign, the effect of Δ can increase or decrease the bowing, depending on the relative magnitude of Δ and δ.

To calculate the total band-gap bowing due to alloy disorder, the shift of E_v, denoted δE_v, also has to be considered. δE_v can be expressed in a form similar to Eq. (5.11.5), except that all the quantities involved are in the Γ_8 representation. The disorder parameters δ and Δ in Table 5.2, the wave function weighing factors $|a|^2$, $|c|^2$, and $2\mathrm{Re}(a^*c)$ for the anion-substituted alloys in Table 5.3 can be used to calculate δE_v from Eq. (5.11.5).

Based on the above analysis, it can be concluded that off-diagonal disorder decreases the band-gap bowing in $Ga_{1-x}In_xAs$, $Ga_{1-x}In_xSb$, $InAs_{1-x}Sb_x$, and $GaAs_{1-x}Sb_x$. This result for $Ga_{1-x}In_xAs$ is consistent with the first finding of Hass *et al.* Note that $InAs_{1-x}Sb_x$ δ and Δ have different signs in the Γ_8 representation but the same sign in the Γ_6 representation. The decrease of the conduction-band bowing is larger than the increase in the valence-band bowing, and a net decrease in bowing due to the off-diagonal disorder is obtained in this case. For $Hg_{1-x}Cd_xTe$, $Hg_{1-x}Zn_xTe$, and $InP_{1-x}As_x$, δ and Δ have different signs and off-diagonal disorders in both Γ_6 and Γ_8 representations reinforce each other to increase the total

TABLE 5.4. Probability Densities $|a|^2$ and $|c|^2$ and Cross-Product $2\mathrm{Re}(a^*c)$ at the Bottom of the Conduction Band of the Cation-Substituted Alloys and at the Top of the Valence Band for the Anion-Substituted Alloys

Alloy	$\|a\|^2$	$\|c\|^2$	$2\mathrm{Re}(a^*c)$
$Ga_xIn_{1-x}As$	0.27	0.73	−0.88
$Ga_xIn_{1-x}Sb$	0.44	0.56	−0.99
$Hg_xCd_{1-x}Te$	0.15	0.85	−0.65
$Hg_xZn_{1-x}Te$	0.15	0.85	−0.72
$InP_{1-x}As_x$	0.99	0.01	0.17
$GaAs_xSb_{1-x}$	0.94	0.06	0.48
$InAs_xSb_{1-x}$	0.96	0.04	0.38

bowing of the gap. It should be noted that the above is aimed at a qualitative discussion of the combined diagonal and off-diagonal order on the bowing parameters b. The size of b depends on the details of band structures and disorder parameters, as will be shown in Chapter 7.

Using the uncertainty principle, the scattering rate R_a of an electron caused by the potential disorder in an alloy is related to the broadening of the energy spectrum, which is given by the imaginary part of the self-energy. Thus, the scattering rate is given by $R_a = -2\ \mathrm{Im}\ \langle\psi|\Sigma(E + i0)|\psi\rangle/\hbar$, where $|\psi\rangle$ is the energy eigenfunction in VCA. With Σ given by Eq. (5.11.1) in the PT and disorder parameters taken from Table 5.2, one can analyze effects of diagonal and off diagonal alloy disorder on the scattering rate for any state using wave functions obtained from VCA. One thing which is clear from the imaginary parts of the self-energies Figs. 5.20 and 5.21 is that the alloy scattering rate is weak for the band-edge states but becomes more important for states away from band edges. However, the effects of alloy scattering on transport depend on details of band structures. A discussion of these effects will be the subject of the next chapter.

5.12. CONCLUDING REMARKS

In this chapter a string of concepts underlying band structure theory is discussed, including formation of bands, band structure methods, features important in semiconductors, Green function and perturbation treatments of alloy disorder, the molecular coherent potential approximation (MCPA), and disorder effects on band structures, particularly band-gap bowing. When our emphasis is on the states near the gap, the fundamental problem of representing many-body excitations by single-particle states with an effective potential is the same whether we have pure semiconductors or disordered alloys. The difference arises from the additional difficulties in diagonalizing the Hamiltonian in disordered alloys.

MCPA is a reasonable approximation within the TB model when alloy disorder is contained in the site diagonal and the first-neighbor off-diagonal matrix elements of the Hamiltonian. This method has been incorporated into an accurate TB Hamiltonian for a systematic calculation of band structures of III–V and II–VI alloys. The calculation and the detailed results are the subject of Chapter 7.

Very little has been said about *ab initio* band theory for disordered semiconductors, because this is a really challenging problem. First one has to obtain an effective potential which accurately described the band gap, say along the line of the Hybertsen and Louie (1987) approximation. Then one has to deal with the real space-disordered potentials. The quasi-random-structure approximation used in conjunction with LDA (Wei *et al.*, 1990) provides a first approximation toward diagonalization of the Hamiltonian using large unit cells. The average Green function method is another approach. For close-pack metal alloys in which muffin-tin potentials are good approximations, the KKR-CPA theory (Gyorffy and Stocks, 1979) has proven to be a good approximation. For semiconductor alloys in which the potentials are highly nonspherical the full potential KKR (Yeh *et al.*, 1991) band theory has to be incorporated in the CPA theory to treat alloy disorder.

REFERENCES

Alibert, C., G. Bordure, A. Laugier, and J. Chevallier (1972), *Phys. Rev. B* **6**, 1301.

Andersen, O.K., O. Jepson, and D. Glotzel (1985), in *Highlights of Condensed Matter Theory*, ed. F. Bassiani *et al.* (North-Holland, Amsterdam).

Bachelet, G.B., D.R. Hamman, and M. Schluter (1982), *Phys. Rev. B* **26**, 4199.

Bardeen, J., and W.H. Brattain (1948), *Phys. Rev.* **74**, 230.

Brattain, W.H., and J. Bardeen (1948), *Phys. Rev.* **74**, 231.

Callaway, J. (1964), *Energy Band Theory* (Academic Press, New York).

Callaway, J. (1974), *Quantum Theory of the Solid State* (Academic Press, New York), Chapter 4.

Chelikowsky, J.R., and M.L. Cohen (1976), *Phys. Rev. B* **14**, 556.

Chen, A.-B. (1973), *Phys. Rev. B* **7**, 2230.

Chen, A.-B., and A. Sher (1982), *Phys. Rev. B* **26**, 6603.

Chen, A.-B., G. Weisz, and A. Sher (1972), *Phys. Rev. B* **5**, 2897.

Cohen, M.L., and D.J. Chadi (1980), in *Handbook on Semiconductors* (North-Holland, Amsterdam) Vol. 2, Chapter 4B.

Cohen, M.L., and J.R. Chelikowsky (1989), *Electronic Structure and Optical Properties of Semiconductors*, 2nd ed. (Springer-Verlag, Berlin).

Gyorffy, B.L. and G.M. Stocks (1979), in *Electronics in Disordered Metals and Metallic Surfaces*, Vol. 42, NATO Advanced Study Inst. Ser. B (Plenum, New York).

Hamann, D.R., M. Schluter, and C. Chiang (1979), *Phys. Rev. Lett.* **43**, 1494.

Harrison, W.A. (1966), *Pseudo-potentials in the Theory of Metals* (W.A. Benjamin, New York).

Hass, K., R.J. Lampert, and H. Ehrenreich (1984a), *Phys. Rev. Lett.* **52**, 77.

Hass, K., B. Velicky, and H. Ehrenreich (1984b), *Phys. Rev. B* **29**, 3697.

Hedin, L., and B.I. Lundquist (1971), *J. Phys. C* **4**, 2064.

Hill, R. (1974), *J. Phys. C: Solid State Phys.* **7**, 521.

Hybertsen, M.S., and S.G. Louie (1987), *Phys. Rev. Lett.* **58**, 1551.

Kane, E.O. (1982), in *Handbook on Semiconductors* (North-Holland, Amsterdam), Vol. 1, Chapter 4A.

Kirpatrick, S., B. Velicky, and H. Ehrenreich (1970), *Phys. Rev. B* **1**, 3250.

Kittel, C. (1986), *Introduction to Solid State Physics*, 6th ed. (Wiley, New York).

Krishnamurthy, S., M.A. Berding, A. Sher, and A.-B. Chen (1988), *Phys. Rev. B* **37**, 4254.

Landolt, H., and R. Bornsetin (1982), in *Numerical Data and Functional Relationships in Science and Technology* (ed. K.H. Hellwidge), Vol. 17, (Springer-Verlag, Berlin).

Landolt, H., and R. Bornstein (1988), in *Numerical Data and Functional Relationships in Science and Technology* (ed. K.H. Hellwidge), Vol. 22, (Springer-Verlag, Berlin).

Lempert, R.J., K.C. Hass, and H. Ehrenreich (1987), *Phys. Rev. B* **36**, 1111.

Ley, L., R.A. Pollak, F.R. McFeely, S.P. Kowalczyk, and A. Shirley (1973), *Phys. Rev. B* **9**, 600.

Madelung, O. (1991), in *Semiconductors: Group IV and III–V Compounds* (Springer-Verlag, New York).

Pauli, W. (1925), *Z. Phys.* **31**, 765.

Shockley, W., and G.L. Pearson (1948), *Phys. Rev.* **74**, 232.

Soven, P. (1967), *Phys. Rev.* **156**, 809.

Spicer, W.E., J.A. Silberman, P. Morgen, I. Lindau, J.A. Wilson, A.-B. Chen, and A. Sher (1982), *Phys. Rev. Lett.* **49**, 948.

Van Hove, L. (1953), *Phys. Rev.* **89**, 1189.

van Schilfgaarde, M., and M. Methfessel (1991), unpublished.

Van Vechten, J.A., and T. K. Bergstresser (1970), *Phys. Rev. B* **1**, 3351.

Velicky, B.S., S. Kirkpatrick, and H. Ehrenreich (1968), *Phys. Rev.* **175**, 747.

Wang, C.S., and B.M. Klein (1981), *Phys. Rev. B* **24**, 3393.

Wei, S.H., L.G. Ferreira, J.E. Bernard, and A. Zunger (1990), *Phys. Rev. B* **42**, 9622.

Williams, E.W., and V. Rehn (1968), *Phys. Rev.* **172**, 798.

Yeh, C.-Y, A.-B. Chen, D.M. Nicholson, and W.H. Buttler (1991), *Phys. Rev. B* **42**, 10976.

6

Transport

Charge transport in semiconductors and alloys will be treated in this chapter with an emphasis on aspects that are strongly influenced by alloys and details of band structures. Alloying not only produces aperiodic potentials that cause extra scattering, it also introduces special features in the band structures that are not often present in the constituent semiconductors, such as the direct and indirect band-gap crossings in some alloys as the alloy compositions are varied.

We will begin with a brief discussion of the master equation and Boltzmann equation (BE). We will then treat low-field transport, generalize Brooks's (unpublished) formula, and apply it to the Si_xGe_{1-x} alloy. After reviewing the important scattering mechanisms, we present an efficient and reliable basis expansion method for solving the Boltzmann equations. These methods are incorporated into full-band-structure calculation of velocity–field characteristics and near ballistic transport, and the results are used to evaluate the merits of materials for high-speed electronics applications. A similar approach is also applied to low-field transport in narrow-gap alloys where Fermi statistics is important.

6.1. MASTER AND BOLTZMANN EQUATIONS

6.1.1. *Master Equation for a Supersystem*

In this section we briefly describe the basic transport theories. Consider an isolated many-body system, one that does not interact with the remainder of the universe. Decompose the Hamiltonian H into two parts

$$H = H_0 + V \tag{6.1.1}$$

where H_0 is a major part for which the eigenvalues E_n and eigenkets $|n\rangle$ are known:

$$H_0|n\rangle = E_n|n\rangle \tag{6.1.2}$$

The V is also time independent and is constructed to be a small part of the Hamiltonian. For a large many-body system, the degeneracy is large, and the eigenenergy spectrum is inevitably continuous. Then the transition rate to a state $|m\rangle$ at t for the system initially in the state $|n\rangle$ at t_0 for a sufficient time interval $t - t_0$ is governed by Fermi's golden rule:

$$W(m/n) = \frac{2\pi}{\hbar} |\langle m|\hat{T}|n\rangle|^2 \delta(E_n - E_m) \tag{6.1.3}$$

where the transition operator is given by $\hat{T} = V + V(E_n + i\eta - H)^{-1}V$, with η being an infinitesimal positive energy. The condition required for the validity of Fermi's golden rule is that $t - t_0$ be larger than a time t_c characteristic of an unperturbed single particle's motion. For example, in lattice vibrations a typical oscillation period is such a characteristic time. Equation (6.1.3) shows that the microscopic reversibility $W(n/m) = W(m/n)$ holds.

The master equation is a gain–loss equation governing the rate of change of the probability $P_n(t)$ of finding the system in any of the eigenstates of H_0, say $|n\rangle$. It reads

$$\frac{dP_n}{dt} = \sum_m [W(n/m)P_m(t) - W(m/n)P_n(t)] \tag{6.1.4}$$

This equation can be derived from the Schrödinger equation provided that the initial many-body state is incoherent and that the golden rule holds (Van Hove, 1955, 1957; Sher and Primakoff, 1960, 1963). Equation (6.1.4) is a statistical–mechanical equation; it does not have the time-reversal symmetry of the Schrödinger equation. In fact the master equation leads to an increase of entropy in time, known as the Boltzmann H theorem (Reif, 1965). Therefore it cannot describe a coherent quantum state. However, for an incoherent initial state—i.e., a state with a large number of accessible eigenstates of H_0 occupied—the master equation is sound. Considerable literature (e.g., Van Hove, 1955, 1957; Sher and Primakoff, 1960, 1963) has been devoted to this subject.

Given microscopic reversibility, $W(n/m) = W(m/n)$, in Eq.(6.1.3), the equilibrium solution to Eq. (6.1.4) is the *a priori* distribution $P_n = 1/\mathcal{N}(E)$ for every state with an eigenenergy on the energy shell $E_n = E$. Here $\mathcal{N}(E)$ is the degeneracy, or the total number of accessible states. This is the expected result, a microcanonical ensemble.

6.1.2. Master Equation for a System in a Heat Bath

Next, consider the situation in which the supersystem consists of two subsystems A and B. Let $H_0 = H_A + H_B$ and $V = V_{AB}$, where H_A and H_B contain only A- and B-system variables, respectively, and the interaction term V_{AB} has all the terms that contain variables from both systems. We designate A as the subsystem of interest and B as the "surroundings," or a heat bath in statistical terms. The state $|n\rangle$ of the supersystem can be specified by ε_n, α_n, η_n, and β_n, where ε_n is the energy of A, and α_n represents the set of quantum numbers of A other than ε_n when the supersystem is in state n; η_n and β_n have similar meanings for B. The probability $P_n(t)$ in Eq. (6.1.4) is now a joint probability specified by these four quantum numbers: $P_n(t) = P(\varepsilon_n, \alpha_n, \eta_n, \beta_n; t)$. The probability of finding the A system in a state specified by ε_n and α_n while the supersystem has an energy, $E = \varepsilon_n + \eta_n$, is then given by

$$P_A(\varepsilon_n, \alpha_n; t) \equiv \sum_{\eta_n \beta_n (\varepsilon_n + \eta_n = E)} P(\varepsilon_n, \alpha_n, \eta_n, \beta_n; t) \tag{6.1.5}$$

Performing this sum on Eq. (6.1.4) and defining an effective transition rate

$$W_A(\varepsilon_n, \alpha_n/\varepsilon_m, \alpha_m; t) \equiv \sum_{\eta_n \beta_n (\varepsilon_n + \eta_n = E) \eta_m \beta_m} W(n/m)P_m(t) \Big/ \sum_{\eta_m \beta_m (\varepsilon_m + \eta_m = E)} P_m(t) \tag{6.1.6}$$

we find

$$\frac{dP_A(\varepsilon_n, \alpha_n; t)}{dt} = \sum_{\varepsilon_m \alpha_m} [W_A(\varepsilon_n, \alpha_n/\varepsilon_m, \alpha_m; t) P_A(\varepsilon_m, \alpha_m; t) - W_A(\varepsilon_m, \alpha_m/\varepsilon_n, \alpha_n; t) P_A(\varepsilon_n, \alpha_n; t)] \tag{6.1.7}$$

Except for the time dependence of W_A, Eq. (6.1.7) looks like a master equation for the system of interest A. The time dependence is removed if B is large compared with A and remains effectively in equilibrium even when A is not. The condition that B is in equilibrium is that the joint probability distributions are given by

$$P(\varepsilon_n, \alpha_n, E - \varepsilon_n, \beta_n; t) = P(\varepsilon_n, \alpha_n, E - \varepsilon_n, \beta_n'; t) \tag{6.1.8}$$

for all β_n and β_n'. The summation over η_m in both the numerator and denominator of Eq. (6.1.6) replaces the argument η_m in P_m by $E - \varepsilon_m$. Then P_m can be pulled out of the sum and divided out to yield a t-independent rate

$$W_A(\varepsilon_n, \alpha_n/\varepsilon_m, \alpha_m) = \sum_{\beta_n \beta_m} \frac{W(\varepsilon_n, \alpha_n, E - \varepsilon_n, \beta_n/\varepsilon_m, \alpha_m, E - \varepsilon_m, \beta_m)}{\mathcal{N}_B(E - \varepsilon_m)} \tag{6.1.9}$$

where $\mathcal{N}_B$ is the degeneracy of the B system

$$\mathcal{N}_B(\eta_m) = \sum_{\beta_m} 1 \tag{6.1.10}$$

With this t-independent transition rate, we then have a master equation for the A system:

$$\frac{dP_A(\varepsilon_n, \alpha_n; t)}{dt} = \sum_{\varepsilon_n \alpha_m} [W_A(\varepsilon_n, \alpha_n/\varepsilon_m, \alpha_m) P_A(\varepsilon_m, \alpha_m; t) - W_A(\varepsilon_m, \alpha_m/\varepsilon_n, \alpha_n) P_A(\varepsilon_n, \alpha_n; t)] \tag{6.1.11}$$

Note that the two W_A's have the ratio

$$\frac{W_A(\varepsilon_n, \alpha_n/\varepsilon_m, \alpha_m)}{W_A(\varepsilon_m, \alpha_m/\varepsilon_n, \alpha_n)} = \frac{\mathcal{N}_B(E - \varepsilon_n)}{\mathcal{N}_B(E - \varepsilon_m)} = \frac{e^{-\varepsilon_n/kT}}{e^{-\varepsilon_m/kT}} \tag{6.1.12}$$

The last step is obtained by the usual argument in statistical mechanics that the heat reservoir B is much larger than the system of interest A, so $\ln \mathcal{N}_B(E - \varepsilon) \approx \ln \mathcal{N}_B(E) - [d \ln \mathcal{N}_B(E)/dE]\varepsilon = \ln \mathcal{N}_B(E) - \varepsilon/kT$, and $\mathcal{N}_B(E - \varepsilon) \approx \mathcal{N}_B(E) e^{-\varepsilon/kT}$. The condition in Eq. (6.1.12) ensures that Eq. (6.1.11) has the equilibrium solution expected of a canonical ensemble, namely

$$P_A(\varepsilon_n, \alpha_n) = e^{-\varepsilon_n/kT}/Z_A \tag{6.1.13}$$

where $Z_A = \sum \mathcal{N}_A(\varepsilon_n) e^{-\varepsilon_n/kT}$ is the partition function.

6.1.3. Single-Particle States

A reduction procedure analogous to that proceeding from the supersystem master equation to that for the system of interest interacting with a heat bath can be used to obtain a similar equation for the single-particle states. For degenerate electrons, the single-particle master equation becomes

$$\frac{df(\mathbf{k},s;t)}{dt}=\sum_{\mathbf{k}'s'}(w(\mathbf{k},s/\mathbf{k}',s')[1-f(\mathbf{k},s;t)]f(\mathbf{k}',s';t)$$
$$-w(\mathbf{k}',s'/\mathbf{k},s)[1-f(\mathbf{k}',s';t)]f(\mathbf{k},s;t)) \quad (6.1.14)$$

In Eq. (6.1.14), $\mathbf{k}$, s are respectively the wave vector and spin index, and $f(\mathbf{k},s;t)$ is the average number of electrons in state $\mathbf{k}$, s at time t. A derivation of this equation for electrons interacting with phonons, which serve as a heat bath and remain in thermal equilibrium, is presented in Section 6.2. The single-particle transition probability rates in Eq. (6.1.14) still have to obey the relation

$$\frac{w(\mathbf{k}',s'/\mathbf{k},s)}{w(\mathbf{k},s/\mathbf{k}',s')}=\frac{e^{-\varepsilon(\mathbf{k}',s')/k_BT}}{e^{-\varepsilon(\mathbf{k},s)/k_BT}} \quad (6.1.15)$$

where the Boltzmann constant is now denoted k_B. This condition and Eq. (6.1.14) lead to the Fermi function being the equilibrium solution

$$f_0(\mathbf{k},s)=1/\left(e^{[\varepsilon(\mathbf{k},s)-\varepsilon_F]/k_BT}+1\right) \quad (6.1.16)$$

where the Fermi energy ε_F is determined from the requirement of charge neutrality in the solid.

For nondegenerate electrons (e.g., in lightly doped semiconductors at room temperature), Eq. (6.1.15) reduces to a form similar to Eq. (6.1.7):

$$\frac{df(\mathbf{k},s;t)}{dt}=\sum_{\mathbf{k}'s'}[w(\mathbf{k},s/\mathbf{k}',s')f(\mathbf{k}',s';t)-w(\mathbf{k}',s'/\mathbf{k},s)f(\mathbf{k},s;t)] \quad (6.1.17)$$

The corresponding equilibrium solution is then the Boltzmann distribution in Eq. (6.1.13), as is expected in this limit.

6.1.4. The Boltzmann Equation

The Boltzmann equation, when applied to electrons in a solid, is a semiclassical theory which permits treatment of transport in inhomogeneous solids, e.g., in p–n junctions where donor and acceptor concentrations vary in space. In this approach, the distribution function is allowed to have a spatial dependence; i.e., $f=f(\mathbf{k},s,\mathbf{r};t)$. Strictly speaking, this is not permitted in quantum mechanics, since momentum and position are not compatible observables. How can this dual $\mathbf{k}$ and $\mathbf{r}$ dependence make sense? The way it works is to artificially divide the crystal into a set of small, but still macroscopic, pieces, each of which interacts with its neighbors and the heat bath. Each piece is small enough to be treated as homogeneous, but the properties of the pieces may change continuously from one to the next. The electrons in each piece are governed by an equation of the form of (6.1.14), but $f(\mathbf{k}, s, \mathbf{r}; t)$ may vary continuously as $\mathbf{r}$ moves from one piece to the next.

To derive the Boltzmann equation, first write the total time derivative of $f(\mathbf{k}, s, \mathbf{r}; t)$ as the sum of partial derivatives:

$$\frac{df}{dt} = \frac{\partial f}{\partial t} + \frac{d\mathbf{k}}{dt} \cdot \nabla_{\mathbf{k}} f + \frac{d\mathbf{r}}{dt} \cdot \nabla f \tag{6.1.18}$$

In a crystal, the average value of $d\mathbf{r}/dt$ for an electron in a state of wave vector $\mathbf{k}$ is the group velocity $\mathbf{v}_{\mathbf{k}} = \nabla_{\mathbf{k}}\varepsilon(\mathbf{k})/\hbar$, and $\hbar(d\mathbf{k}/dt)$ can be identified with the Lorentz force $\mathbf{F}_{\text{ext}} = -e(\mathbf{E} + \mathbf{v} \times \mathbf{B}/c)$. Equation (6.1.18) should be zero if electrons do not experience any collision, because particles are not removed from or added to their trajectories. With collisions, the left side of Eq. (6.1.18) can be set equal to the df/dt given by Eq. (6.1.14) or Eq. (6.1.17), depending on whether we have degenerate or nondegenerate electrons. These df/dt expressions will be denoted $(df/dt)_{\text{coll}}$. Combining the above results, we have the Boltzmann equation

$$\frac{\partial f}{\partial t} - e[\mathbf{E} + (\mathbf{v} \times \mathbf{B}/c)] \cdot \nabla_{\mathbf{k}} f + \mathbf{v} \cdot \nabla f = \left(\frac{df}{dt}\right)_{\text{coll}} \tag{6.1.19}$$

Note that this is in general an integrodifferential equation, because $(df/dt)_{\text{coll}}$ in Eqs. (6.1.14) or (6.1.17) involves a summation over all the wave vectors $\mathbf{k}$ and spins s. The solution and application of this equation to charge transport in semiconductors is the main task of this chapter.

6.2. ELECTRON–PHONON INTERACTION AND SINGLE-PARTICLE MASTER EQUATION

The purpose of this section is to display the structure of the scattering rates of electrons by electron–phonon interactions and to present a derivation of the master equation, Eq. (6.1.14), for single-particle states.

Let us go back to Section 6.1.2. The system of interest A is now composed of electrons and phonons. Let the Hamiltonian of A be $H_{\text{A}} = H_{\text{el}} + H_{\text{ph}} + H_{\text{in}}$, where $H_{\text{el}} = \sum \varepsilon_k c_k^\dagger c_k$ is for electrons only, $H_{\text{ph}} = \sum \hbar\omega_q b_q^\dagger b_q$ is for phonons, and H_{in} is the electron–phonon interaction Hamiltonian, which takes the form

$$H_{\text{in}} = \sum_k \sum_q (V_q b_q c_{k+q}^\dagger c_k + V_q^* b_q^\dagger c_k^\dagger c_{k-q}) \tag{6.2.1}$$

In the above, b_q and $b_q^\dagger$ are phonon destruction and creation operators, and c_k and $c_k^\dagger$ are those for the electrons. For simplicity, the quantum number for the single-particle states $k = (\mathbf{k}, s)$ represents both the wave vector $\mathbf{k}$ and spin index s of electrons. Similarly $q = (\mathbf{q}, \alpha)$ stands for both the wave vector and polarization of phonons. V_q in Eq. (6.2.1) is a coupling constant, assumed to be a function of q.

The eigenstates $|m\rangle$ of the unperturbed Hamiltonian $H_0 = H_{\text{el}} + H_{\text{ph}}$ can be expressed in terms of sets of occupation numbers of electrons n_k and phonons N_q: $|m\rangle = |\{n_k\}, \{N_q\}\rangle$, where n_k for a given k takes a value of 0 or 1, while N_q can have any positive integer values. The average number of electrons in state k is then

$$\langle n_k \rangle = \sum_m n_k P_A(m; t) \tag{6.2.2}$$

where the sum means over all possible combinations of the sets $\{n_k\}$, $\{N_q\}$ for all k and q. Taking the time derivative of Eq. (6.2.2) and using Eq. (6.1.11), we obtain

$$\frac{d\langle n_k(t)\rangle}{dt} = \sum_{mm'} \sum n_k[W_A(m/m')\, P_A(m';t) - W_A(m'/m)\, P_A(m; t)] \tag{6.2.3}$$

Under the condition that phonons remain in thermal equilibrium even when the electrons are not, we can write the joint probability as a product of distributions for electrons and phonons:

$$P_A(m; t) \simeq P_{el}(\{n_k\}; t) P_{ph}(\{N_q\}) \tag{6.2.4}$$

Using this expression for P_A in Eq. (6.2.3) and knowing that $|m\rangle$ stands for $|\{n_k\}, \{N_q\}\rangle$ and $|m'\rangle$ stands for $|\{n'_k\}, \{N'_q\}\rangle$, we can carry out the sum over the phonon variables to reduce Eq. (6.2.3) to

$$\frac{d\langle n_k(t)\rangle}{dt} = \sum_{\{n_l\}} \sum_{\{n'_l\}} n_k \,[w(\{n_l\}/\{n'_l\})\, P_{el}(\{n'_l\}; t)$$

$$- w(\{n'_l\}/\{n_l\})\, P_{el}(\{n_l\}; t)] \tag{6.2.5}$$

where the effective transition rate w between the electronic states is given by

$$w(\{n_l\}/\{n'_l\}) = \sum_{\{N_q\}} \sum_{\{N'_q\}} W_A(\{n_l\},\{N_q\}/\{n'_l\}, \{N'_q\})\, P_{ph}(\{N'_q\}) \tag{6.2.6}$$

Now using the golden rule to calculate the transition rate to the first order in H_{in} and using the properties of the creation and destruction operators, we obtain

$$\sum_{\{n_l\}} \sum_{\{n'_l\}} n_k[w(\{n'_l\}/\{n_l\}) P_{el}(\{n_l\};t)] = \sum_{k'} \langle (1 - n_{k'}) n_k^2 \rangle w(k'/k) \tag{6.2.7}$$

where $w(k'/k) = w_{ab}(k'/k) + w_{em}(k'/k)$ with the subscripts ab and em denoting the absorption and emission of one phonon respectively:

$$w_{ab}(k'/k) = \frac{2\pi}{\hbar} \sum_q |V_q|^2 \langle N_q \rangle \delta(\varepsilon_{k'} - \hbar\omega_q - \varepsilon_k) \delta_{k',k+q} \tag{6.2.8}$$

$$w_{em}(k'/k) = \frac{2\pi}{\hbar} \sum_q |V_q|^2 (\langle N_q \rangle + 1) \delta(\varepsilon_{k'} + \hbar\omega_q - \varepsilon_k) \delta_{k',k-q} \tag{6.2.9}$$

The equilibrium value $\langle N_q \rangle$ is the Bose–Einstein function $\langle N_q \rangle = 1/(e^{\hbar\omega_q/k_BT} - 1)$. The average $\langle (1 - n_{k'}) n_k^2 \rangle$ in Eq. (6.2.7) means

$$\langle(1-n_{k'})n_k^2\rangle = \sum_{\{n_l\}}\sum_{\{n_l'\}}(1-n_{k'})n_k^2 P_{el}(\{n_l'\};t) \tag{6.2.10}$$

Because $n_k^2 = n_k$, if we assume that the occupation numbers are statistically independent, we have $\langle(1-n_{k'}(t))n_k^2(t)\rangle \simeq (1-\langle n_{k'}(t)\rangle)\langle n_k(t)\rangle$. By definition $\langle n_k(t)\rangle$ is the distribution function $f(\mathbf{k}, s; t)$ appearing in Section 6.1.3. Combining these results, Eq. (6.2.5) reduces to the master equation, Eq. (6.1.14), for single-particle distributions. Finally, we note that the scattering rates in Eq. (6.2.8) and (6.2.9) do satisfy the canonical relation in Eq. (6.1.15).

6.3. LOW-FIELD TRANSPORT FOR NONDEGENERATE ELECTRONS IN COLLISION–TIME APPROXIMATIONS

In this section we discuss examples of solutions to the Boltzmann equation, Eq. (6.1.18), for nondegenerate electrons in a uniform semiconductor at low E field using various collision–time approximations. Four relaxation times will be introduced in this chapter: the particle relaxation time τ_R, the momentum relaxation time τ_m, the velocity relaxation time τ_v, and the energy relaxation time τ_E.

The first of these, τ_R, is the reciprocal of the total collision rate. It is used in approximate transport expressions because it is easier to calculate than the other collision times. This approximation is not accurate in transport calculations because it assigns equal weights to forward and wide-angle scattering. The momentum relaxation time τ_m is an improvement over τ_R because it weighs wide-angle scattering more than it does forward scattering. However, the current density driven by an external electric field is directly proportional to the carriers' average group velocity rather than to their average crystal momentum $\mathbf{k}$. Thus, the proper relaxation time to characterize the transport is velocity relaxation time τ_v. If all the carriers have parabolic bands, then the group velocity and momentum are proportional, so $\tau_m = \tau_v$. However, in hot-electron transport or in narrow-gap semiconductors, where the parabolic band is not a good approximation, these two relaxation times can differ substantially. Such cases are treated in Sections 6.5, 6.8, 6.9, and 6.10. The fourth relaxation time, τ_E, is related to the rate at which energy is exchanged between the electron system and its heat bath (usually phonons). In general, one finds $\tau_R < \tau_m$ and $\tau_v < \tau_E$.

As mentioned in Section 6.1.4, the equilibrium f_0 is a Boltzmann distribution, it takes the explicit expression

$$f_0(\mathbf{k}, s, \mathbf{r}) = \frac{nV}{2z}\, e^{-\varepsilon(\mathbf{k},s)/k_BT} \tag{6.3.1}$$

where n is the density of electrons (so $n/2$ is the density per spin), V is the volume of the solid, and z is the single-particle partition function.

Under the influence of a weak static $\mathbf{E}$ field and zero $\mathbf{B}$ field, the steady-state ($\partial f/\partial t = 0$) distribution for either spin in a nonmagnetic semiconductor satisfies the following equation:

$$-e\mathbf{E}\cdot\frac{\nabla_{\mathbf{k}} f}{\hbar} = \sum_{\mathbf{k}'}\left(w(\mathbf{k}/\mathbf{k}')\,[f(\mathbf{k}')-f_0(\mathbf{k}')] - w(\mathbf{k}'/\mathbf{k})\,[f(\mathbf{k})-f_0(\mathbf{k})]\right) \tag{6.3.2}$$

In (6.3.2) f_0 has been inserted for further manipulation. This insertion does not alter the equation, because the w's and f_0's satisfy the principle of detailed balance:

$$w(\mathbf{k}/\mathbf{k}')\, f_0(\mathbf{k}') = w(\mathbf{k}'/\mathbf{k}) f_0(\mathbf{k}) \tag{6.3.3}$$

Let $f(\mathbf{k}) = f_0(\mathbf{k}) + \delta f(\mathbf{k})$ and retain only linear terms in E. Then Eq. (6.3.2) becomes

$$-e\mathbf{E}\cdot\frac{\nabla_{\mathbf{k}} f_0(\mathbf{k})}{\hbar} = \sum_{\mathbf{k}'} w(\mathbf{k}/\mathbf{k}')\,\delta f(\mathbf{k}') - \frac{\delta f(\mathbf{k})}{\tau_{\mathrm{R}}(\mathbf{k})} \tag{6.3.4}$$

where the particle relaxation time $\tau_{\mathrm{R}}(\mathbf{k})$ is defined as

$$[\tau_{\mathrm{R}}(\mathbf{k})]^{-1} \equiv \sum_{\mathbf{k}'} w(\mathbf{k}'/\mathbf{k}) \tag{6.3.5}$$

In the relaxation time approximation, the gain term in Eq. (6.3.4) is neglected. Then the solution is

$$\delta f(\mathbf{k}) = \tau_{\mathrm{R}}(\mathbf{k}) e\mathbf{E}\cdot\frac{\nabla_{\mathbf{k}} f_0(\mathbf{k})}{\hbar} = \tau_{\mathrm{R}}(\mathbf{k}) e\mathbf{v}(\mathbf{k})\cdot\mathbf{E}\frac{\partial f_0(\mathbf{k})}{\partial\varepsilon(\mathbf{k})} \tag{6.3.6}$$

The current density is

$$\mathbf{j} = -e\sum_{\mathbf{k}s}\frac{\mathbf{v}(\mathbf{k}) f(\mathbf{k})}{V} = -2e\sum_{\mathbf{k}}\frac{\mathbf{v}(\mathbf{k}) f(\mathbf{k})}{V} \tag{6.3.7}$$

Because $\sum \mathbf{v}(\mathbf{k}) f_0(\mathbf{k}) = 0, f(\mathbf{k})$ in Eq. (6.3.7) can be replaced by $\delta f(\mathbf{k})$. Using the solution of δf from Eq. (6.3.6), Eq. (6.3.7) yields

$$\mathbf{j} = -2e^2\sum_{\mathbf{k}}\frac{\partial f_0(\mathbf{k})}{\partial\varepsilon(\mathbf{k})}\,\frac{\tau_{\mathrm{R}}(\mathbf{k})\mathbf{v}(\mathbf{k})[\mathbf{v}(\mathbf{k})\cdot\mathbf{E}]}{V} \tag{6.3.8}$$

Using the definition $j_\alpha = \sum_\beta \sigma_{\alpha\beta} E_\beta$, the electrical conductivity tensor can be identified as

$$\sigma_{\alpha\beta} = -2e^2\sum_{\mathbf{k}}\frac{\partial f_0(\mathbf{k})}{\partial\varepsilon(\mathbf{k})}\,\frac{\tau_{\mathrm{R}}(\mathbf{k}) v_\alpha(\mathbf{k}) v_\beta(\mathbf{k})}{V} \tag{6.3.9}$$

For a cubic crystal, the off-diagonal components can be shown to be zero, so $\sigma_{\alpha\beta} = \sigma\delta_{\alpha\beta}$ with σ given by

$$\sigma = -2e^2\sum_{\mathbf{k}}\frac{\partial f_0(\mathbf{k})}{\partial\varepsilon(\mathbf{k})}\,\frac{\tau_{\mathrm{R}}(\mathbf{k}) v_\alpha(\mathbf{k}) v_\alpha(\mathbf{k})}{V}$$

$$= -\frac{2e^2}{3V}\sum_{\mathbf{k}}\frac{\partial f_0(\mathbf{k})}{\partial\varepsilon(\mathbf{k})}\,\tau_{\mathrm{R}}(\mathbf{k}) v^2(\mathbf{k}) \tag{6.3.10}$$

To proceed further we must obtain an explicit expression for $\tau_{\mathrm{R}}(\mathbf{k})$ corresponding to all the scattering mechanisms, the band structure $\varepsilon(\mathbf{k})$, and associated group velocity $\mathbf{v}(\mathbf{k})$, and perform the Brillouin zone integration arising from the conversion $\Sigma_{\mathbf{k}} = [V/(2\pi)^3]\int d^3k$. For a parabolic band structure $\varepsilon(\mathbf{k}) = \hbar^2k^2/2m^*$ with a constant collision time τ and f_0 given by Eq. (6.3.1), Eq. (6.3.9) can be integrated to yield the familiar formula

$$\sigma = ne^2\tau/m^* \tag{6.3.11}$$

The most unsatisfactory approximation made so far is the neglect of the gain terms in Eq. (6.3.4). By neglecting these terms, the relaxation time approximation produces a mobility $\mu = \sigma/n$ that is too small. In Section 7 we shall consider ways to solve Eq. (6.3.2) numerically. An approximate way to incorporate the gain terms is to use the momentum relaxation time $\tau_m(\mathbf{k})$, to be discussed next.

Let us multiply $\mathbf{k}$ on both sides of Eq. (6.3.1) and sum over $\mathbf{k}$ to get

$$\begin{aligned} -\sum_{\mathbf{k}} \mathbf{k}e\mathbf{E}\cdot\frac{\nabla_{\mathbf{k}}f}{\hbar} &= \sum_{\mathbf{k}}\sum_{\mathbf{k}'} \mathbf{k}\,(w(\mathbf{k}/\mathbf{k}')\,\delta f(\mathbf{k}') - w(\mathbf{k}'/\mathbf{k})\delta f(\mathbf{k})) \\ &= \sum_{\mathbf{k}}\left(\sum_{\mathbf{k}'} (\mathbf{k}' - \mathbf{k})w(\mathbf{k}'/\mathbf{k})\right)\delta f(\mathbf{k}) \end{aligned} \tag{6.3.12}$$

To go from the first equality to the second in Eq. (6.3.12) we have interchanged $\mathbf{k}$ and $\mathbf{k}'$ in the first term. Now we can define a momentum relaxation time $\tau_m(\mathbf{k})$ as

$$\frac{\mathbf{k}}{\tau_m(\mathbf{k})} \equiv -\sum_{\mathbf{k}'} (\mathbf{k}' - \mathbf{k})w(\mathbf{k}'/\mathbf{k}) \tag{6.3.13}$$

Taking the dot product of the unit vector $\hat{\mathbf{k}}$ on both sides of this equation and dividing by k produces the result

$$\frac{1}{\tau_m(\mathbf{k})} = \sum_{\mathbf{k}'} (1 - \frac{k'}{k}\cos\theta)w(\mathbf{k}'/\mathbf{k}) \tag{6.3.14}$$

where $\cos\theta = \hat{\mathbf{k}}\cdot\hat{\mathbf{k}}'$. Using Eq. (6.3.13) in Eq. (6.3.12) and linearizing the equation, we find

$$\sum_{\mathbf{k}} \mathbf{k}\,[e\mathbf{E}\cdot\frac{\nabla_{\mathbf{k}}f_0}{\hbar} - \frac{\delta f(\mathbf{k})}{\tau_m(\mathbf{k})}] = 0 \tag{6.3.15}$$

Clearly the sum vanishes if each term in brackets is zero; however, it not evident that this condition is necessary. In other words, setting each bracketed term to zero gives one solution, but it may not be unique. However, taking this to be the case, the solution for δf has exactly the same functional form as that in Eq. (6.3.5) with $\tau_R(\mathbf{k})$ replaced by $\tau_m(\mathbf{k})$.

In Eq. (6.3.14) the scattering rate in the backward directions, i.e., $\theta > \pi/2$, is enhanced while that in the forward scattering directions ($\theta < \pi/2$) is reduced, so τ_m and τ_R can behave quite differently. Qualitatively, the momentum relaxation time approximation is more appropriate to use for electrical transport, because scattering at $\theta = 0$ does not reduce the current, while a 180° scattering is the most effective in reducing the current density.

A good example to show the different behavior of τ_R and τ_m is the scattering of free electrons by polar optical phonons governed by the Frohlich (1937) Hamiltonian, which is

Eq. (6.2.1) with the coupling V_q given by $-\lambda/\sqrt{qV}$ [see Section 6.6 and Eq. (6.6.9)]. The single-particle scattering rate is the sum of w_{em} and w_{ab} of Eqs. (6.2.8) and (6.2.9), respectively. Because the optical phonons have a very small dispersion in the frequencies, the phonon frequency can be set equal to a constant value ω_0. Using the free-electron band structure $\varepsilon(\mathbf{k}) = \hbar^2k^2/(2m^*)$ and defining $k_0 \equiv (2m^*\omega_0/\hbar)^{1/2}$, we can carry out the summation over q for w_{em} and w_{ab} explicitly to yield

$$w_{ab}(k'/k) = \frac{2\pi}{\hbar}\frac{\lambda^2}{V\hbar\omega_0}\langle N_0\rangle \frac{\delta[k'^2 - (k^2 + k_0^2)^{1/2}]}{[2|\mathbf{k}' - \mathbf{k}|^2(k^2 + k_0^2)^{1/2}]} \tag{6.3.16}$$

$$w_{em}(k'/k) = \frac{2\pi}{\hbar}\frac{\lambda^2}{V\hbar\omega_0}\langle N_0 + 1\rangle \frac{\delta[k'^2 - (k^2 - k_0^2)^{1/2}]\theta(k - k_0)}{[2|\mathbf{k}' - \mathbf{k}|^2(k^2 - k_0^2)^{1/2}]} \tag{6.3.17}$$

where θ is the Heaviside step function and $\langle N_0\rangle = 1/(e^{\hbar\omega_0/k_BT} - 1)$. The particle collision time can then be calculated from

$$\frac{1}{\tau_R} = \frac{V}{(2\pi)^3}\int d^3k' \; [w_{ab}(k'/k) + w_{em}(k'/k)] \tag{6.3.18}$$

The angular part of the integration can be carried out analytically:

$$\int \frac{d\Omega'}{|\mathbf{k}' - \mathbf{k}|^2} = \frac{2\pi}{kk'}\cosh^{-1}\left|\frac{k'^2 + k^2}{k'^2 - k^2}\right| \tag{6.3.19}$$

Then the radial integral can also be obtained in a closed form to give

$$\frac{1}{\tau_R} = 2\alpha\omega_0\left[\frac{\langle N_0\rangle \sinh^{-1}(k/k_0)}{(k/k_0)} + \frac{\langle N_0 + 1\rangle \cosh^{-1}(k/k_0)\theta(k - k_0)}{(k/k_0)}\right] \tag{6.3.20}$$

where α is related to λ by $\alpha = \lambda\sqrt{k_0}/2\sqrt{\pi}\hbar\omega_0$. Similarly, the integrals can also be done for the momentum relaxation time τ_m to yield

$$\frac{1}{\tau_m} = \alpha\omega_0\left\{\langle N_0\rangle\left(\frac{k}{k_0}\left[\left(\frac{k}{k_0}\right)^2 + 1\right]^{1/2} - \sinh^{-1}\left(\frac{k}{k_0}\right)\right)\right.$$

$$\left. + \left(\langle N_0 + 1\rangle\left(\frac{k}{k_0}\right)\left[\left(\frac{k}{k_0}\right)^2 + 1\right]^{1/2} - \cosh^{-1}\left(\frac{k}{k_0}\right)\right)\theta(k - k_0)\right\}/(k/k_0)^3 \tag{6.3.21}$$

A direct numerical calculation shows that $\tau_m > \tau_R$ for all energies (or k), and in the extreme limits we have the ratios

$$\frac{\tau_m}{\tau_R} = \begin{cases} 12/5, & k \ll k_0 \\ \ln(2k/k_0), & k \gg k_0 \end{cases} \tag{6.3.22}$$

Thus the two relaxation times can have significant differences. However, both relaxation times will give the same temperature dependence of the carrier mobility $\mu \equiv \sigma/(ne)$ in both the low-T and high-T limits. To see this we first note that the expression for the mobility of nondegenerate electrons with a parabolic band in both relaxation time approximations has compact form

$$\mu = \frac{4e}{3\sqrt{\pi}m^*} \int d\eta \, \eta^{1/2}\tau(k)e^{-\eta} \tag{6.3.23}$$

where the integral variable is given by $\eta = \varepsilon(\mathbf{k})/(k_B T)$. At low temperature T in the limit $T << T_0$, where $k_B T_0 \equiv \hbar\omega_0$, the small k limit can be used for τ_R and the associated mobility can be approximated by

$$\mu_R \simeq \frac{e}{2m^*\alpha\omega_0} e^{T_0/T} \tag{6.3.24}$$

Similar approximation can be made for τ_m, so $\mu_m = 12\mu_R/5$ according to Eq. (6.3.22). At high temperature, $T >> T_0$, if the approximation $\ln(2k/k_0) \simeq 1$ is made, the mobility, using both relaxation times, becomes

$$\mu_m \simeq \mu_R \simeq \frac{2e}{3\sqrt{\pi}m^*\alpha\omega_0}\left(\frac{T_0}{T}\right)^{1/2} \tag{6.3.25}$$

Thus, although μ_R and μ_m have a similar qualitative temperature dependence, their magnitudes can be substantially different.

In analogy with the momentum relaxation time, a velocity relaxation time τ_v can be defined by replacing the $1 - \mathbf{k}'\cdot\mathbf{k}/k^2$ factor in Eq. (6.3.14) by $1 - \mathbf{v}_{\mathbf{k}'}\cdot\mathbf{v}_{\mathbf{k}}/v_{\mathbf{k}}^2$, where $\mathbf{v}_{\mathbf{k}}$ is the group velocity.

Finally, also similar to Eq. (6.3.13), one can define an energy relaxation time τ_E for electrons encountering inelastic collisions such as in the electron–phonon interaction

$$\frac{\varepsilon(\mathbf{k})}{\tau_E(\mathbf{k})} \equiv -\sum_{\mathbf{k}'} [\varepsilon(\mathbf{k}') - \varepsilon(\mathbf{k})]w(\mathbf{k}'/\mathbf{k}) \tag{6.3.26}$$

For the rates given by Eq. (6.2.8) and (6.2.9) for the electron–phonon interaction, this equation becomes

$$\frac{\varepsilon(k)}{\tau_E(k)} \equiv -\sum_{k'} [\hbar\omega_q w_{ab}(k'/k) - \hbar\omega_q w_{em}(k'/k)] \tag{6.3.27}$$

The energy relaxation time characterizes the rate at which energy is transferred between electrons and phonons (or other heat bath). It plays an important role in hot-electron transport.

6.4. MOBILITIES IN ALLOYS: EXAMPLE, Si_xGe_{1-x}

One major difference in the transport properties between alloys and the constituent crystals is that electrons in disordered alloys suffer scattering from the aperiodic potentials in addition to all the other scattering mechanisms, such as electron–phonon and electron–impurity scattering encountered by the electrons in a constituent compound. Many years ago Nordheim (1931) and Brooks (unpublished) obtained an expression for the alloy scattering limited electron mobilities μ_A (assuming the alloy scattering is the only scattering mechanism) in metals and semiconductors, respectively. Brooks's well-known formula reads

$$\mu_A = \frac{(2\pi)^{1/2} e\hbar^4 N_0}{3x(1-x)m^{*5/2}(\Delta E)^2 (kT)^{1/2}} \tag{6.4.1}$$

where N_0 is the number of atoms per unit volume, m^* is the band-edge effective mass, x is the fractional concentration of the alloy, and ΔE is an energy parameter characterizing the alloy potential fluctuations. Although this formula has been widely used, the identification of the alloy disorder parameter ΔE remains uncertain. Various suggestions have been made for ΔE, e.g., the band-edge discontinuity (Harrison and Hauser, 1976), or band-gap differences. Any of these simple choices is bound to fail when Eq. (6.4.1) is applied to alloy systems, where one encounters conduction-band minima transferring between two states of different symmetry, e.g., between X and L in Si_xGe_{1-x}. For example, if ΔE is taken to be the difference in the corresponding band edges, then one finds that $\Delta E \simeq 0.1$ eV for the X (Δ) valley and about 1.2 eV for the L valley. The values that fit the experiment are about 0.6 eV for L and about 0.5 eV for X (Harrison and Hauser, 1976). There is also a problem with the mass m^* that enters Eq. (6.4.1), particularly for an anisotropic valley.

Recall from Chapter 5 that the CPA formalism of the alloy problem requires one to define the potential disorder. Once a given alloy model is defined, the alloy disorder parameter is given. In the tight-binding CPA model (TBCPA) the alloy disorder is characterized by the difference between the two constituent compounds in the diagonal and off-diagonal matrix elements of the TB Hamiltonian. To estimate the alloy disorder limited mobility in $Hg_{1-x}Cd_xTe$ using TBCPA, Hass *et al.* (1983) defined ΔE to be $f_s\delta\varepsilon_s$, where f_s is the ratio of the cation s partial density of states to the whole density of states, and $\delta\varepsilon_s$ is the difference between the s atomic term values of Hg and Cd atoms. We will derive this result and extend the formula to treat more complex alloy bands with indirect gaps and multiple band minima.

Start with Eq. (6.3.10) and apply it to the alloy scattering limited conductivity. If the **k** dependence of the collision time is only through the energy, then the electrical conductivity can be written as

$$\sigma = -\int \frac{\partial f}{\partial \varepsilon} \sigma(\varepsilon)\, d\varepsilon \tag{6.4.2}$$

where the energy-dependent $\sigma(\varepsilon)$ is given by

$$\sigma(\varepsilon) = 2Ne^2 \mathrm{v}^2(\varepsilon)\tau(\varepsilon)\rho(\varepsilon)/3V \tag{6.4.3}$$

In Eq. (6.4.3), N is the total number of unit cells in the alloy, V is the volume, ρ is the density of states per unit cell per spin, and $\mathrm{v}^2(\varepsilon)$ is the average:

$$\mathrm{v}^2(\varepsilon) = \frac{1}{N} \sum_{\mathbf{k}} \frac{\mathrm{v}^2(\mathbf{k})\delta[\varepsilon - \varepsilon(\mathbf{k})]}{\rho(\varepsilon)} \tag{6.4.4}$$

Equations (6.4.2) and (6.4.3) can also be obtained within CPA alloy theory of the electrical conductivity (Chen *et al.*, 1972; particularly Eqs. (E16) and (E18)).

The alloy scattering lifetime in Eq. (6.4.3) is related to the broadening of the energy level $\Delta(\varepsilon)$ by $\tau(\varepsilon) = \hbar/2\Delta$. In the CPA alloy theory, $\Delta(\varepsilon)$ is the absolute value of the imaginary part of the self-energy in the averaged Green's function. In the weak-scattering limit, $\Delta(\varepsilon)$ in a single-band model is given by Eq. (5.9.13), which in present terms is

$$\Delta(\varepsilon) = \pi x(1-x)(\Delta E)^2 \rho(\varepsilon) \tag{6.4.5}$$

Then the mobility is $\mu_A = \sigma/(ne)$, with the electron density n given by

$$n = 2\frac{N}{V}\int f(\varepsilon)\rho(\varepsilon)\, d\varepsilon \tag{6.4.6}$$

For a nondegenerate semiconductor, $f(\varepsilon)$ is the Boltzmann distribution, $f(\varepsilon) \simeq e^{-(\varepsilon-\varepsilon_F)/(kT)}$. Furthermore, for a parabolic band where $\varepsilon(k) = \hbar k^2/(2m^*)$ and $\rho(\varepsilon) = (2m^*)^{3/2}\varepsilon^{1/2}/(4\hbar^2\pi^2)$, Eq. (6.4.2) can been integrated explicitly to give the Brooks formula in Eq. (6.4.1).

For a real semiconductor alloy in a TB description, the alloy scattering can arise from both the term value differences and from fluctuations of the off-diagonal matrix elements. For the case in which the disorder in the off-diagonal matrix elements is negligible, the alloy disorder is characterized by two parameters $\delta\varepsilon_s$ and $\delta\varepsilon_p$, which are respectively the differences in the s- and p-term values between the two species of substituted atoms. An effective alloy broadening may be defined as

$$\Delta(\varepsilon) = (\Delta_s\rho_s + \Delta_p\rho_p)/\rho \tag{6.4.7}$$

where ρ_s, ρ_p are partial densities of states (PDOS) projected onto the s and p states of the substituted sites, and Δ_s, Δ_p are similar to Eq. (6.4.5) with ρ replaced by ρ_s and ρ_p, respectively. For $Hg_{1-x}Cd_xTe$, the s disorder is predominant (Hass *et al.*, 1983; Chen and Sher, 1982a) and one can neglect Δ_p. Defining $\rho_s = f_s\rho$, approximating $\Delta \simeq \Delta_s\rho_s/\rho = f_s\Delta_s$, and using Eq. (6.4.5) for Δ_s, we arrive at

$$\Delta(\varepsilon) = \pi x(1-x)(f_s\delta\varepsilon_s)^2\rho(\varepsilon) \tag{6.4.8}$$

Thus, by comparison with Eq. (6.4.5), $f_s\delta\varepsilon_s$ plays the role of the alloy scattering parameter ΔE in this special case where the p-term value fluctuation $\delta\varepsilon_p$ is small, as pointed by Hass *et al.* (1983).

For an alloy with an indirect-gap minimum, one has to consider both s and p contributions to the alloy scattering rates and the masses that enter ρ and v^2. Again, Eqs. (6.4.2) through (6.4.7) can be combined to yield a Brooks formula:

$$\mu_A = \frac{(2\pi)^{1/2} e\hbar^4 N_0}{3x(1-x)m_c m_d^{3/2}(\Delta E)^2(kT)^{1/2}} \tag{6.4.9}$$

where the effective alloy scattering parameter ΔE is now given as

$$\Delta E^2 = N_v(f_s^2\delta\varepsilon_s^2 + f_p^2\delta\varepsilon_p^2) \tag{6.4.10}$$

with N_v being the number of equivalent minima (e.g., 6 for Si). Note that two kinds of effective masses enter in Eq. (6.4.9): the effective density of states mass $m_d \equiv (m_t^2 m_l)^{1/3}$ and the conductivity mass $m_c \equiv 3(2/m_t + 1/m_l)^{-1}$, where m_t and m_l are respectively the transverse and longitudinal effective masses.

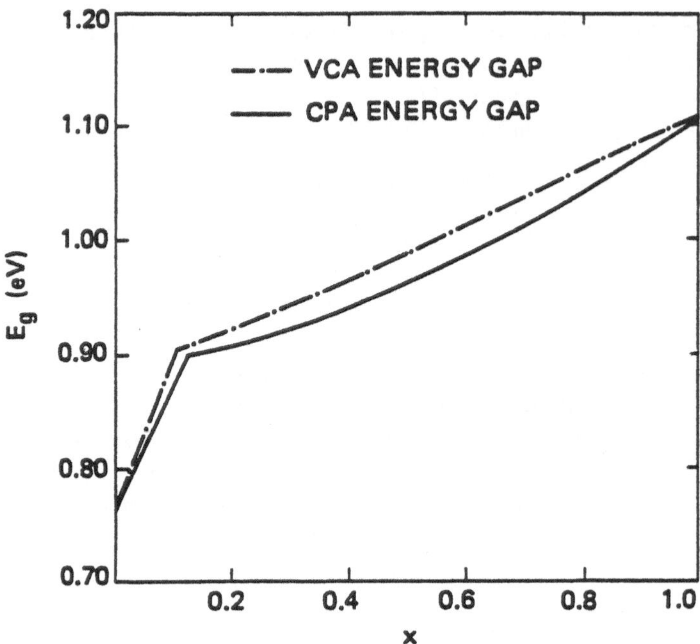

FIGURE 6.1. Variation of the VCA energy gap (dash-dotted line) and the CPA energy gap (solid line) for Si_xGe_{1-x} as a function of x (Krishnamurthy *et al.*, 1986).

Next we consider a still more complicated case where the contributions to the mobility come from more than one band. For example, in Si_xGe_{1-x} the X and L minima cross near $x = 0.15$ (see Fig. 6.1). At the crossover concentration there are two contributions to the net conductivity, so $\sigma = \sum \sigma_i$, where i sums over all X and L valleys. For each σ_i the quantities v_i and ρ_i in Eq. (6.4.3) now take different values for different bands. The effective alloy scattering time $\tau(\varepsilon)$ requires more consideration. The complication comes from the fact that the effective broadening Δ, which is related to the imaginary part of the self-energy, can now come from both bands. For the weak-scattering case, Δ can be calculated from Eq. (5.7.6). If the alloy disorder only comes from the term value differences, Eq. (6.4.7) still holds. In this case the fractions f_s and f_p of the partial densities of states can be obtained from those of the minimum valleys. For example, in Si_xGe_{1-x} alloy at the crossover concentration, the density of states is the sum of those of the two minima X and L: $\rho = \rho_X + \rho_L$, where both ρ_X and ρ_L include all the equivalent pockets, 6 for X and 4 for L. Defining f_{sX} as the fraction of s partial density of states of ρ_X and making a similar definition for f_{sL}, we can write f_s as $f_s = \rho_s/\rho = (f_{sX}\rho_X + f_{sL}\rho_L)/\rho$ and similarly for f_p. Knowing Δ from Eq. (6.4.7) and using $\tau_i = \hbar/2\Delta$, we can still define σ_i and $\mu_i = \sigma_i/n_i$ for each band minimum, with n_i given by Eq. (6.4.6) with ρ replaced by ρ_i. The expression for μ_i is no longer in a simple analytical form like Eqs. (6.4.1) or (6.4.9); however, it can be computed using Eqs. (6.4.2) and (6.4.3).

A simple application of this result was made by Krishnamurthy *et al.* (1985, 1986) to study the x-dependence of the electron mobility in Si_xGe_{1-x} alloys. Figure 6.1 shows their calculated band gap as a function of x for both the VCA and CPA calculations. Table 6.1 lists the quadratic approximations for the band-gap variations of the L and X minima, the

TABLE 6.1. Effective Masses, Band Gaps, Fractions of Partial Densities of States f_{sX} and f_{sL}, and the Term Value Differences $\delta\varepsilon_s$ and $\delta\varepsilon_p$ for the Si_xGe_{1-x} Alloy[a]

$m_l(X)$	$0.97\ m_0$
$m_t(X)$	$0.19 m_0$
$m_l(L)$	$1.64 m_0$
$m_t(L)$	$0.082 m_0$
E_X (eV)	$0.8941 + 0.0421x + 0.1691x^2$
E_L (eV)	$0.7596 + 1.0860 + 0.3306x^2$
f_{sx}	$0.333 + 0.05x\ (0 < x < 0.3)$
	$0.339 + 0.03x\ (0.3 < x < 1.0)$
f_{sL}	$0.632 + 0.13x$
$\delta\varepsilon_s$ (eV)	1.42
$\delta\varepsilon_p$ (eV)	0.20

[a]Krishnamurthy *et al.* (1986)

longitudinal and transverse masses, the term value differences $\Delta\varepsilon_s$ and $\Delta\varepsilon_p$, and the fractions f_{sX} and f_{sL} of the s partial densities of states. Because $\Delta\varepsilon_p$ is much smaller than $\Delta\varepsilon_s$, the contribution from $\Delta\varepsilon_p$ to the alloy scattering can be neglected. These parameters allow us to calculate the alloy scattering limited mobility explicitly within the effective-mass approximation. To compare with the measured mobility, the scattering rate has to include the additional scattering rate $1/\tau_0$ due to impurities and phonons. For a quick first estimate $1/\tau_0$ for the X and L valleys in the alloy were assumed to be the same as those for the pure Si and Ge respectively. Then the mobility for each band is given by

$$1/\mu_i = 1/\mu_i^{\circ} + 1/\mu_i^{A} \tag{6.4.11}$$

where μ_i° are the measured mobilities for Si and Ge (Sze, 1981, p. 22), and μ_i^{A} are the alloy limited mobilities calculated from the above procedure using the CPA band structures and disorder scattering rates. Finally the net electron mobility in the alloy is the average value of those from the two valleys:

$$\mu = \sum_i n_i\mu_i \Big/ \sum_i n_i \tag{6.4.12}$$

The calculated electron mobility is plotted as a function of x in Fig. 6.2.

For $x \leq 0.05$ and $x \geq 0.20$, the energy of the higher valley is much larger than kT, so this valley can be neglected. In this alloy, the s scattering dominates, and the L valley has a much larger s density of states than that of the X valley. This large alloy scattering causes a steep decrease in the mobility for small x values; μ reaches its minimum when $x = 0.14$, the concentration at which the L valley crosses the X-valley edge. When $x > 0.14$, the higher-mobility X valley is the minimum one and becomes the major conduction band, while the L valley gradually leaves the picture, so the mobility increases until it reaches a local maximum at $x = 0.17$. As x further increases, because μ_i is inversely proportional to $x(1-x)$, the total μ starts to decrease until it reaches its minimum at $x = 0.5$. After that, μ increases, following the $1/x(1-x)$ rule reasonably well. These results are also observed in the Hall mobility μ_H (Glicksman, 1955, 1958), also plotted as the dashed curve in Fig. 6.2. Although the calculated drift mobility μ_d is not exactly the same as μ_H, the qualitative x dependence of the calculated drift mobility is seen to correlate well with the experimental μ_H.

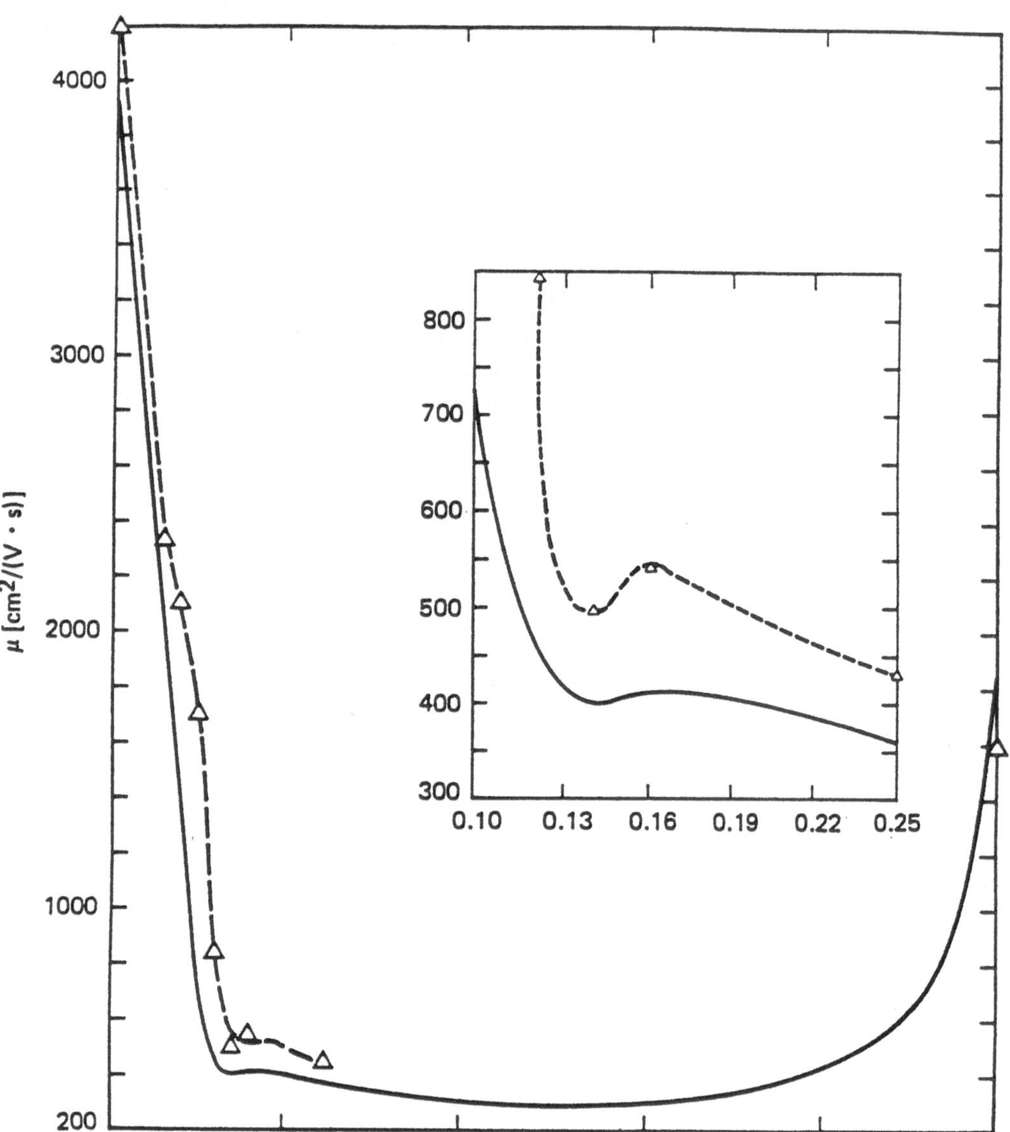

FIGURE 6.2. Si_xGe_{1-x}: Calculated (solid line) electron drift mobility and the experimental Hall mobility (dashed line) from Harrison and Hauser (1976) and Glicksman (1955) are plotted as a function of alloy concentration (Krishnamurthy *et al.*, 1985).

6.5. HOT-ELECTRON v–*E* CHARACTERISTICS: COMPARISON OF MATERIALS' MERITS

In this section we will use the alloy mobility model of Section 4 and a simplified hot-electron theory (Sze, 1981, p. 645) to evaluate the merits of semiconductors and their alloys for high-field-drift transport devices.

Under high-field conditions, the electrons gain energy when they are accelerated from the conduction-band minimum to higher energy levels. The average electron energy in steady state is often considerably higher than the average lattice excitation energy per particle. The electron temperature, T_e, which represents the average electron energy, is therefore much higher than the lattice temperature T. For all direct-gap semiconductors considered here, the mobility in the lower valley, denoted μ_1, is much higher than that in the higher valleys, denoted μ_2. Neglecting μ_2, the drift velocity can be approximated by

$$v_d \simeq \mu_1 E/(1 + Re^{-\Delta E/kT_e}) \tag{6.5.1}$$

where E is the external electric field, R is the ratio between the density of states in the upper valleys and that of the lower valley, and ΔE is the difference between the minima of these two valleys. In steady state, the rate of energy gained by an electron from the field must be equal to the energy lost to the lattice:

$$ev_d E = 3k(T_e - T)/2\tau_E \tag{6.5.2}$$

where τ_E is the electron–lattice energy relaxation time. Equations (6.5.1) and (6.5.2) can be solved simultaneously to yield the velocity–field (v–E) characteristic curves, from which the peak velocity v_p and the corresponding threshold field E_T, i.e., the E field for the maximum v, can be obtained. These values have the approximate forms $v_p \simeq \sqrt{\mu_1 \Delta E/\tau_E}$ and $E_T \simeq \sqrt{R\Delta E/(\tau_E \mu_1)}$.

Alloy effects enter not only from the concentration dependence of the band parameters R and ΔE but also from the effect of alloy scattering on the mobility μ_1. A simple estimate of μ_1 is given by Eq. (6.4.11), modified to include two different temperatures:

$$1/\mu_1 = 1/\mu_0(T) + 1/\mu_A(T_e) \tag{6.5.3}$$

In Eq. (6.5.3), μ_0 is the mobility limited by all scattering mechanisms except alloy scattering and is taken as the concentration-weighed average of the experimental values of the two constituent compounds at the lattice temperature T, which in general is different from T_e. The alloy scattering limited mobility μ_A can be calculated using the generalized Brooks formula discussed in Section 4. Finally, the energy relaxation time τ_E needed in Eq. (6.5.2) can be estimated using a formula given by Conwell (1967, Eq. (3.6.26)), assuming that electrons lose energy to the lattice mainly through interaction with polar optical phonons.

Thus the calculation of the v–E characteristics in the present model requires the following input: the low-field mobility μ_1 and the effective mass m^* for the lower valley, the intervalley energy spacing ΔE, and the ratio R between the density of states of the upper valley to that of the lower valley. Table 6.2 lists the important band parameters of five III–V compounds studied by Krishnamurthy *et al.* (1987). Also listed are the band gaps E_g and the impact ionization energy parameters E_p; E_p is the threshold energy for a hot electron to cause impact ionization of a valence-band electron, i.e., the creation of an electron–hole pair. It is related to the gap by $E_p = E_g(1 + 2\alpha)/(1 + \alpha)$ (Handbook, 1972), where α is the ratio of electron mass to the heavy-hole mass. The reason for considering E_p is that it is also an important parameter in determining which material has the highest practical peak velocity v_p and the smallest threshold electric field E_T. Since $v_p \simeq \sqrt{\mu_1 \Delta E/\tau_E}$, v_p is limited by the largest energy an electron can gain, as characterized by ΔE in a wide-gap semiconductor. In a small-gap semiconductor with $\Delta E > E_p$, impact ionization will take place before electrons are accelerated to their top speed. This limits the suitability of such materials for hot-electron devices. This information and the experimental low-field mobilities are sufficient for the

TABLE 6.2. Important Band Parameters of Five III–V Compounds Used in the v–E Characteristics Study[a]

Parameters	GaAs	InAs	InP	GaSb	InSb
E_g (eV)	1.52	0.42	1.42	0.86	0.25
m^*/m	0.067	0.026	0.072	0.050	0.018
ΔE (eV)	0.282	1.11	0.82	0.27	0.73
R	42.95	186.02	45.02	200.10	1493.14
E_p (eV)	1.70	0.44	1.34	0.98	0.26

[a]Krishnamurthy *et al.* (1987)

calculation of v–E curves for the constituent compounds. For alloys, the molecular CPA theory described in Chapter 5 using a band structure scheme similar to that described in Chapter 7 was carried out to obtained the band structure parameters and the alloy scattering limited mobilities μ_A by Krishnamurthy *et al.* (1987) in their systematic evaluation of the v_p and the associated E_T for semiconductors and their alloys.

6.5.1. Example: $In_{1-x}Ga_xAs$

To illustrate the various effects on the v–E characteristic curve in the present model, the results for $In_{1-x}Ga_xAs$ are discussed here in some detail. Figure 6.3 displays the calculated v–E curves at $T = 300$ K for several x values. Although the assumption $\mu_2 = 0$ for the upper valley of the present model invalidates the v–E curves for $E > E_T$, these curves serve for qualitative comparisons between alloys of different x values. Since the mobility in InAs is larger than that in GaAs, the peak value v_p decreases with x while the corresponding threshold electric field E_T increases with x. This predicated behavior has been observed experimentally (Nag, 1984). At $x = 0.47$, the v_p value is about 50% larger than that of GaAs (Windhorn *et al.*, 1982a,b; Kowalsky and Schlacketzki, 1985). Figure 6.4 shows that v_p in a given alloy is reduced by temperature, because the mobility μ_1 is also reduced. Figure 6.5 shows a comparison of the calculated v–E curves using three different approximations. Curves (a) and (c) are calculated using the effective-mass approximation (EMA), while curve (b) is the result obtained using the actual alloy band structure to calculate the alloy scattering limited mobility. The affect of alloy scattering is seen to be reduced using the detailed band structure. Figure 6.6 summarizes the dependence of v_p on alloy concentration x (the dashed curve) at $T = 77$ K and 300 K. Also plotted are the values of ΔE and E_p. As mentioned earlier, if ΔE is larger than E_p, avalanche breakdown is expected to occur before v_p is reached. Thus the v_p versus x plots are only meaningful for the range of x where $\Delta E < E_p$. In the present case, this range is $x > 0.47$. The one experimental datum (the circle) (Windhorn *et al.*, 1982a,b) shown in the figure lies slightly below the calculated value.

6.5.2. Merits of Other Alloys Compared with GaAs

The v–E characteristic curves were also deduced by Krishnamurthy *et al.* (1987) for the direct-gap ternary alloys GaInSb, InAsP, InAsSb, GaAsSb, and InPSb. The results are summarized in Fig. 6.7, where the ratios of v_p and $(E_T)^{-1}$ to those of GaAs are plotted across the full range of alloy concentrations. Also plotted are the values of ΔE and E_p. Only for those alloys which satisfy the condition $E_p > \Delta E$ can the peak values v_p shown in the plot be reached before avalanche breakdown takes place. The circles in the figure are the experimental values (Windhorn *et al.*, 1982a,b; Kowalsky and Schlachetzki, 1985). Considering the simplicity of the model, the agreement with experiment is remarkable. The ratio of the peak velocity to the energy dissipated by an electron traveling at this speed is $(eE_T)^{-1}$; accordingly $(E_T)^{-1}$ measures the efficiency of an alloy relative to GaAs in hot-electron-device applications. The calculated values of $(E_T)^{-1}$ also agree very well with the experimental data. The results summarized in Fig. 6.5 suggest that InP-based alloys are even more promising candidates for high-speed devices than GaAs. However, a more rigorous calculation in Section 7 suggests that $Ga_{1-x}In_xAs$ alloys remain superior.

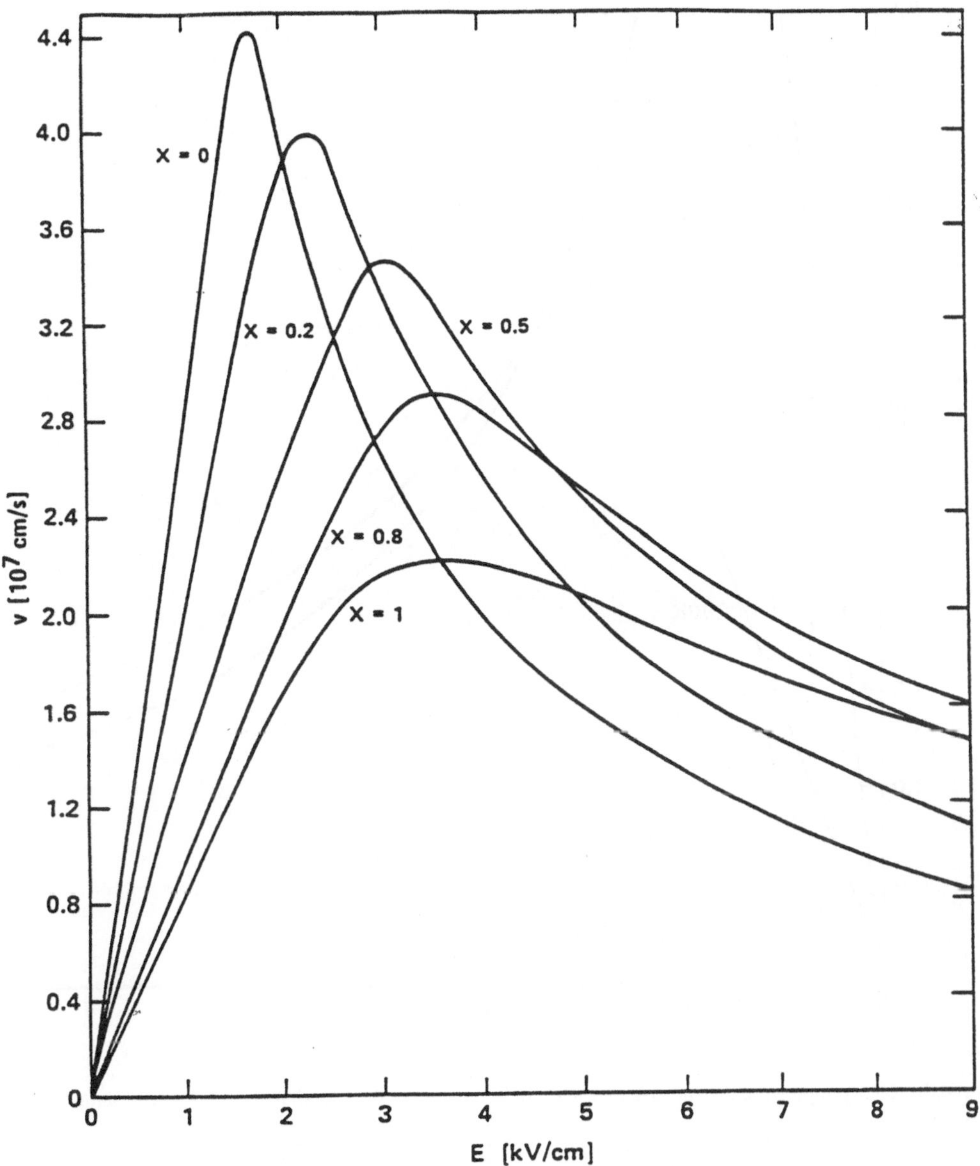

FIGURE 6.3. Velocity field characteristics of $Ga_xIn_{1-x}As$ alloys (Krishnamurthy *et al.*, 1987).

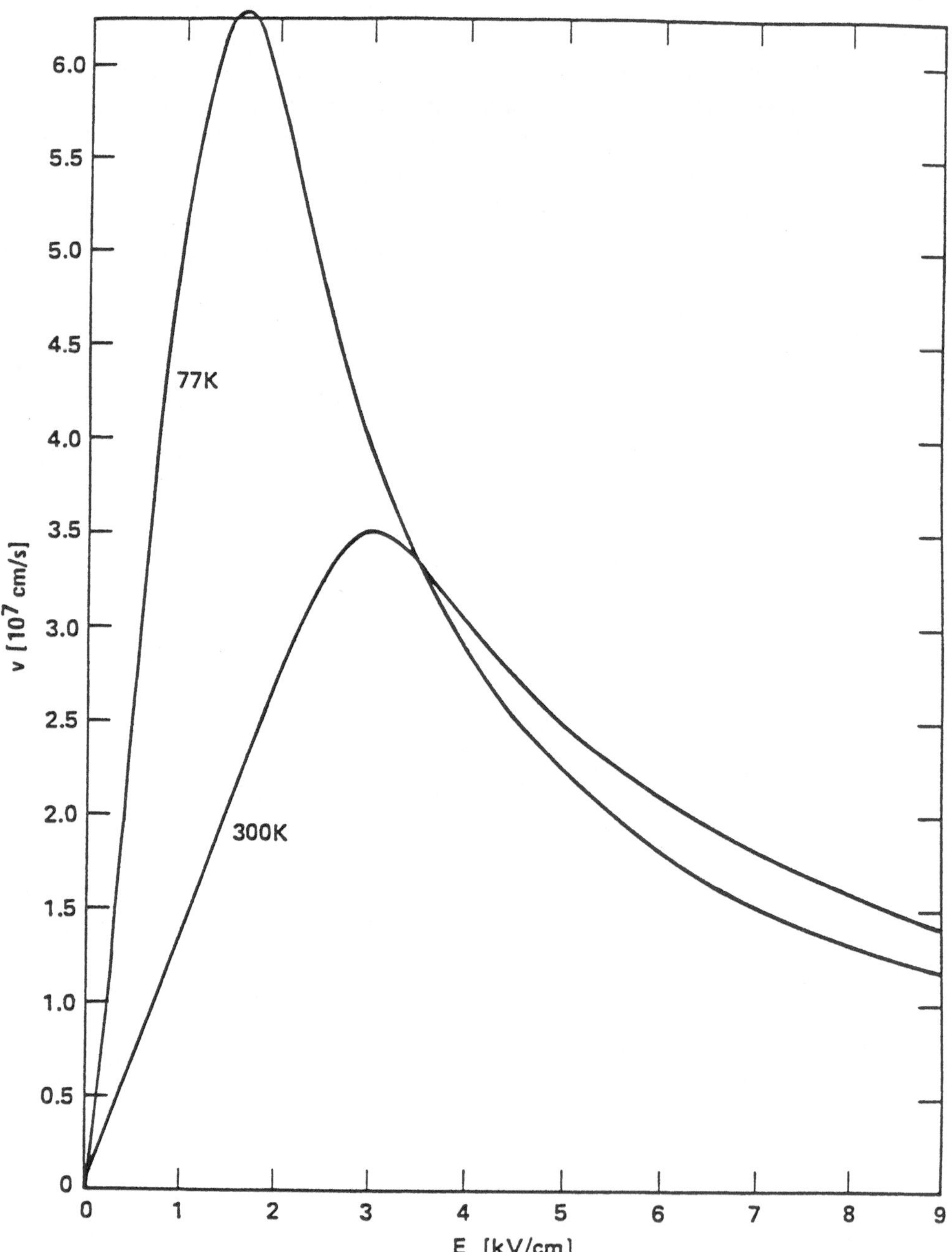

FIGURE 6.4. Temperature dependence of v – E characteristics of the $Ga_{0.5}In_{0.5}As$ alloy (Krishnamurthy *et al.*, 1987).

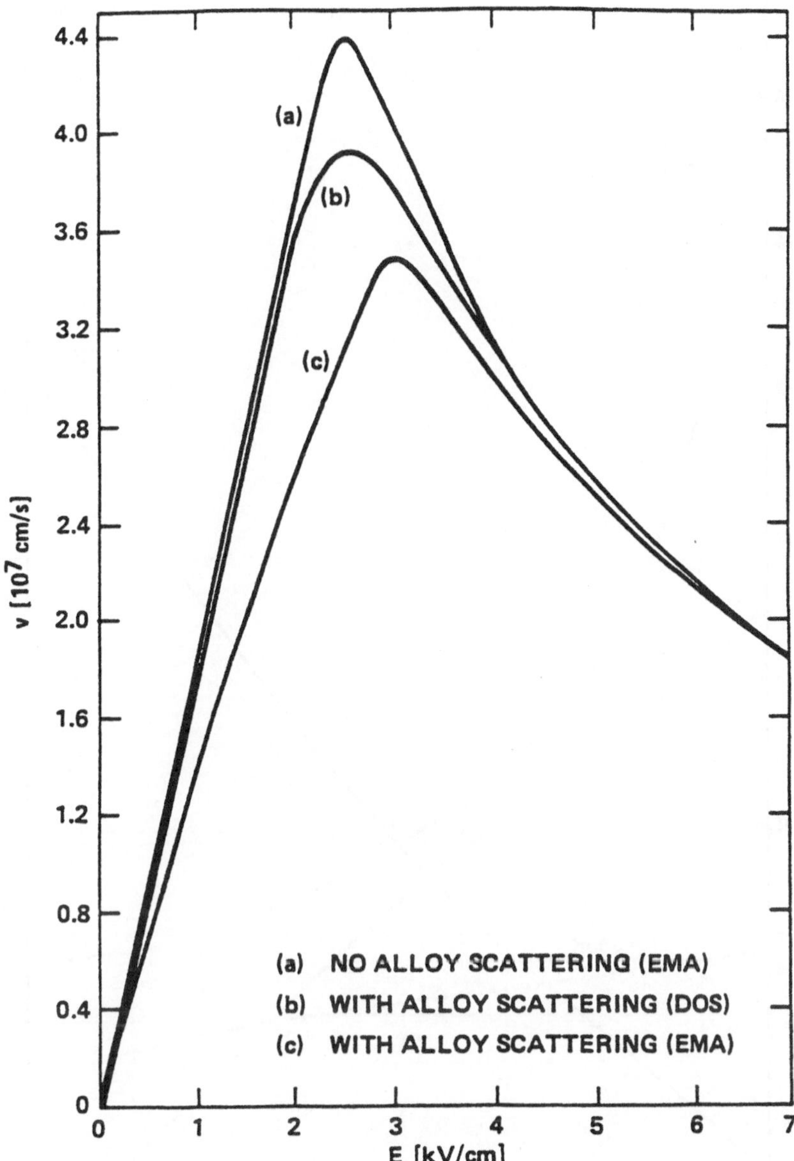

FIGURE 6.5. Effect of band structure and alloy scattering on v – *E* characteristics for $Ga_{0.5}In_{0.5}As$ at 300K. Results of the calculation (a) in EMA without alloy scattering, (b) in EMA with alloy scattering, and (c) with the exact DOS and alloy scattering (Krishnamurthy *et al.*, 1987).

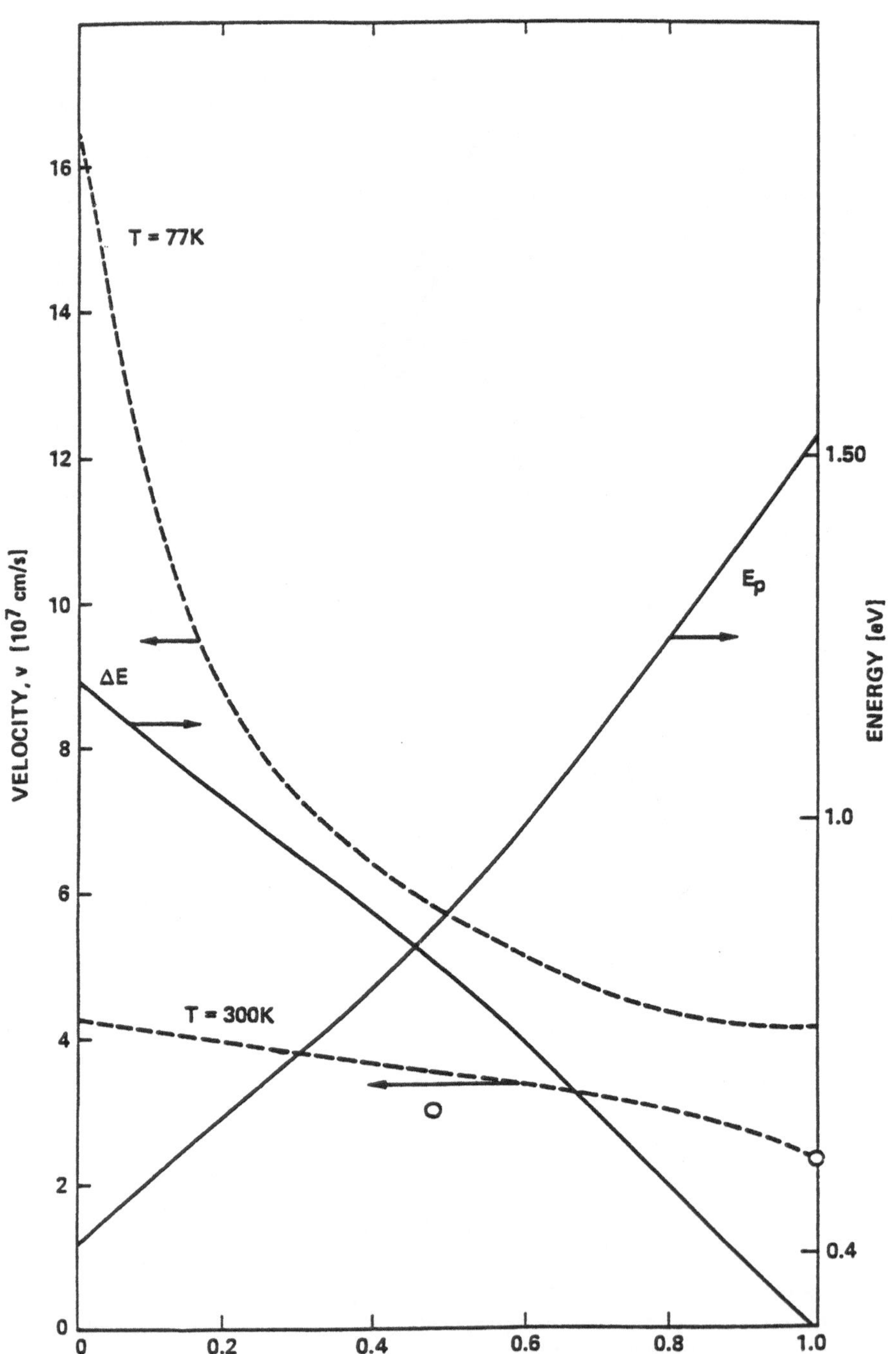

FIGURE 6.6. Variation of ΔE, E_p, v_p as a function of GaAs concentration x in $Ga_xIn_{1-x}As$ alloys. Experimental values (open circles) of v_p are from Windhorn *et al.* (1982). (Krishnamurthy *et al.*, 1987).

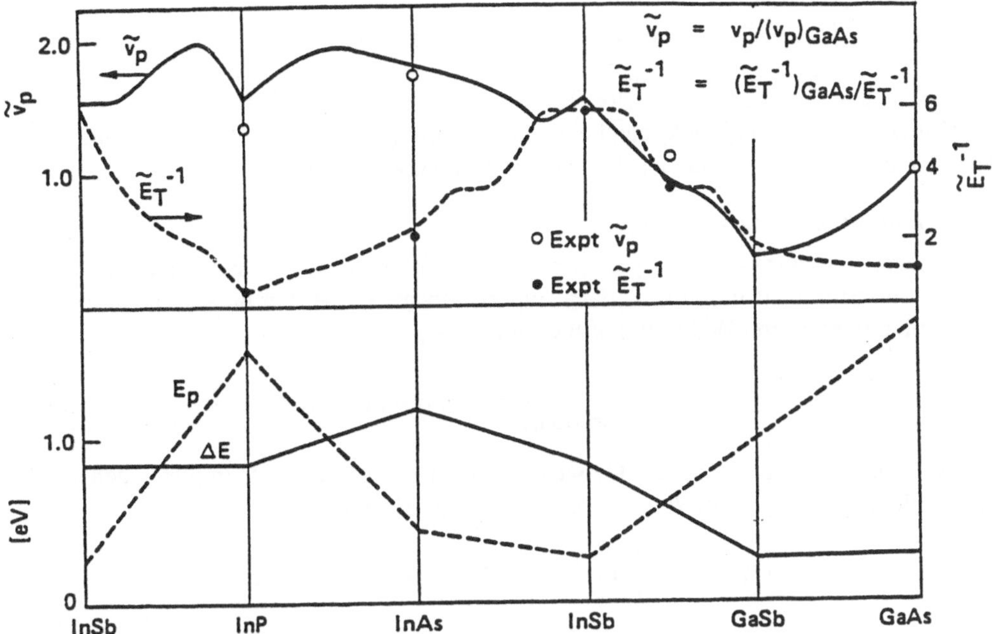

FIGURE 6.7. Variations of ΔE (solid line in lower half), E_p (dotted line in lower half), $\tilde{v}_p$ (solid line in upper half), and $\tilde{E}_T^{-1}$ (dotted line in upper half) are shown. $\tilde{E}_T^{-1}$ are peak velocity and threshold field in the unit of respective GaAs values. The experimental values of peak velocity (open circles) and threshold fields (filled circles) are also shown (Krishnamurthy *et al.*, 1987).

6.6. SCATTERING MECHANISMS

The calculation of the transition rates $w(k'/k)$ is at the heart of transport theory. Before we proceed to discuss more rigorous calculations using realistic band structures, we consider here several important scattering mechanisms of electrons in semiconductors and alloys. These mechanisms are (1) electron–phonon interactions including scattering by the deformation Hamiltonian and by polarization field induced by longitudinal optical phonons in polar semiconductors, (2) ionized impurity scattering, (3) alloy scattering, and (4) the electron–electron interaction. There are excellent books discussing the electron–phonon interactions and impurity scattering (Conwell, 1967; Ridley, 1988).

One emphasis in this chapter is to incorporate realistic band structures in the transport calculations. Within the framework of the long-range tight-binding method of Chapter 7, the general procedure for calculating the matrix elements is as follows. After diagonalizing the TB Hamiltonian, one obtains the band energies and the associate eigenfunctions $|n\mathbf{k}\rangle$, which can be expressed in terms of the Bloch basis functions $|\mathbf{k}j\alpha\rangle$ constructed from orthonormal local basis functions $|Lj\alpha\rangle$ built of Slater or Gaussian orbitals (Chen and Sher 1982b; also see Section 7.1):

$$|n\mathbf{k}\rangle = \sum A_{j\alpha}|\mathbf{k}j\alpha\rangle$$

The Bloch basis functions can be further expanded in terms of plane waves $|\mathbf{k}+\mathbf{G}\rangle$:

$$|\mathbf{k}j\alpha\rangle = \sum c_{j\alpha}(\mathbf{k}+\mathbf{G})|\mathbf{k}+\mathbf{G}\rangle$$

where **G** is a reciprocal lattice vector. Thus, once the scattering Hamitonians V are defined, the matrix elements $\langle n'\mathbf{k}'|V|n\mathbf{k}\rangle$ can be evaluated through the local orbitals or plane waves. Then the transition rates can be calculated from the golden rule.

6.6.1. *Ionized Impurity Scattering*

The most important impurity scattering of electrons in a doped semiconductor often is due to ionized dopant impurities. Each impurity located at **R** can be assigned a screened Coulomb potential (in MKS units) of the form

$$V_{\mathrm{I}}(r) = \frac{Ze^2}{4\pi\kappa_\infty\varepsilon_0|\mathbf{r}-\mathbf{R}|}\exp(-\lambda|\mathbf{r}-\mathbf{R}|) \tag{6.6.1}$$

where $\lambda = n_{\mathrm{I}}e/(\kappa_\infty\varepsilon_0 k_{\mathrm{B}}T)^{1/2}$ is the Debye screening length, with n_{I} the impurity density. The matrix element of $V_{\mathrm{I}}(r)$ with R at the origin between two plane waves is

$$\begin{aligned} v(\mathbf{k}'-\mathbf{k}) &\equiv \frac{1}{V}\int e^{-i\mathbf{k}'\cdot\mathbf{r}}\frac{Ze^2}{4\pi\kappa_\infty\varepsilon_0|\mathbf{r}|}\exp(-\lambda|\mathbf{r}|)\, e^{i\mathbf{k}\cdot\mathbf{r}}\, d^3r \\ &= \frac{Ze^2}{V\kappa_\infty\varepsilon_0(|\mathbf{k}'-\mathbf{k}|^2+\lambda^2)} \end{aligned} \tag{6.6.2}$$

Then the matrix elements between the eigenstates are

$$\begin{aligned} \langle n'\mathbf{k}'|V_{\mathrm{I}}|n\mathbf{k}\rangle &= \sum_{j\alpha j'\alpha'}\sum A^*_{j'\alpha',n'}A_{j\alpha,n} \\ &\times \sum_{GG'}\sum C^*_{j\alpha}(\mathbf{k}'+\mathbf{G}')C_{j\alpha}(\mathbf{k}+\mathbf{G})v(\mathbf{k}'+\mathbf{G}'-\mathbf{k}-\mathbf{G}) \end{aligned} \tag{6.6.3}$$

This scattering is elastic, so in perturbation theory the transition rate is given by

$$w(\mathbf{k}'/\mathbf{k}) = \frac{2\pi}{\hbar}|\langle n'\mathbf{k}'|V|n\mathbf{k}\rangle|^2\delta[\varepsilon_{n'}(\mathbf{k}')-\varepsilon_n(\mathbf{k})] \tag{6.6.4}$$

6.6.2. *Bare Electron–Phonon Interaction*

In terms of the tight-binding Hamiltonian described in Chapter 7, the most direct way to implement the electron–phonon interaction due to deformation is to express the scattering Hamiltonian H_{ep} in the orthonormal local orbitals basis $|\mathrm{L}j\alpha\rangle$:

$$\langle \mathrm{L}'j'\alpha'|H_{\mathrm{ep}}|\mathrm{L}j\alpha\rangle = \frac{\partial\mathcal{V}(\mathbf{d})}{\partial d}\hat{\mathbf{d}}\cdot(\boldsymbol{\eta}_{\mathrm{L}'j'\alpha'}-\boldsymbol{\eta}_{\mathrm{L}j\alpha}) \tag{6.6.5}$$

where $\mathcal{V}$ represents the matrix elements of the electronic Hamiltonian between the two orthogonal local orbitals $|\mathrm{L}j\alpha\rangle$ and $|\mathrm{L}'j'\alpha'\rangle$ separated by a displacement $\mathbf{d} = (\mathbf{L}+\boldsymbol{\tau}_j+\boldsymbol{\eta}_{\mathrm{L}j\alpha}-\mathbf{L}'-\boldsymbol{\tau}_{j'}-\boldsymbol{\eta}_{\mathrm{L}'j'\alpha'})$. The derivative in Eq. (6.6.5) is evaluated at equilibrium positions, where the $\boldsymbol{\eta}$'s are set to zero.

The small displacement $\boldsymbol{\eta}$ in the phonon field is given by

$$\eta_{Lj\alpha} = \sum_{q\lambda} \left(\frac{\hbar}{2NM_j\omega_{\lambda q}} \right)^{1/2} [\hat{\mathbf{e}}_{\lambda q}\, e^{i\mathbf{q}\cdot(\mathbf{L}+\boldsymbol{\tau}_j)} b_{\lambda q} + \hat{\mathbf{e}}^*_{\lambda q}\, e^{-i\mathbf{q}\cdot(\mathbf{L}+\boldsymbol{\tau}_j)}\, b^\dagger_{\lambda q}] \tag{6.6.6}$$

where the summation is over the phonon wave vectors **q** and the polarization index λ, M_j is the mass of the atom at $\mathbf{L} + \boldsymbol{\tau}_j$, $\hat{e}_{\lambda q}$ is the eigenvector of a phonon mode, and $b_{\lambda q}$ and $b^\dagger_{\lambda q}$ are respectively the destruction and creation operators of phonons. Explicit expressions for the matrix elements are

$$\langle N_{\lambda q} - 1, \mathbf{k}'j'\alpha' | H_{\text{ep}} | N_{\lambda q'}, \mathbf{k}j\alpha \rangle = \delta_{\mathbf{k}',\mathbf{k}+\mathbf{q}} \left(\frac{\hbar}{2N\omega_{\lambda q}} \right)^{1/2} V^{\alpha'\alpha}_{j'j}(\mathbf{k}+\mathbf{q}, \mathbf{k}; \lambda) \sqrt{N_{\lambda q}} \tag{6.6.7}$$

where

$$V^{\alpha'\alpha}_{j'j}(\mathbf{k}+\mathbf{q}, \mathbf{k}, \lambda) = \sum_{\mathbf{d}} \frac{\partial \mathcal{V}(\mathbf{d})}{\partial d}\, \hat{\mathbf{d}}\cdot\hat{\mathbf{e}}_{\lambda q} \left(\frac{e^{-i\mathbf{k}\cdot\mathbf{d}}}{\sqrt{M_{j'}}} - \frac{e^{-i\mathbf{k}'\cdot\mathbf{d}}}{\sqrt{M_j}} \right) \tag{6.6.8}$$

What we now need are matrix elements between the energy eigenstates, i.e., $\langle N_{\lambda q} - 1, n'\mathbf{k}' | H_{\text{ep}} | N_{\lambda q'}, n\mathbf{k} \rangle$, which can be obtained from Eq. (6.6.7) with $V^{\alpha'\alpha}_{j'j}(\mathbf{k}+\mathbf{q}, \mathbf{k}, \lambda)$ replaced by

$$V_{n'n}(\mathbf{k}+\mathbf{q}, \mathbf{k}, \lambda) = \sum_{j\alpha j'\alpha'} \sum A^*_{j'\alpha',n'} A_{j\alpha,n} V^{\alpha'\alpha}_{j'j}(\mathbf{k}+\mathbf{q}, \mathbf{k}, \lambda) \tag{6.6.9}$$

The transition rates can now be calculated using Eqs. (6.2.8) and (6.2.9), where the summation runs over both λ and **q**, and V_q is replaced by $V_{n'n}(\mathbf{k}+\mathbf{q}, \mathbf{k}, \lambda)$.

6.6.3. *Polar Optical Phonon Scattering*

In addition to the bare electron–phonon interactions, the electrons also suffer scattering from the polarization field induced by longitudinal optical phonons in a polar semiconductor. This induced interaction can be described by the Frohlich (1937) Hamiltonian (Callen, 1949; Ehrenreich, 1957)

$$V_{\text{ep}} = i \left(4\pi e^2 \frac{\hbar\omega_0}{2V} \left(\frac{1}{\kappa_\infty} - \frac{1}{\kappa_0} \right) \right)^{1/2} \sum_q \frac{1}{q} (b_q e^{i\mathbf{q}\cdot\mathbf{r}} + b^\dagger_q e^{i\mathbf{q}\cdot\mathbf{r}}) \tag{6.6.10}$$

where κ_0 and κ_∞ are the zero and infinite frequency dielectric constants, ω_0 is the LO phonon frequency, which is taken as a constant because of small dispersion. A direct calculation of the V_{ep} using the band structure wave functions yields the transition rates in the forms given by Eqs. (6.2.8) and (6.2.9) with the $|V_q|^2$ replaced by $|A|^2 |\langle n'\mathbf{k}' | \exp(i\mathbf{q}\cdot\mathbf{r}) | n\mathbf{k} \rangle|^2$, where A is the combined constant factor before the summation in Eq. (6.6.9).

We note that all the scattering rates due to electron–phonon interactions discussed in this chapter are based on first-order perturbation theory, which only causes the emission or absorption of one phonon going from the initial to the final states. Higher-order scattering will involve multiphonon processes. However, Fales (1975) has shown, in the case of polar optical phonon scattering, there are sequential cancellations between successive orders which leave first-order scattering to be the dominant term.

6.6.4. Alloy Scattering

The basic ideas of alloy scattering were discussed in Section 4. Here we describe how to incorporate these ideas into an actual calculation. Within the tight-binding scheme, alloy disorder can be described by the fluctuations in the site-diagonal matrix elements and the first few neighbor interactions. One can start with a periodic unperturbed Hamiltonian in the virtual crystal approximation (VCA) and define the fluctuations as ΔV for a given alloy configuration. In most semiconductors, ΔV is weak and can be treated within perturbation theory (see Section 5.7). One can then calculate the matrix elements $\langle n'\mathbf{k}'|\Delta V|n\mathbf{k}\rangle$ between the eigenstates of the unperturbed Hamiltonian. Then the alloy scattering rates are calculated using the golden rule:

$$w(k'/k) = \frac{2\pi}{\hbar}\overline{|\langle n'\mathbf{k}'|\Delta V|n\mathbf{k}\rangle|^2}\delta\,[\varepsilon_{n'}(\mathbf{k}') - \varepsilon_n(\mathbf{k})] \tag{6.6.11}$$

where the overbar means the average over all alloy configurations.

If the fluctuation in the alloy Hamiltonian is large, multiple-scattering theory of Chapter 5 has to be used to calculate the scattering matrix. Within the general CPA theory, the self-energy operator Σ associated with the self-consistently determined effective Green function in the alloy is proportional to the averaged $\mathbf{k}$-dependent alloy scattering rate

$$R(\mathbf{k}) = 2|\mathrm{Im}\langle n\mathbf{k}|\Sigma|n\mathbf{k}\rangle|/\hbar \tag{6.6.12}$$

This is equivalent to assuming that the scattering rates are isotropic in $\mathbf{k}$-space so the following expression for $w(\mathbf{k}'/\mathbf{k})$ can be used in the Boltzmann equation:

$$w(\mathbf{k}'/\mathbf{k}) = R(\mathbf{k})\delta[\varepsilon_{n'}(\mathbf{k}') - \varepsilon_n(\mathbf{k})]/\mathcal{D}(\varepsilon) \tag{6.6.13}$$

where $\mathcal{D}(\varepsilon)$ is the density of states per spin at $\varepsilon = \varepsilon_n(\mathbf{k})$.

6.6.5. Electron–Electron Scattering

The scattering of electrons by other conduction electrons can be important in a semiconductor when the doping concentration is moderately high (i.e., $n > 10^{17}$ 1/cm^3). The electron–electron scattering rates can be estimated from the q-dependent dielectric function for the phonon-plasmon-coupled system, which is given by

$$\varepsilon(q,\omega) = \varepsilon_\infty(\omega^2 - \omega_l^2)/(\omega^2 - \omega_t^2) + \chi(q,\omega) \tag{6.6.14}$$

where ω_l and ω_t are respectively the longitudinal and transverse phonon frequencies, and $\chi(q,\omega)$ is the electron linear density response function. In the so-called random phase approximation for free electrons, $\chi(q,\omega)$ is the Linhard function; $\chi = \chi_1 + i\chi_2$ contains real and imaginary parts. The effective scattering rates can then be calculated from

$$w(\mathbf{k}'/\mathbf{k}) = 8\pi e^2\int d\omega\,\left|\mathrm{Im}\left(\frac{1}{\varepsilon(q,\omega)}\right)\right|\delta\,[\varepsilon_n(\mathbf{k}) - \varepsilon_{n'}(\mathbf{k}\pm\mathbf{q}) \pm \hbar\omega] \tag{6.6.15}$$

where $\mathbf{k}'$ is $\mathbf{k}\pm\mathbf{q}$ and the $\pm$ choice depends on whether it is an absorption or emission process.

6.7. EXPANSION SOLUTION OF THE BOLTZMANN EQUATION

Analytical solutions of the Boltzmann transport equation rarely exist except for a few simple and usually unrealistic cases. Besides the approximate solutions based on the collision–time approximation discussed in the previous sections, numerical solutions can be obtained by iterative techniques (Budd, 1967; Rees, 1969). The most popular numerical technique is the Monte Carlo method (Fawcett, 1970, 1974; Hauser *et al.*, 1976; Littlejohn *et al.*, 1978, 1982), which is extremely time consuming, particularly when a full band structure is used. There has been a continuing effort to develop faster numerical techniques (Butcher and Fawcett, 1966; Conwell and Vassel, 1966; Aubert *et al.*, 1984). Here we describe a fast numerical method that uses full band structures and includes all scattering mechanisms to yield converged population distributions at high field (Krishnamurthy *et al.*, 1989).

The steady-state Boltzmann equation in the presence of an applied constant electric field for nondegenerate electrons is given by

$$-e\mathbf{E}\cdot\frac{\nabla_{\mathbf{k}}f}{\hbar}=\sum_{\mathbf{k}'}(w(\mathbf{k}/\mathbf{k}')\,f(\mathbf{k}')-w(\mathbf{k}'/\mathbf{k})\,f(\mathbf{k})) \tag{6.7.1}$$

We can expand $f(\mathbf{k})$ in terms of a set of basis function, $\phi_i(\mathbf{k})$, such that

$$f(\mathbf{k})=\sum_i C_i\phi_i(\mathbf{k}) \tag{6.7.2}$$

Substituting $f(\mathbf{k})$ in this form in Eq. (6.7.1), multiplying $\phi_j(\mathbf{k})$ on both sides of the equation, and summing over all $\mathbf{k}$, we obtain a matrix equation

$$\sum_i M_{ji}C_i=0 \tag{6.7.3}$$

where the matrix elements are given by

$$M_{ji}=\sum_{\mathbf{k}}\frac{e}{\hbar}\phi_j(\mathbf{k})\nabla_{\mathbf{k}}\phi_i(\mathbf{k})\cdot\mathbf{E}$$

$$+\sum_{\mathbf{k}\mathbf{k}'}\sum\phi_j(\mathbf{k})\,(w(\mathbf{k}/\mathbf{k}')\,\phi_i(\mathbf{k}')-w(\mathbf{k}'/\mathbf{k})\,\phi_i(\mathbf{k})) \tag{6.7.4}$$

Equation (6.7.3) means that the eigenvector corresponding to the zero eigenvalue of the matrix M gives us the required coefficients to obtain the distribution function, from which the drift velocity can be computed

$$\mathbf{v}_{\mathrm{d}}=\sum_{\mathbf{k}}\frac{\mathbf{v}(\mathbf{k})f(\mathbf{k})}{N} \tag{6.7.5}$$

If the chosen basis set is complete, the zero eigenvector is ensured, and the distribution function can be accurately obtained. However, to be complete, a basis set usually contains a very large number of functions. Then a numerically difficult diagonalization of a very large matrix must be carried out. To facilitate the calculation we need to devise a basis set that is finite and yet accurate for describing the distribution function even at high field strength. It was found that for all direct-gap semiconductors each conduction-band valley only needs simple s-like and p-like basis functions to yield a satisfactory solution (Krishnamurthy *et al.*, 1989, to be referenced as KSC). The functional forms of the basis functions are

$$\phi_s(k) = N_s e^{-\alpha(\varepsilon_k - \varepsilon_0)} \tag{6.7.6}$$

$$\phi_p(k) = N_p \nabla_{\mathbf{k}} \varepsilon_k \cdot \mathbf{E} e^{-\beta(\varepsilon_k - \varepsilon_0)} \tag{6.7.7}$$

where ε_0 is the minimum energy of the appropriate valley, and N_s and N_p are the normalization constants. In an actual calculation, the electric field is chosen to be in the z direction. Then all four L valleys are equivalent and can use the same basis functions. For the X valleys, two different sets are required, one for those whose $\mathbf{k}$ is perpendicular to z and the other set with $\mathbf{k}$ parallel to z. Including the two for Γ, there are a total of eight independent basis functions, so M is an 8×8 matrix. Because the basis is incomplete, there is no assurance that M will have a zero eigenvalue. To obtain an optimal solution, the exponents α and β are also varied. It was found that a simplified set with $\alpha = \beta$ for all valleys yielded accurate answers for low-to-moderate electric field strengths. Physically, this corresponds to assigning them the same effective hot-electron temperature even though the distribution function is not a Maxwellian. In all cases to be discussed, the nearly zero eigenvalues were obtained in less than 10 iterations, with each iteration taking little time (a few seconds on a workstation).

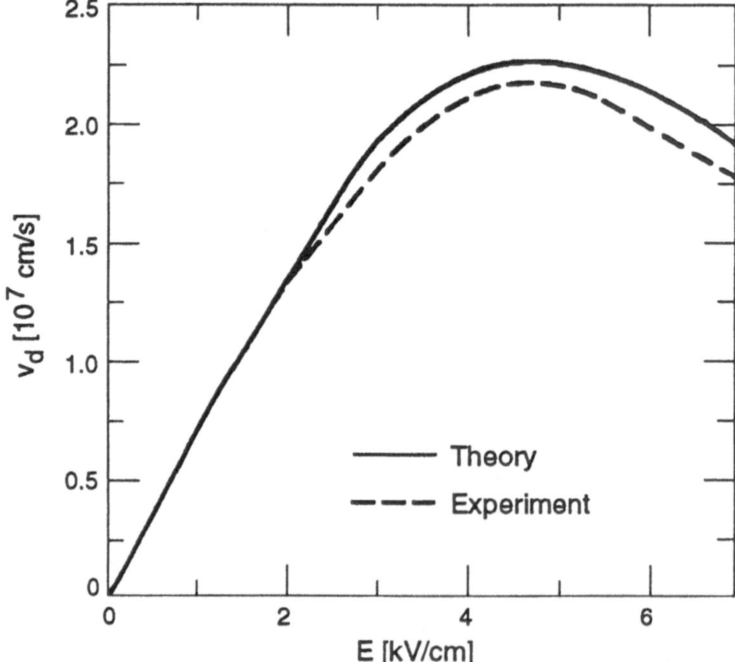

FIGURE 6.8. Calculated (solid line) and experimental (dashed line) v – E curves for bulk GaAs at a lattice temperature of 300 K and an impurity concentration of 5×10^{16} cm^{-3}, respectively (Krishnamurthy *et al.*, 1988a).

Figure 6.8 shows a v–E curve (the solid line) for GaAs calculated from this method by KSC and the corresponding experimental curve (the dashed line) (Masselink *et al.*, 1987). The scattering mechanisms treated in this calculation include only ionized impurity scattering and the polar optical phonon scattering discussed in Section 6. The lattice temperature used is 300 K, and the impurity concentration is $5 \times 10^{16}/\text{cm}^3$. This value of impurity concentration was an adjusted value selected so the calculation produces the experimental value of 7500 cm^2 V-s for the low-field electron mobility (Masselink *et al.*, 1987). The calculated peak velocity v_p of 2.3×10^7 cm/s is slightly larger ($\approx$10%) than the experimental value. However, the threshold field E_T of 4.5 kV/cm obtained in the calculation is in excellent agreement with experiment. While this comparison is not sufficient to suggest further theoretical improvement, we should point out that the present calculation is only accurate for low- to medium-field strengths. For high fields, more basis functions are needed than the ones used in the present calculation.

It is instructive to examine the energy distribution function $g(\varepsilon)$, which can be obtained from the solution $f(\mathbf{k})$ by

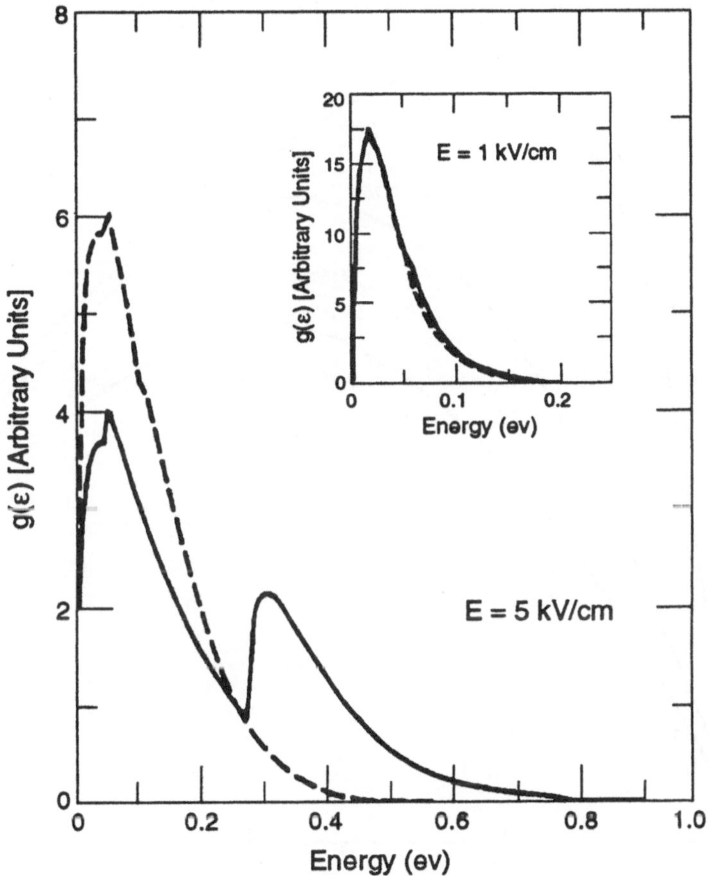

FIGURE 6.9. Our $g(\varepsilon)$ (solid line) is compared with that calculated using the drifted Maxwellian distribution (dashed line) at a 5-kV/cm electric field in GaAs at 300 K. Corresponding curves are shown in the inset for 1 kV/cm of electric field strength (Krishnamurthy *et al.*, 1989).

$$g(\varepsilon) = \frac{1}{N} \sum_{\mathbf{k}} f(\mathbf{k}) \delta[\varepsilon - \varepsilon(\mathbf{k})] \tag{6.7.8}$$

Figure 6.9 shows a comparison between the calculated distribution (solid line) based on the basis expansion method and a displaced Maxwellian (dashed line) that gives the same peak velocity. Curves at both 1-kV/cm and 5-kV/cm field strengths are presented. At 1 kV/cm, the displaced Maxwellian does not differ much from the calculated distribution function. However, at 5 kV/cm the displaced Maxwellian differs substantially from the solution to BE, which shows a bimodal behavior linked to the Γ and L valleys, respectively.

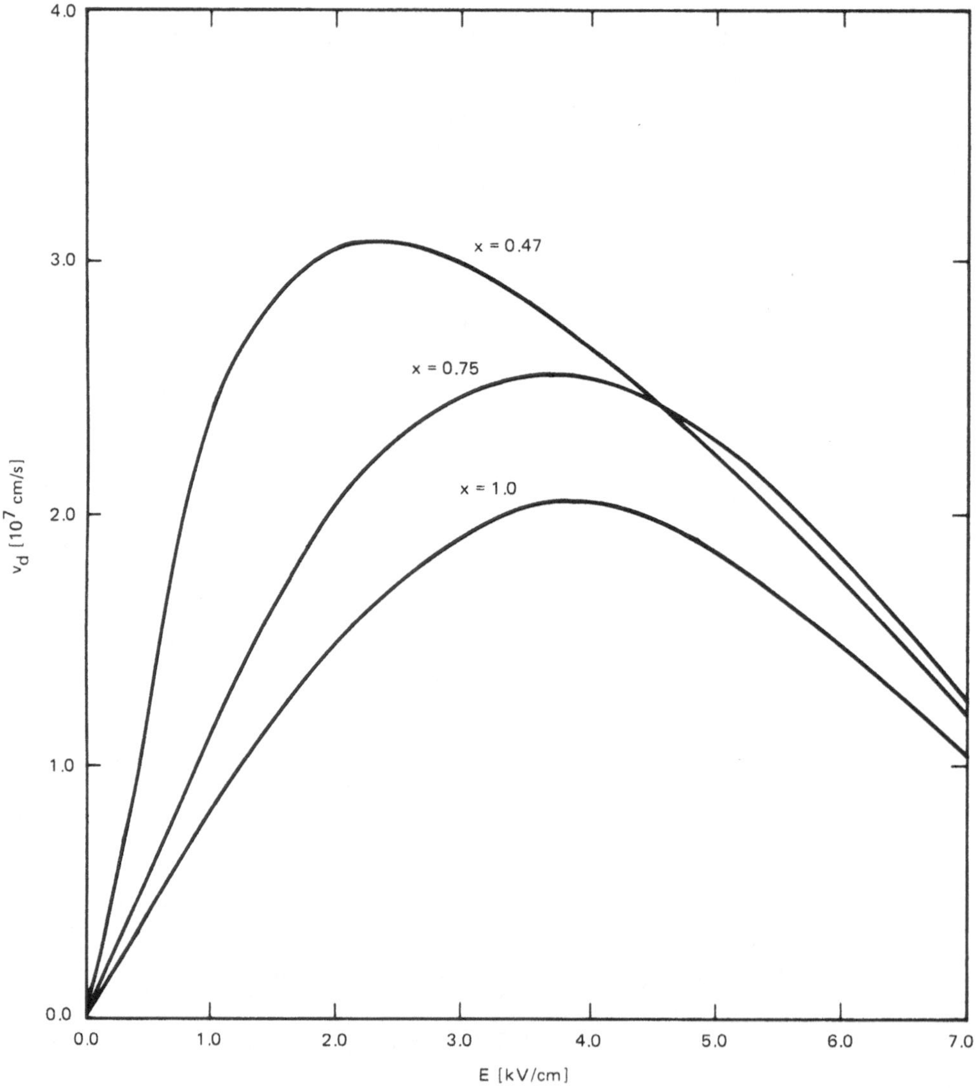

FIGURE 6.10. Velocity–field characteristics of GaAs-based alloys (Krishnamurthy *et al.*, 1987).

The above v–E characteristics calculation was also extended to alloys by including alloy disorder scattering calculated from CPA theory. The calculated v–E curves for $Ga_xIn_{1-x}As$ at $x = 0.47$ and 0.75 are compared with the pure GaAs result in Fig. 6.10. The $x = 0.47$ case is seen to have the highest v_p and smallest E_T values. The curves for the alloys, when compared with those in Fig. 6.3 based on the simple v–E model, are not as sharp and have lower values for the peak velocity. The calculated values of $v_p = 3.2 \times 10^7$ cm/s and $E_T = 3.0$ kV/cm for the $Ga_{0.47}In_{0.53}As$ certainly lie within the experimental ranges of v_p, between 2.2×10^7 and 3.2×10^7 cm/s, and E_T, between 2.7 kV/cm and 4.6 kV/cm (Marsh *et al.*, 1980; Gammel *et al.*, 1981; Bandy *et al.*, 1981; Windhorn *et al.*, 1982a,b; Hase *et al.*, 1987). These calculated values are very close to the measured results $v_p = 3.1 \times 10^7$ cm/s and $E_T = 2.9$ kV/cm of

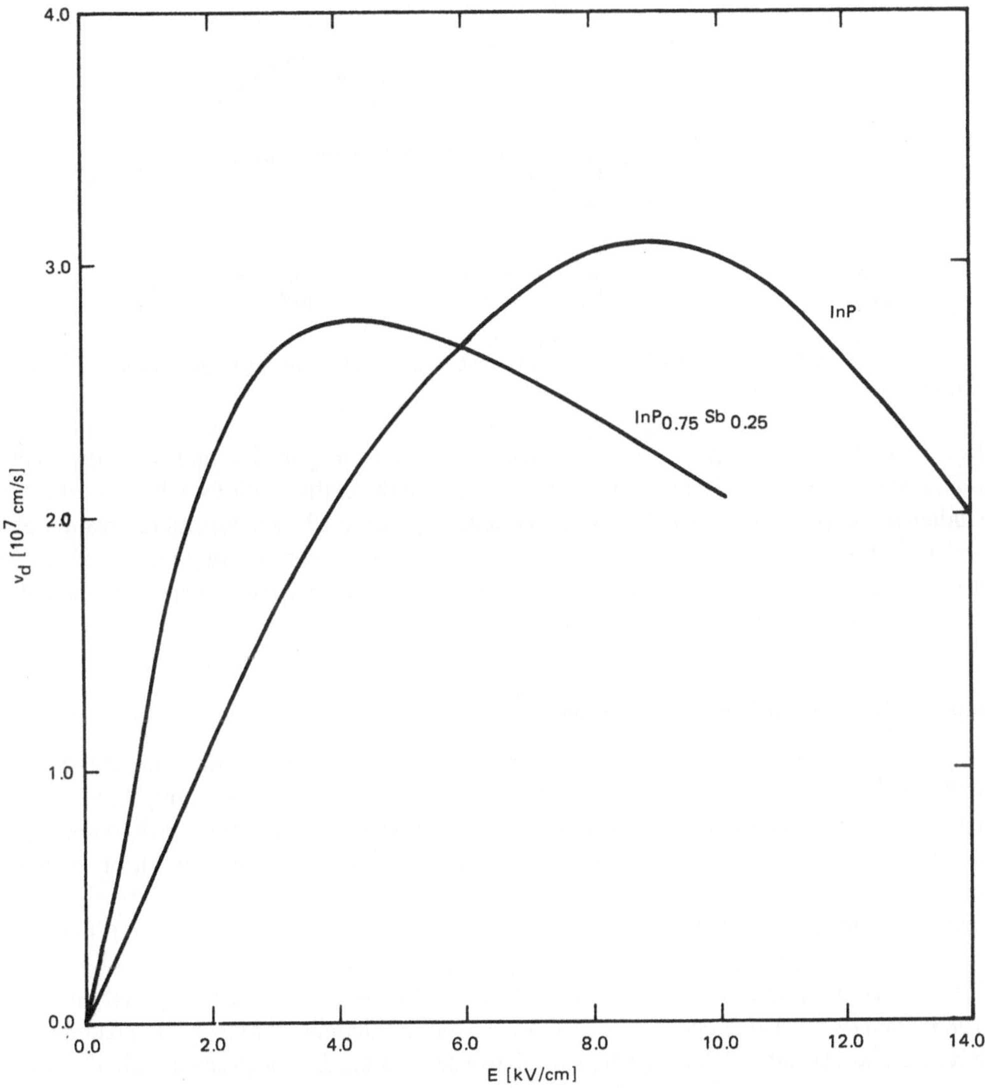

FIGURE 6.11. Velocity–field characteristics of InP-based alloys (Krishnamurthy *et al.*, 1987).

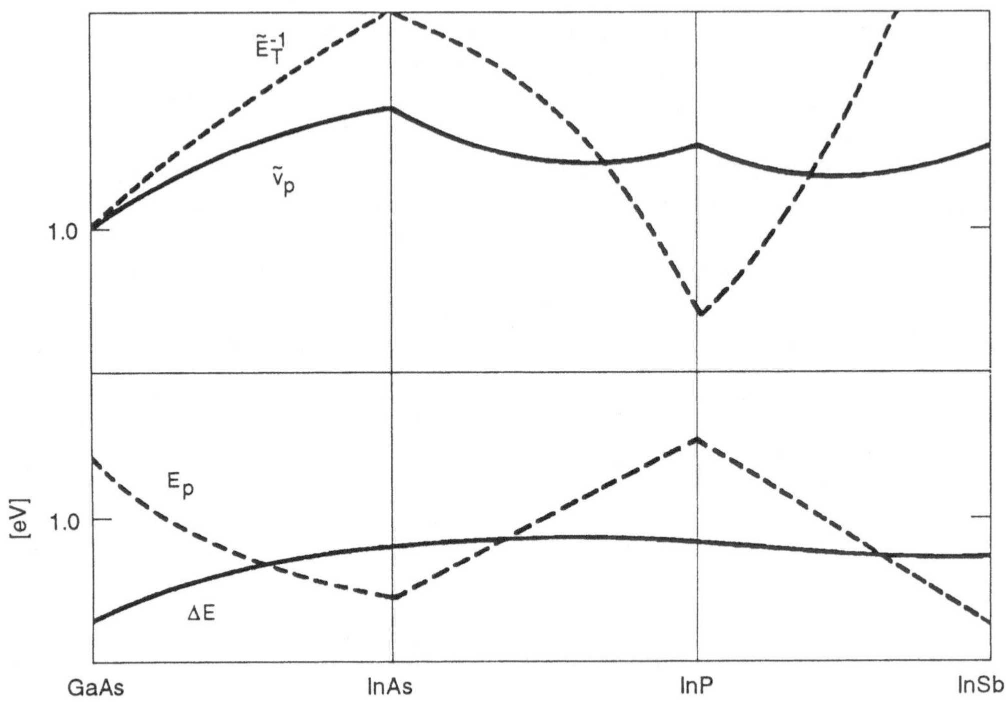

FIGURE 6.12. Comparative evaluation of various semiconductor alloys for high-speed device applications (Krishnamurthy *et al.*, 1987).

Marsh *et al.* (1981). Figure 6.11 shows the v–E curves for pure InP and the alloy system $InP_{0.75}Sb_{0.25}$. These results are comparable to those of the GaInAs alloys, whereas the model studies in Section 5 favor the InP-based systems. Figure 6.12 summarizes the results of v_p and $E_{\bar{T}}^{-1}$ for several alloys in terms of their ratios to the corresponding GaAs values. We note again that the values are meaningful only in the regions where the intervalley separation is smaller than the impact ionization energy; i.e., $\Delta E < E_p$.

6.8. NEAR-BALLISTIC TRANSPORT

In an ideal ballistic-transport device, the charge carriers will traverse the active region with no collisions. In reality, the presence of lattice imperfections, impurities, lattice vibrations, and boundaries between the injector and collector limits such collisionless transport. However, the effects of scattering can be minimized if the length of the active region is small when compared with the mean free path of the charge carriers. Under this condition, electron velocities in the ballistic regime can be on the order of 10^8 cm/s. The measured electron mean free path for most high-quality III–V compounds and alloys ranges from a few hundred to a few thousand angstroms (Levi *et al.*, 1985, 1986; Heiblum *et al.*, 1985; Ohasi *et al.*, 1985; Hasse et al., 1985). Current semiconductor fabrication technologies have been advanced to a point where submicron devices can be constructed with dimensions comparable to these attainable mean free paths. Devices that exhibit ballistic transport have been demonstrated. Using a hot-electron transistor depicted in Fig. 6.13 with a base width

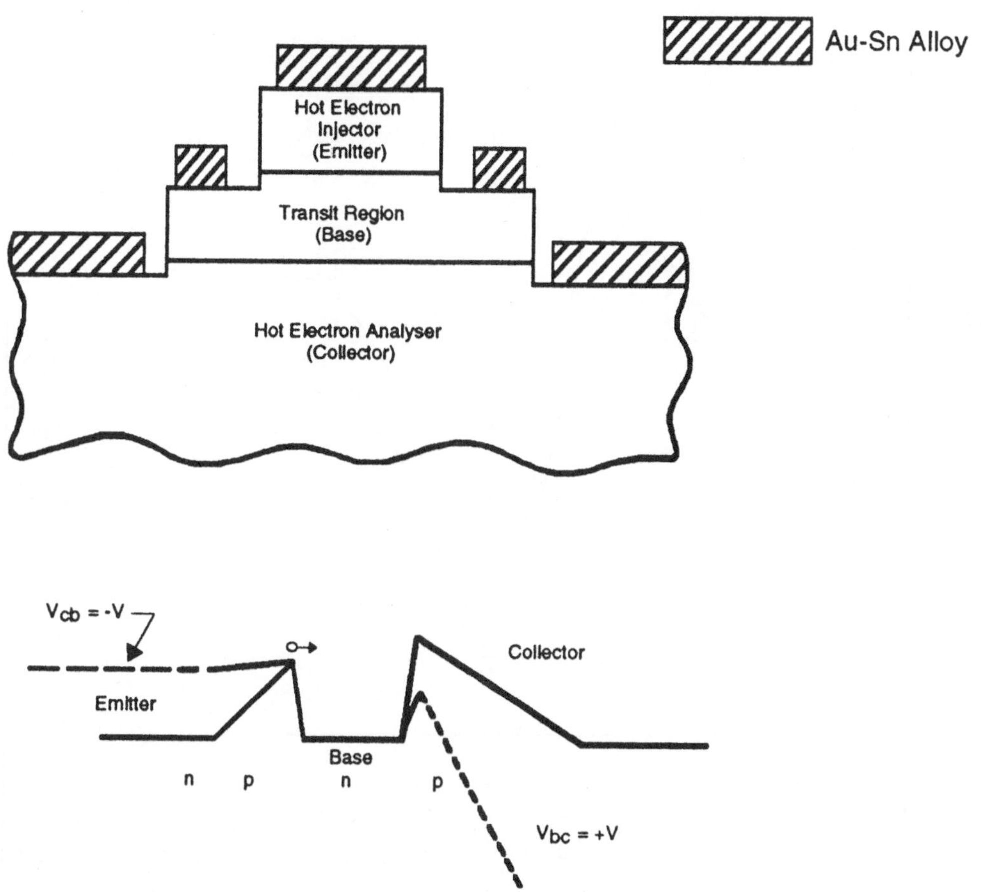

FIGURE 6.13. Schematic for hot-electron transistor (Levi *et al.*, 1985).

as large as 850 Å, Levi *et al.* (1985, 1986) observed near-ballistic transport at 4.2 K in GaAs samples. For a base width as small as 310 Å, Heiblum *et al.* (1986) observed that about 50% of the injected electrons transport across GaAs ballistically. However, with the addition of a barrier at the emitter side, Yokoyama *et al.* (1984) observed ballistic transport of electrons in GaAs samples with 1000-Å base width. An I–V measurement by Ohashi *et al.* (1985) provides evidence of near-ballistic transport at 300 K across 2500 to 5000 Å in $Ga_{0.47}In_{0.53}As$.

There are many issues that need to be addressed in modeling a high-speed device based on ballistic transport. These issues include effective injection of the electron into the active region, the transport of charge carriers in the active region and across the boundaries between different device regions, and collection of the charge carriers. If waves matching across different regions in the device can be optimized and the additional scattering due to surfaces and interfaces can be minimized, then the performance of the device is determined by the transport properties in the active region. Because charge carriers traverse the active region in the shortest time possible at the peak group velocity, the optimal strategy for high-speed devices is to inject the carriers into the active region in the direction and at the energy that

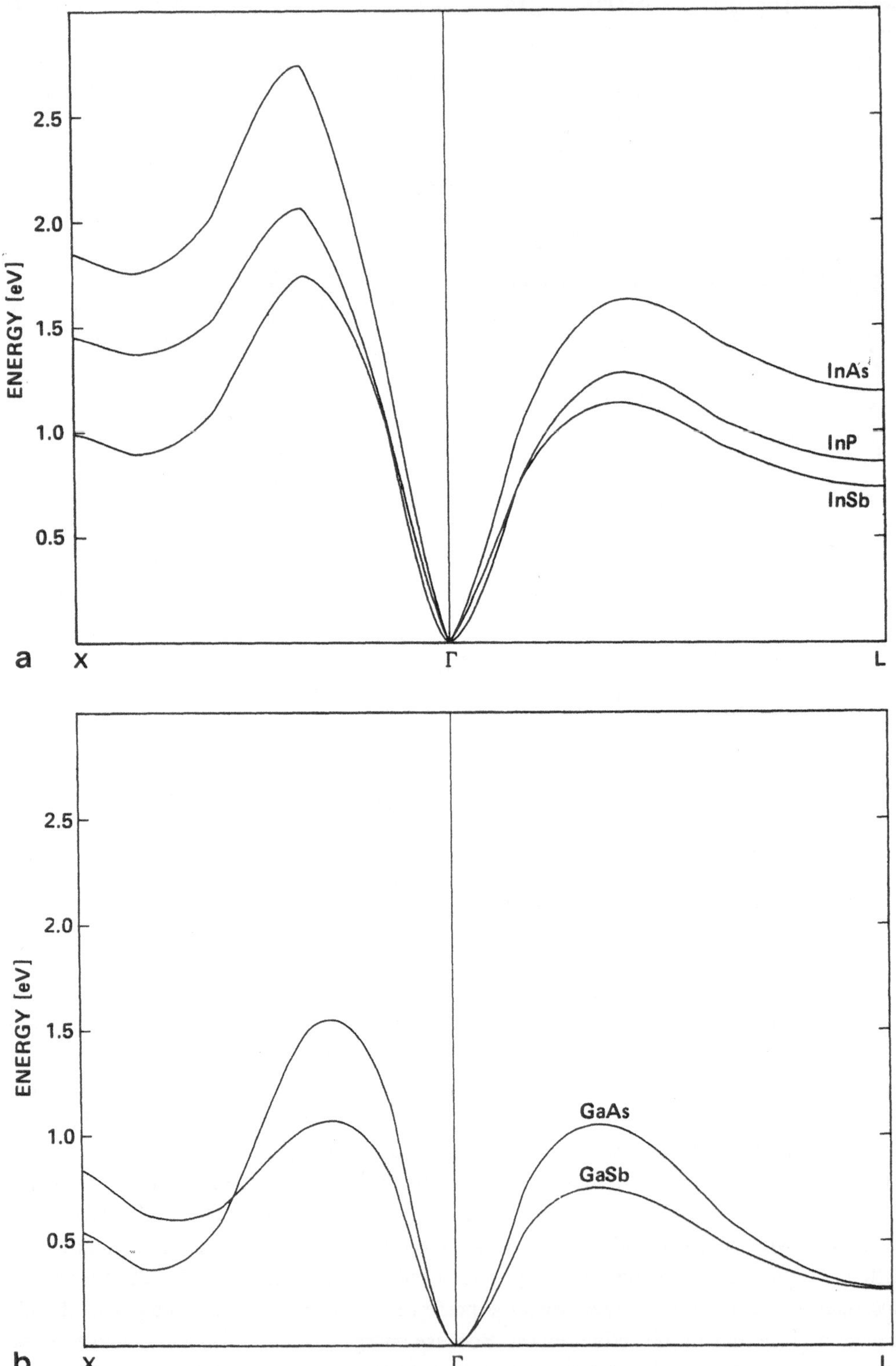

FIGURE 6.14. Lowest conduction band of (a) InP, InAs, InSb, and (b) GaAs, GaSb compounds. All energies measured in eV with respect to their respective conduction-band minima (Krishnamurthy *et al.*, 1988b).

gives them their peak velocity and longest lifetime. An overall evaluation of the group velocity and mean free path for several III–V compounds and alloys for possible ballistic transport devices was performed by Krishnamurthy *et al.* (1988a,b) based on detailed calculations using realistic band structures similar to those presented in Chapter 7. This section is devoted to a discussion of these results.

The group velocity $\mathbf{v}(\mathbf{k})$ of electrons in a semiconductor is calculated from the energy-band dispersion $\varepsilon(\mathbf{k})$ by

$$\mathbf{v}(\mathbf{k}) = \nabla_{\mathbf{k}}\varepsilon(\mathbf{k})/\hbar \tag{6.8.1}$$

To compare the group velocities among semiconductors for ballistic transport, we plot in Fig. 6.14 the lowest conduction bands along Γ–L and Γ–X for the five direct-gap III–V semiconductors to be considered. These band structures were obtained by Krishnamurthy *et al.* (1988a,b); they are quite similar to those discussed in Chapter 7. Because there is a strong $\mathbf{k}$ dependence of these bands, the associated group velocities have even a larger variation with $\mathbf{k}$. For example, Fig. 6.15 shows the value of $\mathbf{v}(\mathbf{k})$ along the (111) (Γ–L) direction and (100) (Γ–X) directions. The maximum value occurs along the Γ–X direction and not too far from the bottom of the conduction band. The group velocity along L also has a peak, but the value is smaller than that along Γ–X. Note that the group velocity becomes negative (has direction opposite to $\mathbf{k}$) as $\mathbf{k}$ moves further away from Γ. These patterns are similar for all the III–V direct-gap compounds and alloys studied in this section. In Fig. 6.16, the group velocities along (100) for several useful compounds and alloys are plotted as a function of injected electron energy. The maximum velocity ranges from 1.1 to 1.4×10^8 cm/s. However,

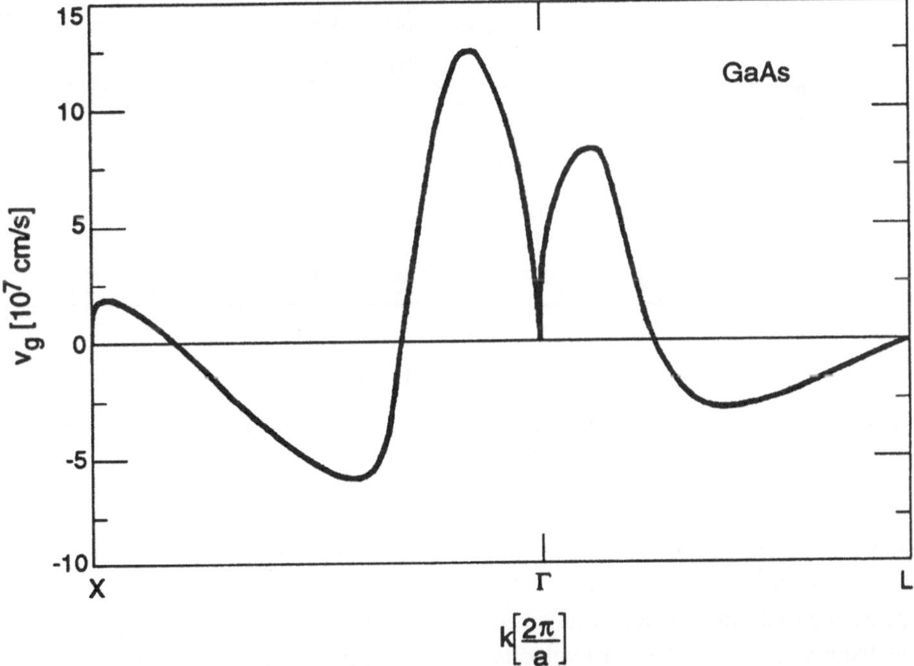

FIGURE 6.15. Group velocity of electrons in GaAs in the (100) and (111) directions (Krishnamurthy *et al.*, 1988b).

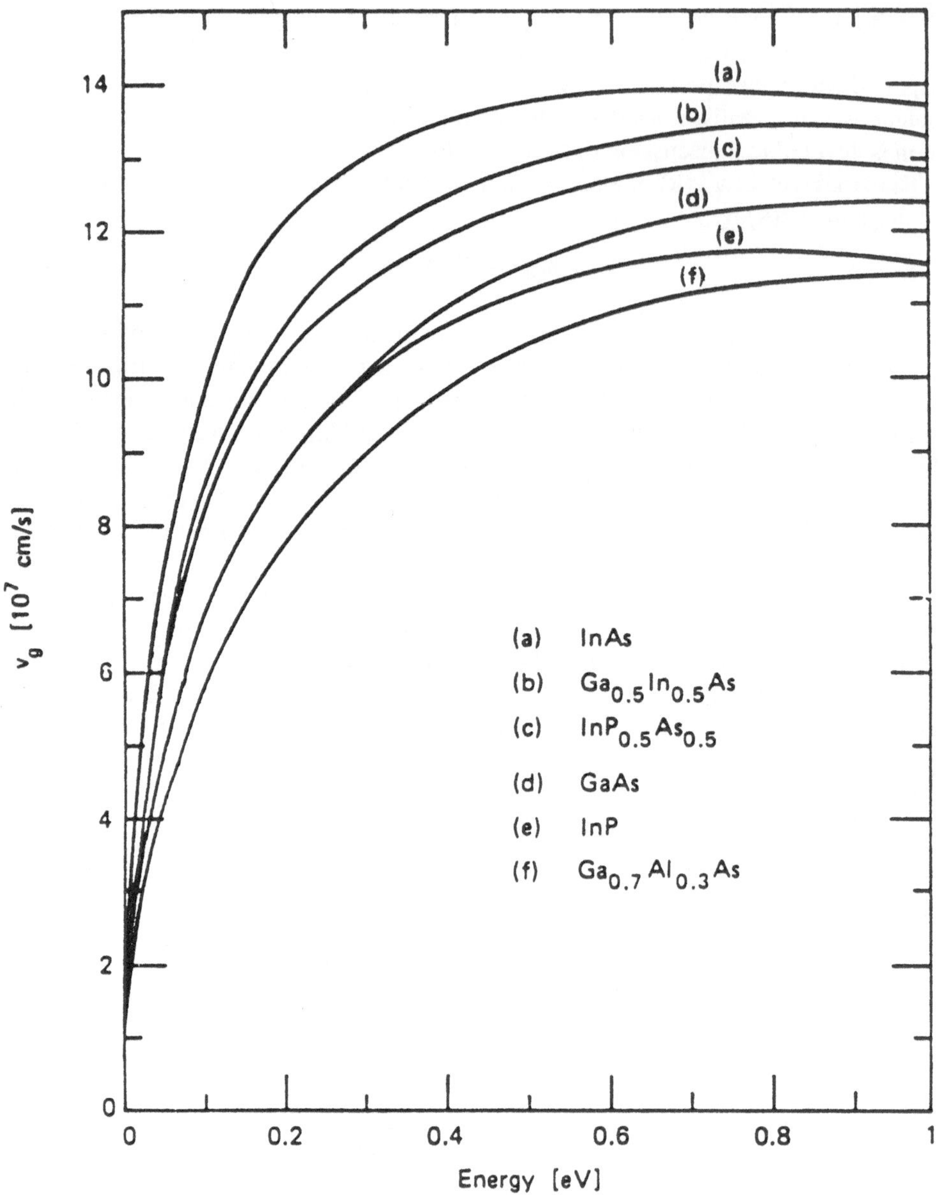

FIGURE 6.16. Variation of group velocity with electron energy in the (100) direction. (Krishnamurthy *et al.*, 1988b).

in using this figure one has to keep in mind that the injection energy cannot exceed either the first intervalley energy spacing ΔE or the impact ionization energy E_p discussed in Section 5.

Next consider the mean free path. Since group velocity is the most relevant quantity, it is appropriate to define a velocity mean free path Λ as $n = n_0 e^{-d/\Lambda}$, where n_0 is the number of injected electrons with forward velocity v_0 and n is the number of electrons retaining v_0 after traveling a distance d in the base region. This mean free path can be assumed to be the

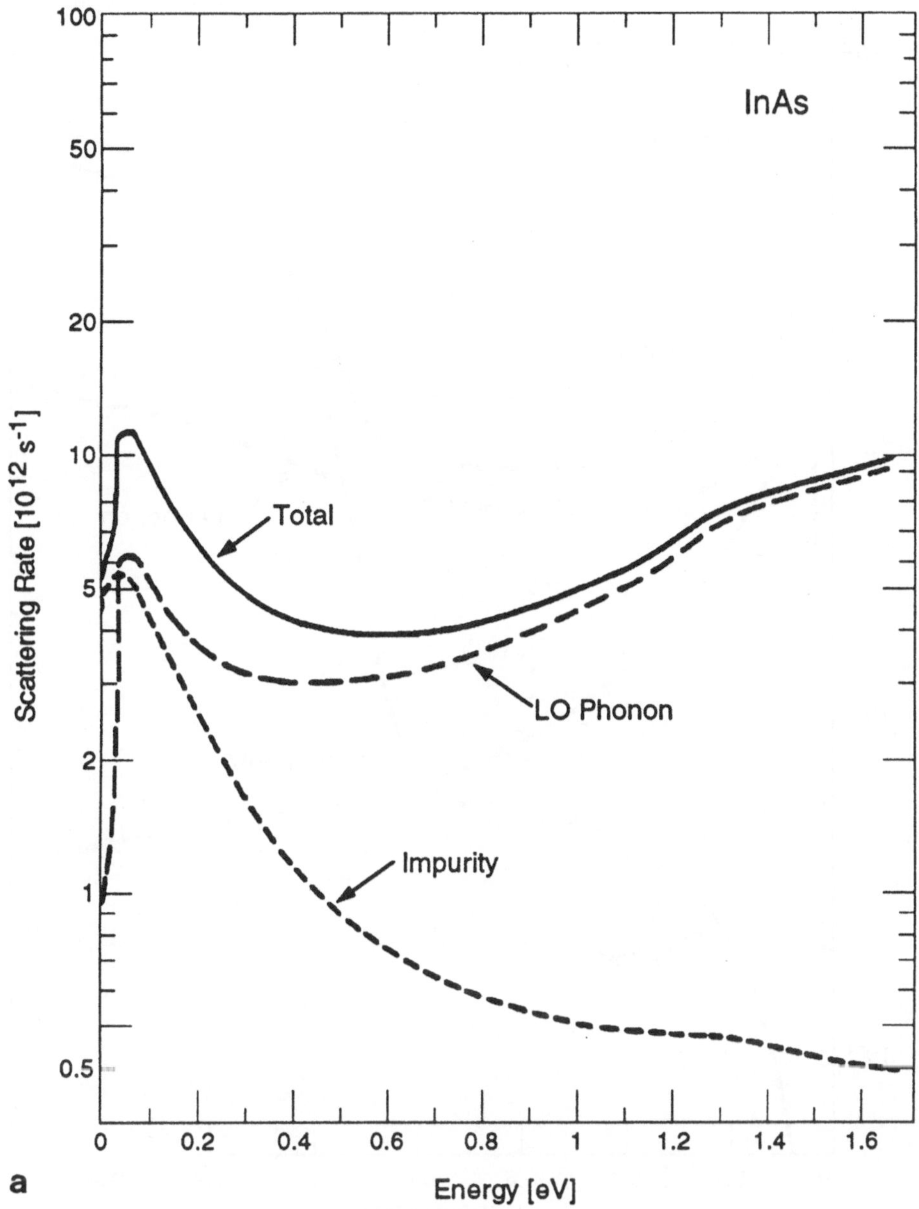

FIGURE 6.17. Phonon (long dash), 10^{18}cm^{-3} impurities (short dash), alloy disorder (dash dot), and effective total scattering rate (solid) as a function of energy in the (100) direction for (a) InAs; (b) $Ga_{0.5}In_{0.5}As$; (c) GaAs; (d) $InP_{0.5}As_{0.5}$; (e) InP; and (f) $Ga_{0.7}Al_{0.3}As$ (Krishnamurthy *et al.*, 1988b).

product of v_0 and a velocity relaxation time τ_v. Similar to the momentum relaxation time considered in Section 3, τ_v can be calculated from Eq. (6.3.14) with the factor $1-(\mathbf{k}'/\mathbf{k})\cos\theta$ replaced by $1-\mathbf{v}(\mathbf{k})\cdot\mathbf{v}(\mathbf{k}')/|\mathbf{v}(\mathbf{k})|^2$.

Because crystals oriented in the (100) direction offer the largest electron group velocity, the calculations of the mean free paths by KSC were carried out in that direction only. The

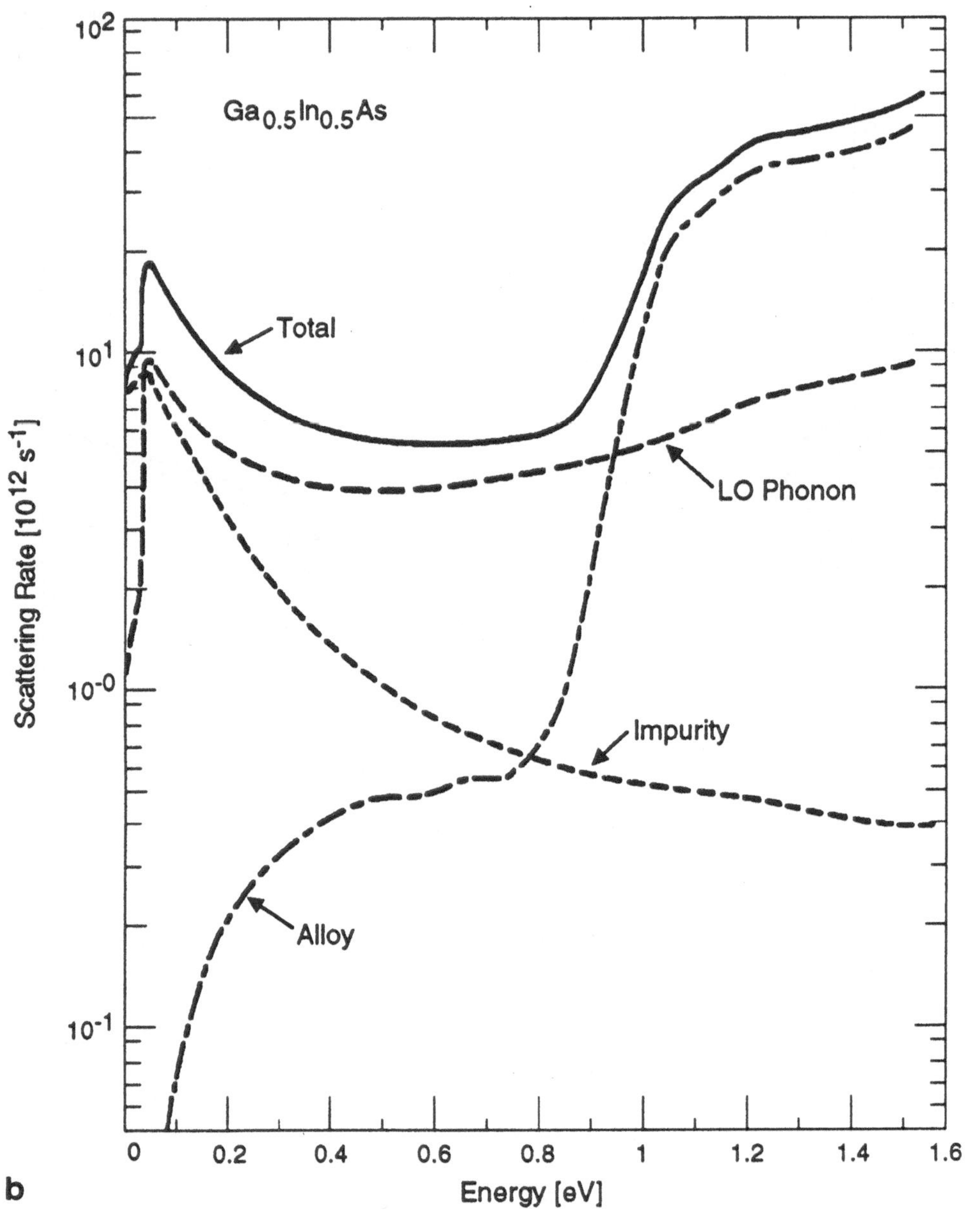

FIGURE 6.17. (continued)

velocity relaxation times due to LO phonons τ_p, ionized impurities τ_i, and alloy disorder τ_a were calculated separately. Then the total scattering rate $1/\tau_v$ is obtained from the sum

$$1/\tau_v = 1/\tau_p + 1/\tau_i + 1/\tau_a \tag{6.8.2}$$

from which the mean free path is then computed, $\Lambda = v\tau_v$. These scattering rates are functions of the injection energy ε or group velocity v_0. The calculated scattering rates due to different

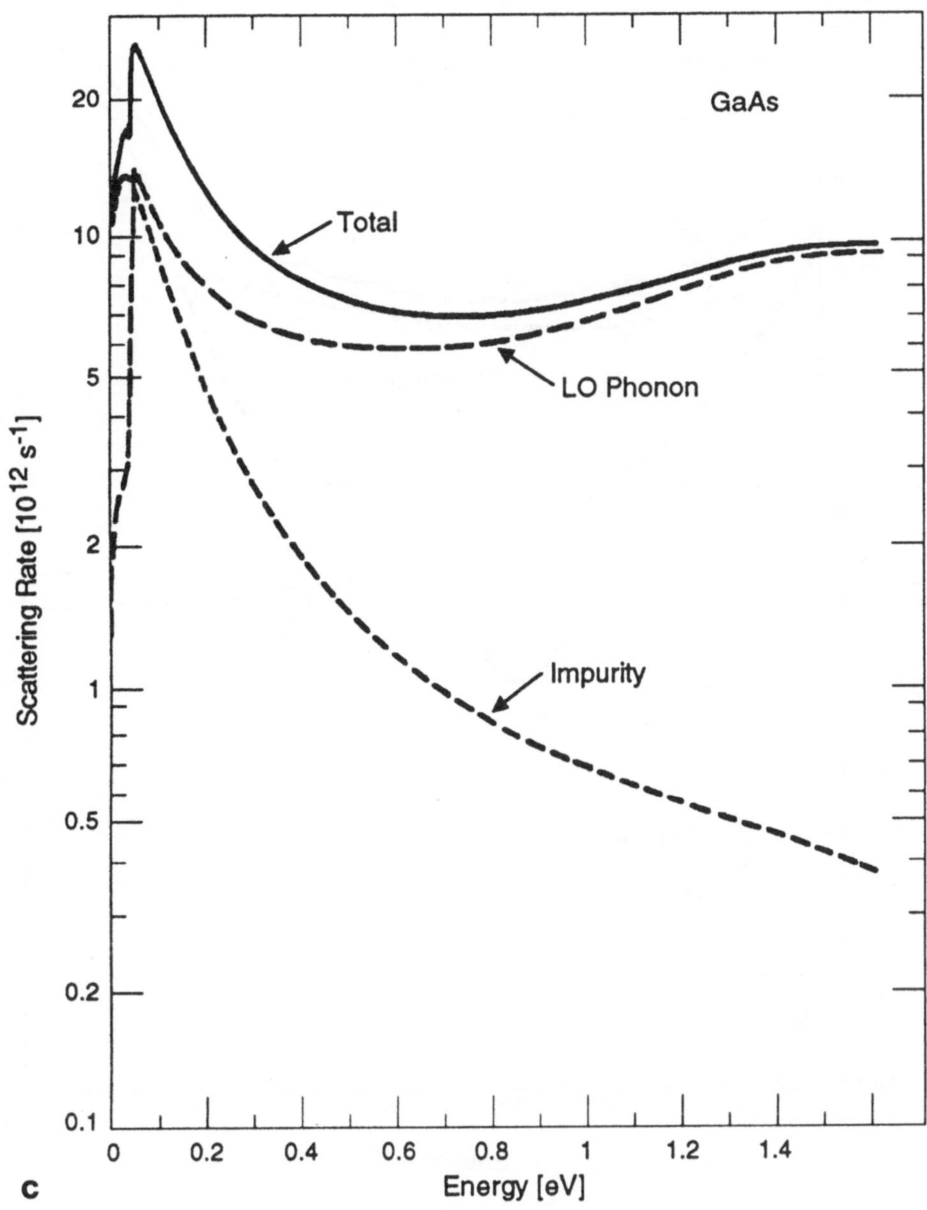

FIGURE 6.17. (continued)

mechanisms and the total rates as a function of the injection energy for several III–V compounds and alloys are shown in Figs. 6.17a–f. With an assumed impurity density of 10^{18} $1/cm^3$, impurity scattering dominates at very low injection energies. Because the effective mass is largest in $Ga_{0.7}Al_{0.3}As$ followed by InP, GaAs, $InP_{0.5}As_{0.5}$, $Ga_{0.5}In_{0.5}As$, and InAs respectively, the impurity scattering rates follow the same order (see Table 6.2 for the masses used).

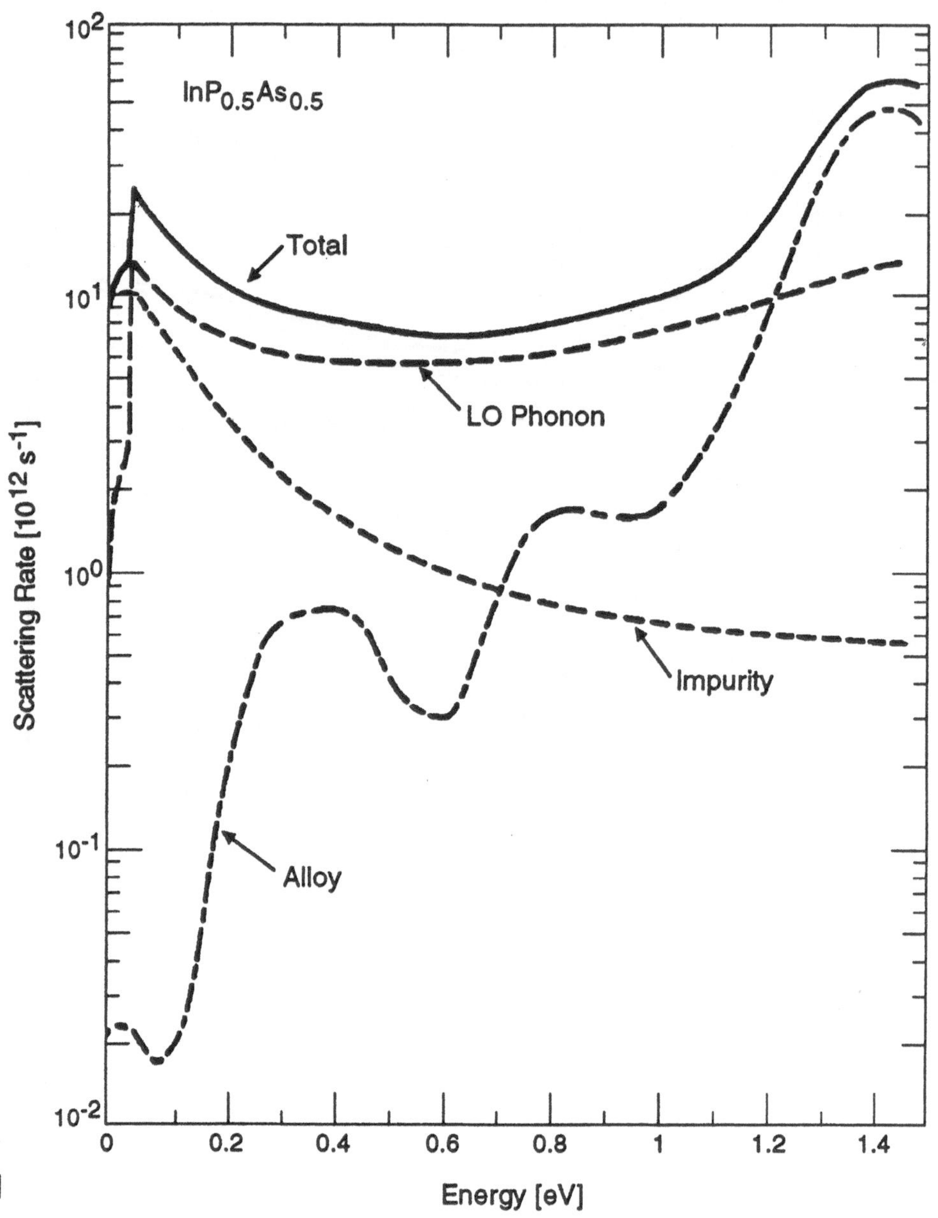

FIGURE 6.17. (continued)

The average matrix element for impurity scattering decreases with energy, whereas the density of states (DOS) increases with energy. These two competing mechanisms give rise to a peak in the impurity scattering rate. At higher energies the decrease in the matrix element dominates. Hence, the contribution from impurities to the total scattering rate becomes negligible. Phonon scattering at low energies can be analyzed by the number of phonons N_p available at that energy and the electron effective mass m^*. Larger values of N_p and m^* imply

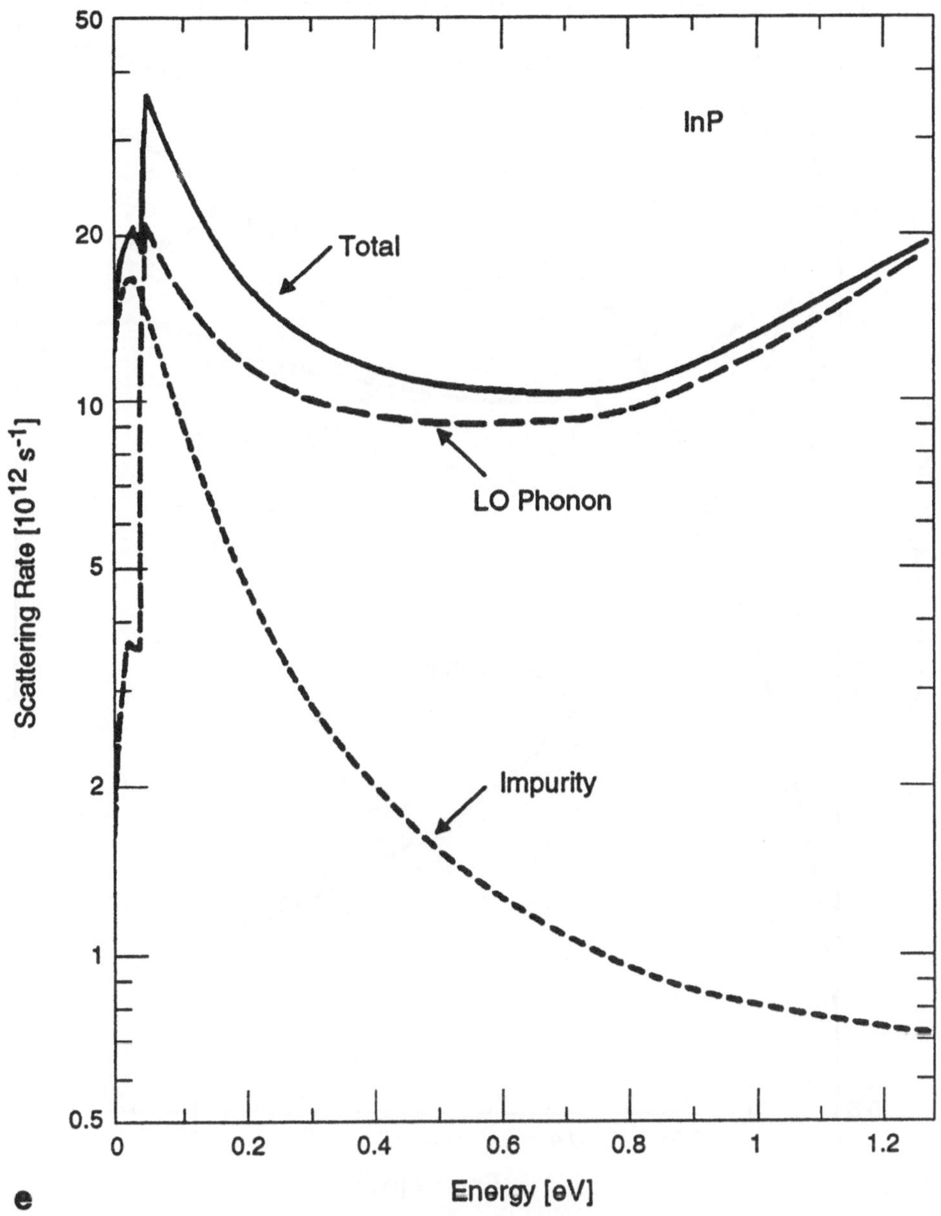

FIGURE 6.17. (continued)

more effective phonon scattering. For energies greater or equal to the LO phonon energy $\hbar\omega_0$, electrons can also interact with the lattice by emitting phonons, which is reflected in the sudden increase in the scattering at $\varepsilon = \hbar\omega_0$. Phonon scattering is less effective than impurity scattering for really low injection energies, because of the small number of phonons available at low **q** and the low-electron DOS. However, for injection energies larger than $\hbar\omega_0$, phonon scattering is a dominant mechanism. For alloys, disorder scattering arises from

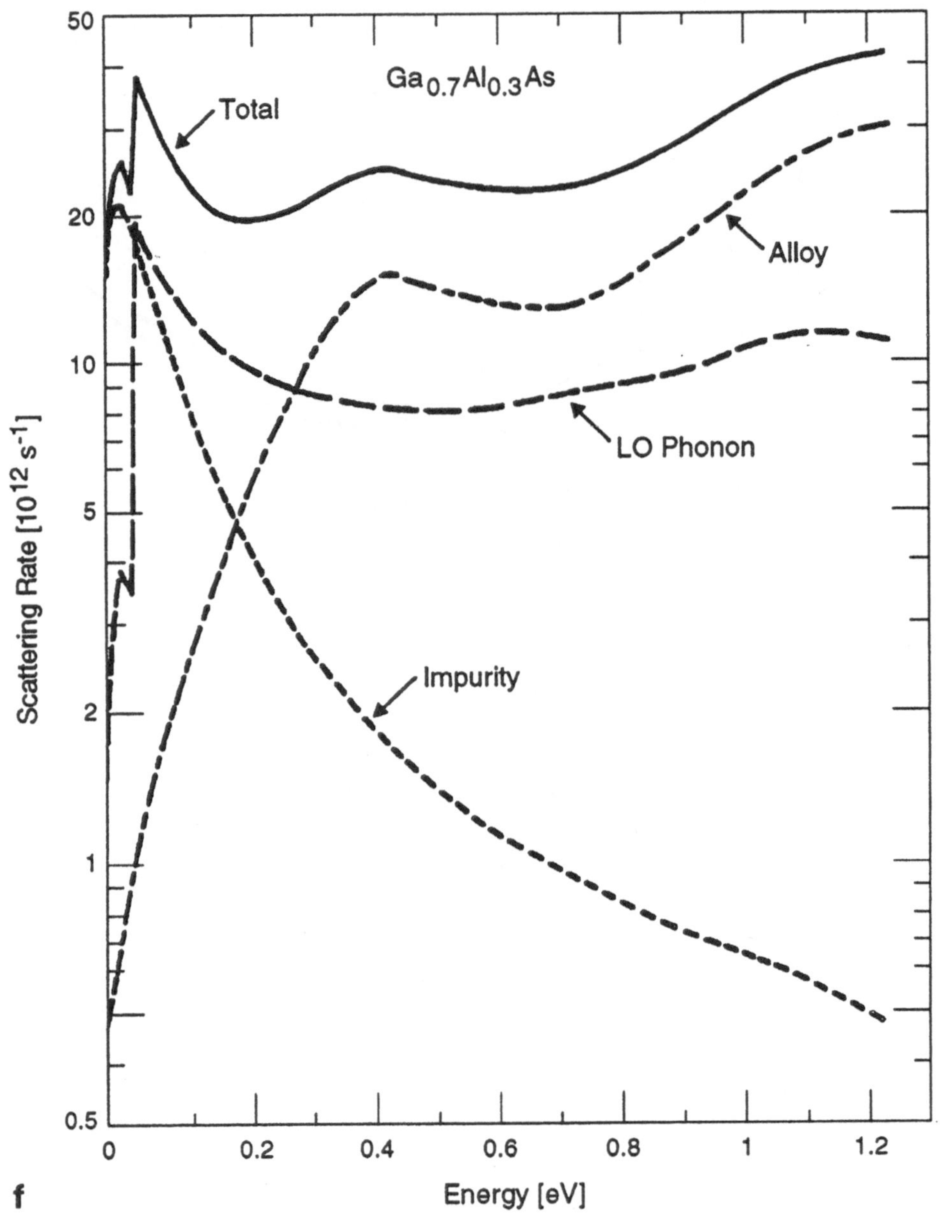

FIGURE 6.17. (continued)

fluctuations in both the site-diagonal and off-diagonal Hamiltonian matrix elements. This is treated in CPA theory. The alloy scattering rate increases rapidly with increase in the ejection energy, because the electron density does so. The alloy disorder scattering rate can exceed the LO phonon scattering rate at high injection energies.

The variation of Λ in the (100) direction as a function of injection energy is shown in Fig. 6.18. Again, to use this figure one has to check first to determine if the injection energy

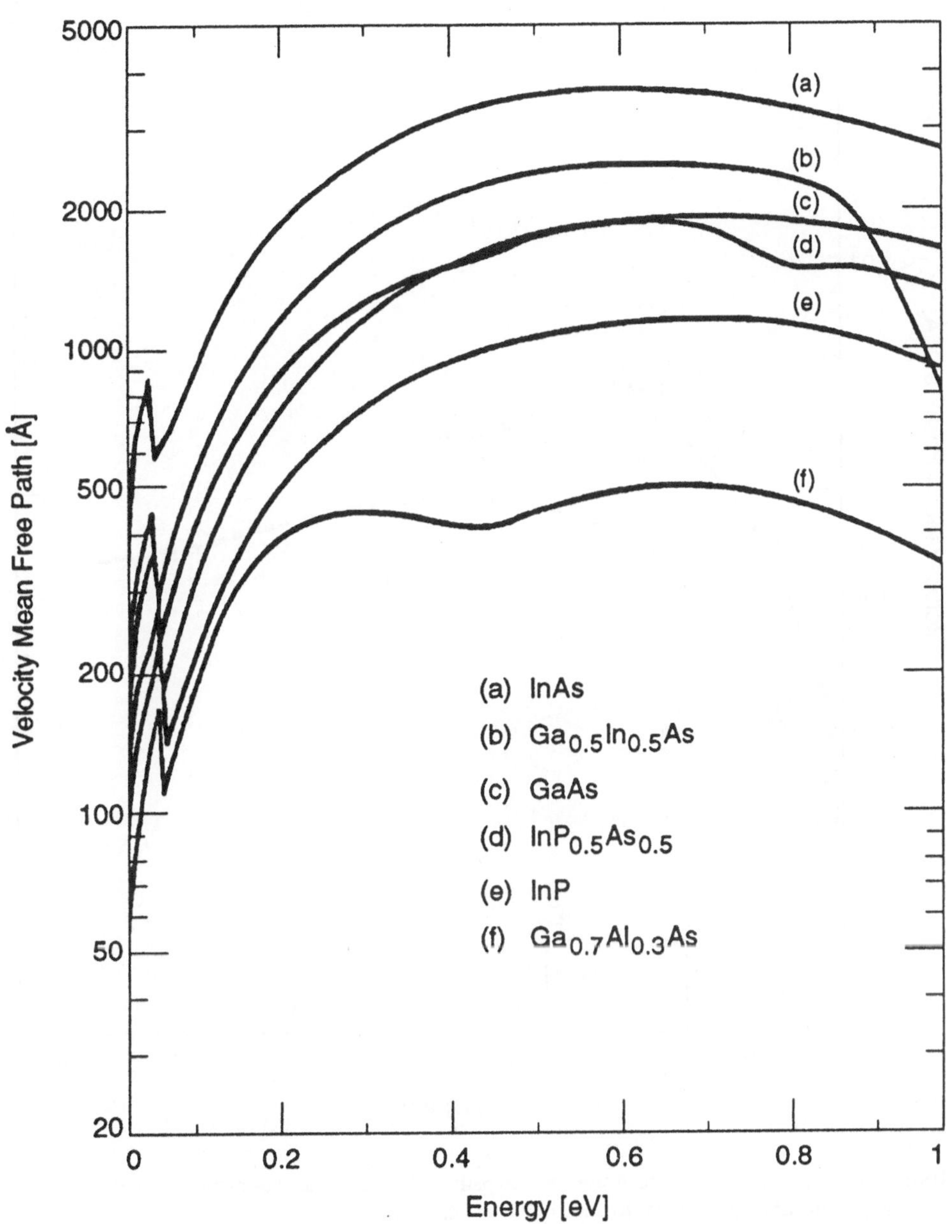

FIGURE 6.18. Velocity mean free paths in the (100) direction at 300 K (Krishnamurthy *et al.*, 1988b).

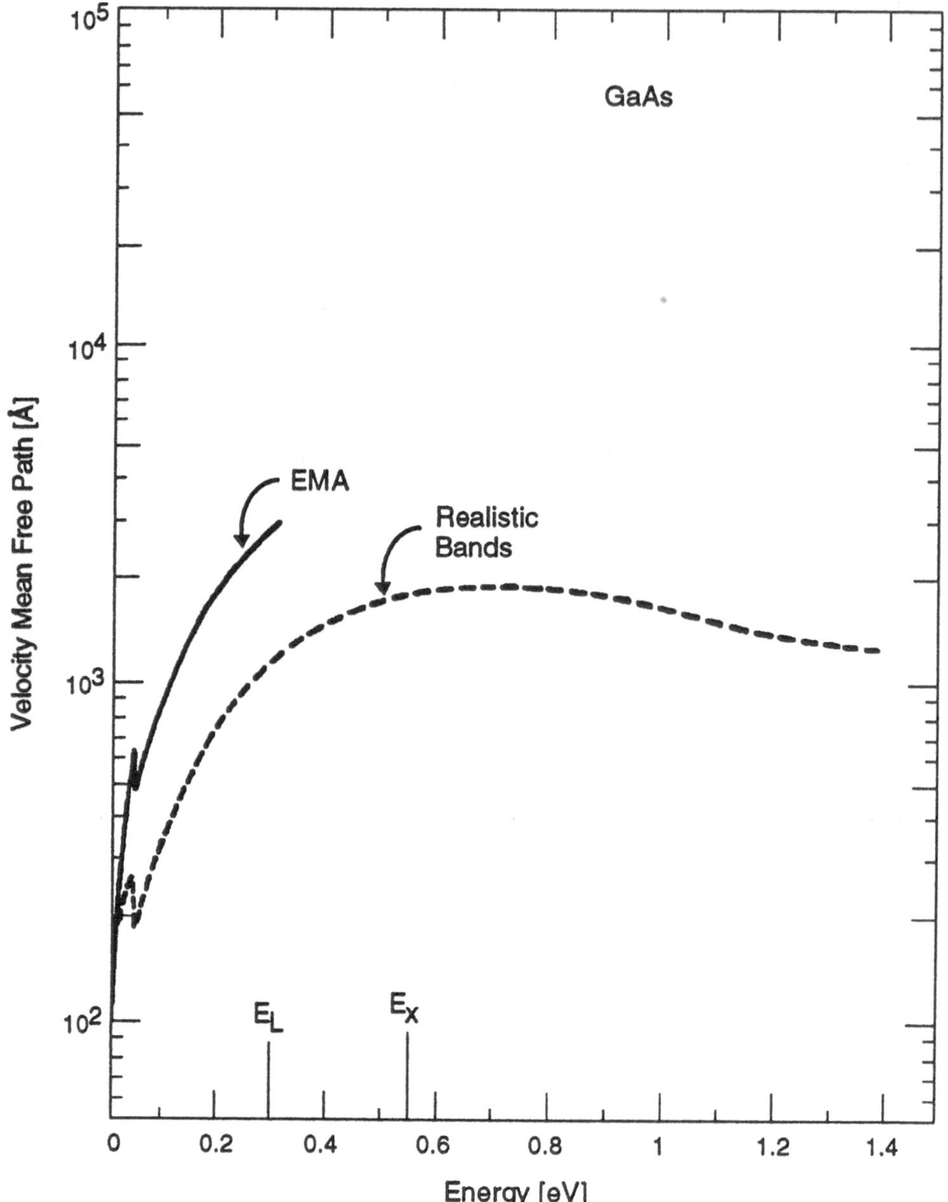

FIGURE 6.19. Comparison of the effective mean free path in GaAs calculated using the realistic-band structure (dashed line) and the EMA (solid line) (Krishnamurthy *et al.*, 1988b).

exceeds either the intervalley separation ΔE or the impact ionization energy E_p. In Fig. 6.19 a comparison is made between the calculation using the full band structures and an effective-mass approximation for GaAs. Similar comparisons have also been found for the other materials. The Λ calculated from EMA is not shown at energies above the second conduction-band minimum E_L or ΔE, because of uncertainties involved in the calculation of

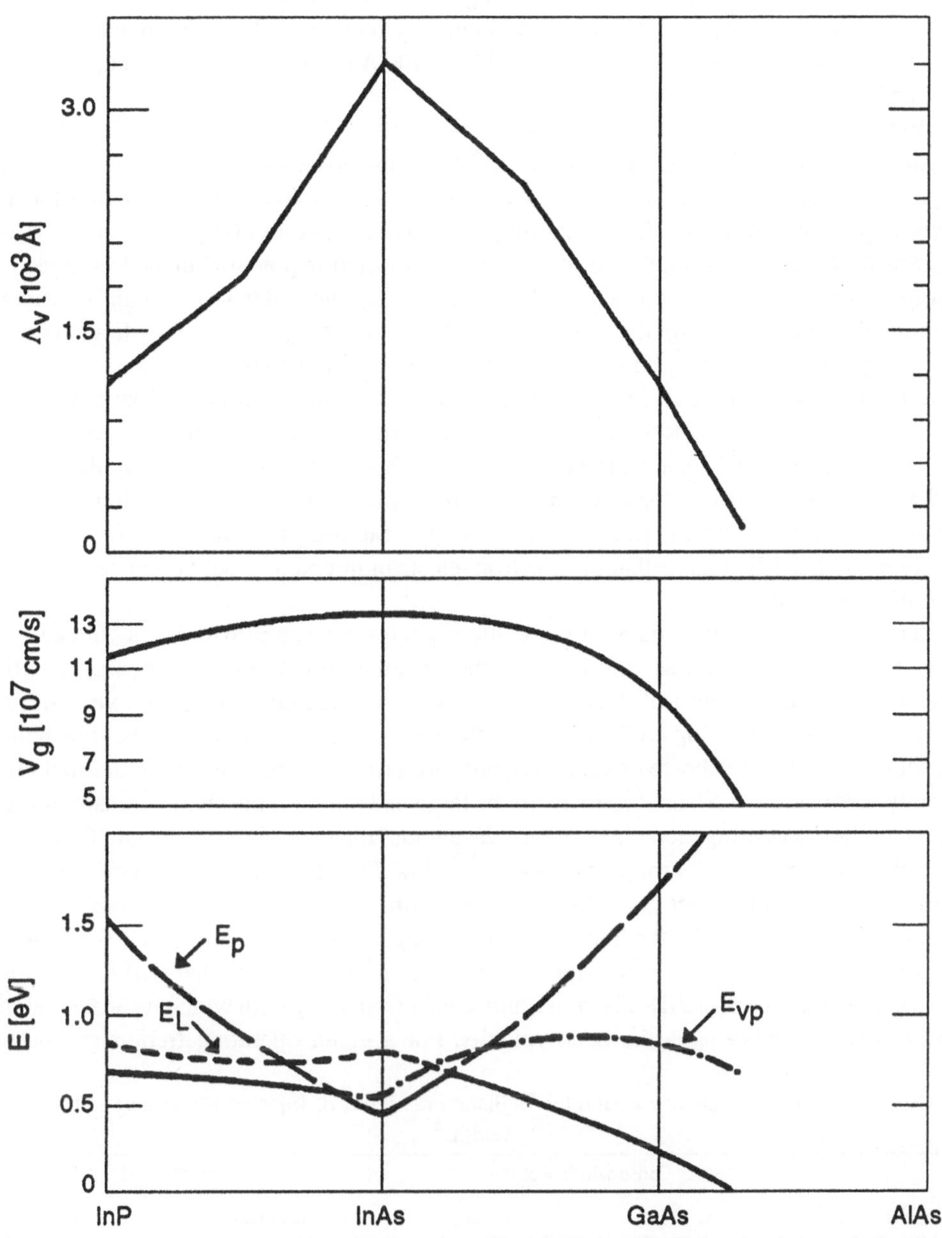

FIGURE 6.20. In the bottom portion of this figure E_L (short dashed), E_p (long dashed), and energy at peak v_g values, E_{vp}, (dash dot) and near-optimal injection energy (solid) are plotted as functions of alloy concentration. Corresponding to near-optimal injection energy, v_g and Λ_v are plotted in the middle and top portions of this figure, respectively (Krishnamurthy *et al.*, 1988b).

intervalley scattering using this method. At low energy, up to 25 meV, both methods yield the same Λ. Again, we see a sudden change of Λ at low energy due to the onset of phonon–emission scattering. As the injection energy increases, the difference between the two calculations becomes larger. At E_L the Λ from EMA is three times the calculation using full band structures.

Before we use the calculated group velocity and mean free path Λ as a basis to compare the materials' merits for high-speed near-ballistic transport devices, we need to discuss the range of validity of the calculation. The scattering rates in Eq. (6.8.2) do not include the bare electron–phonon interaction (deformation potential) (see Section 6.6.2) and the electron–electron electron interaction. While scattering by deformation potential in the low-injection-energy regime is not as important as the scattering by the polar LO phonon considered previously, it contributes significantly to the scattering at an injection energy larger than the intervalley energy ΔE through intervalley scattering. This will be treated in the next section. Thus the calculated values of Λ for $\varepsilon > \Delta E$ may not be accurate. Several previous calculations (Levi *et al.*, 1985, 1986; Rorison and Herbert, 1986) showed that the electron–electron scattering rate was 10^{13} 1/s at an injection energy of 250 meV. This rate is roughly the same as that from all the other scattering mechanisms combined. However, we found that the electron–electron interaction only gives a velocity scattering rate of 1.5×10^{12} 1/s at this injection energy. This means that the electron–electron interaction only contributes 15% or so to the total scattering rate.

Thus with a 15% adjustment of the scattering rates, we can utilize the calculated v and Λ to evaluate the materials' merits. We choose an optimal injection energy ε which corresponds to the smallest of the three energies: ε_{vp} at the maximum group velocity; ΔE, the first intervalley spacing ($\Delta E = E_L$ for all the III–V studied); and E_p for the threshold of impact ionization. The choice of ε_{vp} is obvious, because we want to have the optimal group velocity. Choosing $\varepsilon < \Delta E$ is done to avoid the large and unfavorable charge transfer through the intervalley scattering, and choosing $\varepsilon < E_p$ avoids impact ionization. Figure 6.20 shows plots of mean free path Λ, group velocity v_g, and E_{vp}, E_L, and E_p for InP, InAs, GaAs AlAs, and their alloys with direct gaps. For Λ, straight lines are used to connect the calculated values. InAs turns out to be the material with the largest mean free path and group velocity.

In a device with base width L, the fraction of injected electrons that will be ballistic is approximately $\alpha = \exp(-L/\Lambda)$. The maximum cutoff frequency with which it can be operated is $\nu_c = v_p/2\pi L$, where v_p is the peak velocity. For a given operating frequency, the best

TABLE 6.3. Cutoff Frequency ν_c and the Ballistic Fraction α of Injected Electrons for Two Base Widths[a]

	Base width = 500 Å		Base width = 100 Å	
Materials	ν_c (THz)	α	ν_c (THz)	α
GaAs	3.2	0.63	15.6	0.91
$Ga_{0.5}In_{0.5}As$	4.1	0.82	20.7	0.96
InAs	4.3	0.86	21.5	0.97
$InP_{0.5}As_{0.5}$	4.1	0.76	20.7	0.95
InP	3.7	0.64	18.6	0.91
$Ga_{0.7}Al_{0.3}As$	1.7	0.04	8.3	0.54

[a]Krishnamurthy *et al.* (1988a)

material will be the one with the largest α. Table 6.3 lists the calculated values of ν_c and α using the calculated Λ for base widths of 100 and 500 Å. With the larger base width, InAs and $Ga_{0.5}In_{0.5}As$ are clearly preferred for high-frequency operation. However, to achieve an operational frequency of 10 THz when the base width is about 100 Å, no material offers a significant advantage over GaAs.

6.9. INTERVALLEY SCATTERING

One uncertainty in the scattering rate calculation in the preceding section is the neglect of the bare electron–phonon interaction. This interaction is believed to play an important role in hot-electron transport, because it is effective in causing intervalley scattering of hot electrons from the Γ minimum to the L minimum. There have been controversies over the reliability of theoretical calculations of the coupling constant of this intervalley scattering. The experimentally deduced coupling constants $D_{\Gamma L}$ for GaAs also vary greatly, with quoted numbers lying between 1.5×10^8 and 10×10^8 eV/cm (Littlejohn *et al.*, 1977; Collins and Yu, 1984; Kash *et al.*, 1983; Shah *et al.*, 1987). The value 7×10^8 eV/cm is frequently used in the literature. In this section, a parameter-free theory of the intervalley scattering coupling constant (Krishnamurthy *et al.*, 1988c) based on a full-band-structure calculation is presented. Then all the scattering rates are included in the velocity relaxation time calculation to explain the observed ratio of the emitter-to-collector current (Hase *et al.*, 1987).

The formal expressions for the scattering rates associated with the bare electron–phonon interaction are given in Section 6.6.2. For the intervalley scattering from Γ to L, the matrix element can be calculated at a fixed phonon wave vector $\mathbf{q} = (1/2, 1/2, 1/2)(2\pi/a)$. The phonon frequencies involved, denoted ω_{iv}, are the LA and LO phonons at this $\mathbf{q}$. Thus the effective matrix element of Eq. (6.6.8), now denoted $V_{\Gamma L}$, for a given phonon mode (LA or LO) is a constant. Then the transition rates of Eqs. (6.2.8) and (6.2.9) can be used to calculate the intervalley scattering rate. When the band structure at L is described by the effective-mass approximation, the scattering rate can be obtained in a closed form:

$$\frac{1}{\tau_{iv}} = \frac{m_l^{1/2} m_t}{2^{1/2} \pi \hbar^3 \rho \omega_{iv}} |V_{\Gamma L}|^2 [N_q (\varepsilon + \hbar\omega_{iv})^{1/2} + (N_q + 1)(\varepsilon - \hbar\omega_{iv})^{1/2}] \tag{6.9.1}$$

where ε is the injection energy measured from the L-valley minimum and the emission term does not exist until ε is larger than $\hbar\omega_{iv}$. By comparing Eq. (6.9.1) with the form of deformation potential intervalley scattering rate (Conwell, 1967), the intervalley coupling constant $D_{\Gamma L}$ can be identified to be $|V_{\Gamma L}|^2$.

To actually calculate the matrix elements in Eq. (6.6.8), we have to know how the tight-binding matrix elements vary with interatomic distances d. Krishnamurthy *et al.* (1988b) assumed that the TB matrix elements scale as $1/d^n$ and a pair repulsive energy scales as $u(d) = \eta/d^{2n}$. The equilibrium bond length d_0 and the bulk modulus were used to fix the power n and the parameter η. The calculated value of $D_{\Gamma L}$ for GaAs is 2.05×10^8 eV/cm for the LA phonon scattering and 2.40×10^8 eV/cm for the LO phonon scattering. If we combine the scattering rates from the two phonon branches in Eq. (6.9.1) and use only one effective $D_{\Gamma L}$ and one effective intervalley phonon frequency $\omega_{iv} = (\omega_{LA} + \omega_{LO})/2$, then the value for the effective $D_{\Gamma L}$ for GaAs is 3.2×10^8 eV/cm. The corresponding values in InP and InAs are 2.4 and 2.8×10^8 eV/cm, respectively.

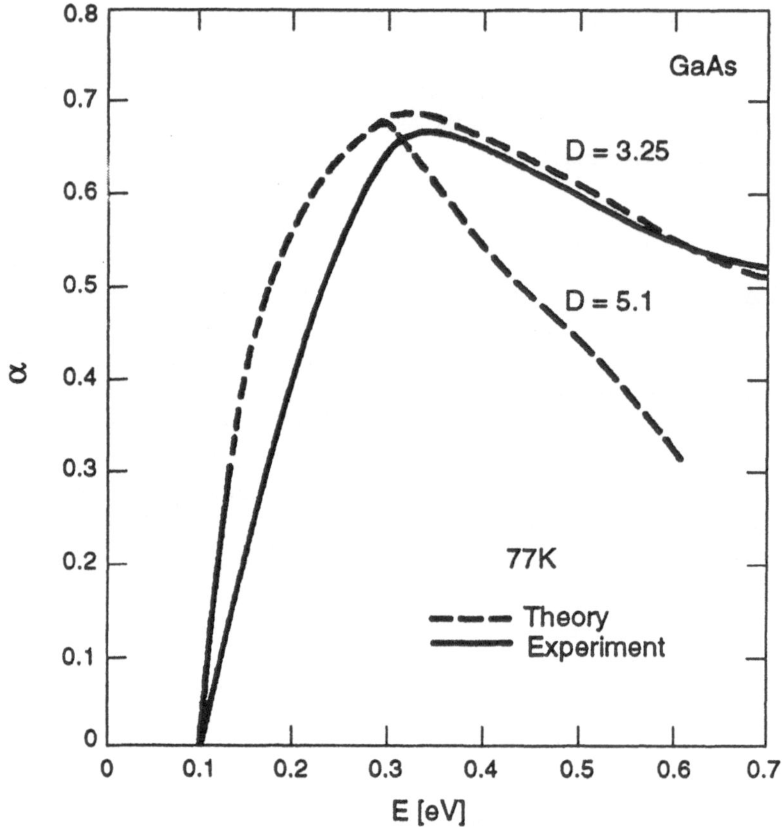

FIGURE 6.21. Calculated (dashed line) and experimental (solid line) values of α as a function of injection energy in a GaAs hot-electron transistor at 77 K (Krishnamurthy *et al.*, 1988c).

These intervalley scattering rates were included along with the other scattering mechanisms considered in the preceding section by Krishnamurthy *et al.* (1988c) for a calculation of the velocity mean free path Λ. The ballistic ratio $\exp(-L/\Lambda)$ is used to estimate the ratio α of the collector-to-emitter current in a hot-electron transistor (HET) with a base length L. The calculated α for GaAs is plotted in Fig. 6.21 as a function of the injection energy for two values of $D_{\Gamma L}$. The solid line in the figure is the experimental α by Hase *et al.* (1987). The agreement between the experiment and theory is good when the present value $D_{\Gamma L} = 3.2 \times 10^8$ eV/cm is used in the calculation, whereas the calculation using the previously deduced value $D_{\Gamma L} = 5.1 \times 10^8$ eV/cm gives a poor result for injection energies higher than the intervalley separation E_L, implying the coupling constant is too large. There is also a substantial difference between theory and experiment at low injection energy. This can be attributed to the nonmonoenergetic nature of the injected electrons in the experiment. In the calculation, it is assumed that all electrons originate from the Fermi energy. However, all electrons with energies above the conduction-band minimum but below the Fermi energy can tunnel into the base. At low injection energies, the average group velocity of the electrons will be much smaller than their group velocity at the Fermi energy. Correspondingly, experimental values of α are smaller than that predicated by the theory. This correction becomes negligible at high injection energies where the group velocity peaks and is nearly independent of the injection energy.

In summary, the intervalley coupling constants associated with scattering by phonons are calculated using full band structures. The calculated values of $D_{\Gamma L}$ are found to be smaller than previous results. Using the present coupling constant, the calculated ballistic fraction α is in reasonable agreement with the experimental collector-to-emitter carrier ratio in GaAs. However, we note that the previous $D_{\Gamma L}$ values were deduced based on the EMA. They may be appropriate within EMA, but not for a full band structure calculation.

6.10. NARROW-GAP MATERIALS

In this section the expansion solution method is generalized to solve the Boltzmann equation (BE) in cases where Fermi statistics must be retained along with a nonparabolic band and all important scattering mechanisms. This enables accurate solutions for the transport properties of semiconductors with narrow band gaps. The temperature and carrier concentration variations of the mobility of $Hg_{0.78}Cd_{0.22}Te$ (Krishnamuthy and Sher, 1994) will be used as an example.

Start from the BE (6.1.19) and (6.1.14). For simplicity the spin index s is suppressed. In steady state with the absence of the **B** field, $-e\mathbf{E}\cdot\nabla_{\mathbf{k}}f = (df/dt)_{\text{coll}}$. In the following, a recursive method for solving the BE will be derived. First write the distribution function f as the sum of the equilibrium Fermi function f_0 given by Eq. (6.1.16) plus a deviation δf:

$$f(\mathbf{k}) = f_0(\mathbf{k}) + \delta f(\mathbf{k}) \tag{6.10.1}$$

Noting that $\Sigma\delta f(\mathbf{k}) = 0$ and using the detailed balance condition

$$\sum_{\mathbf{k}'} (w(k/k')[1 - f_0(k)]f_0(k') - w(k'/k)[1 - f_0(k')]f_0(k)) = 0$$

we have

$$-e\mathbf{E}\cdot\nabla_{\mathbf{k}}f = \sum_{k'} \{(\mathcal{W}(k/k')\delta f(k') - \mathcal{W}(k'/k)\delta f(k) + [w(k'/k) - w(k/k')]\delta f(k)\delta f(k'))\} \tag{6.10.2}$$

where $\mathcal{W}$ is a renormalized transition rate,

$$\mathcal{W}(k/k') = w(k/k')\frac{1 - f_0(k)}{1 - f_0(k')} \tag{6.10.3}$$

Care must be taken in the numerical evaluation of $\mathcal{W}$ when $f_0(k')$ is close to unity.

Note that for elastic scattering, $\mathcal{W}$ and w are equal. However, for inelastic scattering the relative size of $\mathcal{W}$ to w depends on the energies at k and k'. If both initial and final energies are larger (or smaller) than the Fermi energy ε_F, then only small corrections to w occur. However, if the initial state is above and the final state is below ε_F, then for that scattering event $\mathcal{W}$ is suppressed. This tends, for example, to decrease the contribution of inelastic scattering events involving phonon emission.

To facilitate the power series expansion in terms of the applied field, we write the electric field as $\lambda\mathbf{E}$ and δf as

$$\delta f(k) = \sum_{n=1}^{\infty} f_n \lambda^n \tag{6.10.4}$$

where λ is an ordering index that will be set to unity at the end of the calculation. Equation (6.10.4) begins with the first power because that is the leading term, as can be seen from Eq. (6.10.2). Using these expressions in Eq. (6.10.2) and collecting the coefficients of the same power of λ on both sides, we obtain a series of equations. The lowest two orders are:

$$-e\mathbf{E}\cdot\nabla_{\mathbf{k}} f_0(k) = \sum_{\mathbf{k}'} [\mathcal{W}(k/k')f_1(k') - \mathcal{W}(k'/k)f_1(k)] \tag{6.10.5}$$

$$-e\mathbf{E}\cdot\nabla_{\mathbf{k}'} f_1(k) = \sum_{\mathbf{k}} \{(\mathcal{W}(k/k')f_2(k') - \mathcal{W}(k'/k)f_2(k)$$

$$+ [w(k'/k) - w(k/k')]f_1(k)f_1(k')\} \tag{6.10.6}$$

From f_0, Eq. (6.10.5) is solved for f_1, and this solution is used in Eq. (6.10.6) for f_2, and so on. This procedure can be continued to the required accuracy. Such a procedure is needed for hot-electron transport. For a low electric field, the lowest-order solution f_1 is sufficient.

This formalism is essential to an accurate study of electron transport in narrow-gap materials like $Hg_{0.78}Cd_{0.22}Te$. The band gap as a function of alloy concentration in this range is given in Fig. 7.52. With this small band gap, the conduction band deviates strongly from a parabolic band. It is well known that, even in large-gap materials, the constant EMA is valid only near the band edge (Kane, 1957) (within about $E_g/10$). The parabolic approximation is particularly poor for narrow-gap materials, and nonparabolic corrections calculated in the **k**·**p** formalism are often used (Schmidt, 1970; Meyer and Bartoli, 1982; Bartoli *et al.*, 1982). The calculated conduction band in Chapter 7 for the narrow-gap material can be well reproduced by the hyperbola

$$\varepsilon(\mathbf{k}) = (\gamma k^2 + c^2)^{1/2} - c \tag{6.10.7}$$

If the two parameters γ and c are adjusted to fit the band gap and the effective mass, Eq. (6.10.7) reduces to the same expression obtained in Kane's $k\cdot p$ method (Meyer and Bartoli, 1982). However, these two parameters can also be adjusted to obtain the best fit to a full band structure $\varepsilon(k)$ over the range of energy of interest. Figure 6.22 shows a comparison of bands obtained from different procedures. The heavy lines are the band structures obtained by the diagonalization of the Hamiltonian discussed in Chapter 7. The thin line is the fitted conduction band using Eq. (6.10.7) with γ and c chosen to be 48.3 and 0.058 respectively, which is seen to be really close to the actual band in this energy range. The thin dashed line is based on Kane's $k\cdot p$ method, which corresponds to Eq. (6.10.7) with γ and c equal to 41.2 and 0.05, respectively. The $k\cdot p$ band is reasonable but not as accurate as the thin line. On the other hand, the constant-mass band (heavy dashed line) deviates considerably from the actual band.

Two qualitative features of the band structure in Fig. 6.22 that impact transport properties should be noted. First, for energies $\varepsilon(\mathbf{k}) - c$ greater than 50 meV where the energy is linear in k, the group velocity is a constant independent of k. Then the DOS increases proportionally to ε rather than $\varepsilon^{1/2}$ as in the case of a parabolic band. The first feature, namely

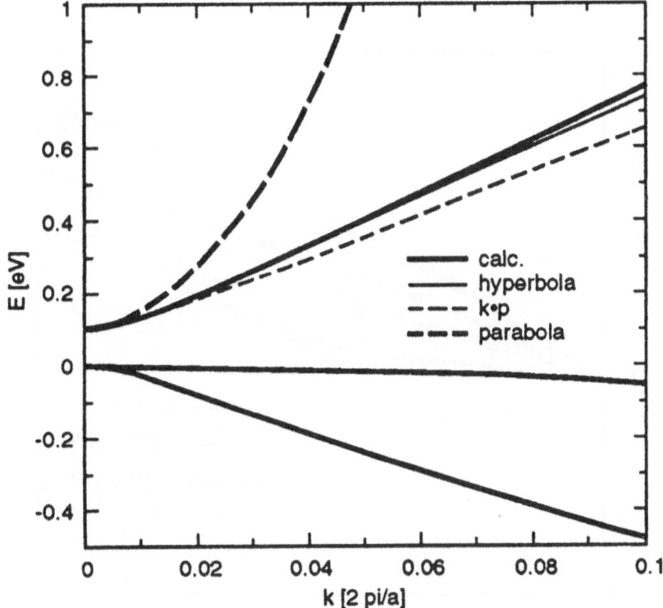

FIGURE 6.22. Approximation to the 77 K near band-edge electronic structure of $Hg_{0.78}Cd_{0.22}Te$ (Krishnamurthy and Sher, 1994).

a constant group velocity, means there is no change in the drift velocity with an increase in electron temperatures. The second feature increases the scattering rates, because for a given **k** the DOS into which the scattering can occur is higher than that in a parabolic band. This effect decreases the drift velocity and mobility.

For a quantitative calculation, the first step is to determine the Fermi level ε_F as a function of temperature and donor concentration n_D. A knowledge of temperature-dependent gaps $E_g(T)$ is essential, for which an empirical expression (in eV) (Brice, 1986)

$$E_g = 0.0954 + 0.327T/1000 \tag{6.10.8}$$

can be used. The ε_F can be calculated from the charge neutrality condition that the number of electrons n in the conduction band is equal to the number of holes p in the valence band plus the number of ionized donors n_D^+; i.e., $n = p + n_D^+$. Explicitly, $n = 2\Sigma f_0(\mathbf{k})$ for all $\varepsilon(\mathbf{k}) \geq \varepsilon_c = E_g$, $p = 2\Sigma[1 - f_0(\mathbf{k})]$ for all $\varepsilon(\mathbf{k}) \leq \varepsilon_v = 0$, and

$$n_D^+ = \frac{n_D}{1 + \exp(\varepsilon_F - \varepsilon_D)/(k_B T)} \tag{6.10.9}$$

The calculation of n uses the conduction-band structure given by Eq. (6.10.7). For p the valence-band structure is based on the results to be discussed in Chapter 7, which corresponds to using a DOS with an effective hole mass of 0.65. Figure 6.23a,b show the calculated ε_F as a function of temperature for two donor levels $\varepsilon_D - \varepsilon_C = 0$ and 30 meV respectively. In each figure, several donor concentrations are considered. The dashed line is E_g versus T given by Eq. (6.10.8).

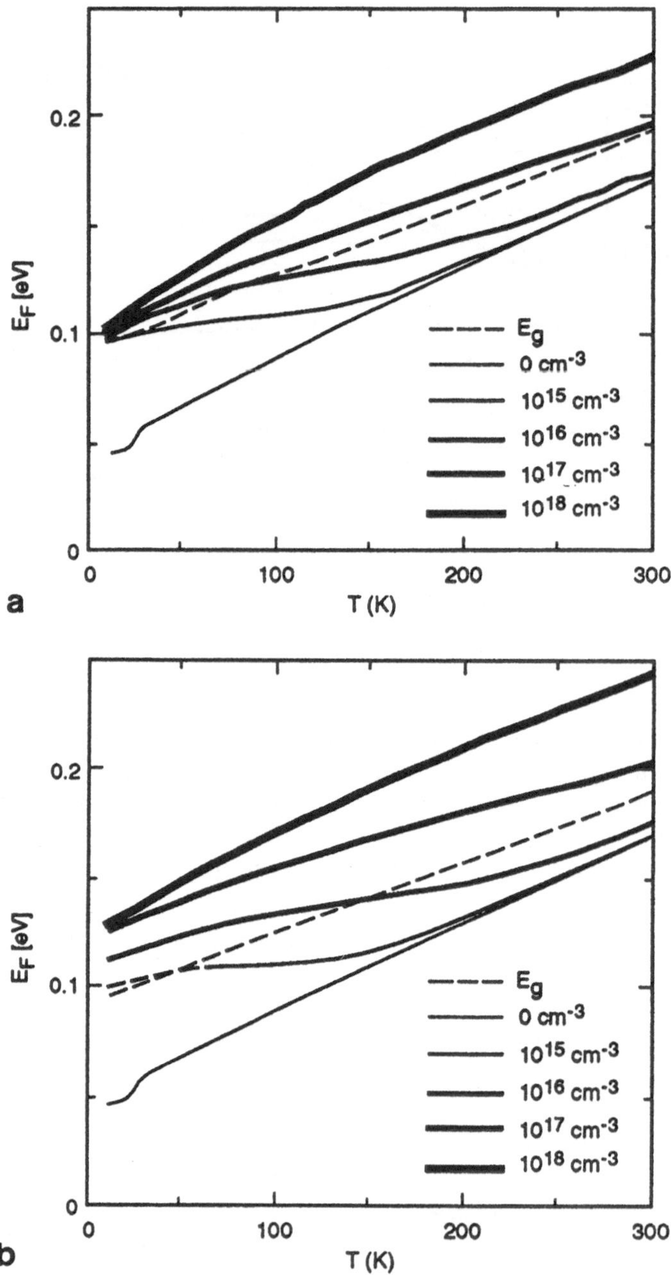

FIGURE 6.23. Fermi energy as a function of temperature for several donor concentrations: (a) $E_D - E_C = 0$ eV; (b) $E_D - E_C = 30$ meV (resonant in the conduction band) (Krishnamurthy and Sher, 1994).

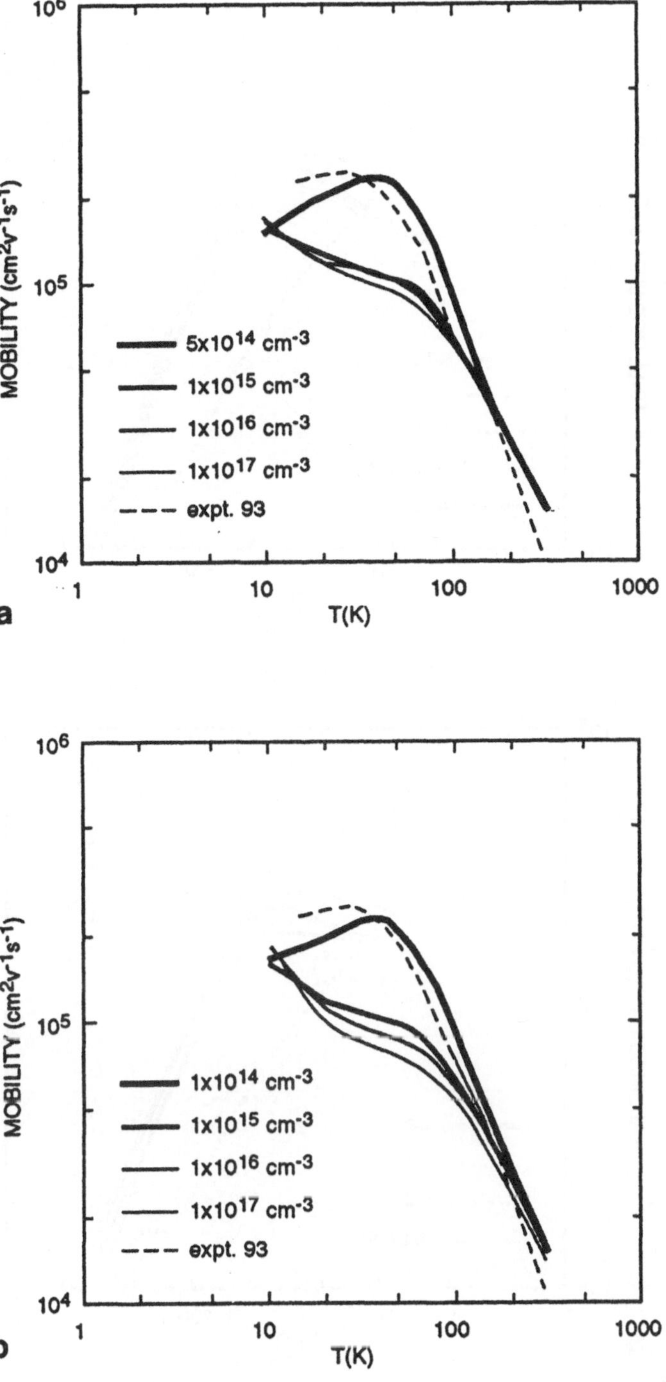

FIGURE 6.24. Drift mobility as a function of temperature for several donor concentrations: (a) $E_D - E_C = 0$ eV; (b) $E_D - E_C = -30$ meV (Krishnamurthy and Sher, 1994).

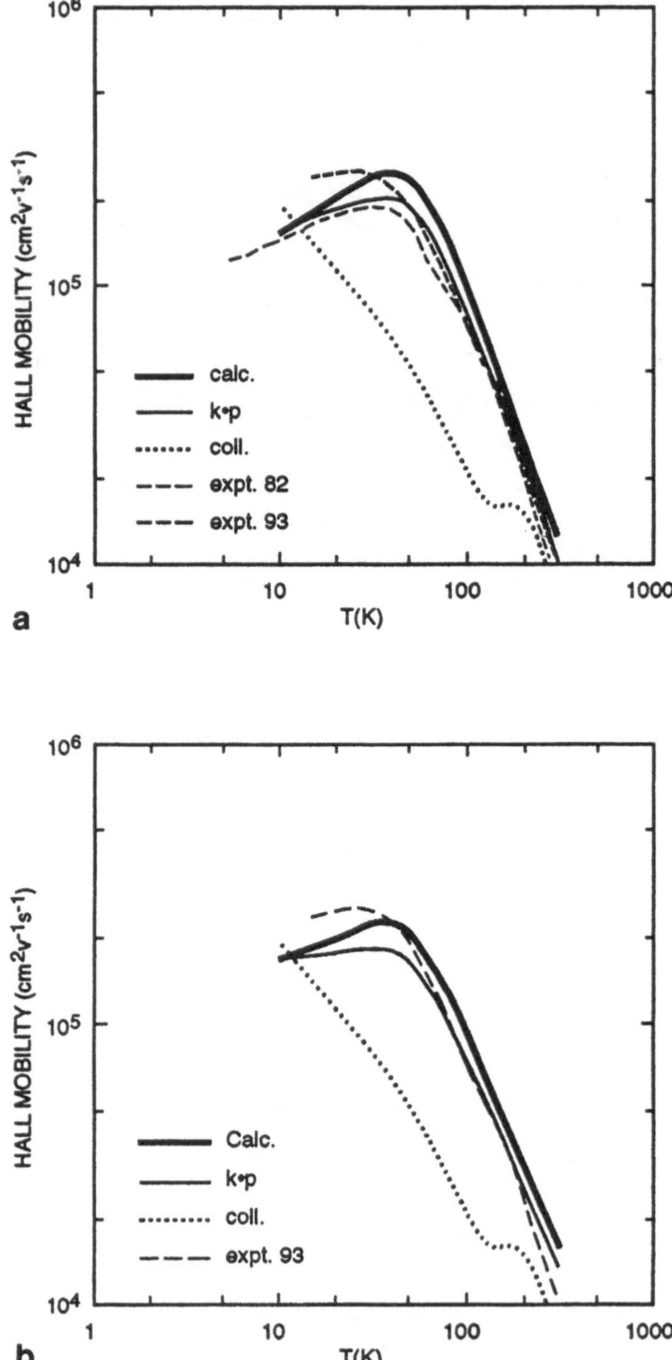

FIGURE 6.25. Hall mobility as a function of temperature with various approximations: (a) $E_D - E_C = 0$ eV; (b) $E_D - E_C = 30$ meV (Krishnamurthy and Sher, 1994).

In the calculation of mobility, only two of the most important scattering mechanisms, ionized impurity scattering and polar optical phonon scattering as described in Section 6.6 (see also Fig. 6.17), are included. The basis expansion method of Section 6.7 is used to solve f_1 from Eq. (6.10.5). Only two basis functions, $\exp(-\beta[\varepsilon(\mathbf{k})-\varepsilon_c])$ and $\mathbf{k}\cdot\hat{E}\exp(-\beta[\varepsilon(\mathbf{k})-\varepsilon_c])$, for f_1 are needed to obtain converged results. The calculated mobilities for several impurity concentrations as functions of the temperature (solid lines) are shown in Figs. 6.24a,b for the two donor levels considered. A peak in the mobility exists for a low-impurity concentration ($n_D = 10^{14}/\text{cm}^3$), in agreement with experiment (Bajaj, 1993) (dashed line). The peak is the result of the competition between increase in the screening of the impurity potential by free carriers and the increase in the numbers of ionized impurities and phonons as temperature increases. The increase of screening reduces the scattering rates by ionized impurities, while the increase in the numbers of ionized impurities and phonons enhances the scattering. This effect is only seen at low donor concentration, indicating that the change of screening with T in the highly doped materials is not as effective as that in the lightly doped systems.

In Fig. 6.25a,b, the drift mobilities calculated from different approaches are compared with the experimental Hall mobilities (Bajaj, 1983; Bajaj *et al.*, 1982). The calculations are done at the experimental donor concentration ($n_D = 5\times10^{14}\ 1/\text{cm}^3$). Both the full-band-structure calculation and the $k\cdot p$ method produce the peak structure of the experiment while the collision–time approximation fails to do so. The $k\cdot p$ method yields smaller mobilities than the full-band-structure calculation because a smaller γ value in Eq. (6.10.7) is used in the former. A smaller γ means a larger DOS and, hence, lower mobility. Note that a small peak near 200 K in the curve calculated using the collision–time approximation is due to a sudden change in the Fermi energy.

In summary, this section describe the procedure for a detailed calculation of the low-field mobilities in a narrow-gap semiconductor. The emphasis is the solution of the Boltzmann equation using Fermi statistics and accurate band structures. The results show that a straight calculation can explain the observed peak structure in the mobility (with T) at low carrier concentration without resorting to additional scattering mechanisms or parameters. However, the results can be improved using better T-dependent band structures and including other scattering mechanisms left out in the present calculation.

6.11. CONCLUDING REMARKS

This chapter presents the most important aspects of charge transport theory in semiconductor alloys. After a brief discussion of the master and Boltzmann equations, methods are presented to solve the Boltzmann equation for steady-state single-particle distribution in both low and high electric fields and for both degenerate and nondegenerate electrons. A consistent theme is the application of realistic band structures and wave functions. One very effective basis-expansion method has been employed in the solution of the Boltzmann equation. We show that alloying affects transport not only through alloy disorder scattering but also through the modification of the band structures. In fact, the latter plays a more important role in most cases.

A systematic comparative study of materials' merits for ballistic-transport devices shows that GaInAs and InPAs alloys are better than GaAs for devices with large base width ($D \geq$

500 Å), but with a thin base width ($D \simeq 100$ Å) no material offers a significant advantage over GaAs.

One key element in our theory is the availability of the realistic band structures (compiled in Chapter 7). Although it is very difficult to prove theoretically how accurate the wave functions based on our minimum basis empirical model are for the calculation of transition rates for transport coefficients, comparison between experiments and our calculations indicates that they are exceedingly good. They are accurate enough to serve as an engineering design tool. Considering the fundamental difficulties associated with the *ab initio* band theory for disordered semiconductor alloys discussed in Section 5.12, we cannot foresee in the near future the emergence of a practical *ab initio*–based theory capable of predicting transport coefficients for semiconductors as done in this chapter. However, many aspects of the present theory can be improved. For example, more accurate treatments of impurity potentials, electron–phonon interactions, and electron–electron scattering should be considered. The theory should be extended to transport across interfaces so it can model a real device.

Linear response theory has not been discussed in this chapter, because fortunately the alloy scattering for the systems that we considered have weak alloy scattering that does not warrant such a treatment. For strong-scattering alloys, the more rigorous Kubo (1957) and Greenwood (1958) conductivity formula based on the linear response should be used for the low-field transport. This theory has already been incorporated into the CPA Green function formalism (Velicky, 1969; Chen *et al.*, 1972). When alloy scattering is strong, new phenomena not included in this chapter may arise. Perhaps the best known is Anderson localization (Mott and Davis, 1979). There is no evidence for Anderson localization due to alloy scattering in the usual semiconductors, but it may occur in some of the highly strained systems people are trying to prepare, e.g., $Si_{1-x-y}Ge_xC_y$. Finally for high-field transport, the response theory has to be extended to higher orders.

REFERENCES

Aubert, J.P., J.C. Vaissiere, and J.P. Nougier (1984), *J. Appl. Phys.* **56**(4), 1128.

Bajaj, J. (1993), unpublished.

Bajaj, J., S.H. Shin, G. Bostrup, and D.T. Cheung (1982), *J. Vac. Sci. Technol.* **21**, 244.

Bandy, S., C. Nishimoto, S. Hyder, and C. Hooper (1981), *Appl. Phys. Lett.* **38**, 817.

Bartoli, F.J., J.R. Meyer, R.E. Allen, and C.A. Hoffman (1982), *J. Vac. Sci. Technol.* **21**, 241.

Brice, J.C. (1986), *Properties of HgCdTe*, EMIS Data Review Series No. 3, p. 103.

Brooks. H. (unpublished). A discussion of this formula can be found in L. Makowski and M. Glicksman (1973), *J. Phys. Chem. Solids* **34**, 487.

Budd, H. (1967), *Phys. Rev.* **158**, 798.

Butcher, P.N., and W. Fawcett (1966), *Phys. Lett.* **21**, 489.

Callen, H (1949), *Phys. Rev.* **76**, 1394.

Chen, A.-B., G. Weisz, and A. Sher (1972), *Phys. Rev. B* **5**, 2897.

Chen, A.-B., and A. Sher (1982a), *J. Vac. Sci. Technol.* **21**(1), 138.

Chen, A.-B., and A. Sher (1982b), *Phys. Rev. B* **26**, 6603.

Collins C.L., and P.Y. Yu (1984), *Phys. Rev. B* **30**, 4501.

Conwell, E.M. (1967), *High Field Transport in Semiconductors*, Solid State Physics Suppl., Vol. 9 (Academic Press, New York and London).

Conwell, E.M., and M.O. Vassel (1966), *J. Phys. Soc. Jpn.* (suppl.) **21**, 527.

Ehrenreich, H. (1957), *J. Phys. Chem. Solids* **2**, 131.

Fales, C.L. (1975), *Electron-Electron Resonances*, Ph.D. thesis, The College of William and Mary.

Fawcett, W., A.D. Boardman, and S. Swain (1970), *J. Phys. Chem. Solids* **31**, 1963.
Fawcett, W., and D.C. Herbert (1974), *J. Phys. C* **7**, 1641.
Frohlich, H. (1937), *Proc. R. Soc. A* **160**, 230.
Gammel, H. Ohno, and J.M. Ballantyne (1981), *IEEE J. Quantum Electron* **QE-17**, 267.
Glicksman, M. (1955), *Phys. Rev.* **100**, 1146.
Glicksman, M. (1958), *Phys. Rev.* **111**, 125.
Handbook of Physics (1972), 3rd ed. (McGraw-Hill, New York).
Harrison, J.W., and J. R. Hauser (1976), *J. Appl. Phys.* **47**, 292.
Hase, I., H. Kawai, S. Imanaga, K. Kaneko, and N. Watanabe (1987), *J. Appl. Phys.* **62**, 2558.
Hass, K.C., H. Ehrenreich, and B. Velicky (1983), *Phys. Rev. B* **27**, 1088.
Hasse, M.A., V.M. Robbins, N. Tabatabaie, and G.E. Stillman (1985), *J. Appl. Phys.* **57**(6), 2295.
Hauser, J.R., M.A. Littlejohn, and T.H. Glisson (1976), *Appl. Phys. Lett.* **28**, 458.
Heiblum, M., E. Calleja, I.M. Anderson, W.P. Dumke, C.M. Knoedler, and L. Osterling (1986), *Phys. Rev. Lett.* **56**, 2854.
Heiblum, M., M.I. Nathan, Dd.C. Thomas, and C.M. Knoedler (1985), *Phys. Rev. Lett.* **55**, 2200.
Kane, E.O. (1957), *J. Phys. Chem. Solids* **1**, 249.
Kash, K., P.A. Wolf, and W.A. Bonner (1983), *Appl. Phys. Lett.* **42**, 173.
Kowalsky, W., and A. Schlacketzki (1985), *Solid-State Electron.* **28**, 299.
Krishnamurthy, S., A. Sher, and A.-B. Chen (1985), *Appl. Phys. Lett.* **47**, 160.
Krishnamurthy, S., A. Sher, and A.-B. Chen (1986), *Phys. Rev. B* **33**, 1026.
Krishnamurthy, S., A. Sher, and A.-B. Chen (1987), *J. Appl. Phys.* **61**, 1475.
Krishnamurthy, S., A. Sher, and A.-B. Chen (1988a), *Appl. Phys. Lett.* **52**, 468.
Krishnamurthy, S., A. Sher, and A.-B. Chen (1988b), *J. Appl. Phys.* **63**(9), 4540.
Krishnamurthy, S., A. Sher, and A.-B. Chen (1988c), *Appl. Phys. Lett.* **53**, 1853.
Krishnamurthy, S., A. Sher, and A.-B. Chen (1989), *Appl. Phys. Lett.* **55**, 1002.
Krishnamurthy, S., and A. Sher (1994), *J. Appl. Phys.* **75**, 7904.
Levi, A.F.J., J.R. Hayes, and R. Bhat (1986), *Appl. Phys. Lett.* **48**, 1609.
Levi, A.F.J., J.R. Hayes, P.M. Platzman, and W. Wiegman (1985), *Phys. Rev. Lett.* **55**, 2071.
Littlejohn, M., J.R. Hauser, and T.M. Glisson (1977), *J. Appl. Phys.* **48**, 4587.
Littlejohn, M.A., J.R. Hauser, and T.H. Glisson (1982), *Appl. Phys. Lett.* **30**, 242.
Littlejohn, M.A., J.R. Hauser, T.H. Glisson, D.K. Ferry, and J.W. Harrison (1978), *Solid-State Electron.* **21**, 107.
Marsh, J.H., P.A. Houston, and P.N. Robson (1980), in *Gallium Arsenide and Related Compounds*, ed. H.W. Thim (Institute of Physics, Bristol).
Masselink, W.T., N. Braslau, W.I. Wang, and S.L. Wright (1987), *Appl. Phys. Lett.* **51**(9), 1533.
Meyer J.R., and F.J. Bartoli (1982), *J. Vac. Sci. Technol.* **21**, 237.
Molowski, L., and M. Glicksman (1973), *J. Phys. Chem. Solids* **34**, 487.
Nordheim, L. (1931), *Ann Phys.* **9**, 607; *ibid*, 641.
Nag, B.R. (1984), *Pramana* **23**, 411.
Ohashi, T., M.I. Nathan, S.D. Mukherjee, G.W. Wicks, G. Rubino, and L.F. Eastman (1985), Inst. Phys. Conf. Ser. No. 74, 293.
Rees, H.D. (1969), *J. Phys. Chem. Solids* **30**, 643.
Reif, F. (1965), *Statistical Thermal Physics* (McGraw-Hill, New York), p. 624.
Ridley, B.K. (1988), *Quantum Processes in Semiconductors* (Oxford Science).
Rorison, J.M., and D.C. Herbert (1986), *J. Phys. C* **19**, 3991.
Schmidt, J.L. (1970), *J. Appl. Phys.* **41**, 2876.
Shah, J., B. Deveaud, T.C. Damen, W.T. Tsang, A.C. Gossard, and P. Lugli (1987), *Phys. Rev. Lett.* **59**, 2222.
Sher, A., and H. Primakoff (1960), *Phys. Rev.* **119**, 178.
Sher, A., and H. Primakoff (1963), *Phys. Rev.* **130**, 1267.
Sze, S.M (1981), *Physics of Semiconductor Devices* (Wiley, New York).
Van Hove, L. (1955), *Physica* **XXI**, 517.
Van Hove, L. (1957), *Physica* **XXIII**, 441.
Windhorn, T.H., L.W. Cook, and G.E. Stillman (1982a), *IEEE Electron. Dev. Lett.* **DL-3**(1), 18.
Windhorn, T.H., L.W. Cook, and G.E. Stillman (1982b), *J. Electron. Mater.* **11**, 1065.
Yokoyama, N., K. Imamura, T. Pshima, H. Nishi, S. Muto, K. Kondo, and S. Hiyamizu (1984), *Jpn. J. Appl. Phys.* **23**, L311.

7

Band Structures of Selected Semiconductors and Their Alloys

This chapter is devoted to quantitative calculations of band structures of semiconductors and their alloys using a hybrid pseudopotential and tight-binding band model. This method is capable of producing accurate band structures for the pure constituent semiconductors and allows the execution of the alloy theory discussed in Chapter 5. Detailed band structures and Hamiltonian parameters are derived for a number of semiconductors, and the results are used in an extensive study of alloy systems.

7.1. HYBRID PSEUDOPOTENTIAL AND TIGHT-BINDING MODEL (HPT)

The emphasis here is to seek a simple but quantitative band model applicable to both the constituent semiconductors and their alloys. To this end, we first review some options now available for pure constituent semiconductors. Although one *ab initio* method, the GW method (Hyberson and Lowie, 1987), has been demonstrated to be able to incorporate an excitation self-energy potential into the one-electron Hamiltonian in the LDA calculation to obtain accurate band gaps for semiconductors, the calculation is still so time-consuming that it makes comprehensive studies impractical. As mentioned, the bands obtained from the empirical pseudopotential method (EPM), such as those obtained by Chelikowsky and Cohen (1976) and shown in Figs. 5.6 through 5.9, are the most accurate available. A question which so far is unanswered is, how can these results be incorporated into alloy calculations?

In contrast, the empirical tight-binding method (ETB) described in Section 5.10 provides a framework for the molecular CPA calculation discussed in Chapter 5. However, a first-neighbor or even second-neighbor ETB does not produce accurate results for some important band quantities. For example, the effective masses at the band edges, which require the inclusion of longer-range interactions, are not reproduced well. The hybrid empirical pseudopotential and tight-binding method (HPT) (Chen and Sher, 1980) encompasses the merits of EMP and ETB. HPT produces band structures for the pure semiconductors to within current experimental certainty. It also provides a means to implement the alloy calculation.

The key idea of HPT is to use the empirical pseudopotentials to construct a long-range universal tight-binding Hamiltonian H_0 and a short-range tight-binding Hamiltonian H_1 to distinguish different constituent crystals' potentials and, hence, the potential fluctuations in

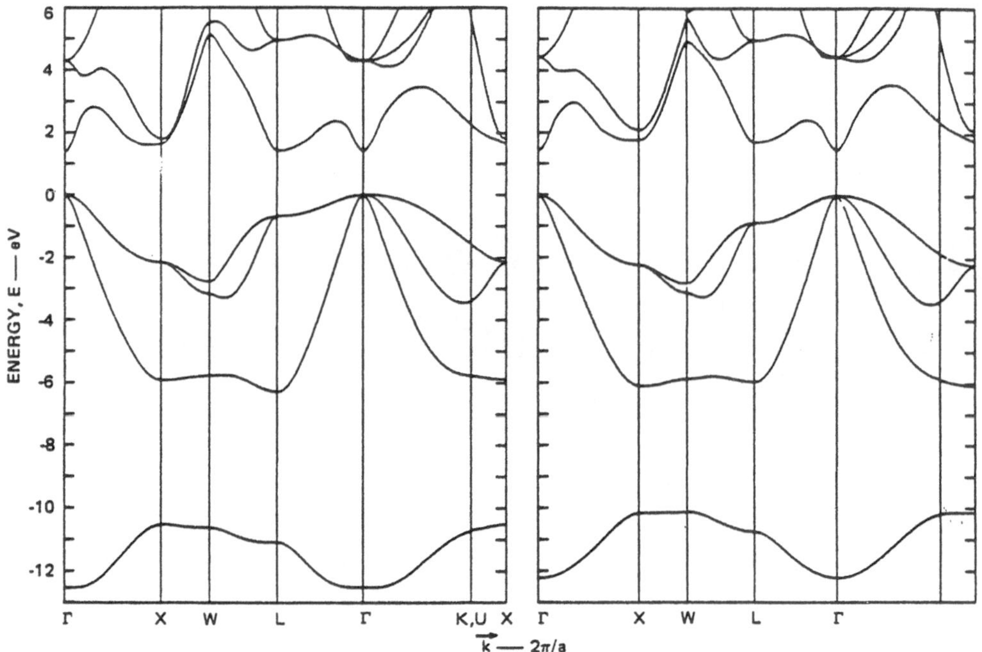

FIGURE 7.1. Comparison between the band structures of GaAs obtained from (a) Gaussian orbitals with $B_A = 0.3$ and $B_C = 0.2$ and (b) converged plane waves.

an alloy. The reason for using the word "universal" will become clear later. A major difference between the present method and usual TB, or the Slater–Koster (1954) model, is that now all the basis functions are explicitly defined. The local basis functions to be considered are either Gaussian orbitals of the type $\phi_{lm}(r) = r^n e^{-\gamma r^2} Y_{lm}$ or Slater orbitals $\phi_{lm}(r) = r^n e^{-\gamma r} Y_{lm}$, where Y_{lm} are real spherical harmonics and n is an interger. When an empirical pseudopotential of the form of Eq. (5.3.8) is used, it is straightforward to calculate the band structures from these orbitals following the LCAO procedure described in Section 5.2.

It is useful to point out that, while a converged calculation requires more than 10 orbitals per atom (Chadi, 1977), a minimum basis set of 4 orbitals per atom consisting of one s and three p orbitals can produce the lowest six bands fairly accurately. As an example, Fig. 7.1 (Chen and Sher, 1980) compares the band structures of GaAs for (a) an LCAO calculation using a minimum basis Gaussian set with $n = 0$ for the s and $n = 1$ for the p orbitals, and (b) a converged plane-wave calculation. Both calculations used the same empirical local pseudopotential (Cohen and Bergstresser, 1966). The LCAO basis functions used are Gaussian orbitals with two exponential parameters $\beta_a = 0.3$ and $\beta_c = 0.2$, respectively, for the anion and cation orbitals, where β is related to the actual exponential parameter γ by $\gamma = \beta(2\pi/a)^2$. The quality of the band structure produced by a minimum set of Slater orbitals with the corresponding exponential parameters $\beta_a = 1.8$, $\beta_c = 1.5$, and $n = 1$ for both s and p orbitals is similar to those in Fig. 7.1. The bands plotted in Fig. 7.1 take the maximum of the valence bands as the zero energy. On this energy scale, the bands from the two calculations can barely be distinguished from each other. On an absolute scale, however, the energies of the LCAO bands are about 1 eV higher than the corresponding energies in the converged bands. However, this rigid shift of energy does not play a role in the electrical and transport properties that are governed by relative energies. The main point in Fig. 7.1 is to show that accurate relative band structures of semiconductors can be obtained from a minimum basis set.

To obtain an accurate description of these bands in the important energy range around the band gap and in the valence bands is the main purpose of this chapter. One reason for this success using the minimum orbital set is that, unlike the usual tight-binding calculations, the present LCAO calculation includes all matrix elements between local orbitals at all distances. A straight application of the above LCAO calculation using the minimum basis set and the available empirical local pseudopotential, however, is not capable of producing the band structures within experimental accuracy. The main sources of error are truncation of the basis functions and errors of the pseudopotential itself. As mentioned, a nonlocal pseudopotential is needed to attain the accuracy of the band structures in Figs. 5.6 to 5.9. These effects can be incorporated in the present method to produce band structures of similar quality.

To simplify the description, we will use local orbitals that are normalized and mutually orthogonal, the orthonormal local orbitals (OLO). Although there is no intrinsic difficulty in using nonorthogonal basis functions in the Green function formalism (Lohez and Lannoo, 1983), the theory in Chapter 5 has to be modified. To directly use the theory in Chapter 5, let us convert the nonorthogonal Gaussian or Slater orbitals into OLO (Chen and Sher, 1982). Let $|u_{\mathrm{L}j\alpha}\rangle$ be a nonorthogonal orbital (NOLO) centered at the jth atom of the unit cell specified by a lattice vector $\mathbf{L}$, and let $|\phi_{kj\alpha}\rangle$ be the corresponding Bloch functions, as described by Eq. (5.2.3). These Bloch functions are not mutually orthonormal for the same $\mathbf{k}$. Let $S(\mathbf{k})$ be the overlap matrix, defined by $S_{j\alpha,j'\alpha'} = \langle\phi_{kj\alpha}|\phi_{kj'\alpha'}\rangle$, and let U be a unitary matrix that transforms $S(\mathbf{k})$ into a diagonal form $S_0(\mathbf{k})$ by $U^{+}(\mathbf{k})S(\mathbf{k})U(\mathbf{k}) = S_0(\mathbf{k})$. Then the matrix $B(\mathbf{k}) = US_0^{-1/2}U^{+}$ transforms the overlap matrix S into an identity, $B^{+}SB = I$, and changes $|\phi_{k\alpha}\rangle$ into a set of orthonormal Bloch functions: $|\mathbf{k}j'\alpha'\rangle = \sum B_{j\alpha j'\alpha'}|\phi_{kj\alpha}\rangle$. The inverse transformation of Eq. (5.2.3) then yields a set of OLO $|\mathrm{L}j\alpha\rangle$:

$$|\mathrm{L}j\alpha\rangle = \frac{1}{\sqrt{N}}\sum_{k} e^{-i\mathbf{k}\cdot(\mathbf{L}+\boldsymbol{\tau}_j)}\,|\mathbf{k}j\alpha\rangle \tag{7.1.1}$$

where $\boldsymbol{\tau}_j$ is the position vector of the jth atom in the $\mathbf{L}$ cell relative to $\mathbf{L}$. Figure 7.2 shows the s and p_x types of OLO constructed from the minimum Gaussian orbital (GO) set and comparison with the original Gaussians. The OLO are seen to spread more than the GO, but the range of the Hamiltonian matrix elements between the OLOs is found to be comparable to that between the original GO. The calculation of the wave functions for these OLO orbitals and the Hamiltonian matrix elements between these orbitals requires a k space integration within the first Brillouin zone. However, the Hamiltonian matrix $H_0(\mathbf{k})$ in the Bloch basis $|\mathbf{k}j\alpha\rangle$ for each $\mathbf{k}$ is readily obtained from the original LCAO Hamiltonian H_{N} by the transformation $H_0(\mathbf{k}) = B^{+}(\mathbf{k})H_{\mathrm{N}}(\mathbf{k})B(\mathbf{k})$. These OLOs can be regarded as a set of mutually orthonormal local s and p orbitals. The procedure just presented provides a means for constructing the long-range TB Hamiltonian H_o in the OLO basis.

The band structure produced by H_o, as shown in Figure 7.1, does not include the spin-orbit interaction. The spin-orbit interaction can be effectively treated in this model by adding a spin-orbit Hamiltonian H_{so}. The simplest way to represent H_{so} is to use OLO in the Γ representation $|\mathrm{L}j\gamma\rangle$ given by Eq. (5.10.10) (also discussed in Appendix 7A). In this

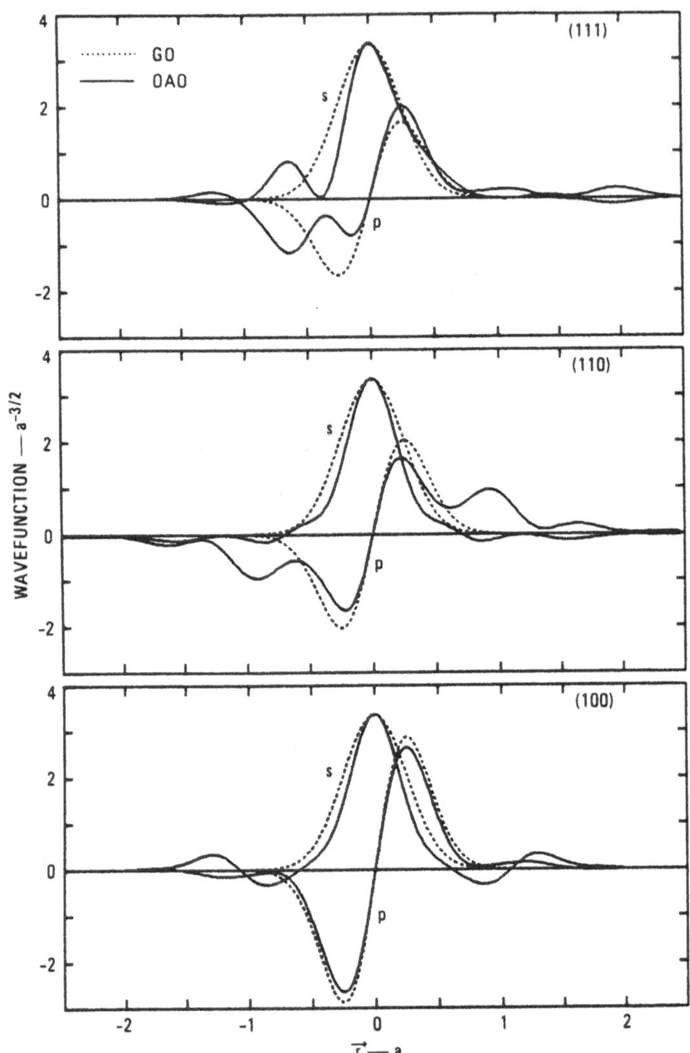

FIGURE 7.2. Comparison between the A_1 and T_2 orthonormal atomic-like orbitals (OAOs) and the s- and p_x-type Gaussian orbitals (GOs) along three directions.

representation H_{so} is purely diagonal with the values 0, $-2\lambda^j$, and λ^j for Γ_6, Γ_7, and Γ_8 representations respectively for each site (see Eq. (7A.2)). For a zinc blende or wurtzite semiconductor, only two parameters, λ^A for the anions and λ^C for the cations, are needed. The value of λ should be close to the atomic spin-orbit coupling constant, as determined by the difference $3\lambda = \varepsilon_{p3/2} - \varepsilon_{p1/2}$ for the two p-levels listed in Table 2.1.

To obtain band structures in agreement with experiment, a tight-binding Hamiltonian H_1 with only the first- and second-neighbor interaction parameters can be added to fine-tune the band structures. The TB parameters in H_1 are adjusted to fit the desired band energies. A comparison of the detailed band structures to be discussed later (Figs. 7.3 through 7.11) with those calculated from the empirical pseudopotentials (Fig. 5.7) demonstrates the capability of the present method to incorporate experimental data to produce accurate band structures for the constituent semiconductors.

The next step is to show how this method is extended to alloy calculations. In doing so we not only have to pay attention to the accuracy of the band structures of the constituents, but we also must obtain meaningful Hamiltonians for alloys, particularly terms relating to the disorder parameters. In this connection, a systematic investigation of the LCAO Hamiltonian matrix elements for the III–V zinc blende compounds (Chen and Sher, 1980) using empirical pseudopotentials shows that the long-range matrix elements, i.e., third neighbor and beyond, for all systems are nearly the same if crystal units are used. In crystal units, the dimensionless energy ε is related to the energy E in Rydberg by $E = \varepsilon(2\pi/a)^2$, where a is the lattice constant in units of Bohr radii. This result indicates that, because the bond-length variations among III–V compounds are only a few percent, there is a common "universal" Hamiltonian H_0 which provides the long-range matrix elements for all III–V semiconductors and alloys. The same idea can also be extended to the IV–IV and II–VI materials groups. The variations from one system to another within a group and the potential disorder in their alloys are then confined to the short-range matrix elements, which are obtained by fitting each constituent's band structure.

Specifically, the parametrization procedure starts with a common set of pseudopotentials in crystal units. Then, the following procedure is used to obtain the band structures and Hamiltonians: (1) For a given group such as the III–V compounds, a common set of pseudopotential form factors in crystal units is obtained from the average of the available empirical values. These are then used to generate a common zeroth-order tight-binding Hamiltonian matrix H_0 in the OLO basis constructed from Slater or Gaussian orbitals with a fixed pair of the decay parameters β_a and β_c. The empirical pseudopotentials are usually written as the symmetrical and antisymmetric combinations $V_s(g) = (V^A(g) + V^C(g))/2$ and $V_a(g) = (V^A(g) - V^C(g))/2$, where $V^A(g)$ and $V^C(g)$ are the potential form factors of the anion and cation respectively at the wave vector with a magnitude g. The only values of $V_s(g)$ needed are at $g^2 = 3$, 8, and 11 in units of $(2\pi/a)^2$, and for $V_a(g)$ at $g^2 = 3$, 4, and 11. Table 7.1 lists these form factors in crystal units for the III–V systems. (2) A local perturbation TB Hamiltonian matrix H_1 and a spin-orbit Hamiltonian H_{so} are adjusted to fit the band structures for the constituent systems. H_1 only contains the term value corrections $\delta\varepsilon_s^A$, $\delta\varepsilon_s^C$, $\delta\varepsilon_p^A$, $\delta\varepsilon_p^C$; the first-neighbor interactions h_{ss}^{AC}, h_{sx}^{AC}, h_{xs}^{AC}, h_{xx}^{AC}, and h_{xy}^{AC}; and the second-neighbor interactions of the two-center forms v_{ss}, $v_{sp\sigma}$, $v_{pp\sigma}$, and $v_{pp\pi}$ among anions, and similar forms among second-neighbor cations. As mentioned, H_{so} only contains the two parameters λ^A and λ^C. These parameters are adjusted to produce the experimental spin-orbit splitting of the top of the valence band Δ_0. The most important band features, those which have been accurately determined experimentally, are fitted exactly. The other less important or less accurately determined quantities are obtained with a least-square fit. Examples of important band quantities are the energies for the Γ_{15v}, Γ_{1c}, X_{1c}, and L_{1c} states, the spin-orbit splitting Δ_0, and

TABLE 7.1. Common Pseudopotential Form Factors $v(g)$ in Crystal Units at Several g^2 Values Used to Construct the Unperturbed Hamiltonian H_0 for the III–V Semiconductors

$g^2[(2\pi/a)^2]$	3	4	8	11
V_s	–0.7059	0	0.0376	0.1562
V_a	0.2228	0.1620	0	0.0492

the effective mass at the bottom of the conduction band. Because the number of parameters involved is large and there are considerable uncertainties in the experimental band quantities for some compounds, the parameters determined from the fitting are not expected to be unique. The diagonal and first-neighbor interactions contain most of the variations from one system to another. The smaller second-neighbor parameters are used to fine tune the band structures. The differences in the term values and the first-neighbor interactions then define the alloy disorder potential. The alloy theory discussed in Chapter 5 is then applied. Appendix 7A provides a detailed description of band calculation using the HPT model.

7.2. BAND STRUCTURES AND HAMILTONIAN PARAMETERS FOR III–V CONSTITUENT COMPOUNDS

Among III–V compounds, GaAs has the most extensive and accurate experimental data base and thus provides a good place to start the parametrization of the empirical Hamiltonian.

TABLE 7.2. Lattice Constants a (in Bohr radii), Matrix Element Corrections $\delta\varepsilon$, h, and v for H_1 (in crystal units), Spin-Orbit Coupling Constants λ (in eV), Valence-Band Offsets E_v (in eV) Relative to GaAs, and Resulting Term Values (in eV) for the Nine Constituent III–V Zinc Blende Semiconductors

	AlP	AlAs	AlSb	GaP	GaAs	GaSb	InP	InAs	InSb
a	10.331	10.656	11.595	10.293	10.685	11.516	11.091	11.448	12.244
$\delta\varepsilon_s^A$	−1.6198	−1.9393	−2.1796	−1.6308	−1.9465	−1.9490	−1.9566	−2.1677	−2.3890
$\delta\varepsilon_s^C$	−1.9612	−1.9437	−1.9991	−2.1001	−2.1153	−2.0457	−1.9352	−2.1309	−1.8185
$\delta\varepsilon_p^A$	−2.1371	−2.0894	−2.2442	−2.1822	−2.0900	−2.2677	−2.2021	−2.1320	−2.3380
$\delta\varepsilon_p^C$	−2.1096	−2.0645	−2.0800	−2.1611	−1.9964	−2.1473	−1.8913	−1.7478	−1.9399
h_{ss}^{AC}	−0.1415	−0.1961	−0.0980	−0.1596	−0.1773	0.1207	0.1464	−0.0862	0.3240
h_{sx}^{AC}	−0.1377	−0.2261	−0.0847	−0.0684	−0.0766	−0.0165	−0.1134	−0.0440	0.1911
h_{xs}^{AC}	0.0519	0.0607	0.2504	0.0321	0.0562	0.1002	0.1140	0.2335	0.1204
h_{xx}^{AC}	−0.0121	0.0774	−0.0129	−0.0272	−0.0150	−0.2114	−0.0179	−0.1573	−0.3193
h_{xy}^{AC}	0.0465	0.0823	−0.3970	−0.0289	0.0855	0.0122	−0.0546	0.0666	−0.1478
v_{ss}^A	0.0189	0.0195	−0.0300	−0.0150	0.0100	−0.0463	−0.0320	−0.0117	−0.0603
$v_{sp\sigma}^A$	0.0110	−0.0111	−0.0055	0.0204	−0.0079	0.0067	0.0175	0.0066	0.0354
$v_{pp\sigma}^A$	0.0073	−0.0020	−0.0045	0.0166	0.0200	0.0345	0.0101	0.0055	0.0508
$v_{pp\pi}^A$	−0.0029	0.0114	0.0238	0.0035	0.0024	0.0065	0.0122	0.0070	0.0081
v_{ss}^C	−0.0125	−0.0114	0.0351	0.0051	−0.0133	0.0257	0.0162	−0.0054	0.0198
$v_{sp\sigma}^C$	0.0110	−0.0111	−0.0055	0.0204	−0.0079	0.0067	0.0175	0.0066	0.0354
$v_{pp\sigma}^C$	0.0053	−0.0383	−0.0093	0.0246	−0.0046	0.0273	0.0120	0.0117	0.0338
$v_{pp\pi}^C$	0.0138	0.0100	0.0025	0.0140	0.0170	0.0083	0.0013	0.0003	−0.0005
λ^A	0.025	0.123	0.254	0.025	0.120	0.256	0.031	0.128	0.284
λ^C	0.000	0.000	0.000	0.050	0.050	0.050	0.092	0.092	0.092
E_v	−0.794	−0.550	−0.260	−0.540	0.000	0.140	−0.340	0.070	0.260
E_s^A	−5.052	−6.260	−6.247	−5.145	−6.260	−5.399	−5.854	−6.360	−6.353
E_s^C	1.272	1.279	0.859	0.577	0.464	0.682	1.217	0.341	1.417
E_p^A	−0.836	−0.560	−1.092	−1.071	−0.560	−1.202	−1.009	−0.660	−1.315
E_p^C	3.454	3.460	2.861	3.219	3.762	2.627	3.951	4.296	3.067

The term values in H_1 of both Ga and As are allowed to vary freely to obtain the most accurate bands in comparison with experimental results. The fitting then yields the term values for both Ga and As measured with respect to the top of the valence band E_v of GaAs, which can be taken as the reference energy and set equal to zero: $E_v(\mathrm{GaAs}) = 0$. The valence-band offsets—i.e., the differences in E_v between semiconductors—are used to set the E_v values for other semiconductors. Since the valence-band offsets among semiconductors are not strictly additive, we do not expect to have a unique set of E_v which will yield accurate band offsets. However, the set given in Table 7.2, which are adopted from those of Bauer *et al.* (1987) with the exception of AlAs, serve to give a first estimate of E_v. Using the results for GaAs as a guide, we then proceed to other systems. Although it is desirable to have fixed term values for each element to be used in the MCPA alloy calculation, in reality the potential of each atom is modified when the atom is in different crystal environments, as manifested by the need for small variations in the term values in order to achieve the desired accuracy of the fitted band structures in the present scheme. The resulting Hamiltonian parameters are listed in Table 7.2. Appendix 7A describes how to use these parameters in the band structure calculation. These Hamiltonian parameters will be used later in the alloy calculations. The calculated band structures are described system by system and are compared with experiments and other calculations. A comprehensive compilation of the band structure results for the III–V compounds is available in the semiconductor data book edited by Madelung (1991), which is the main source of the experimental and other theoretical results used here for comparison.

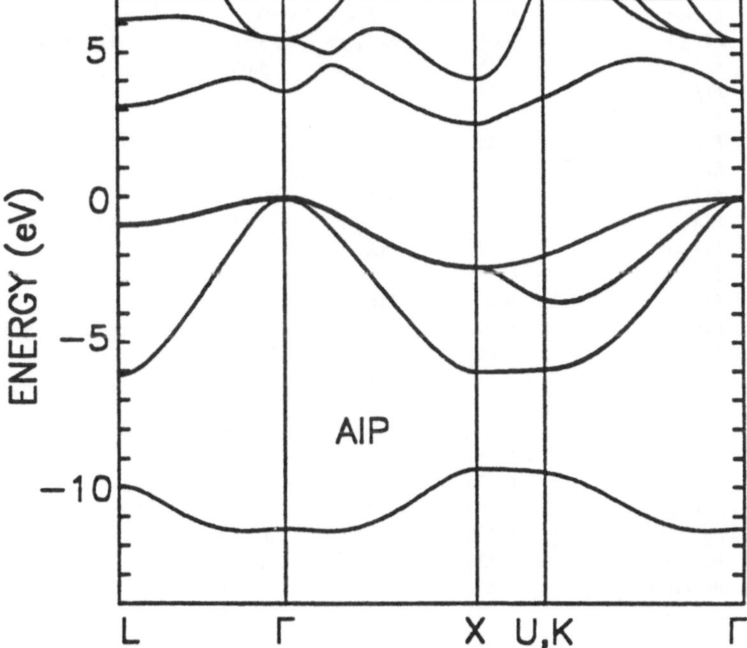

FIGURE 7.3. Calculated band structure energy in eV versus crystal wave vector of AlP along several symmetry directions.

7.2.1. AlP

AlP is not studied as commonly as other III–V semiconductors considered in this section. As a consequence, the only experimental data available that can be used to assign the Hamiltonian parameters are the indirect gap $E(X_{6c} - \Gamma_{8v}) = 2.51$ eV and the gap $E(\Gamma_{6c} - \Gamma_{8v}) = 3.63$ eV. The parametrization of the band structure for this system starts with the term values and spin-orbit constants of Al and P derived from AlAs and GaP respectively and the mean tight-binding parameters of these two systems. Then two of the parameters are modified to fit the two experimental gaps. The other parameters are only slightly modified, so valence-band energies are not too different from the LCAO calculation (Huang and Ching, 1985). The calculated bands are shown in Fig. 7.3. The band energies at three symmetry k points and several important effective masses are listed in Table 7.3 along with the LCAO calculation. We note that while most of the band energies and effective masses of these two band structures are qualitatively similar, the location of the conduction-band minimum E_c is different. The present band has its E_c located exactly at X and has a moderate longitudinal

TABLE 7.3. Calculated Band Structure of AlP from the Present HPT Method Compared with an LCAO Calculation (Huang and Ching, 1985) and Experiment[a]

	Theory			
	HPT		LCAO	Experiment
$E(\Gamma_{6v})$	−11.42	$E(\Gamma_{1v})$	−11.82	
$E(\Gamma_{7v})$	−0.068			
$E(\Gamma_{8v})$	0	$E(\Gamma_{15v})$	0	
$E(\Gamma_{6c})$	3.63	$E(\Gamma_{1c})$	3.74	3.63[b] (4K)
$E(\Gamma_{7c})$	5.45	$E(\Gamma_{15v})$	5.09	
$E(\Gamma_{8c})$	5.46			
$E(X_{6v})$	−9.36	$E(X_{1v})$		
$E(X_{6v})$	−6.04	$E(X_{3v})$		
$E(X_{6v})$	−2.45			
$E(X_{7v})$	−2.42	$E(X_{5v})$	−2.27	
$E(X_{6c})$	2.51	$E(X_{1c})$	2.51	2.505[b] (4K)
$E(X_{7c})$	4.07	$E(X_{3c})$	4.30	
$E(L_{6v})$	−9.95			
$E(L_{6v})$	−6.14	$E(L_{1v})$		
$E(L_{6v})$	−0.99			
$E(L_{4,5v})$	−0.95	$E(L_{3v})$	−0.80	
$E(L_{6c})$	3.13	$E(L_{1c})$	3.57	
$E(L_{6c})$	6.15	$E(L_{3c})$		
$E(L_{4,5c})$	6.16			
$m_l(X_{6c})$	0.756		3.67	
$m_t(X_{6c})$	0.226		0.212	
$m_c(\Gamma_{6c})$	0.187			
$m_{hh}[100]$	0.496		0.513	
$m_{hh}[111]$	1.228		1.372	
$m_{lh}[100]$	0.252		0.211	
$m_{lh}[111]$	0.193		0.145	
m_{soh}	0.339			

[a]The energies (in eV) are relative to the top of the valence band. The effective masses are multiples of free-electron mass.
[b]Monemar (1973).

effective mass $m_l = 0.756m_o$, where m_o is the free-electron mass. The LCAO band has its E_c located on the Δ-axis (Γ–X) and has a large $m_l = 3.67m_o$.

7.2.2. *AlAs*

Although GaAlAs has been a very important alloy in quantum-well structures, the experimental information for the band structure of AlAs is rather sketchy, mainly because good bulk crystals of AlAs are hard to grow. We have to rely on a combination of experimental and estimated results (Adachi, 1985), previous calculated bands (Huang and Ching, 1985), and extrapolated values from the GaAlAs alloy as input in our band structure fitting. The resulting band structure is plotted in Fig. 7.4. The symmetry-point energies and important effective masses are tabulated in Table 7.4. The minimum gap is an indirect gap of 2.23 eV with the conduction-band minimum (CBM) located slightly off the symmetry point X. This is the so-called camel's back structure. The calculated CBM is shifted by $k_o = 0.069(2\pi/a)$ away from the X point with a corresponding energy shift $\Delta E = 0.35$ meV below the X_{6c} level. The longitudinal mass associated with a camel back is usually large; our calculated value is 10.96. The calculated transverse mass at CBM is 0.234 times the free-electron value. These numbers should be compared with the values $k_o = 0.042(2\pi/a)$, $\Delta E = 0.2$ meV, $m_l = 10.7$, and $m_t = 0.227$ extrapolated from GaP and used in a two-band model analysis of this structure (Kopylov, 1985). The 2.95 eV for the Γ_{6c} level is an input in our calculation. This value is smaller than the experimental value of 3.13 eV. However, this lower value is needed to explain the band gaps of $Ga_{1-x}Al_xAs$ alloys, as discussed later. The values of L_{6c} and $L_{4,5v}$ are chosen so that they give values in accord with the E1 optical

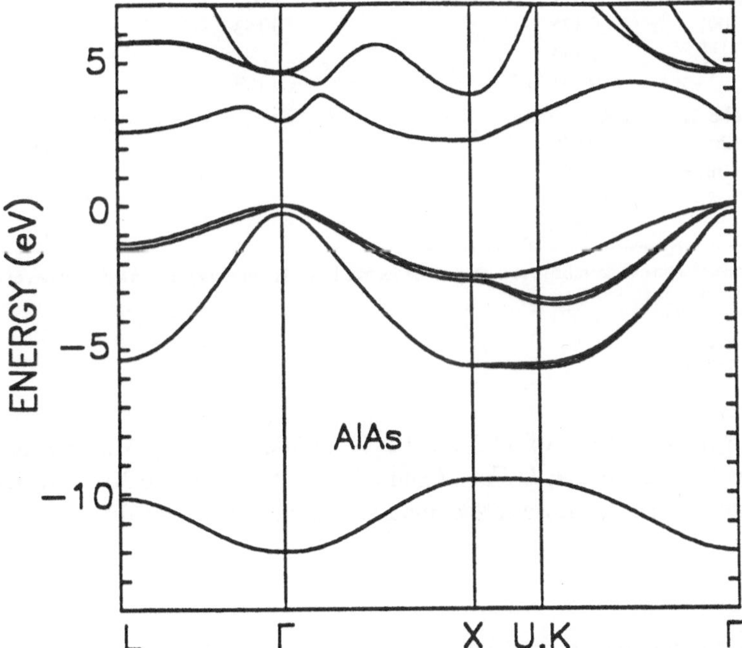

FIGURE 7.4. Calculated band structure energy in eV versus crystal wave vector of AlAs along several symmetry directions.

TABLE 7.4. Comparison of the Calculated Band Structure of AlAs from the Present HPT Method with an LCAO Calculation (Huang and Ching, 1985), and the Results Compiled by Madelung (1991)[a]

	Theory			Experimental and estimated values
	HPT		LCAO	
$E(\Gamma_{6v})$	−12.00	$E(\Gamma_{1v})$	11.95	
$E(\Gamma_{7v})$	−0.30			−0.30[b] (300K)
$E(\Gamma_{8v})$	0	$E(\Gamma_{15v})$	0	
$E(\Gamma_{6c})$	2.95	$E(\Gamma_{1c})$	2.79	3.13[c] (4K)
$E(\Gamma_{7c})$	4.60	$E(\Gamma_{15v})$	4.48	
$E(\Gamma_{8c})$	4.67			
$E(X_{6v})$	−9.56	$E(X_{1v})$	−9.63	
$E(X_{6v})$	−5.63	$E(X_{3v})$	−5.69	
$E(X_{6v})$	−2.68			
$E(X_{7v})$	−2.50	$E(X_{5v})$	−2.38	
$E(X_{6c})$	2.23	$E(X_{1c})$	2.37	2.229[c] (4K)
$E(X_{7c})$	3.84	$E(X_{3c})$	3.84	
$E(L_{6v})$	−10.16			
$E(L_{6v})$	−5.35	$E(L_{1v})$	−5.95	
$E(L_{6v})$	−1.49			
$E(L_{4,5v})$	−1.30	$E(L_{3v})$	−0.88	
$E(L_{6c})$	2.59	$E(L_{1c})$	2.81	
$E(L_{6c})$	5.68	$E(L_{3c})$	5.86	
$E(L_{4,5c})$	5.74			
$m_c(\Gamma_{6c})$	0.135		0.150	
m_{hh} [100]	0.421		0.409	
m_{hh} [111]	0.817		1.022	
m_{lh} [100]	0.175		0.153	
m_{lh} [111]	0.145			
m_{soh}	0.261		0.109	
Camel back structure				
k_0 $(2\pi/a)$	0.069			0.042[d]
ΔE (meV)	0.35			0.2[d]
m_l^*	10.96			10.7[d]
m_t	0.234			0.227[d]

[a]The energies (in eV) are relative to the top of the valence band. The effective masses are multiples of the free electron mass.
[b]Adachi (1985).
[c]Monemar (1973).
[d]Kopylov (1985).

transition energy $E(L_{6c}–L_{4,5v})$ of 3.83 eV at 300 K (Adachi, 1985). The spin-orbit splitting at the top of valence band $\Delta_0 = E(\Gamma_{8v} - \Gamma_{7v})$ in the present band is 0.30 eV, which is also an input. The calculated hole masses are comparable to those calculated by Huang and Ching (1985).

7.2.3. *AlSb*

The experimental data base for the band structure of AlSb is only slightly better than that for AlP. The first four pieces of experimental data listed in Table 7.5 are sightly adjusted to give the zero-temperature values for the band structure parametrization. In the absence of

TABLE 7.5. Comparison of the Calculated Band Structure of AlSb from the Present HPT Method with an LCAO Calculation (Huang and Ching, 1985), and the Results Compiled by Madelung (1991)[a]

	Theory			Experimental and estimated values
	HPT		LCAO	
$E(\Gamma_{6v})$	–11.77	$E(\Gamma_{1v})$	–11.77	
$E(\Gamma_{7v})$	–0.676			–0.673[b] (295 K)
$E(\Gamma_{8v})$	0	$E(\Gamma_{15v})$	0	
$E(\Gamma_{6c})$	2.39	$E(\Gamma_{1c})$	2.05	2.384[b] (25 K)
$E(\Gamma_{7c})$	3.50	$E(\Gamma_{15v})$	3.50	
$E(\Gamma_{8c})$	3.59			
$E(X_{6v})$	–8.98	$E(X_{1v})$	–8.72	
$E(X_{6v})$	–6.40	$E(X_{3v})$	–5.44	
$E(X_{6v})$	–1.63			
$E(X_{7v})$	–1.21	$E(X_{5v})$	–2.31	
$E(X_{6c})$	1.69	$E(X_{1c})$	2.08	1.686[c] (27 K)
$E(X_{7c})$	2.94	$E(X_{3c})$	3.02	
$E(L_{6v})$	–10.24			
$E(L_{6v})$	–4.87	$E(L_{1v})$		
$E(L_{6v})$	–1.35			
$E(L_{4,5v})$	–0.90	$E(L_{3v})$	–0.90	
$E(L_{6c})$	2.33	$E(L_{1c})$	1.94	2.327[c] (35 K)
$E(L_{6c})$	3.84	$E(L_{3c})$		
$E(L_{4,5c})$	3.88			
$m_c(\Gamma_{6c})$	0.185			
m_{hh} [100]	1.00		0.336	
m_{hh} [111]	1.34		0.872	
m_{lh} [100]	0.136		0.123	
m_{lh} [111]	0.131		0.091	
m_{soh}	0.273			
Camel back structure				
k_0 $(2\pi/a)$	0.120			0.101[d]
ΔE (meV)	7.88			7.4[d]
m_l^*	2.13			1.8[d]
m_t^*	0.37			0.259[d]

[a]The energies (in eV) are relative to the top of the valence band. The effective masses are multiples of the free electron mass.
[b]Joullie *et al.* (1982).
[c]Alibert *et al.* (1983).
[d]Kopylov (1985).

more experimental data, the valence-band states of the LCAO calculation (Huang and Ching, 1985) are used, The Hamiltonian parameters derived from GaSb are used as a start. They are then modified by replacing the term values and the spin-orbit coupling constant of Ga by the appropriate values for Al and adjusting four other tight-binding parameters to fit the four experimental data points just mentioned and several LCAO valence-band states. The calculated band structure is shown in Fig. 7.5. The calculated band energies at the symmetry points and several important effective masses are tabulated in Table 7.5 and compared with the LCAO calculation. The band energies calculated from the LCAO method differ from the

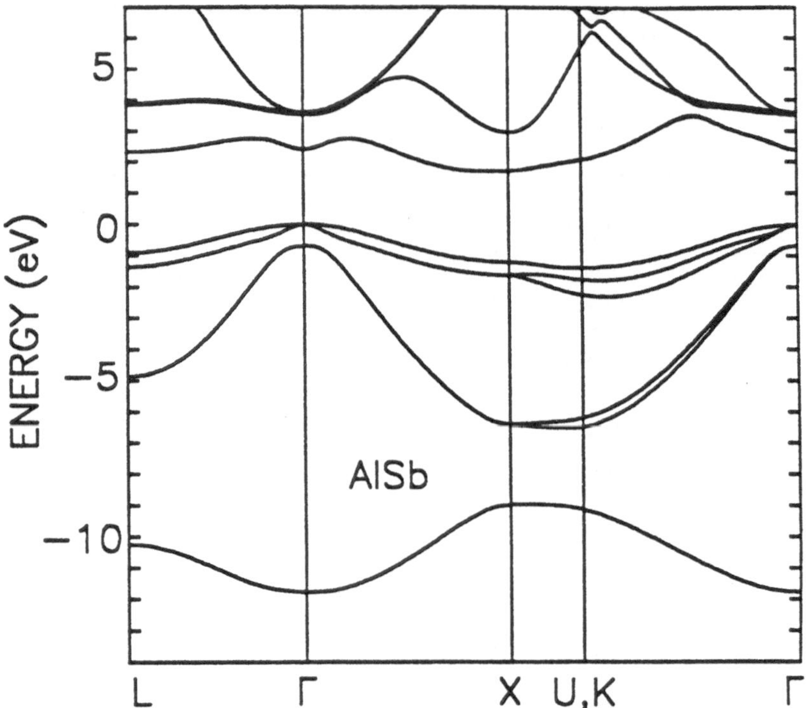

FIGURE 7.5. Calculated band structure energy in eV versus crystal wave vector of AlSb along several symmetry directions.

present results by several tenths of an electron volt for most energies and about 20% for most masses. The conduction-band minimum occurs along the Δ-axis slightly off the symmetry X point. The camel-back structure is similar to that in AlAs and to that deduced from a **k**·**p** analysis (Kopylov, 1985).

7.2.4. GaP

There are sufficient data to allow a full empirical parametrization of the band structure of GaP using the preset HPT method. Figure 7.6 shows the calculated band structure. Table 7.6 lists the calculated band energies at the symmetry points along with the empirical pseudopotential method (EPM) (Cohen and Chelikowsky, 1989) and the compiled experimental results (Madelung, 1991). Column A contains the band energies derived from angle-resolved photoemission spectra (Solal *et al.*, 1984), while column B contains those deduced from x-ray photoemission spectra (Ley *et al.*, 1974) and other experiments and studies. Note that the band energies from the EMP calculation and the angle-resolved photoemission experiment did not take into account the small spin-orbit splittings, for which the nonrelativistic labels of states are also given. Our calculated band energies follow those of the EMP calculation except for small adjustments to fit the quoted experimental values closely. The 2.350 eV for the X_{6c} state is used to produce the value deduced from low-temperature wavelength modulation exciton data (Humphreys *et al.*, 1978). However, the calculated conduction-band minimum E_c does not fall at X but at $\Delta k = 0.034(2\pi/a)$ away from X and on the Δ-axis and with an energy $\Delta E = 0.55$ meV below the value of the X_{6c} level.

TABLE 7.6. Comparison of Band Structure of GaP from the Present Calculation with Results Compiled by Madelung (1991) and Empirical Pseudopotential Calculation (EMP) (Cohen and Chelikowsky, 1989)[a]

	Theory			Experiment	
	HPT	EMP		A	B
$E(\Gamma_{6v})$	−12.99	−12.99	$E(\Gamma_{6v})$	−12.3	−13.2
$E(\Gamma_{7v})$	−0.081				
$E(\Gamma_{8v})$	0	0	$E(\Gamma_{1v})$		
$E(\Gamma_{6c})$	2.866	2.88	$E(\Gamma_{1c})$		
$E(\Gamma_{7c})$	5.24	5.24	$E(\Gamma_{15v})$		
$E(\Gamma_{8c})$	5.39				
$E(X_{6v})$	−9.80	−9.46	$E(X_{1v})$		−9.6
$E(X_{6v})$	−6.88	−7.07	$E(X_{3v})$	−6.80	
$E(X_{6v})$	−2.71				
$E(X_{7v})$	−2.70	−2.73	$E(X_{5v})$	−3.00	−2.7
$E(X_{6c})$	2.35	2.16	$E(X_{1c})$		
$E(X_{7c})$	2.89	2.71	$E(X_{3c})$		
$E(L_{6v})$	−10.86	−10.60	$E(L_{1v})$	−10.8	−10.6
$E(L_{6v})$	−6.61	−6.84	$E(L_{1v})$	−6.8	−6.9
$E(L_{6v})$	−1.26				
$E(L_{4,5v})$	−1.20	−1.10	$E(L_{3v})$	−0.9	
$E(L_{6c})$	2.65	2.79	$E(L_{1c})$		
$E(L_{6c})$	5.62	5.74			
$E(L_{4,5c})$	5.71				
$m_l\,(\Delta_{6c})$	4.88				4.8[b]
$m_t(\Delta_{6c})$	0.254				0.254[b]
$m_c(\Gamma_{6c})$	0.158				
m_{hh} [100]	0.488				
m_{hh} [111]	1.059				0.67[c]
m_{lh} [100]	0.204				
m_{lh} [111]	0.167				0.17[c]
m_{soh}	0.296				0.465[d]
Camel back structure					
$k_0\ (2\pi/a)$	0.034				0.025[e]
ΔE (meV)	0.55				3.5[e]
m_l^*	4.88				10.9[e]
m_t^*	0.254				0.25[e]

[a]The energies are in eV relative to the top of the valence band. The effective masses are multiples of the free-electron mass.
[b]Miura *et al.* (1983a).
[c]Schwerdtfeger (1972).
[d]Sharma *et al.* (1983).
[e]Miura *et al.* (1983b).

This is the camel-back structure mentioned earlier. We have used the second-neighbor parameters $v^A_{sp\sigma}$ and $v^C_{sp\sigma}$ to bring the longitudinal mass at this minimum close to the experimental value of 4.8. We found that both Δk and ΔE are very sensitive to these parameters. The transverse mass turns out to agree with the experimental value of 0.254 without being fitted. Note these values are different from the $\Delta k = 0.025(2\pi/a)$, $\Delta E = 3.5$ meV, and the large longitudinal mass of 10.9 obtained in a **k**·**p** analysis of this structure (Miura *et al.*, 1983b). A small adjustment of the band energy is made for the $E(L_{6c}) = 2.650$

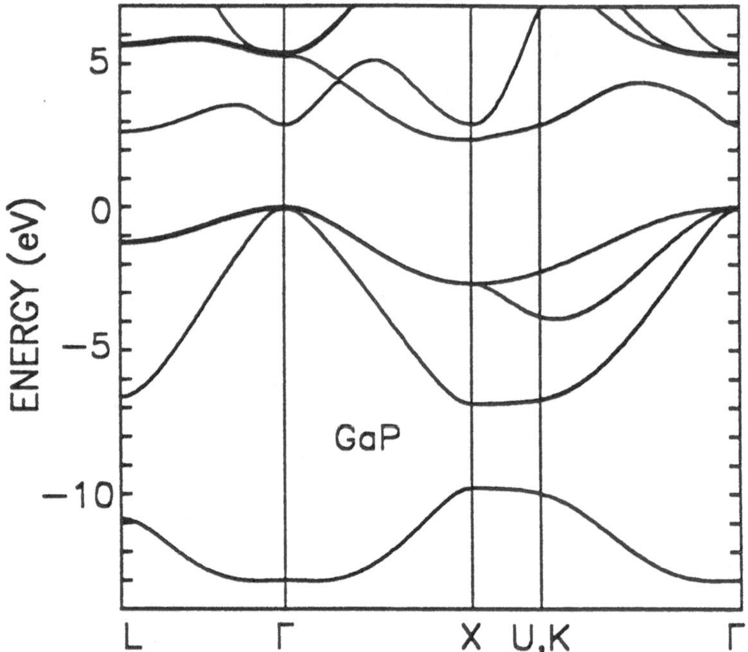

FIGURE 7.6. Calculated band structure energy in eV versus crystal wave vector of GaP along several symmetry directions.

eV to yield the 2.637-eV $E(L_{6c} - \Gamma_{6c})$ energy spacing at 78 K deduced from electroabsorption (Kopylov and Pikhin, 1977). The value $E(\Gamma_{6c}) = 2.866$ eV adopted is slightly lower than the 2.895 eV extrapolated from the photoconductivity data (Nelson *et al.*, 1964). A small spin-orbit splitting at the top of the valence band $\Delta_o = 0.08$ eV was also observed (Takizawa, 1983) and has been adopted in the present band. Although the calculated light-hole mass $m_{lh}[111] = 0.167$ is very close to the experimental value of 0.17, the calculated heavy-hole mass $m_{hh}[111] = 1.059$ is considerably larger than the experimental value of 0.67. In contrast, our calculated effective mass for the spin-orbit split-hole state (Γ_{7v}), $m_{soh} = 0.296$, is smaller than that obtained from **k**·**p** theory.

7.2.5. *GaAs*

The calculated band structure is shown in Fig. 7.7. As mentioned earlier, the band structure looks similar to that in Fig. 5.7. Table 7.7 is a comparison of the present band energies and effective masses with experiment and other calculations. Column A contains the band energies derived from an angle-resolved photoemission experiment (Chiang *et al.*, 1980). Column B is for other experimental results. The minimum band gap is a direct gap $\Gamma_{6c} - \Gamma_{8v} = 1.52$ eV, which is an input in the calculation. Although a different value of 1.63 eV has been assigned to the gap in angular photoemission data, the 1.52-eV gap at $T = 0$ is a well-established value, because it yields a correct prediction of the exciton luminescence peak position at low temperature. The Γ_{7c}, X_{6c}, L_{6c}, X_{7v}, and $L_{4,5v}$ are chosen so that they produce the correct critical transition energies identified from optical data (Aspnes and Studna, 1973). The spin-orbit splitting at the top of the valence band $\Delta_o = \Gamma_{8v} - \Gamma_{7v} = 0.341$

TABLE 7.7. Comparison of Band Structure of GaAs Calculated from the Hybrid Pseudopotential Tight-Binding Method (HPT) with That from the Empirical Pseudopotential Calculation (Cohen and Chelikowsky, 1989), and with the Experimental and Estimated Results[a]

	Theory			Experiment and estimate	
	HPT	EMP		A[b]	B
$E(\Gamma_{6v})$	−13.10	−12.55	$E(\Gamma_{6v})$	−13.1	
$E(\Gamma_{7v})$	−0.341	−0.35			−0.341[c] (4.2 K)
$E(\Gamma_{8v})$	0	0	$E(\Gamma_{1v})$	0	
$E(\Gamma_{6c})$	1.52	1.51	$E(\Gamma_{1c})$	1.632	1.519[d] (1.8 K)
$E(\Gamma_{7c})$	4.72	4.55	$E(\Gamma_{15v})$	4.716	
$E(\Gamma_{8c})$	4.89	4.71			
$E(X_{6v})$	−10.58	−9.83	$E(X_{1v})$	−10.75	
$E(X_{6v})$	−6.66	−6.88	$E(X_{3v})$	−6.70	
$E(X_{6v})$	−3.04	−2.99			
$E(X_{7v})$	−2.89	−2.89	$E(X_{5v})$	−2.80	
$E(X_{6c})$	2.03	2.03	$E(X_{1c})$	2.18	
$E(X_{7c})$	2.60	2.38	$E(X_{3c})$	2.54	
$E(L_{6v})$	−11.16	−10.60	$E(L_{1v})$	−11.24	
$E(L_{6v})$	−6.65	−6.83	$E(L_{1v})$	−6.70	
$E(L_{6v})$	−1.51	−1.42			
$E(L_{4,5v})$	−1.30	−1.20	$E(L_{3v})$	−1.30	
$E(L_{6c})$	1.82	1.82	$E(L_{1c})$	1.85	
$E(L_{6c})$	5.40				
$E(L_{4,5c})$	5.53	5.52			
m_c	0.067				0.067[e]
m_{hh} [100]	0.380				m_h 0.51[e]
m_{hh} [111]	0.847				
m_{lh} [100]	0.089				m_l 0.082[e]
m_{lh} [111]	0.079				
m_{soh}	0.169				0.154[e]

[a]Column A is deduced from angle-resolved photoemission experiment, and column B is from other experiments and estimates. The energies are in eV relative to the top of the valence band. The effective masses are multiples of the free-electron mass.
[b]Chiang *et al.* (1980).
[c]Aspnes and Studna (1973).
[d]Skromme and Stillman (1984).
[e]Blakemore (1982).

eV is also fitted exactly. The other spin-orbit splittings $\Delta_1 = L_{6v} - L_{4,5} = 0.210$ eV and $\Delta_0' = \Gamma_{8c} - \Gamma_{7c} = 0.180$ eV are in good agreement with experiment (Aspnes and Studna, 1973). The rest of the band energies listed in Table 7.7 are seen to agree with the experimental results or the EMP calculation. The effective electron mass $m_c = 0.0670$ is also an input. m_c in the present method can be fitted exactly and conveniently using the second-neighbor parameters $V_{sp\sigma}$, because these parameters control m_c but do not affect the energies listed in Table 7.7. The calculated heavy-hole masses are rather anisotropic, with $m_h[100] = 0.360$ and $m_h[111] = 0.850$. The experimental average effective heavy-hole mass $m_h = 0.51$ falls between these two values. The calculated light-hole masses from the present bands are $m_{lh}[100] = 0.089$ and $m_l[111] = 0.078$, which also compare well with the average experimental light-hole mass of 0.082. Finally the mass for the spin-orbit split hole from the present calculation is $m_{soh} =$

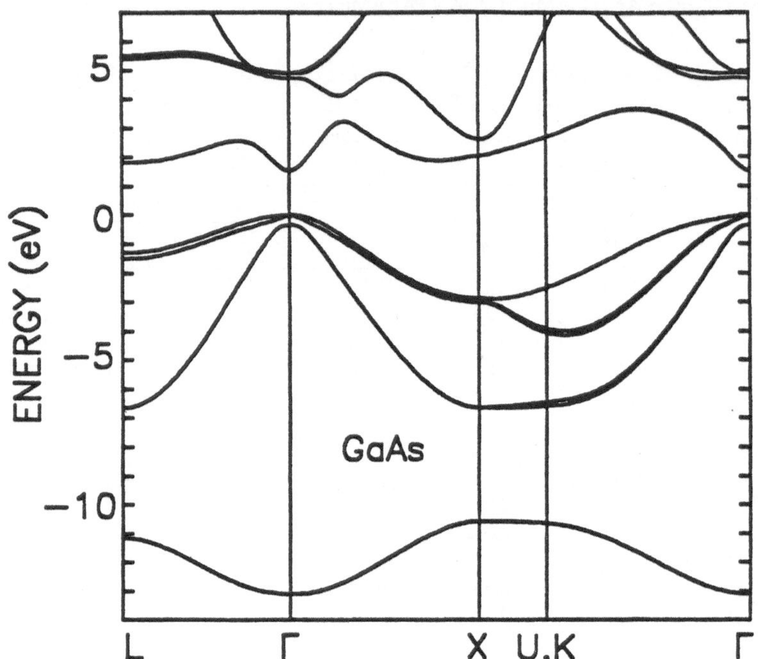

FIGURE 7.7. Calculated band structure energy in eV versus crystal wave vector of GaAs along several symmetry directions.

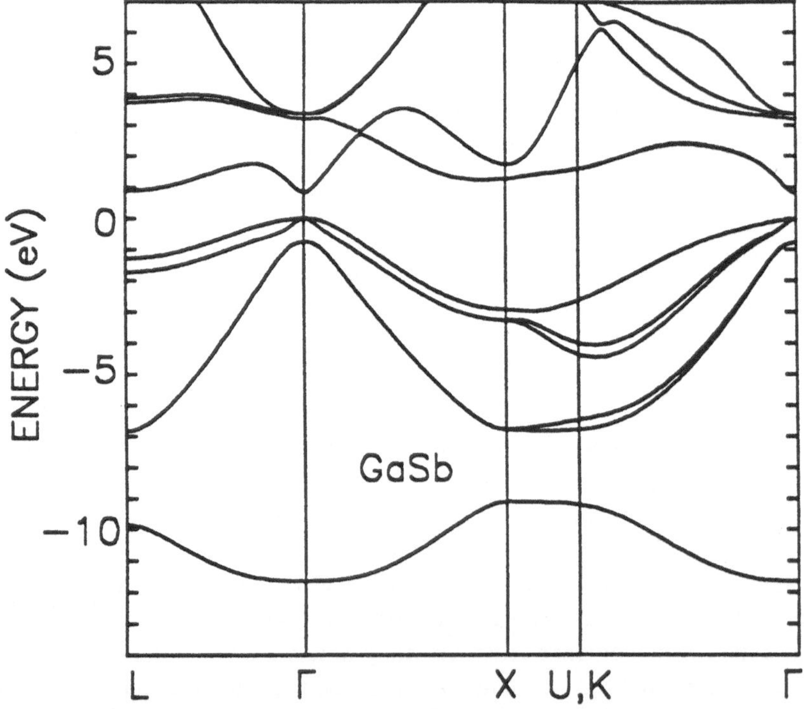

FIGURE 7.8. Calculated band structure energy in eV versus crystal wave vector of GaSb along several symmetry directions.

TABLE 7.8. Comparison of Band Structure of GaSb from the Present Calculation with Results Compiled by Madelung (1991) and Empirical Pseudopotential Calculation (Cohen and Chelikowsky, 1989)[a]

	Theory		Experiment	
	HPT	EMP	A[b]	B
$E(\Gamma_{6v})$	−11.64	−12.00	−11.64	
$E(\Gamma_{7v})$	−0.759	−0.76	−0.82	−0.756[c]
$E(\Gamma_{8v})$	0	0		
$E(\Gamma_{6c})$	0.822	0.86		0.822[c]
$E(\Gamma_{7c})$	3.20	3.44		3.191[c]
$E(\Gamma_{8c})$	3.36	3.77		3.404[c]
$E(X_{6v})$	−9.11	−9.33	−9.62	
$E(X_{6v})$	−6.81	−6.76	−6.90	
$E(X_{6v})$	−3.30	−2.61	−3.10	
$E(X_{7v})$	−2.96	−2.37	−2.86	
$E(X_{6c})$	1.25	1.72		
$E(X_{7c})$	1.72	1.79		
$E(L_{6v})$	−9.85	−10.17	−10.06	
$E(L_{6v})$	−6.83	−6.25	−6.60	
$E(L_{6v})$	−1.72	−1.45	−1.55	−1.530[c]
$E(L_{4,5v})$	−1.28	−1.00	−1.10	
$E(L_{6c})$	0.90	1.22		1.095[c]
$E(L_{6c})$	3.72	4.43		4.36[c]
$E(L_{4,5c})$	3.88	4.59		4.59[c]
m_c	0.0412			0.0412[d] (4.2 K)
m_{hh} [100]	0.296			m_h 0.28[e]
m_{hh} [111]	0.709			
m_{lh} [100]	0.048			m_l 0.050[e]
m_{lh} [111]	0.044			
m_{soh}	0.135			

[a]The energies are in eV relative to the top of the valence band. The effective masses are multiples of the free-electron mass.
[b]Chiang and Eastman (1980).
[c]Aspnes *et al.* (1976).
[d]Hill and Scherdtfeger (1974).
[e]Heller and Hamerly (1985).

0.168, compared with the experimental value of 0.154. From these comparisons, it can be concluded that the present HPT method has produced an accurate band structure for GaAs.

7.2.6. GaSb

There is also sufficient experimental information for a suitable fitting of the band structure of GaSb using the present method. The calculated band structure is shown in Fig. 7.8. The energies at three symmetry points and the calculated effective masses are compared with the EMP calculation and experimental data in Table 7.8. Column A is for the angle-resolved photoemission results, and Column B for other experiments. Except for modifications of some band-edge states to produce the experimental values, our band energies are very close to the EMP results. The modified energies include the 0.822 eV for the band gap $E(\Gamma_{6c} - \Gamma_{8v})$, the values of $E(X_{6c}) = 1.250$ eV and $E(L_{6c}) = 0.900$ eV. This $E(L_{6c})$ value and the −1.28 eV for the $E(L_{4,5v})$ give an energy spacing of 2.18 eV, which is close to the 2.185-eV

E1 optical transition energy found experimentally (Alibert *et al.*, 1983). An even lower value (0.871 eV) for $E(L_{6c})$ has also been deduced experimentally (Alibert *et al.*, 1983). The present value of $E(L_{6c})$ is needed in order to bring the band structures of alloys of GaSb (e.g. $Ga_{1-x}Al_xSb$; see next section) into agreement with experiment. The X_{6c} energy of 1.250 eV produces an energy spacing $E(X_{6c}) - E(\Gamma_{6c}) = 428$ meV, which agrees with the value deduced from the low-temperature (10 K) electroreflectance data (Aspnes *et al.*, 1976). The large spin-orbit splitting $\Delta_o = 0.758$ eV is an input based on the experimental value. The calculated spin-orbit splittings of the topmost valence states at other symmetry points also agree reasonably well with the experimental values. The conduction-band mass at Γ_{6c} reproduces the experimental value. The calculated light-hole masses of 0.048 and 0.044 compare well

TABLE 7.9. Comparison of Band Structure of InP from Present Calculation with Results Compiled by Madelung (1991) and Empirical Pseudopotential Calculation (Cohen and Chelikowsky, 1989)[a]

	Theory		
	HPT	EMP	Experiment
$E(\Gamma_{6v})$	−11.40	−11.420	
$E(\Gamma_{7v})$	−0.108	−0.21	−0.108[b]
$E(\Gamma_{8v})$	0	0	0
$E(\Gamma_{6c})$	1.42	1.50	1.42[b]
$E(\Gamma_{7c})$	4.80	4.64	4.8[c]
$E(\Gamma_{8c})$	5.06	4.92	4.87[d]
$E(X_{6v})$	−8.77	−8.91	
$E(X_{6v})$	−6.20	−6.01	−5.9[e]
$E(X_{6v})$	−2.21	−2.09	
$E(X_{7v})$	−2.20	−2.06	−2.2[e]
$E(X_{6c})$	2.38	2.44	2.38[b]
$E(X_{7c})$	3.06	2.97	
$E(L_{6v})$	−9.66	−9.67	
$E(L_{6v})$	−5.70	−5.84	
$E(L_{6v})$	−1.20	−1.09	−1.23[c]
$E(L_{4,5v})$	−1.12	−0.94	−1.12[c]
$E(L_{6c})$	2.03	2.19	2.03[b]
$E(L_{6c})$	5.60	5.58	
$E(L_{4,5c})$	5.77	5.70	
m_c	0.0765		0.0765[f] (4.2 K)
m_{hh} [100]	0.516		0.56[g]
m_{hh} [111]	0.987		0.60[g]
m_{lh} [100]	0.098		m_l 0.12[g]
m_{lh} [111]	0.090		
m_{soh}	0.177		0.121[g]

[a]The energies are in eV relative to the top of the valence band. The effective masses are multiples of the free-electron mass.
[b]Cammassel *et al.* (1980).
[c]Matatagui *et al.* (1968).
[d]Shaklee *et al.* (1966).
[e]Williams *et al.* (1983).
[f]Helm *et al.* (1985).
[g]Leotin *et al.* (1974).

with the average mass of 0.05 deduced from transport data. While the calculated heavy-hole mass along (100) is the same as the quoted value, the calculated heavy-hole masses in (111) is much larger.

7.2.7. *InP*

With the availability of the EMP band structure and experimental data for the important gaps, as shown in Table 7.9, the parametrization of the band structure of InP using the HPT can be carried out without guessing the input band quantities. It can be seen in this table that the energies of the following states have been chosen to fit the existing values: Γ_{6v}, Γ_{7v}, Γ_{8v}, Γ_{6c}, Γ_{7c}, $L_{4,5v}$, and L_{6c}. The X_{6c} level is chosen so that the energy spacing $E(X_{6c}) - E(\Gamma_{6c}) =$ 965 meV is slightly larger than the experimental value of 960 meV at 8 K (Onton *et al.*, 1972). The X_{7v} level is then set at −2.12 eV so that the energy spacing $E(X_{6c}) - E(X_{7v})$ is about the same as that from the EMP calculation, which has been fitted to the optical data. The calculated band structure in Fig. 7.9, when compared with Fig. 7.7, looks very much like that of GaAs. The main differences are the spin-orbit splittings. In GaAs the spin-orbit splittings in the valence bands are larger than those in the conduction bands, whereas in InP they are reversed. For example, the calculated value for $\Delta_o = E(\Gamma_{8v}) - E(\Gamma_{7v})$ in InP is 0.109 eV, compared with the splitting in the conduction band $\Delta_o' = E(\Gamma_{8c}) - E(\Gamma_{7c}) = 0.271$ eV. The reason for this reversal is that the valence-band *p* states are dominated by the anion component, while the conduction-band states have a larger cation component, and the spin-orbit coupling constant for In is larger than for P. In GaAs, As has a larger coupling constant. The EMP calculation also gives a larger Δ_o' than Δ_o, although not as pronounced

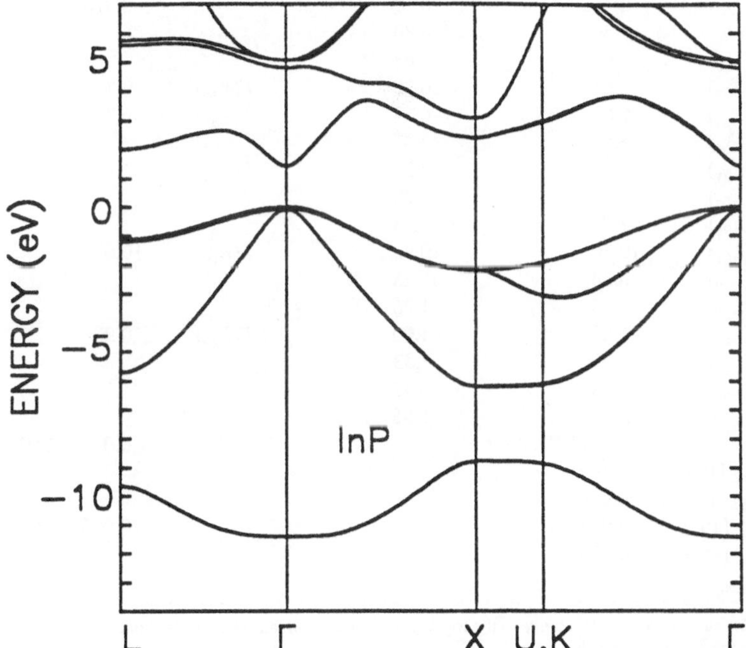

FIGURE 7.9. Calculated band structure energy in eV versus crystal wave vector of InP along several symmetry directions.

as the present calculation. It is surprising that the experimental data shown in Table 7.9 give a very small value of Δ_o'. The effective mass at the conduction-band minimum is fitted to the experimental value. The calculated masses for the heavy hole along the (100) and the light-hole masses are in reasonable agreement with the experimental values, but the heavy-hole mass $m_{hh}[111]$ and the hole mass for the spin-orbit split band m_{soh} from the calculation are considerably heavier than the experimental masses.

7.2.8. InAs

The comparison between the present HPT calculation with the EMP calculation and the compiled experimental and estimated results shown in Table 7.10 for InAs is very similar to that for InP. Most of the band energies follow those of the EMP. Modifications are only made to produce the experimentally deduced values. These include the band gap $E(\Gamma_{6c}) - E(\Gamma_{8v}) = 0.422$ eV, the spin-orbit splitting $\Delta_o = 0.382$ eV, and $E(X_{6v}) = -10.36$ eV. The calculated band structure shown in Fig. 7.10 is very similar to that obtained from the EMP calculation

TABLE 7.10. Comparison of Band Structure of InAs from Present Calculation with Results Cited by Madelung (1991) and Empirical Pseudopotential Calculation (Cohen and Chelikowsky, 1989)[a]

	Theory			
	HST	EMP		Experiment
$E(\Gamma_{6v})$	−12.30	−12.69	$E(\Gamma_{6v})$	−12.3[b]
$E(\Gamma_{7v})$	−0.382	−0.43		−0.38[c] (1.5 K)
$E(\Gamma_{8v})$	0	0		
$E(\Gamma_{6c})$	0.422	0.37		0.418[d] (4.2K)
$E(\Gamma_{7c})$	4.39	4.39		
$E(\Gamma_{8c})$	4.67	4.63		
$E(X_{6v})$	−9.66	−10.20	$E(X_{1v})$	−9.8[b]
$E(X_{6v})$	−6.60	−6.64	$E(X_{3v})$	−6.3[e]
$E(X_{6v})$	−2.52	−2.47		
$E(X_{7v})$	−2.37	−2.37	$E(X_{5v})$	−2.4[e]
$E(X_{6c})$	2.28	2.28		
$E(X_{7c})$	2.79	2.66		
$E(L_{6v})$	−10.36	−10.92	$E(L_{1v})$	−10.6[b]
$E(L_{6v})$	−6.38	−6.23		
$E(L_{6v})$	−1.25	−1.26		
$E(L_{4,5v})$	−1.00	−1.00	$E(L_{3v})$	0.90[c]
$E(L_{6c})$	1.53	1.53		
$E(L_{6c})$	5.38	5.42		
$E(L_{4,5c})$	5.57	5.55		
m_c	0.0212			0.0231[f] (150 K)
m_{hh} [100]	0.399			
m_{hh} [111]	0.954			
m_{lh} [100]	0.027			0.026[c] (20 K)
m_{lh} [111]	0.026			
m_{soh}	0.089			

[a]The energies are in eV relative to the top of the valence band, and the effective masses are the ratios to the free-electron value.
[b]Ley *et al.* (1974).
[c]Pidgeon *et al.* (1967).
[d]Varfolomeeve *et al.* (1975).

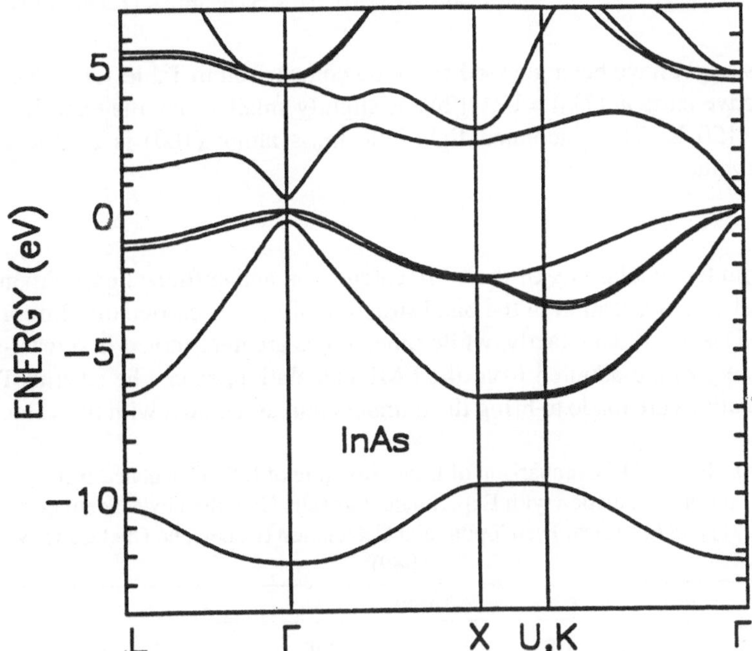

FIGURE 7.10. Calculated band structure energy in eV versus crystal wave vector of InAs along several symmetry directions.

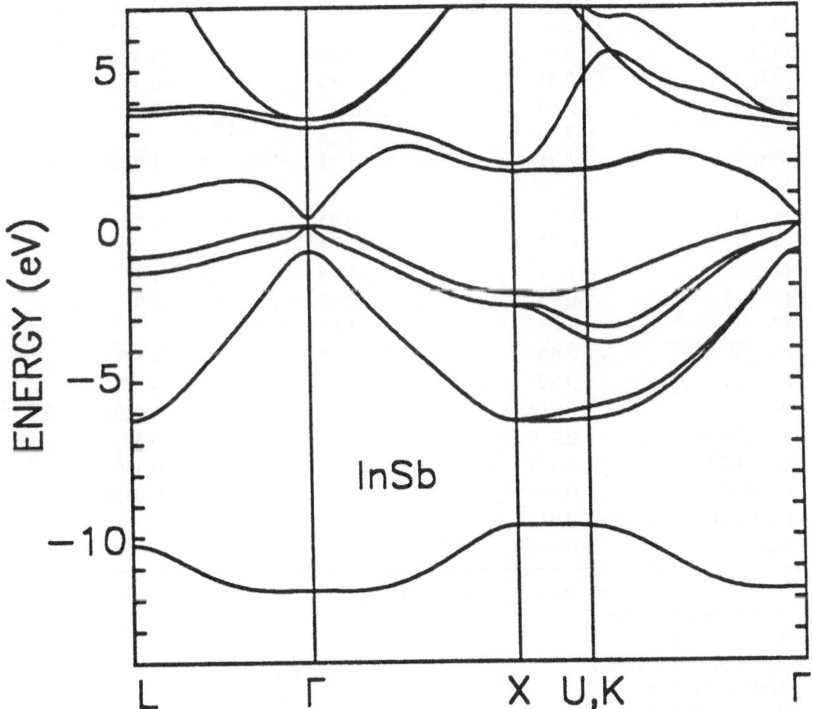

FIGURE 7.11. Calculated band structure energy in eV versus crystal wave vector of InSb along several symmetry directions.

in Fig. 5.7, as might have been guessed from the comparison in Table 7.10. The calculated electron effective mass $m_c(\Gamma_{6c}) = 0.0212m_o$ is slightly smaller than the experimental value of 0.0231 at 150 K. The calculated light-hole mass along (100) is comparable to the experimental value.

7.2.9. *InSb*

Again with the availability of the EMP calculation and sufficient experimental data, a fully empirical parametrization of the band structure of InSb is carried out. Important states at the band edges are fitted exactly, while other states are determined in a least-square fit. Most input energies are adopted from the EMP calculation, as can be seen in Table 7.11. Small modifications are made to bring the numbers into agreement with the values deduced

TABLE 7.11. Comparison of Band Structure of InSb Calculated from Present Calculation with Experimental Results Compiled by Madelung (1991) and Empirical Pseudopotential Calculation (Cohen and Chelikowsky, 1989)[a]

	Theory		
	HPT	EMP	Experiment
$E(\Gamma_{6v})$	-11.70	-11.710	-11.7^b
$E(\Gamma_{7v})$	-0.852	-0.82	-0.850^c
$E(\Gamma_{8v})$	0	0	
$E(\Gamma_{6c})$	0.237	0.25	0.235^d
$E(\Gamma_{7c})$	3.14	3.16	3.141^c
$E(\Gamma_{8c})$	3.42	3.59	3.533^c
$E(X_{6v})$	-9.66	-9.20	-9.5^b
$E(X_{6v})$	-6.33	-6.43	-6.4^b
$E(X_{6v})$	-2.64	-2.45	-2.4^b
$E(X_{7v})$	-2.24	-2.24	
$E(X_{6c})$	1.71	1.71	1.79^c
$E(X_{7c})$	1.95	1.83	
$E(L_{6v})$	-10.24	-9.95	-10.5^b
$E(L_{6v})$	-6.22	-5.92	
$E(L_{6v})$	-1.49	-1.44	-1.4^b
$E(L_{4,5v})$	-0.96	-0.96	-0.9^b
$E(L_{6c})$	1.03	1.03	
$E(L_{6c})$	3.60	4.30	4.32^c
$E(L_{4,5c})$	3.80	4.53	4.47^c
m_c	0.0135		0.01359^e (4.2 K)
m_{hh} [100]	0.410		0.34^f
m_{hh} [111]	0.971		0.45^f
m_{lh} [100]	0.0145		m_l 0.0158^g
m_{lh} [111]	0.0142		
m_{soh}	0.106		

[a]The energies are in eV relative to the top of the valence band. The effective masses are multiples of the free-electron mass.
[b]Ley et al. (1974).
[c]Logothetidis *et al.* (1985).
[d]Little *et al.* (1983).
[e]Zengen *et al.* (1983).
[f]Bagguley *et al.* (1963).
[g]Seiler *et al.* (1980).

from experiments. These include the following states: $E(\Gamma_{6c}) = 0.237$ eV, $\Delta_o = 0.850$ eV, $E(\Gamma_{7c}) = 3.141$ eV, and $E(X_{6c}) = 1.79$ eV. The calculated band structure is shown in Fig. 7.11, which is similar to that calculated from EMP. The only appreciable difference is that the spin-orbit splitting for the low-lying valence band along the Σ axis (Γ–K) in the present calculation is larger. The electron mass $m_c(\Gamma_{6c}) = 0.0135$ is fitted to the low-temperature experimental value. Again, while the calculated light-hole masses and the heavy-hole mass along (100) are in good accord with the experiment, the calculated heavy-hole mass along the (111) direction is heavier.

7.3. THE HPT MODEL APPLIED TO III–V PSEUDOBINARY ALLOYS

The hybrid pseudopotential and tight-binding model is designed to facilitate alloy calculations. Similar to that used for the pure constituent crystals, the alloy Hamiltonian $\hat{H}$ in the OLO basis consists of three parts:

$$\hat{H} = \hat{H}_0 + \hat{H}_1 + \hat{H}_{so} \tag{7.3.1}$$

where $\hat{H}_o$ in crystal units is taken to the common Hamiltonian H_o minus a diagonal matrix D; i.e., $\hat{H}_0 = H_0 - D$, where the matrix elements of D have the following values: $D_S^A = 0.616$, $D_p^A = 1.971$, $D_S^C = 2.214$, and $D_p^C = 2.796$. The removal of these diagonal matrix elements makes $\hat{H}_0$ a universal matrix for the constituent compounds and alloys. Both $\hat{H}_1$ and $\hat{H}_{so}$ contain alloy disorder. $\hat{H}_1$ is a short-range perturbation Hamiltonian that contains varying term values E_α, first-neighbor matrix elements h, and second-neighbor matrix elements v. H_{so} is the spin-orbit Hamiltonian, which is site diagonal with the coupling constants λ varying with the alloying atoms. These energy parameters, except the second-neighbor parameters v, can be derived from those of the constituent compounds listed in Table 7.2. Since the v's are small, they will be treated in the virtual crystal approximation (VCA). Appendix 7B contains a detailed description of the matrix elements entering these Hamiltonians.

From Eq.(7.3.1) we can write the alloy Hamiltonian as

$$H_{alloy} = \overline{H} + U \tag{7.3.2}$$

where $\overline{H}$ is the concentration averaged $\hat{H}$ of Eq. (7.3.1) and U is the disorder part. $\overline{H}$ will be referred to as the VCA Hamiltonian.

To be specific, let us consider a pseudobinary alloy $A_{1-x}B_xC$, such as $Ga_{1-x}In_xAs$. U can be decomposed into three parts:

$$U = U_o + U_1 + U_{so} \tag{7.3.3}$$

The first term arises from the term value fluctuations. It takes the form

$$U_o = \sum_{L} \sum_{\alpha} |L\alpha\rangle \delta E_\alpha^L \langle L\alpha| \tag{7.3.4}$$

where L designates both the position L and the kind of alloying atoms (A or B), and α labels the type of orbitals; i.e. s, p_x, etc. δE_α^L takes the value $\delta E_\alpha^A = E_\alpha^A - \overline{E}_\alpha$ if site L is occupied by an A atom, and δE_α^B by a B atom. Here $\overline{E}_\alpha$ is the averaged term value $\overline{E}_\alpha = (1-x)E_\alpha^A + xE_\alpha^B$.

The term values of the III–V systems relative to the top of the valence band of GaAs are listed in Table 7.2.

The second term U_1 arises from the fluctuations $\delta h = h - \bar{h}$ in the first-neighbor interactions h. It may be written as

$$U_1 = \sum_{\mathrm{L}\alpha} \sum_{j\beta} \left(|\mathrm{L}\alpha\rangle \delta h^{\mathrm{L}}_{\alpha,j\beta} \langle \mathrm{L}j\beta| + |\mathrm{L}j\beta\rangle \delta h^{\mathrm{L}}_{j\beta,\alpha} \langle \mathrm{L}\alpha| \right) \tag{7.3.5}$$

The labels L and α have the same meaning as those in U_0. The index j runs over the four C atoms surrounding the alloying atom at site **L**, and β over the orbital types. δh^{L} takes the value of δh^{A} if the **L** site is occupied by an A atom, and δh^{B} if occupied by a B atom. Since the first-neighbor bond lengths d in an alloy differ from the value d° in the pure constituent crystal, the value of h in general is different from the pure-crystal value h_o listed in Table 7.2. We may assume h in crystal units scales as $h = h_o(\bar{d}/d)^2$, where d is the actual length for the bond that connects the two orbitals under consideration, and $\bar{d}$ is the VCA bond length. Recall also that the bond lengths d_{AC} and d_{BC} in a pseudobinary alloy $A_{1-x}B_xC$ follow a bimodal distribution (see Fig. 1.1). Assuming $d^{\circ}_{\mathrm{BC}} > d^{\circ}_{\mathrm{AC}}$, then, according to the results discussed in Chapter 1, the effective bond lengths in the alloy can be approximated by $d_{\mathrm{AC}} = \bar{d} - 3x\delta/4$ and $d_{\mathrm{BC}} = \bar{d} + 3(1-x)\delta/4$, with $\bar{d}$ being the average, $\bar{d} = (1-x)d^{\circ}_{\mathrm{AC}} + xd^{\circ}_{\mathrm{BC}}$, and δ the difference $\delta = d^{\circ}_{\mathrm{BC}} - d^{\circ}_{\mathrm{AC}}$ of the constituents' bond lengths.

Finally, U_{so} describes the fluctuation in the spin-orbit interaction. It may be conveniently written as

$$U_{\mathrm{so}} = \sum_{\mathrm{L}} \left(|\mathrm{L}\Gamma_8\rangle \delta\lambda^{\mathrm{L}} \langle \mathrm{L}\Gamma_8| - 2|\mathrm{L}\Gamma_7\rangle \delta\lambda^{\mathrm{L}} \langle \mathrm{L}\Gamma_7| \right) \tag{7.3.6}$$

in the Γ-OLO basis. $\delta\lambda^{\mathrm{L}}$ is the difference between the spin-orbit coupling constant of the atom at site **L** and the VCA value.

Note that while $\overline{\mathrm{H}}$ is most conveniently obtained from Tables 7.1 and 7.2 in crystal units, the disorder energy parameters must be entered in the usual units of eV or Rydbergs. Recall that the energy E in Rydbergs is related to the dimensionless crystal units ε by $E = \varepsilon(2\pi/a)^2$, with a being the lattice constant in Bohr radii. We remind the reader that Appendix 7B provides a detailed description of the alloy Hamiltonian.

Despite the complicated appearance of the disorder Hamiltonian, it is in a form enabling a direct application of the alloy theory discussed in Chapter 5. First we recognize that U can be written as the sum of molecular contributions:

$$U = \sum_L U_L \tag{7.3.7}$$

In principle, this allows a CPA calculation using the IATA iteration method discussed in Chapter 5. However, each U_{L} involves 40 orbitals including the spins: 8 from the center atom on site **L** and 32 from the four surrounding atoms. In practice, this demands very lengthy computation, because we need to calculate a 40×40 Green function matrix for many energies in each iteration, and each matrix element requires a complicated integration inside the first Brillouin zone [see Eq. (5.10.8)]. This complexity is greatly reduced if a further approximation is made to construct a model suitable for direct application of molecular CPA (MCPA).

This extra approximation is to neglect some disorder matrix elements in U_1. When the atomlike OLOs are transformed into the Γ-OLO basis, as in those described by Eq. (5.10.2), U_1 not only contains interactions among the Γ-OLO of the same center, but also nonzero matrix elements between neighboring centers. However, these off-center interactions are small. If we keep only the on-center U, we have the molecular model described in Section 5.10. Then the CPA self-energy becomes block diagonal into eight 2×2 matrices, and the MCPA calculation can be carried out with only a modest computational effort. It is convenient to define a scattering Hamiltonian W to be the difference between the constituents' Hamiltonians. In the molecular model using the Γ-OLO basis, W is site diagonal: $W = \Sigma w_L$. Each w_L is block diagonal into eight 2×2 matrices, characterized by a diagonal disorder parameter δ and an off-diagonal disorder parameter Δ, like that given in Eq. (5.11.2). In terms of w_L, the disorder potential for an A atom at site L is $U_A = xw_L$, and for a B atom $U_B = -(1 - x)w_L$. From a practical point of view, the non-self-consistent ATA calculation using the molecular model is all we actually need for all the semiconductor alloys considered in this book. For cases where second-order perturbation theory (Section 5.11) is reasonably accurate, ATA is better. For modestly strong scattering cases, ATA is nearly as accurate as CPA. However, ATA only uses the VCA Green function, which can be efficiently and accurately calculated using Eq. (5.10.11), whereas CPA requires the calculation of the Green functions that contain the self-energy operators [see Eq. (5.10.8)]. The computational time for CPA is about 50 times greater than ATA. Unless otherwise stated, the alloy calculations to follow will be based on molecular ATA calculations. Appendix 7B also describes the molecular ATA calculation.

7.4. BAND STRUCTURES OF SELECTED III–V ZINC BLENDE ALLOYS

The most extensively studied band feature in semiconductor alloys $A_{1-x}B_xC$ are the concentration variation of the fundamental band gaps or, more generally, of three conduction-band states—Γ_{6c}, X_{6c}, and L_{6c}—relative to the top of the valence band Γ_{8v}, and the crossing of the direct and indirect band gaps. As mentioned, most band energies are not linear functions of x, and they are often approximated by a quadratic form

$$E = \alpha + \beta x + \gamma x^2 \tag{7.4.1}$$

Equation (7.4.1) can be written in the usual form:

$$E(x) = \bar{E} - x(1 - x)b \tag{7.4.2}$$

in terms of the linearly averaged energy $\bar{E} = (1 - x)E_{AC} + xE_{BC}$ and the bowing parameter b. Note that here b and γ have the same value. A brief comparison of these band features calculated from the present model with those obtained previously is presented next.

7.4.1. $Ga_{1-x}Al_xAs$

Table 7.12 lists the coefficients α, β, and γ of Eq. (7.4.1) for the three lowest conduction-band gaps deduced from the present calculation along with those obtained previously. We can treat Eq. (7.4.2) as an exact expression and calculate b as a function of x to test the validity of the quadratic approximation. The calculated results for b shown in Table 7.4 indicate that the quadratic form is very good for the X and L gaps, but is only approximately

TABLE 7.12. Coefficients α, β, and γ (all in eV) of Eq. (7.4.1) for the Three Lowest Conduction-Band Energies Relative to the Top of the Valence Band for $Ga_{1-x}Al_xAs$

States	α	β	γ	Temperature (K)	Source
$E(\Gamma_{6c})$	1.520	1.162	0.269	0	Present calculation
$E(\Gamma_{6c})$	1.425	1.155	0.37	295	Lee *et al.* (1980)
$E(\Gamma_{6c})$	1.430	1.205	0.26	295	Aubel *et al.* (1985)
$E(\Gamma_{6c})$	1.420	1.087	0.438	295	Saxena (1981)
$E(X_{6c})$	2.030	0.025	0.175	0	Present calculation
$E(X_{6c})$	1.900	0.125	0.143	295	Casey & Panish (1978)
$E(X_{6c})$	1.911	0.005	0.245	295	Lee *et al.* (1980)
$E(X_{6c})$	1.905	0.10	0.16	295	Saxena (1981)
$E(L_{6c})$	1.820	0.570	0.200	0	Present calculation
$E(L_{6c})$	1.708	0.642	0	295	Casey & Panish (1978)
$E(L_{6c})$	1.734	0.574	0.055	295	Lee *et al.* (1980)
$E(L_{6c})$	1.705	0.695	0	295	Saxena (1981)

correct for the Γ gap. However, this increase in b for the Γ gap as a function of x is not as drastic as that suggested by Casey and Panish (1978), who empirically deduced these expressions: $E(\Gamma) = 1.424 + 1.247x$ for $x < 0.45$, and $E(\Gamma) = 1.424 + 1.247x + 1.147(x - 0.45)^2$ for $x > 0.45$, both at room temperature. The calculated values of γ shown in Table 7.12 are the averaged values of b in Table 7.13. The value of b can be decomposed into two contributions, one from upward-bowing b_v of the valence-band top Γ_{8v}, and a second from the downward-bowing b_c of the bottom of the conduction band; i.e., $b = b_c - b_v$. The calculated value for b_v is -0.115 eV, which gives a valence-band offset $\Delta E_v = -0.55x + 0.115x(1-x)$ as a function of x between $Ga_{1-x}Al_xAs$ and GaAs. The bowing parameters can be further decomposed into two contributions: one from the VCA calculation, and the other from the disorder contribution obtained from the ATA calculation. The disorder parameters in $Ga_{1-x}Al_xAs$ in the Γ-OLO basis (see the definition of δ and Δ in Eq. (5.11.2) and Eq. (7B.14)) are listed in Table 7.14. The value of b_v in the present case is found to arise entirely from the VCA contribution, so all the alloy disorder contribution, to be referred to as the extrinsic b_{ex}, comes entirely from the conduction bands.

In Fig. 7.12 these three band gaps are plotted as functions of x. The minimum gap for $x = 0$ (GaAs) is a direct gap of 1.52 eV at Γ. This continues to be the minimum gap as x increases until the Γ gap crosses the X gap at a concentration $x_c = 0.430$ and a gap $E_c = 2.073$ eV. For $x > x_c$, the X gap is the smallest among the three gaps. Although the minimum gap in this x range may not be located precisely at X, the difference between the actual gap and the value

TABLE 7.13. Calculated Values of b (in eV) in Eq. (6.4.2) for $Ga_{1-x}Al_xAs$ at several x Values for the Three Low Conduction-Band States

x	0.1	0.25	0.5	0.7	0.9	Average	b_{ex} [a]
$b(\Gamma)$	0.22	0.24	0.27	0.31	0.36	0.269	0.15
$b(X)$	0.17	0.18	0.18	0.18	0.18	0.175	0.04
$b(L)$	0.18	0.20	0.20	0.20	0.21	0.200	0.11

[a] b_{ex} is the disorder contribution.

TABLE 7.14. Alloy Disorder Parameters δ and Δ defined in Eq. (5.11.2) in the OLO Γ Representations Centered at a Cation Site in $Ga_{1-x}Al_xAs$

Representation	δ (eV)	Δ (eV)
Γ_6	−0.814	0.040
Γ_7	0.202	−0.075
Γ_8	0.352	−0.075

at X should be less than 1 meV, just like that in the pure AlAs case. Table 7.15 lists the values of x_c and E_c from several sources. The calculated values of x_c and E_c for the Γ–X crossing are close to Dingle *et al.*'s results at 2 K. The calculated energies need to be sifted downward by about 0.1 eV in order to be compared with the room-temperature energies listed. The calculated values of x_c and E_c for the Γ–L and X–L crossings are also contained in Table 7.15.

7.4.2. $Ga_{1-x}Al_xSb$

The three coefficients α, β, and γ of Eq. (7.4.1) for the three low conduction-band gaps are listed in Table 7.16. The calculated values of b in Table 7.17 indicate that the quadratic form is not very accurate for the conduction-band edges at Γ and L. The moderate off-diago-

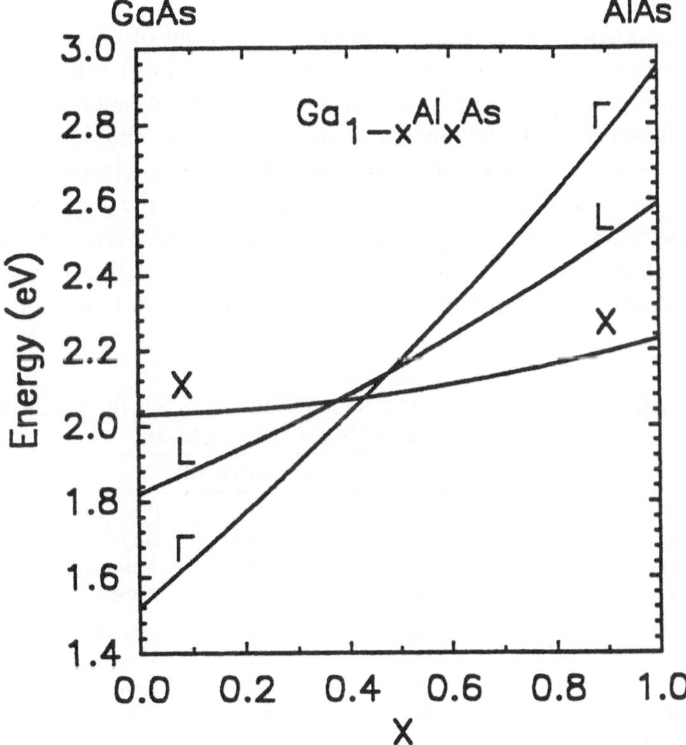

FIGURE 7.12. Conduction-band energy minima in eV at Γ, X, and L relative to the top of this valence band as functions of alloy compositions X in $Ga_{1-x}Al_xAs$.

TABLE 7.15. Crossover Concentration x_c and Energy E_c (in eV) for Three Pairs of Conduction-Band Edges in $Ga_{1-x}Al_xAs$

States	x_c	E_c	Temperature (K)	Source
Γ–X	0.430	2.073	0	Present calculation
Γ–X	0.44	2.04	2	Dingle *et al.* (1977)
Γ–X	0.405	1.953	295	Lee *et al.* (1980)
Γ–X	0.45	1.985	295	Temkin & Keramidas (1980)
X–L	0.375	2.064	0	Present calculation
X–L	0.350	1.942	295	Lee *et al.* (1980)
Γ–L	0.478	2.138	0	Present calculation
Γ–L	0.432	1.992	295	Lee *et al.* (1980)

nal disorder parameters in Table 7.18 make important contributions to the bowing of the Γ and L gaps (see Table 7.17). The calculated bowing parameter for the Γ gap is in reasonable agreement with the values deduced from experiments. While the two experimental bowing parameters for the L gap disagree, the present calculated value is in agreement with one of them. Although the calculated bowing for the X gap is smaller than those of Γ and L, its value is still appreciable, in contrast to the zero bowing deduced experimentally. The calculated valence-band offset relative to GaAs is well approximated by the quadratic form $\Delta E_v = 0.14 - 0.196x - 0.204x^2$.

In Fig. 7.13 these three band gaps are plotted as functions of x. The minimum gap for $x = 0$ (GaSb) is a direct gap of 0.822 eV at Γ. This continues to be the minimum gap as x increases until the Γ gap crosses the L gap at a crossover concentration $x_c = 0.293$ with a gap $E_c = 1.196$ eV. Between $x = 0.293$ and 0.426, the L gap is the minimum gap. At $x = 0.426$, the gap is 1.354 eV. For $x > 0.426$, the X gap is the smallest among the three gaps. Although the minimum gap in this x range may not be precisely at X, just like the camel-back structure in pure AlSb, the difference in energy is only several meV. These results are consistent with experimental findings that the conduction-band minimum switches from Γ to L and then from L to X as x increases. For comparison, Voigt *et al.* (1982) found the Γ–L crossover

TABLE 7.16. Coefficients α, β, and γ (all in eV) of Eq. (7.4.1) for the Three Lowest Conduction-Band States of $Ga_{1-x}Al_xSb$

States	α	β	γ	Temperature (K)	Source
$E(\Gamma_{6c})$	0.822	1.068	0.500	0	Present
$E(\Gamma_{6c})$	0.813	1.097	0.40	4.2	Biryulin *et al.* (1983)
$E(\Gamma_{6c})$	0.81	1.09	0.48	27	Alibert *et al.* (1983)
$E(\Gamma_{6c})$	0.73	1.10	0.47	300	Alibert *et al.* (1983)
$E(X_{6c})$	1.252	0.090	0.348	0	Present
$E(X_{6c})$	1.11	0.577	0	4.2	Biryulin *et al.* (1983)
$E(X_{6c})$	1.12	0.56	0	27	Alibert *et al.* (1983)
$E(X_{6c})$	1.05	0.56	0	300	Alibert *et al.* (1983)
$E(L_{6c})$	0.900	0.784	0.646	0	Present
$E(L_{6c})$	0.893	0.867	0.21	4.2	Biryulin *et al.* (1983)
$E(L_{6c})$	0.86	0.81	0.64	77	Aulombard *et al.* (1985)

TABLE 7.17. Values of b (in eV) in Eq. (6.5.2) at Several x Values for the Three Lowest Conduction-Band States of $Ga_{1-x}Al_xSb$

x	0.1	0.25	0.5	0.7	0.9	Average	b_{ex}
$b(\Gamma)$	0.361	0.417	0.505	0.598	0.742	0.500	0.306
$b(X)$	0.324	0.328	0.349	0.366	0.388	0.348	0.005
$b(L)$	0.548	0.595	0.647	0.714	0.746	0.646	0.182

TABLE 7.18. Alloy Disorder Parameters as Defined in Eq. (5.11.2) in the Γ-OLO Representations Centered at a Cation Site in $Ga_{1-x}Al_xSb$

Representation	δ (eV)	Δ (eV)
Γ_6	−0.177	0.668
Γ_7	−0.333	−0.686
Γ_8	−0.183	−0.686

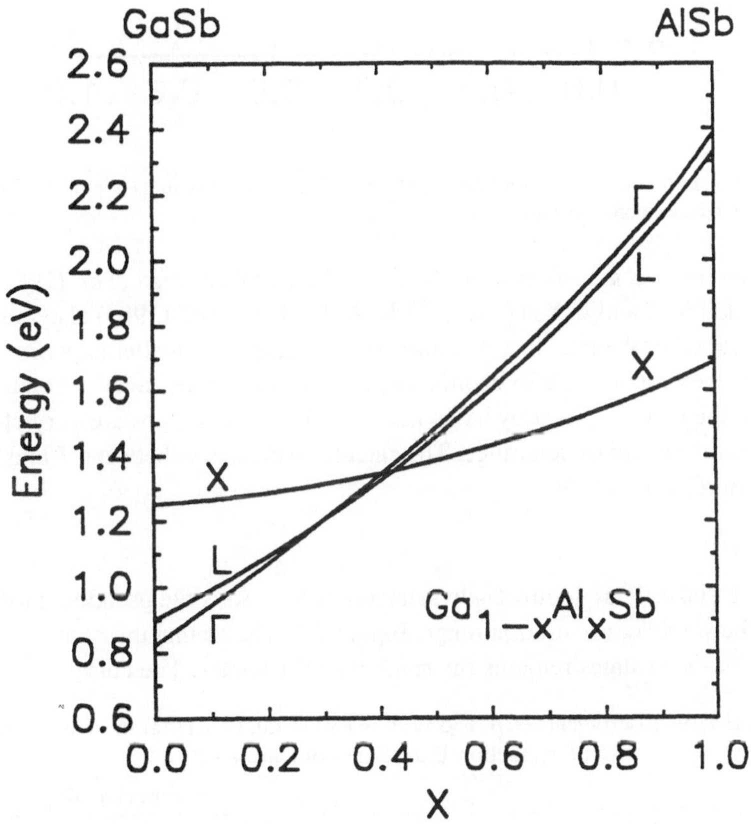

FIGURE 7.13. Conduction-band energy minima in eV at Γ, X, and L relative to the top of this valence band as functions of alloy compositions X in $Ga_{1-x}Al_xSb$.

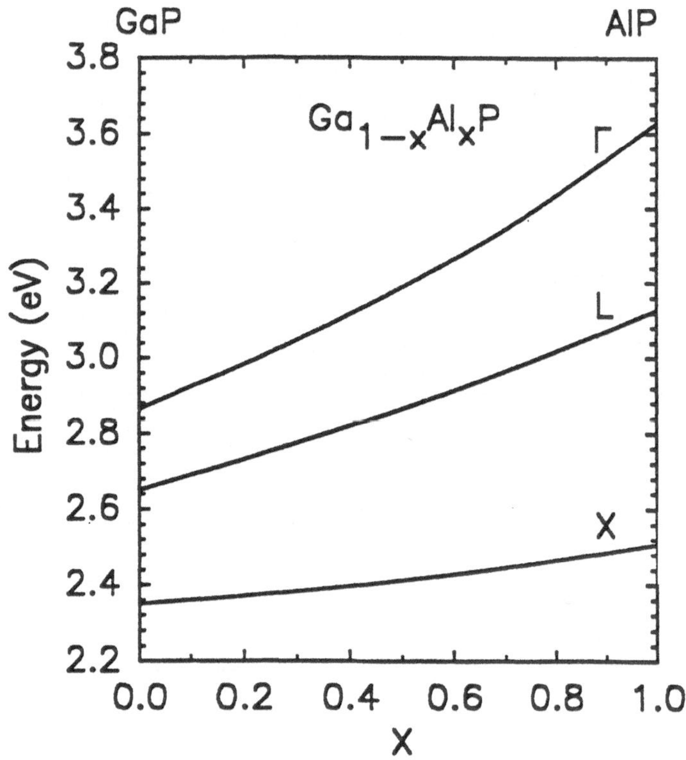

FIGURE 7.14. Conduction-band energy minima in eV at Γ, X, and L relative to the top of this valence band as functions of alloy compositions X in $Ga_{1-x}Al_xP$.

at $x = 0.22$ and the L–X crossover at $x = 0.45$ at 4.2 K, while Alibert *et al.* (1983) found the Γ–L crossover at 0.35 and L–X at 0.46 at 27 K. Aulombard *et al.* (1985) also found the Γ–L crossover at $x = 0.24$. We note that it is necessary to adapt the smaller experimental energy for the L_{6c} (0.90 eV) state in GaSb in order to produce these three stages of conduction-band minimum for the $Ga_{1-x}Al_xSb$ alloy. Also note that the Γ and L gaps are very close to each other over the full concentration range. The calculated crossover between Γ and X occurs at $x_c = 0.406$ with $E_c = 1.347$ eV.

7.4.3. $Ga_{1-x}Al_xP$

Both GaP and AlP are indirect-gap semiconductors, so is the pseudobinary $Ga_{1-x}Al_xP$ throughout the whole concentration range. Figure 7.14 shows that the ordering of the three low conduction-band states remains the same for all x values. The curves are simple, and

TABLE 7.19. Coefficients α, β, and γ (all in eV) in Eq. (7.4.1) for the Three Lowest Conduction-Band States of $Ga_{1-x}Al_xP$

States	α	β	γ	Temperature (K)	Source
$E(\Gamma_{6c})$	2.866	0.524	0.238	0	Present
$E(X_{6c})$	2.350	0.090	0.068	0	Present
$E(L_{6c})$	2.650	0.381	0.096	0	Present

TABLE 7.20. Coefficients α, β, and γ of Eq. (7.4.1) for the Three Lowest Three Low Conduction-Band States of $In_{1-x}Ga_xP$

States	α	β	γ	Temperature (K)	Source
$E(\Gamma_{6c})$	1.420	0.931	0.517	0	Present
$E(\Gamma_{6c})$	1.409	0.695	0.758	2	Onton & Chicotka (1971)
$E(\Gamma_{6c})$	1.345	1.435	0.500	300	Alibert *et al.* (1972)
$E(X_{6c})$	2.380	−0.249	0.220	0	Present
$E(X_{6c})$	2.321	0.017	0.0	2	Onton & Chicotka (1971)
$E(L_{6c})$	2.030	0.314	0.308	0	Present

there is no crossing between any two of the three. Table 7.19 lists the coefficients α, β, and γ for the quadratic approximation of Eq. (7.4.1) for these three conduction-band gaps. The bowing parameters are all small. There is no reliable experimental data for comparison. The calculated valence-band offset is nearly linear and given by $\Delta E_v = -0.541 - 0.256x + 0.003x^2$ relative to GaAs.

7.4.4. $In_{1-x}Ga_xP$

In Table 7.20 the calculated α, β, and γ for the three lowest conduction-band states are compared with those obtained previously. The calculated b shown in Table 7.21 indicates that the quadratic form is a good approximation for all three gaps. The alloy disorder parameters given in Table 7.22 show that there are moderate disorder contributions in both s and p term values between In and Ga, and weak off-diagonal disorder in all three representations. As a result, the extrinsic bowing parameters b_{ex} in Table 7.21 are small. In contrast, the VCA bowing for the Γ gap is large. However, the calculated b, which is close the experimental value of Alibert *et al.* (1983), is smaller than the other two experimental values. The calculated valence-band top has a very small bowing parameter and in eV is given by $\Delta E_v = -0.339 - 0.184x - 0.018x^2$ relative to the top of valence band of GaAs.

The three conduction-band edges are plotted in Fig. 7.15 as functions of x. The Γ gap crosses the X gap at $x_c = 0.695$ and $E_c = 2.314$ eV. For $x > x_c$, the X gap is the smallest among the three gaps. These values are in good agreement with the $x_c = 0.69$ and $E_c = 2.32$ eV for $T < 10$ K found by Joullie and Alibert (1974) and smaller than the $x_c = 0.74$, $E_c = 2.33$ eV values at 2 K estimated by Onton and Chicotka (1971) using a linear x-dependence for the X gap. The calculated X–L crossing occurs at $x_c = 0.571$ and $E_c = 2.309$ eV, and the Γ–L crossing at $x_c = 0.783$ and $E_c = 2.464$ eV.

We note that the calculated direct gap of 2.013 eV at 0 K for the 50–50 ($x = 0.5$) alloy is very close a recently measured value of 1.987 eV at 13 K (Kanatz *et al.*, 1992). We also note that the band gap of alloys also depends on their state of ordering. Alloys with long-range ordering tend to have smaller band gaps than the disordered alloys with the same composi-

TABLE 7.21. Values of b in Eq. (7.4.2) at Several x Values for the Three Lowest Conduction-Band States of $In_{1-x}Ga_xP$

x	0.1	0.3	0.5	0.75	0.9	Average	b_{ex}
$b(\Gamma)$	0.514	0.517	0.518	0.514	0.517	0.517	0.04
$b(X)$	0.250	0.233	0.224	0.208	0.200	0.220	0.07
$b(L)$	0.311	0.309	0.309	0.307	0.303	0.308	0.06

TABLE 7.22. Alloy Disorder Parameters as Defined in Eq. (5.11.2) in the Γ-OLO Representations Centered at a Cation Site in $In_{1-x}Ga_xP$

Representation	δ (eV)	Δ (eV)
Γ_6	0.640	0.101
Γ_7	0.648	0.077
Γ_8	0.774	0.077

tion. A recent experiment by Lee *et al.* (1991) showed that for $Ga_{0.5}In_{0.5}P$ the ordered system has a gap of 1.902 eV while the disordered alloy has a higher value of 1.990 eV at 300 K. This translates into a 0.36-eV difference in the bowing parameter between the ordered and disordered alloys.

It turns out that the reduction of band gaps in the long-range ordered (LRO) semiconductors is quite general (e.g., Gomyo *et al.*, 1987; Kanatz *et al.*, 1992; Jones *et al.*, 1992; and Horner *et al.*, 1993). A review of this topic has been given by Zunger and Mahajan (1993). In the LRO semiconductors such as the CuAuI, chalcopyrite, and CuPt structures discussed in Section 1.4, the lattice periodicity along certain directions is expanded to several times that in the zinc blende structure. Besides, these LRO structures also experience lattice

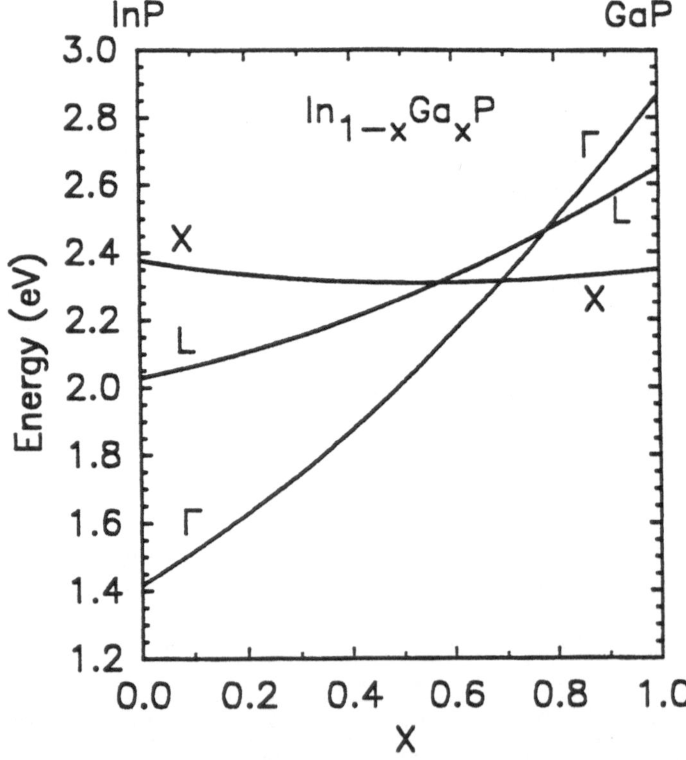

FIGURE 7.15. Conduction-band energy minima in eV at Γ, X, and L relative to the top of this valence band as functions of alloy compositions X in $In_{1-x}Ga_xP$.

TABLE 7.23. Coefficients α, β, and γ of Eq. (7.4.1) for the Least Lowest Conduction-Band States of $Ga_{1-x}In_xAs$

States	α	β	γ	Temperature (K)	Source
$E(\Gamma_{6c})$	1.520	−1.447	0.348	0	Present
$E(\Gamma_{6c})$	1.520	−1.50	0.40	2	Biryulin *et al.* (1983)
$E(\Gamma_{6c})$	1.432	−1.50	0.40	300	Biryulin *et al.* (1983)
$E(X_{6c})$	2.030	0.057	0.190	0	Present
$E(L_{6c})$	1.820	−0.535	0.244	0	Present

distortions, e.g., nonideal *c/a* ratio, and internal interatomic distance relaxations. These lattice changes cause the splitting of the top of the valence band and introduce couplings between the Brillouin zone folded states, which are the origin of the large reduction of the band gaps. A quantitative calculation of these changes has been made by Wei and Zunger (1990).

7.4.5. $Ga_{1-x}In_xAs$

In Table 7.23 the calculated α, β, and γ for the three low conduction-band states are compared with those obtained previously. The calculated *b*'s shown in Table 7.24 indicate that the quadratic form is a reasonable approximation for all three gaps. The alloy disorder parameters given in Table 7.25 show a very small *s* disorder and a moderate *p* disorder in the cation term values, a modestly strong off-diagonal disorder in the Γ_6 representation, and relatively smaller off-diagonal disorder in the Γ_7 and Γ_8 representations. These disorder parameters combine to give rather small extrinsic bowing parameters b_{ex}, as shown in Table 7.15 along with the full contribution. The calculated bowing parameter for the Γ gap is slightly smaller than the experimental results listed in the same table. Note that the bowing parameters deduced from earlier experiments ($b \simeq 0.8$ eV by Abrahams *et al.*, 1959, and $b = 0.60$ eV by Williams and Rhen, 1968) are considerably larger than those listed in Table 7.13. The calculated valence band top has an upward bowing, which is well approximated by the quadratic form $\Delta E_v = 0.199x - 0.128x^2$ (in eV) relative to the top of valence band of GaAs.

In Fig. 7.16 these three band gaps are plotted as functions of *x*. The direct Γ gap is the minimum gap throughout the whole alloy concentration. The ordering $E(\Gamma) < E(L) < E(X)$ follows the same trend. Recently, a number of groups have worked on $Ga_{0.47}In_{0.53}As$ epitaxially grown on a lattice-matched InP substrate. Using the numbers in Table 7.23, our calculated value for the Γ gap at this alloy concentration is 0.851 eV, which is larger than the measured values of 0.813 eV at 2 K and 0.75 eV at 300 K (Goetz *et al.*, 1983). This difference may either imply that the bowing parameters for the Γ gap in Table 7.13 should

TABLE 7.24. Values of *b* in Eq. (7.4.2) at Several *x* Values for the Three Lowest Conduction-Band States of $Ga_{1-x}In_xAs$

x	0.1	0.25	0.5	0.70	0.9	Average	b_{ex}
$b(\Gamma)$	0.364	0.357	0.348	0.339	0.348	0.348	0.03
$b(X)$	0.190	0.190	0.191	0.189	0.189	0.190	0.02
$b(L)$	0.249	0.245	0.244	0.243	0.243	0.244	0.03

TABLE 7.25. Alloy Disorder Parameters as Defined in Eq. (5.11.2) in the Γ-OLO Representations Centered at a Cation Site in $Ga_{1-x}In_xAs$

Representation	δ (eV)	Δ (eV)
Γ_6	0.124	0.420
Γ_7	−0.450	−0.163
Γ_8	−0.576	−0.163

TABLE 7.26. Coefficients α, β, and γ of Eq. (7.4.1) for the Three Lowest Conduction-Band States of $Ga_{1-x}In_xSb$

States	α	β	γ	Temperature (K)	Source
$E(\Gamma_{6c})$	0.822	−0.991	0.406	0	Present
$E(\Gamma_{6c})$	0.813	−0.991	0.413	0	Roth *et al.* (1980)
$E(X_{6c})$	1.252	0.342	0.118	0	Present
$E(L_{6c})$	0.900	−0.098	0.228	0	Present

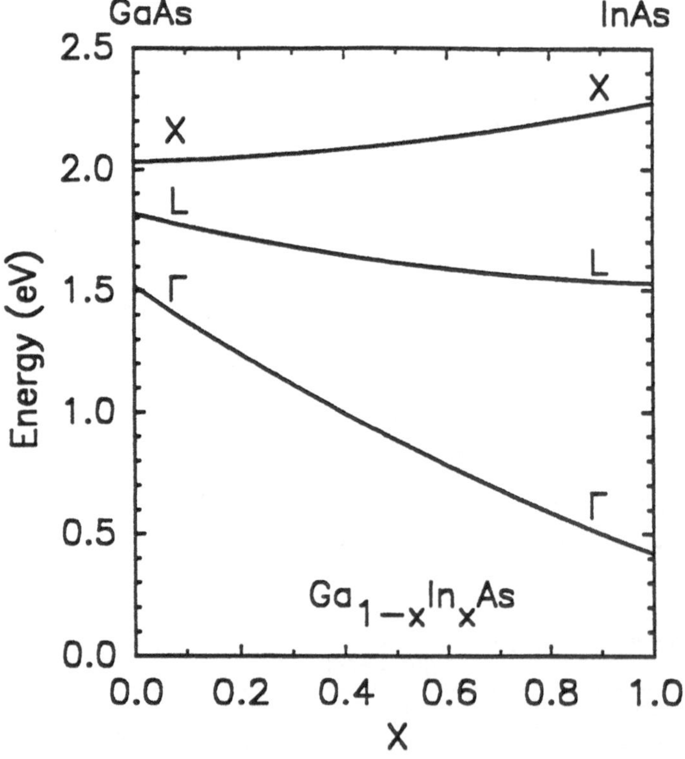

FIGURE 7.16. Conduction-band energy minima in eV at Γ, X, and L relative to the top of this valence band as functions of alloy compositions X in $Ga_{1-x}In_xAs$.

TABLE 7.27. Values of b in Eq. (7.4.2) at Several x Values for the Three Lowest Conduction-Band States of $Ga_{1-x}In_xSb$

x	0.1	0.25	0.5	0.70	0.9	Average	b_{ex}
$b(\Gamma)$	0.426	0.413	0.405	0.398	0.426	0.406	0.043
$b(X)$	0.107	0.115	0.116	0.124	0.144	0.118	0.001
$b(L)$	0.230	0.225	0.228	0.229	0.266	0.228	0.034

be increased by 0.1, or there is structural ordering (short range and long range) in the epitaxial alloys. As discussed earlier, long-range ordering in semiconductor alloys tends to make the band gap smaller than it is in the disordered phase.

7.4.6. $Ga_{1-x}In_xSb$

Table 7.26 compares the calculated α, β, and γ for the three low conduction-band states with those obtained previously. The calculated b's shown in Table 7.27 indicate that the quadratic form is a reasonable approximation for all three gaps. Table 7.28 shows only moderate diagonal disorder and a weak off-diagonal disorder in all three representations. The resulting extrinsic bowing parameters b_{ex}, shown in Table 7.27, are small compared with the VCA contribution. The calculated bowing parameter for the Γ gap is moderately large and is in good agreement with the experiment. The bowing parameters for the other two gaps are smaller, but there is no experimental value to compare against. The three gaps as functions of x are plotted in Fig. 7.17, which are qualitatively similar to those in $Ga_{1-x}In_xAs$ without any crossing between any two gaps. The calculated valence-band top is well approximated by $\Delta E_v = 0.140 + 0.210x - 0.09x^2$ (in eV) relative to that of GaAs.

7.4.7. $In_{1-x}Al_xP$

The three conduction-band edges at Γ, X, and L as functions of the concentration x are plotted in Fig. 7.18. Our calculation shows that the quadratic approximation is very good for all three gaps. The coefficients are given in Table 7.29. The Γ gap has a very large bowing parameter as compared to the L and X gaps. The experimental values listed in the table are not recent; they show no bowing in both Γ and X gaps. The ordering and crossing of these states in Fig. 7.18 are similar to those in $In_{1-x}Ga_xP$. For $x < 0.566$, the direct Γ gap is the minimum gap. The indirect X gap becomes the smallest of the three for $x > 0.566$. The calculated crossover concentrations x_c and energies E_c are tabulated in Table 7.30.

TABLE 7.28. Alloy Disorder Parameters as Defined in Eq. (5.11.2) in the Γ-OLO Representations Centered at a Cation Site in $Ga_{1-x}In_xSb$

Representation	δ (eV)	Δ (eV)
Γ_6	−0.735	−0.146
Γ_7	−0.356	−0.183
Γ_8	−0.482	−0.183

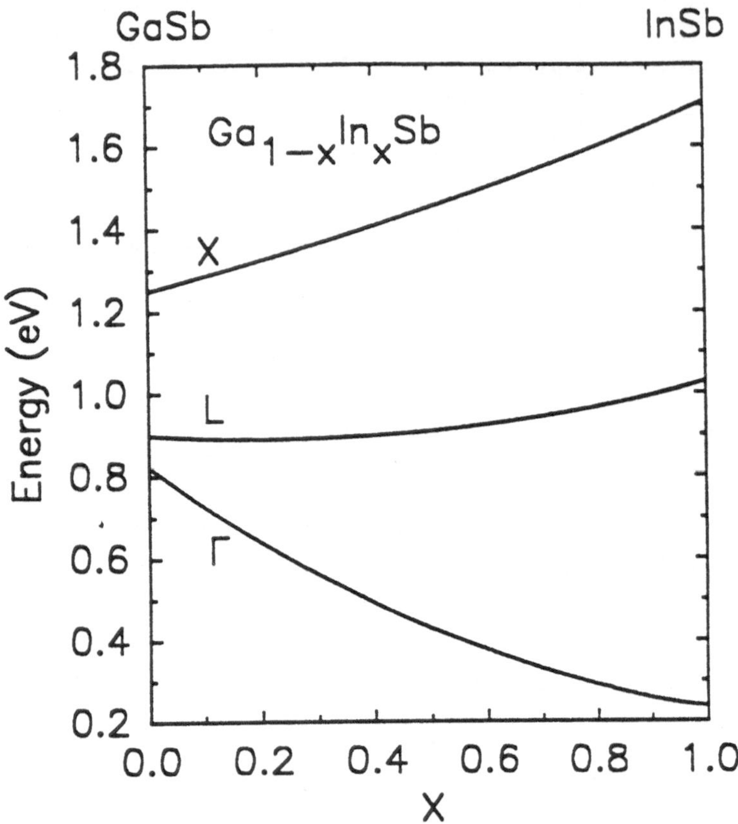

FIGURE 7.17. Conduction-band energy minima in eV at Γ, X, and L relative to the top of this valence band as functions of alloy compositions X in $Ga_{1-x}In_xSb$.

TABLE 7.29. Coefficients α, β, and γ (in eV) of Eq. (7.4.1) for the Lowest Conduction-Band Energies at Γ, X, and L Relative to Valence-Band Top in $In_{1-x}Al_xP$

States	α	β	γ	Temperature (K)	Source
$E(\Gamma_{6c})$	1.420	1.217	0.993	0	Present
$E(\Gamma_{6c})$	1.340	2.23		300	Lucovski and Chen (1970)
$E(X_{6c})$	2.380	0.016	0.113	0	Present
$E(X_{6c})$	2.24	0.18		300	Locovski and Chen (1970)
$E(L_{6c})$	2.030	0.788	0.311	0	Present

TABLE 7.30. Crossover Concentration x_c and Energy E_c (in eV) for the Three Conduction-Band States in $In_{1-x}Al_xP$

States	x_c	E_c	Temperature (K)	Source
Γ–X	0.566	2.424	0	Present calculation
Γ–X	0.44			Onton and Chicotka (1970)
X–L	0.410	2.405	0	Present calculation
Γ–L	0.684	2.713	0	Present calculation

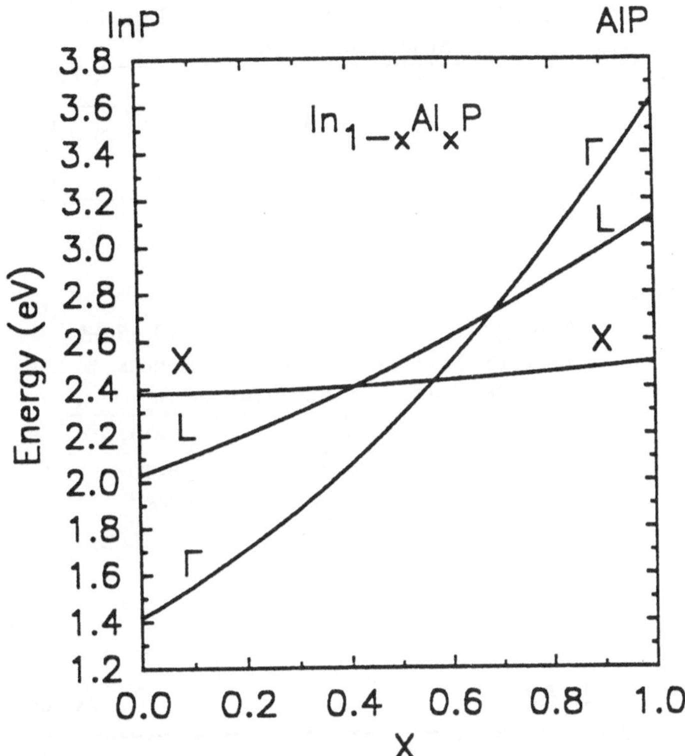

FIGURE 7.18. Conduction-band energy minima in eV at Γ, X, and L relative to the top of this valence band as functions of alloy compositions X in $In_{1-x}Al_xP$.

7.4.8. $In_{1-x}Al_xAs$

Table 7.31 lists the coefficients α, β, and γ of Eq. (7.4.1) for the three low conduction-band edges. The calculated results for b shown in Table 7.32 indicate that the quadratic form is not very accurate for all these band edges. Judging from the moderate disorder parameters in Table 7.33, the extrinsic contribution to the bowing is small compared with the intrinsic VCA contribution. There is only very limited experimental information for the gaps of this alloy. The calculated bowing parameter for the Γ gap turns out to agree reasonably well with the experimentally deduced value of Wakefield *et al.* (1984). There are no good data for the entire X and L gaps as functions of alloy concentration. The calculated top of the valence band yields a valence-band offset $\Delta E_V = 0.07 - 0.115x - 0.506x^2$ relative to GaAs.

TABLE 7.31. Coefficients α, β, and γ of Eq. (7.4.1) for the Three Lowest Conduction-Band States of $In_{1-x}Al_xAs$

States	α	β	γ	Temperature (K)	Source
$E(\Gamma_{6c})$	0.420	1.831	0.698	0	Present
$E(\Gamma_{6c})$	0.33	1.91	0.74	300	Wakeield *et al.* (1984)
$E(X_{6c})$	2.278	–0.747	0.700	0	Present
$E(L_{6c})$	1.529	0.441	0.620	0	Present

TABLE 7.32. Values of b in Eq. (7.4.2) at Several x Values for the Three Lowest Conduction-Band States of $In_{1-x}Al_xAs$

x	0.1	0.25	0.5	0.7	0.9	Average	b_{cx}
$b(\Gamma)$	0.651	0.671	0.702	0.723	0.744	0.698	0.15
$b(X)$	0.722	0.688	0.704	0.693	0.697	0.700	0.04
$b(L)$	0.591	0.599	0.623	0.638	0.647	0.620	0.11

TABLE 7.33. Alloy Disorder Parameters as Defined in Eq. (5.11.2) in the Γ-OLO Representations Centered at a Cation Site in $In_{1-x}Al_xAs$

Representation	δ (eV)	Δ (eV)
Γ_6	−0.938	−0.380
Γ_7	0.652	0.089
Γ_8	0.928	0.089

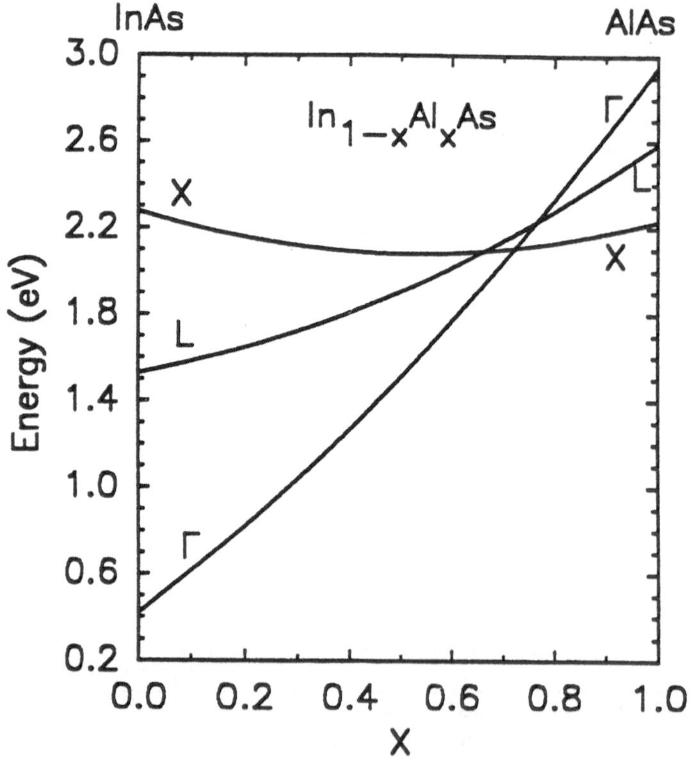

FIGURE 7.19. Conduction-band energy minima in eV at Γ, X, and L relative to the top of this valence band as functions of alloy compositions X in $In_{1-x}Al_xAs$.

Figure 7.19 plots these three band gaps as functions of x. The minimum gap for $x = 0$ (InAs) is a direct gap of 0.42 eV at Γ. This continues to be the minimum gap as x increases until the Γ gap crosses the X gap at a crossover concentration $x_c = 0.724$, and with a gap $E_c = 2.106$ eV. For $x > x_c$, the X gap is the smallest among the three gaps. Although the minimum gap in this x range may not be precisely at X, as evidenced from pure AlAs, the difference in energy is less than 1 meV. This set of values are to be compared with the experimentally deduced values $x_c = 0.68$ and $E_c = 2.05$ at 300 K by Lorenz and Onton (1970). The calculated X and L gaps cross at a smaller $x_c = 0.664$ with an $E_c = 2.092$ eV, and the L gap crosses the Γ gap at $x_c = 0.766$ and $E_c = 2.227$ eV.

7.4.9. $In_{1-x}Al_xSb$

The three conduction-band edges at Γ, X, and L as functions of the concentration x are plotted in Fig. 7.20. The calculation shows that the quadratic approximation is very good for all three gaps. The coefficients are given in Table 7.34. All three gaps have very large bowing parameters. There are no good experimental values for comparison. The ordering and crossing of these states are similar to those in $In_{1-x}Al_xP$. For $x < 0.697$, the direct Γ gap is

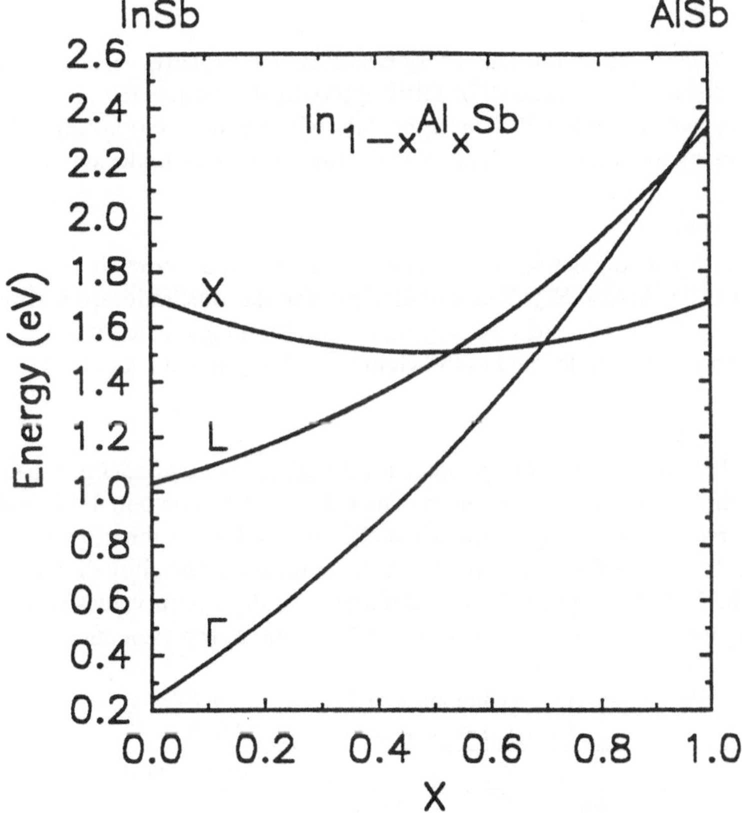

FIGURE 7.20. Conduction-band energy minima in eV at Γ, X, and L relative to the top of this valence band as functions of alloy compositions X in $In_{1-x}Al_xSb$.

TABLE 7.34. Coefficients α, β, and γ of Eq. (7.4.1) for the Three Lowest Conduction-Band States of $In_{1-x}Al_xSb$

States	α	β	γ	Temperature (K)	Source
$E(\Gamma_{6c})$	0.237	1.246	0.907	0	Present
$E(X_{6c})$	1.710	−0.796	0.777	0	Present
$E(L_{6c})$	1.030	0.459	0.841	0	Present

the minimum gap. The indirect X gap becomes the smallest of the three for $x > 0.697$. The calculated crossover concentrations x_c and energies E_c are tabulated in Table 7.35.

7.4.10. $AlAs_{1-x}P_x$

The three lowest conduction-band energies at Γ, X, and L relative to the top of the valence band (Γ_{8v}) as a function of x are shown in Fig. 7.21. All the three curves can be well approximated by quadratic approximations. They all increase monotonically with x. The coefficients for the quadratic forms are given in Table 7.36. Throughout the whole concentration range, the minimum gap is an indirect gap with the conduction-band minimum close to the symmetry point X.

7.4.11. $AlAs_{1-x}Sb_x$

The three lowest conduction band energies at Γ, X, and L relative to the top of the valence band (Γ_{8v}) plotted in Fig. 7.22 can all be fairly approximated as quadratic functions of x with the coefficients given in Table 7.37. Throughout the whole concentration range, the minimum gap is an indirect gap with the conduction-band minimum close to the symmetry point X.

7.4.12. $AlP_{1-x}Sb_x$

The three lowest conduction-band energies at Γ, X, and L shown in Fig. 7.23 are very similar to those in $AlAs_{1-x}Sb_x$. The coefficients for the quadratic approximations as a function of x are given in Table 7.38. Again, the minimum gap is an indirect gap with the conduction-band minimum located in the vicinity of the symmetry X point.

7.4.13. $GaAs_{1-x}P_x$

In Table 7.39 the calculated α, β, and γ for the three lowest conduction-band states are compared with those obtained previously. The calculated b shown in Table 7.40 indicates that the quadratic form is only approximately correct for all three gaps, and disorder contributes only a small fraction to the bowing parameters. The alloy disorder parameters given in Table 7.41 show modestly strong disorder in the s term values of As and P, and moderate diagonal and off-diagonal disorder in the Γ_7 and Γ_8 representations. The calculated

TABLE 7.35. Crossover Concentration x_c and Energy E_c (in eV) for the Three Lowest Conduction-Band States in $In_{1-x}Al_xSb$

States	x_c	E_c	Temperature (K)	Source
Γ–X	0.697	1.538	0	Present calculation
X–L	0.530	1.506	0	Present calculation
Γ–L	0.937	2.196	0	Present calculation

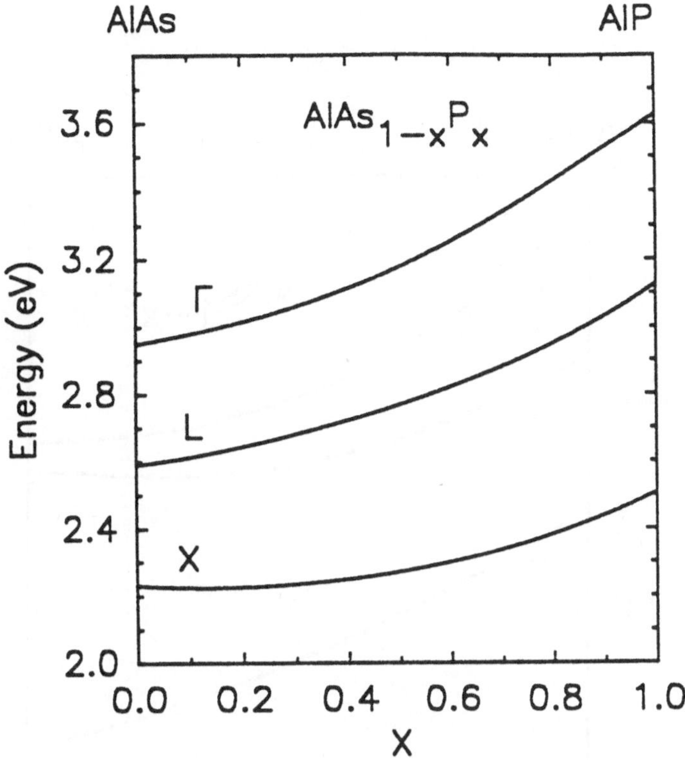

FIGURE 7.21. Conduction-band energy minima in eV at Γ, X, and L relative to the top of this valence band as functions of alloy compositions X in $AlAs_{1-x}P_x$.

TABLE 7.36. Coefficients α, β, and γ (in eV) of Eq. (7.4.1) for the Lowest Conduction-Band Energies at Γ, X, and L Relative to the Valence-Band Top in $AlAs_{1-x}P_x$

States	α	β	γ	Temperature (K)	Source
$E(\Gamma_{6c})$	2.950	0.252	0.427	0	Present
$E(X_{6c})$	2.230	−0.123	0.400	0	Present
$E(L_{6c})$	2.590	0.158	0.380	0	Present

TABLE 7.37. Coefficients α, β, and γ (in eV) of Eq. (7.4.1) for the Lowest Conduction-Band Energies at Γ, X, and L Relative to the Valence-Band Top $AlAs_{1-x}Sb_x$

States	α	β	γ	Temperature (K)	Source
$E(\Gamma_{6c})$	2.950	−1.251	0.691	0	Present
$E(X_{6c})$	2.230	−0.790	0.250	0	Present
$E(L_{6c})$	2.590	−0.734	0.474	0	Present

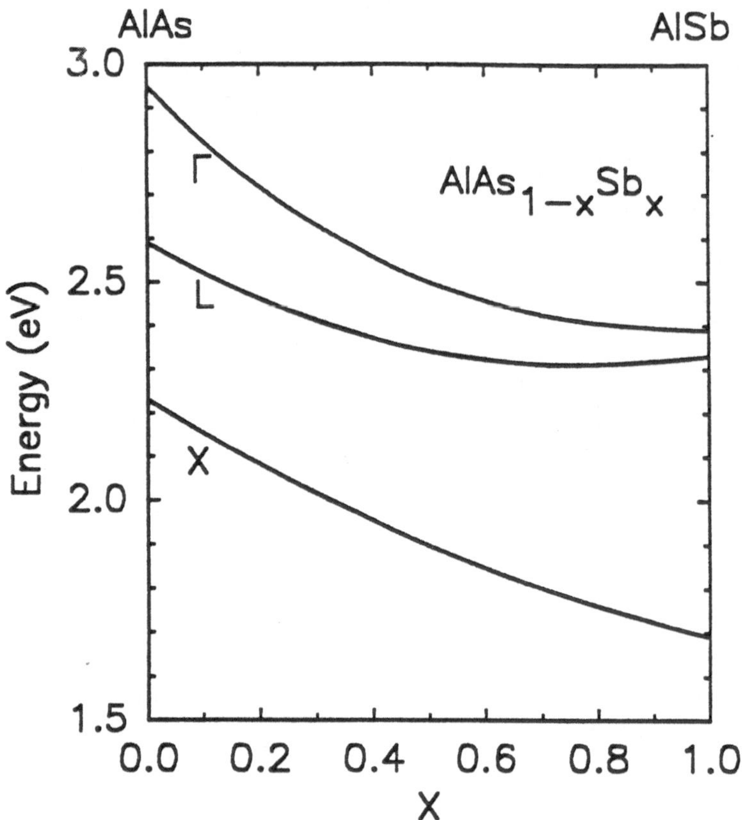

FIGURE 7.22. Conduction-band energy minima in eV at Γ, X, and L relative to the top of this valence band as functions of alloy compositions X in $AlAs_{1-x}Sb_x$.

TABLE 7.38. Coefficients α, β, and γ (in eV) of Eq. (7.4.1) for the Lowest Conduction-Band Energies at Γ, X, and L Relative to the Valence-Band Top $AlP_{1-x}Sb_x$

States	α	β	γ	Temperature (K)	Source
$E(\Gamma_{6c})$	3.630	−2.351	1.111	0	Present
$E(X_{6c})$	2.510	−1.097	0.277	0	Present
$E(L_{6c})$	3.130	−1.556	0.756	0	Present

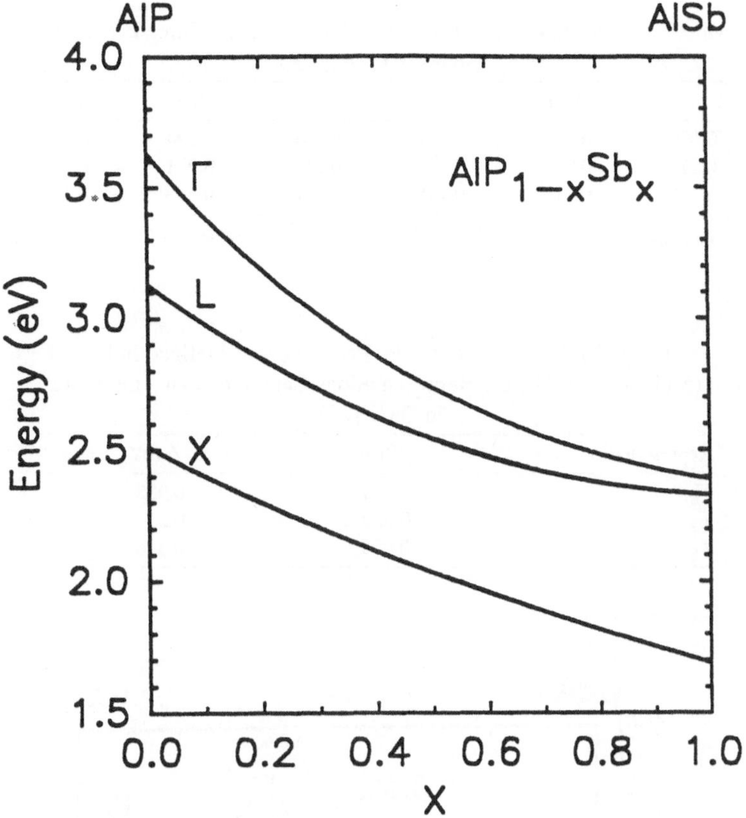

FIGURE 7.23. Conduction-band energy minima in eV at Γ, X, and L relative to the top of this valence band as functions of alloy compositions X in $AlP_{1-x}Sb_x$.

TABLE 7.39. Coefficients α, β, and γ (in eV) for the Lowest Conduction-Band Energies at Γ, X, and L Relative to the Valence-Band Top in $GaAs_{1-x}P_x$

States	α	β	γ	Temperature (K)	Source
$E(\Gamma_{6c})$	1.520	1.231	0.116	0	Present
$E(\Gamma_{6c})$	1.508	1.188	0.174	77	Aspnes (1976)
$E(\Gamma_{6c})$	1.514	1.174	0.186	77	Nelson *et al.* (1976)
$E(\Gamma_{6c})$	1.519	1.155	0.210	6	Onton & Foster (1972)
$E(\Gamma_{6c})$	1.515	1.172	0.186	2	Capizzi *et al.* (1981)
$E(X_{6c})$	2.230	0.076	0.243	0	Present
$E(X_{6c})$	1.971	0.059	0.202	77	Aspnes (1976)
$E(X_{6c})$	2.002	0.070	0.267	6	Onton & Foster (1972)
$E(X_{6c})$	1.971	0.144	0.211	2	Capizzi *et al.* (1981)
$E(L_{6c})$	1.820	0.698	0.133	0	Present
$E(L_{6c})$	1.802	0.77	0.16	77	Aspnes (1976)

TABLE 7.40. Values of b in Eq. (7.4.2) at Several x Values for the Three Lowest Conduction-Band States in $GaAs_{1-x}P_x$

x	0.1	0.30	0.5	0.75	0.9	Average	b_{ex} [a]
$b(\Gamma)$	0.070	0.088	0.111	0.141	0.124	0.116	0.04
$b(X)$	0.217	0.226	0.236	0.263	0.271	0.243	0.02
$b(L)$	0.100	0.115	0.145	0.156	0.163	0.133	0.04

[a] b_{ex} is the disorder contribution.

TABLE 7.41. Alloy Disorder Parameters as Defined in Eq. (5.11.2) in the Γ-OLO Representations Centered at an Anion Site in $GaAs_{1-x}P_x$

Representation	δ (eV)	Δ (eV)
Γ_6	−1.115	0.005
Γ_7	0.320	−0.318
Γ_8	0.606	−0.318

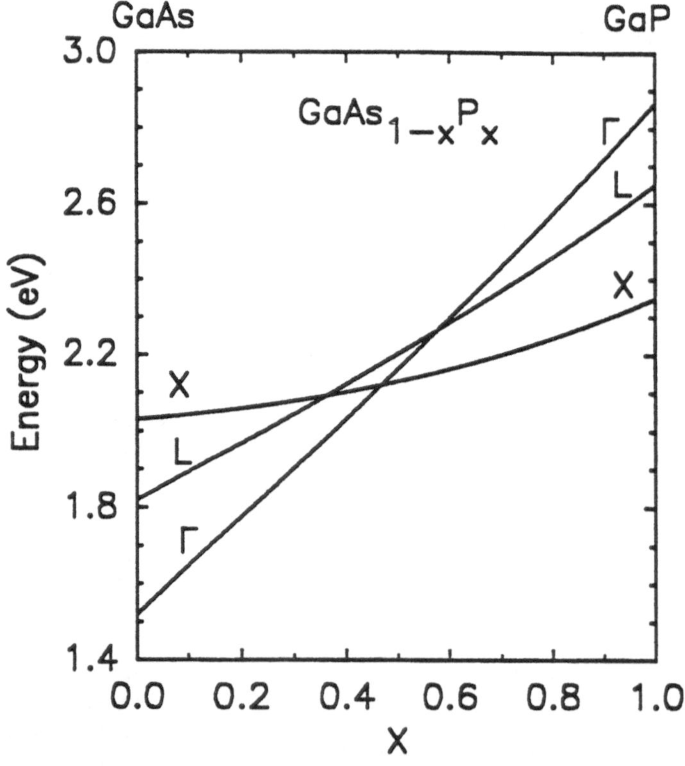

FIGURE 7.24. Conduction-band energy minima in eV at Γ, X, and L relative to the top of this valence band as functions of alloy compositions X in $GaAs_{1-x}P_x$.

valence-band top is a good linear function of x given by $\Delta E_v = -0.54x$ relative to that in GaAs. Figure 7.24 contains plots of these three band gaps as functions of x. The Γ gap crosses the X gap at $x_c = 0.471$ and $E_c = 2.12$ eV. For $x > x_c$, the X gap is the smallest among the three gaps. This x_c value lies between the 0.45 at 77 K measure by Nelsen *et al.* (1976) and the 0.51 value at 30 K by Marciniak and Wittey (1975). The X–L crossing occurs at $x_c = 0.361$ and $E_c = 2.27$ eV, and the Γ–L crossing at $x_c = 0.586$ and $E_c = 2.09$ eV.

7.4.14. $GaAs_{1-x}Sb_x$

This alloy has a direct gap at Γ throughout the whole concentration range. However, because the solid-solution miscibility gap runs into the solidus curve, the range of concentration of the solid phase that can be grown by near-equilibrium growth is limited. Most recent alloy samples have been grown by epitaxial techniques. These samples may contain partial long-range order as discussed in Chapter 1. Since the band gaps of alloys depend on the state of order, it is thus difficult to compare the present calculation, which assumes a random alloy, with the available data. Figure 7.25 shows the three lowest conduction-band states at Γ, X, and L from the present calculation. The direct gap at Γ is the minimum gap through the whole alloy composition. There is no band crossing as x varies. The Γ gap has

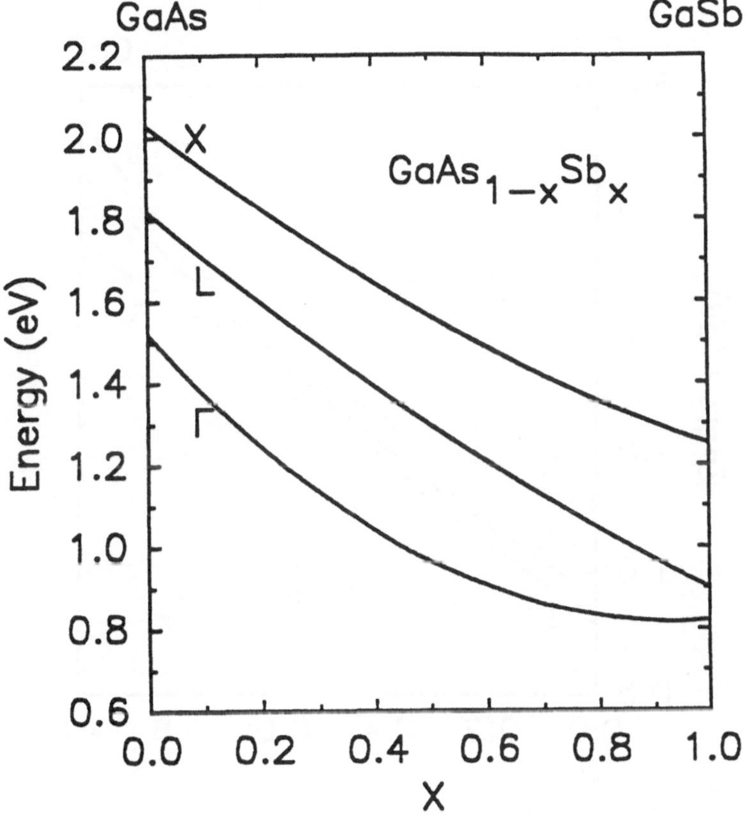

FIGURE 7.25. Conduction-band energy minima in eV at Γ, X, and L relative to the top of this valence band as functions of alloy compositions X in $GaAs_{1-x}Sb_x$.

TABLE 7.42. Coefficients α, β, and γ (in eV) of Eq. (7.4.1) for the Lowest Conduction-Band Energies at Γ, X, and L Relative to the Valence-Band Top $GaAs_{1-x}Sb_x$

States	α	β	γ	Temperature (K)	Source
$E(\Gamma_{6c})$	1.520	−1.528	0.830	0	Present
$E(\Gamma_{6c})$	1.43	−1.9	1.2	300	Nahory *et al.* (1977)
$E(X_{6c})$	2.030	−1.092	0.314	0	Present
$E(L_{6c})$	1.820	−1.168	0.248	0	Present

a very large bowing parameter, as can been seen from the plots and from the coefficients for the quadratic approximations given in Table 7.42. However, the calculated bowing is still smaller than the experimental value. In this respect it is worth noting that the following values of band gaps have been measured for $GaAs_{0.5}Sb_{0.5}$ epitaxially grown on an InP substrate: 0.795 eV at 300 K (Kerr *et al.*, 1986; Chiu *et al.*, 1985), 0.80 and 0.84 eV at 4 K (Cherng *et al.*, 1984). Using the low-T value, the bowing parameter is 1.3 eV. Since this alloy was found to be ordered and the disordered system has a smaller bowing, the calculated value presented here may still be a good representation of the disordered alloy.

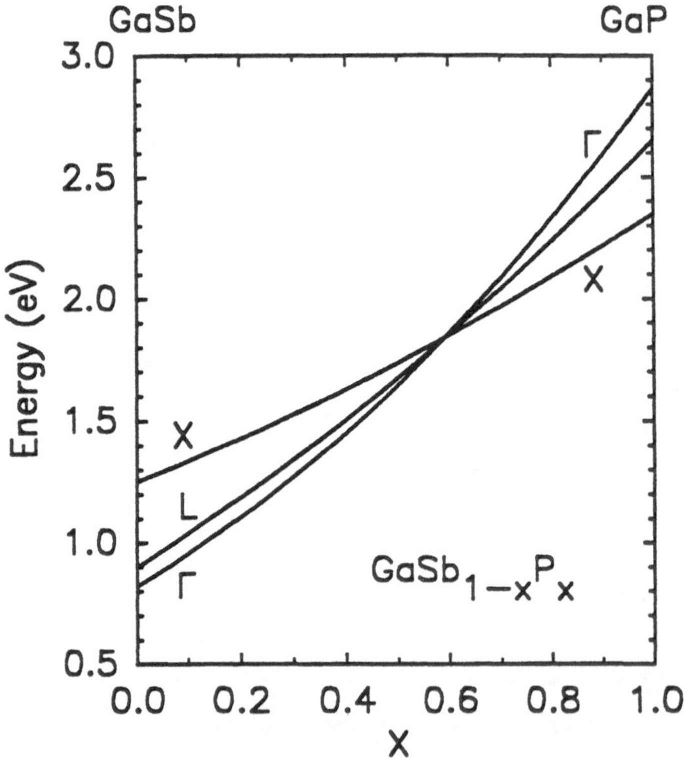

FIGURE 7.26. Conduction-band energy minima in eV at Γ, X, and L relative to the top of this valence band as functions of alloy compositions X in $GaSb_{1-x}P_x$.

TABLE 7.43. Coefficients α, β, and γ (in eV) of Eq. (7.4.1) for the Lowest Conduction-Band Energies at Γ, X, and L Relative to the Valence-Band Top $GaSb_{1-x}P_x$

States	α	β	γ	Temperature (K)	Source
$E(\Gamma_{6c})$	0.822	1.258	0.768	0	Present
$E(X_{6c})$	1.252	0.846	0.252	0	Present
$E(L_{6c})$	0.900	1.358	0.393	0	Present

7.4.15. $GaSb_{1-x}P_x$

This alloy has a very large miscibility gap (see Chapter 4). It can only be grown by nonequilibrium techniques such as MOCVD epitaxial growth. A very interesting point about the band gaps of this alloy is that all the three lowest conduction-band states at Γ, X, and L cross each other at the same concentration $x_c = 0.593$ at 0 K, as shown in Fig. 7.26. The quadratic form of Eq. (7.4.1) is a good representation for all three gaps. The coefficients α, β, and γ are given in Table 7.43. The bowing parameters are comparable to those for $GaAs_{1-x}Sb_x$.

7.4.16. $InAs_{1-x}P_x$

Table 7.44 lists the coefficients α, β, and γ of Eq. (7.4.1) for the three low conduction-band gaps. The calculated results for b shown in Table 7.45 indicate that the quadratic form is a good approximation for all three band edges. The disorder parameters in Table 7.46 are small, so the extrinsic contribution to the bowing is also small compared with the intrinsic VCA contribution. The experimental information is also limited. The calculated bowing parameter for the Γ gap is close to values deduced from experiment (Nicholas *et al.*, 1979). The calculated top of the valence band yields a valence-band offset $\Delta E_v = 0.070 - 0.115x - 0.506x^2$ relative to GaAs. The three band gaps are plotted in Fig. 7.27, which shows no crossing between states. The alloy is a direct-gap semiconductor throughout the entire concentration range.

7.4.17. $InAs_{1-x}Sb_x$

Figure 7.28 shows plots of the three lowest conduction-band energies at Γ, X, and L relative to the top of the valence band Γ_{8v} as functions of the composition x. Qualitatively these gaps are similar to those in $GaAs_{1-x}Sb_x$ except that these gaps are much smaller. The quadratic form of Eq. (7.5.1) is a reasonably good approximation for all three gaps (see Table 7.48), with the coefficients given in Table 7.47. Note that the large bowing of the minimum Γ gap, so that the gaps of the alloys are smaller than that of InSb for a wide range of

TABLE 7.44. Coefficients α, β, and γ (in eV) of Eq. (7.4.1) for the Lowest Conduction-Band Energies at Γ, X, and L Relative to the Valence-Band Top $InAs_{1-x}P_x$

States	α	β	γ	Temperature (K)	Source
$E(\Gamma_{6c})$	0.420	0.712	0.286	0	Present
$E(\Gamma_{6c})$	0.414	0.64	0.36	77	Nicholas *et al.* (1979)
$E(\Gamma_{6c})$	0.356	0.675	0.32	300	Nicholas *et al.* (1979)
$E(X_{6c})$	2.278	−0.086	0.187	0	Present
$E(L_{6c})$	1.529	0.385	0.115	0	Present

TABLE 7.45. Values of b in Eq. (7.4.2) at Several x Values for the Three Lowest Conduction-Band States in $InAs_{1-x}P_x$

x	0.1	0.3	0.5	0.75	0.9	Average	b_{ex} [a]
$b(\Gamma)$	0.278	0.284	0.287	0.288	0.287	0.286	0.044
$b(X)$	0.184	0.185	0.196	0.176	0.177	0.187	0.096
$b(L)$	0.119	0.117	0.116	0.114	0.109	0.115	0.058

[a] b_{ex} is the disorder contribution

TABLE 7.46. Alloy Disorder Parameters as Defined in Eq. (5.11.2) in the Γ-OLO Representations Centered at an Anion Site in $InAs_{1-x}P_x$

Representation	δ (eV)	Δ (eV)
Γ_6	–0.506	–0.497
Γ_7	0.156	–0.230
Γ_8	0.446	–0.230

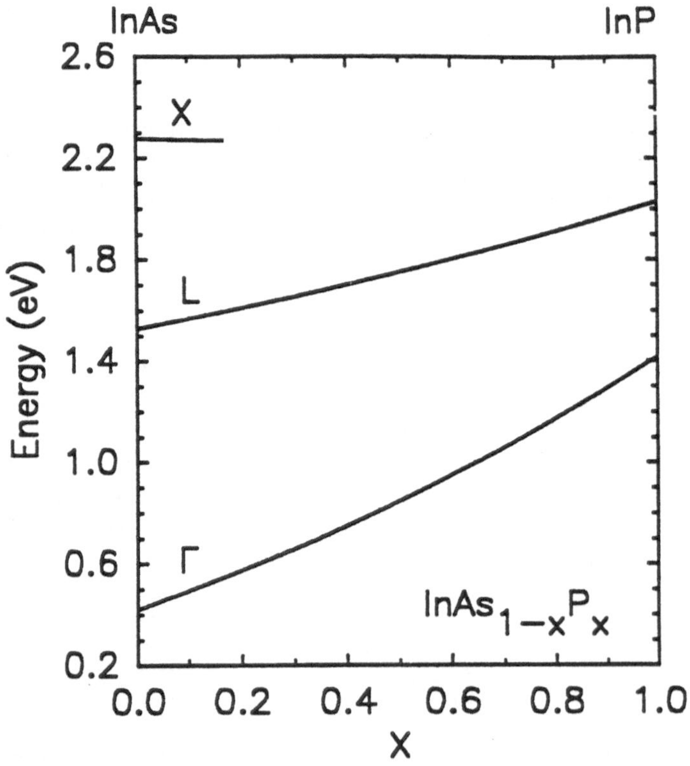

FIGURE 7.27. Conduction-band energy minima in eV at Γ, X, and L relative to the top of this valence band as functions of alloy compositions X in $InAs_{1-x}P_x$.

TABLE 7.47. Coefficients α, β, and γ (in eV) of Eq. (7.4.1) for the Lowest Conduction-Band Energies at Γ, X, and L Relative to Valence-Band Top and the Spin-Orbit Splitting Δ_o in $InAs_{1-x}Sb_x$

States	α	β	γ	Temperature (K)	Source
$E(\Gamma_{6c})$	0.420	−1.064	0.881	0	Present
$E(\Gamma_{6c})$	0.350	−0.850	0.680	300	Berolo and Woolley (1972)
$E(X_{6c})$	2.278	−0.758	0.190	0	Present
$E(L_{6c})$	1.530	−0.681	0.182	0	Present
Δ_o	0.382	0.251	0.219	0	Present
Δ_o	0.40	−0.62	1.02	300	Berolo and Woolley (1972)

TABLE 7.48. Values of b in Eq. (7.4.2) at Several x Values for the Three Lowest Conduction-Band States in $InAs_{1-x}Sb_x$

x	0.1	0.25	0.5	0.70	0.9	Average	b_{ex}
$b(\Gamma)$	0.908	0.896	0.880	0.863	0.844	0.881	−0.080
$b(X)$	0.180	0.188	0.185	0.201	0.199	0.190	−0.052
$b(L)$	0.192	0.184	0.181	0.178	0.180	0.182	−0.060

[a] b_{ex} is the disorder contribution

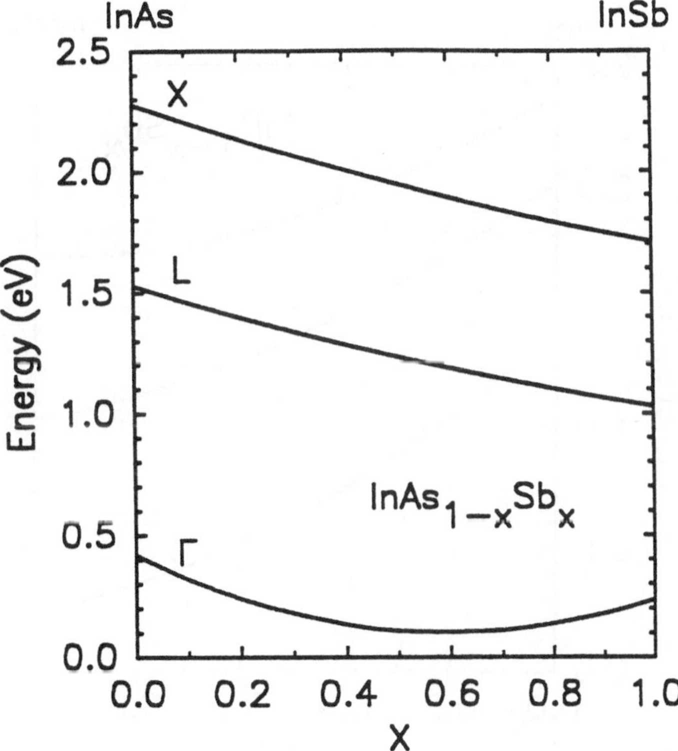

FIGURE 7.28. Conduction-band energy minima in eV at Γ, X, and L relative to the top of this valence band as functions of alloy compositions X in $InAs_{1-x}Sb_x$.

TABLE 7.49. Alloy Disorder Parameters as Defined in Eq. (5.11.2) in the Γ-OLO Representations Centered at an Anion Site in $InAs_{1-x}Sb_x$

Representation	δ (eV)	Δ (eV)
Γ_6	−0.007	0.274
Γ_7	0.967	−0.663
Γ_8	0.499	−0.663

composition, with the smallest gap of 0.1 eV at $x = 0.55$. This result is consistent with an early experiment (Berolo and Woolley, 1972), with the estimated quadratic coefficients listed in Table 7.47. Note that in Table 7.49 the disorder parameters are weak in the Γ_6 representation but are moderately strong in the Γ_7 and Γ_8 representations. These parameters combine to give a negative contribution to the bowing parameters, although their magnitudes are still small compared with the VCA contributions.

There has been considerable discussion of the nonlinear x dependence of the spin-orbit splitting of the top of the valence band $\Delta_o = E(\Gamma_{8v}) - E(\Gamma_{8v})$. Optical data have shown a very large bowing of Δ_o in $InAs_{1-x}Sb_x$, and simple phenomenological theories based on the weakening of the spin-orbit coupling due to disorder mixing of the s states at the valence-

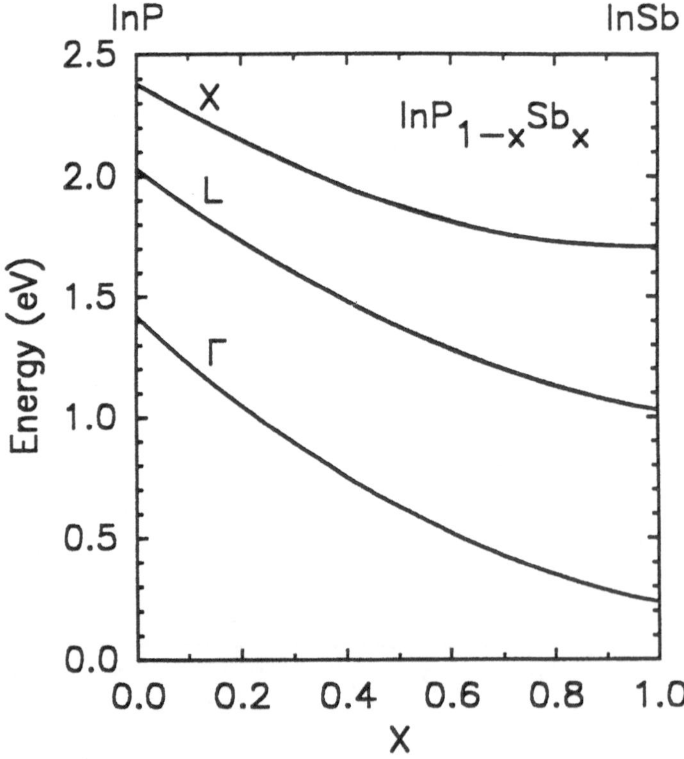

FIGURE 7.29. Conduction-band energy minima in eV at Γ, X, and L relative to the top of this valence band as functions of alloy compositions X in $InP_{1-x}Sb_x$.

TABLE 7.50. Coefficients α, β, and γ of Eq. (7.4.1) for the Lowest Conduction-Band Energies at Γ, X, and L Relative to Valence-Band Top and the Spin-Orbit Splitting Δ_o in $InP_{1-x}Sb_x$

States	α	β	γ	Temperature (K)	Source
$E(\Gamma_{6c})$	1.420	−1.987	0.804	0	Present
$E(X_{6c})$	2.380	−1.343	0.673	0	Present
$E(L_{6c})$	2.030	−1.618	0.618	0	Present

band top (Chadi, 1977) have been used to interpret the data. The calculated results for Δ_o can be well approximated by the quadratic form with the coefficients also listed in Table 7.47. It is seen that the calculated bowing parameter, which is large when compared with other alloy systems, is still considerably smaller than the experimental value. This discrepancy deserves further study.

7.4.18. $InP_{1-x}Sb_x$

This alloy can only be grown by nonequilibrium techniques. The three conduction-band states as functions of alloy composition plotted in Fig. 7.29 behave similarly to those in $GaAs_{1-x}Sb_x$; namely, it is a direct-gap semiconductor throughout the whole x range, the gaps do not cross, and they decrease monotonically with x. Despite the large bowing, all three gaps are well approximated by the quadratic form with the coefficients given in Table 7.50.

7.5. BAND STRUCTURES AND HAMILTONIAN PARAMETERS FOR II–VI ZINC BLENDE COMPOUNDS

The II–VI compounds to be studied include ZnS, ZnSe, ZnTe, CdS, CdSe, CdTe, HgSe, and HgTe. Although ZnS, ZnSe, CdS, and CdSe are normally grown in the wurtzite structure, they can also be grown in the zinc blende structure. Our discussion in this section and the next will be restricted to the zinc blende structure because the present alloy theory has not yet been extended to the wurtzite structure. One feature in the band structures of these II–VI compounds that is different from the III–VI compounds studied so far is the influence of d bands. These bands originate from the valence d states of the Zn, Cd, and Hg, which in their II–VI compounds cut into the bottom part of the sp valence bands. The dashed curves in Fig. 7.30 show the band structure of CdTe, including the d bands obtained from a second-neighbor tight-binding calculation (Lai-Hsu, 1990). The flattened curves at the bottom are the d bands. Since the d states are still about 10 eV below the top of the valence band, the effect of these d bands can be treated by contracting the s, p, and d orbitals* into "effective" s and p basis functions. This is evident from Fig. 7.30, where the solid curves are the bands based on these contracted s and p orbitals. The last several valence bands and the first two conduction bands are seen to be reproduced accurately. This justifies the application of the minimum-basis

* Contraction here means the reduction of the size of Hamiltonian matrix. For example, eigenvalue equations $ax + by = \varepsilon x$ and $cx + dy = \varepsilon y$ for a 2×2 matrix can be reduced to the single equation $ax + [bc/(\varepsilon - d)]x = \varepsilon x$. If d is far away from the energy ε of interest, the factor $1/(\varepsilon - d)$ can be approximated by a constant $1/(\varepsilon_0 - d)$, where ε_0 is a typical value in the energy range of interest.

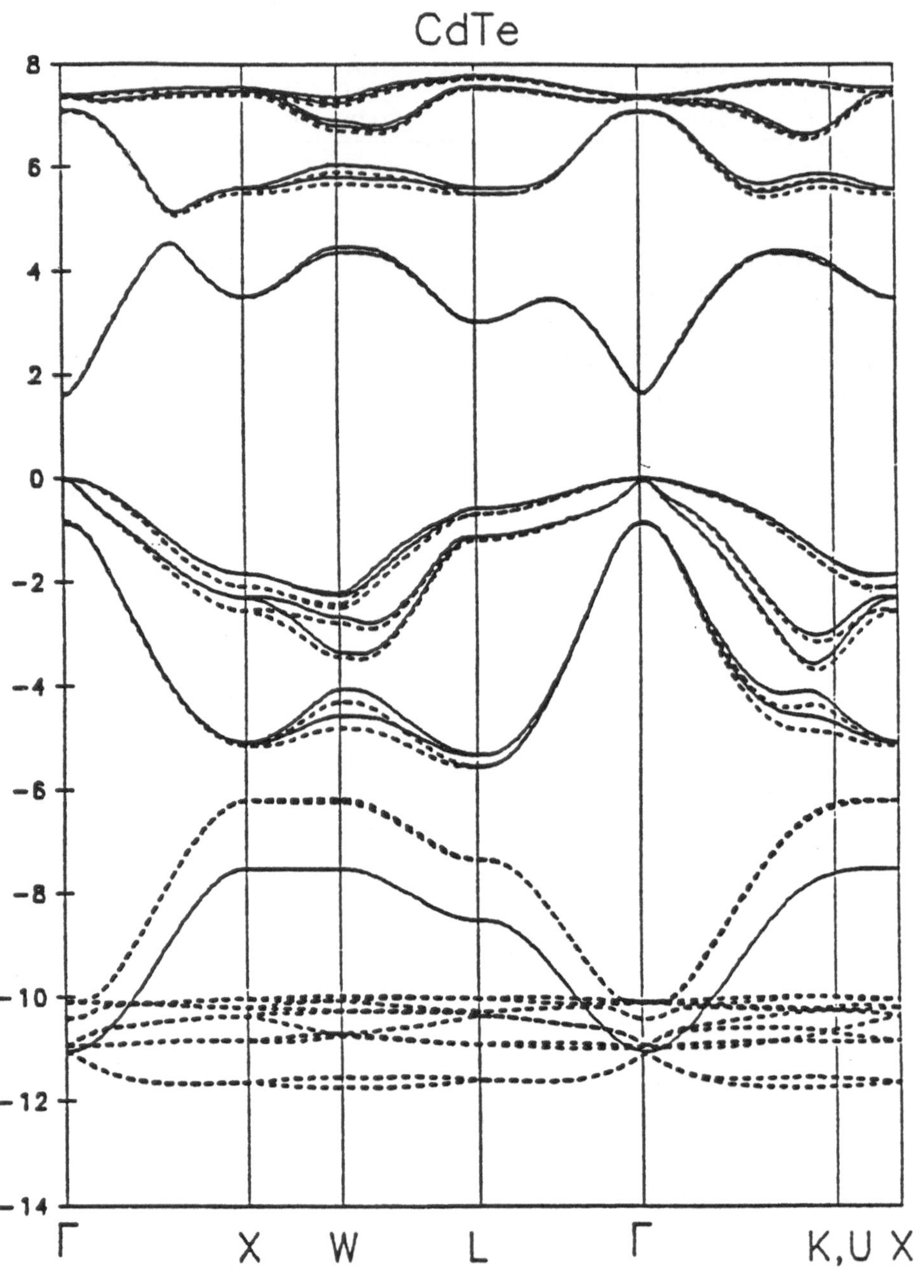

FIGURE 7.30. Comparison between the band structures of CdTe calculated with a second-neighbor tight-binding Hamiltonian including the d-orbitals (the dashed lines) and those without.

TABLE 7.51. Lattice Constants a (in Bohr radii), Matrix Elements $\delta\varepsilon$, h, and v for H_1 (in crystal units), Spin-Orbit Coupling Constants λ (in eV), Valence-Band Offsets E_v (in eV) Relative to GaAs, and Resulting Term Values E^i_α (in eV) for Eight II–VI Zinc Blende Compounds

	ZnS	ZnSe	ZnTe	CdS	CdSe	CdTe	HgSe	HgTe
a	10.216	10.713	11.533	11.026	11.433	12.242	11.499	12.208
$\delta\varepsilon_s^A$	−2.4004	−2.3597	−2.1147	−2.4428	−2.3873	−2.6325	−2.0553	−2.6325
$\delta\varepsilon_s^C$	−1.4769	−1.6804	−1.9959	−1.3215	−1.6605	−1.8382	−1.8801	−2.2481
$\delta\varepsilon_p^A$	−1.9236	−2.0613	−2.0075	−2.0892	−2.0513	−2.1699	−2.0513	−1.9734
$\delta\varepsilon_p^C$	−1.8919	−1.5594	−1.7953	−1.6173	−1.5001	−1.5041	−1.7400	−1.5141
h_{ss}^{AC}	0.0671	−0.1886	−0.2943	0.0280	−0.0294	−0.2837	0.0864	−0.1182
h_{sx}^{AC}	−0.0381	−0.1043	0.2469	0.2932	−0.3031	−0.1555	0.2322	−0.3223
h_{xs}^{AC}	0.4305	0.2575	0.2233	−0.0299	0.4590	0.2654	0.1590	0.3160
h_{xx}^{AC}	0.3614	0.2259	0.0402	0.3732	0.2445	0.1651	−0.1676	0.3074
h_{xy}^{AC}	0.4576	0.2691	0.3097	0.1908	0.2237	0.2152	−0.0385	0.2628
v_{ss}^A	−0.0223	0.0095	0.0181	−0.0235	−0.0344	0.0168	−0.0227	0.0168
$v_{sp\sigma}^A$	0.0250	0.0390	0.0331	0.0376	0.0244	0.0149	0.0244	0.0149
$v_{pp\sigma}^A$	−0.0380	−0.0380	−0.0292	0.0025	−0.0313	0.0153	−0.0313	0.0183
$v_{pp\pi}^A$	−0.0001	0.0108	−0.0028	−0.0001	0.0033	−0.0030	0.0082	−0.0021
v_{ss}^C	−0.0088	−0.0256	−0.0130	−0.0351	−0.0160	−0.0257	−0.0305	−0.0257
$v_{sp\sigma}^C$	0.0250	0.0390	0.0331	0.0376	0.0246	0.0149	0.0246	0.0149
$v_{pp\sigma}^C$	−0.0119	0.0092	0.0207	−0.0041	−0.0947	−0.0474	0.0480	−0.0453
$v_{pp\pi}^C$	0.0072	0.0000	−0.0084	0.0104	0.0624	0.0142	0.0224	0.0142
λ^A	0.020	0.157	0.356	0.020	0.153	0.323	0.153	0.327
λ^C	0.077	0.077	0.077	0.050	0.050	0.050	0.323	0.320
E_v	−1.870	−1.050	−0.430	−1.390	−0.980	−0.530	−0.100	−0.180
E_s^A	−9.184	−8.161	−6.052	−8.071	−7.279	−7.227	−5.847	−7.268
E_s^C	3.794	2.497	0.881	3.943	2.275	1.347	1.356	−0.123
E_p^A	0.244	−0.423	−0.147	−0.522	−0.330	−0.713	−0.326	−0.009
E_p^C	4.653	5.788	4.041	5.207	5.325	4.630	4.290	4.620

HPT model to these compounds, because producing these bands accurately is the prime goal of the HPT model.

Following the parametrization procedure in Section 7.3, the Hamiltonian and band structures for these II–VI compounds have been obtained starting with the same common Hamiltonian H_0 used for the III–V compounds. The resulting parameters for the adjustable TB Hamiltonian H_1 are listed in Table 7.51. These results will be used later for alloy calculations. Next we shall discuss the band structures calculated from these Hamiltonians.

7.5.1. ZnS

The experimental information about band structures of II–VI compounds is not in general as complete as that for the III–V compounds. We have combined theoretical and

TABLE 7.52. Comparison of Calculated Band Structure of ZnS (column HPT) with Experiments and Other Calculations

	Theory						
	HPT		EM-OPW[a]	EMP[b]	LCAO[c]	LDA[d]	Experiments and estimates
$E(\Gamma_{6v})$	−13.50	$E(\Gamma_{1v})$			−12.27	−13.06	−13.5[e]
$E(\Gamma_{7v})$	−0.097						
$E(\Gamma_{8v})$	0	$E(\Gamma_{15v})$	0	0	0		
$E(\Gamma_{6c})$	3.80	$E(\Gamma_{1c})$	3.7	3.7	3.81		3.80[f]
$E(\Gamma_{7c})$	8.35	$E(\Gamma_{15v})$	9.0	8.3	9.22		
$E(\Gamma_{8c})$	8.54						
$E(X_{6v})$	−11.04	$E(X_{1v})$			−10.80	−11.88	−12.0[e]
$E(X_{6v})$	−4.80	$E(X_{3v})$	−3.3		−3.95	−4.80	−5.5[e]
$E(X_{6v})$	−2.27						
$E(X_{7v})$	−2.25	$E(X_{5v})$	−1.1	−1.6	−1.57	−2.30	−2.5[e]
$E(X_{6c})$	4.45	$E(X_{1c})$	5.6	5.1	5.76		
$E(L_{6v})$	−11.54				−11.18	−12.17	
$E(L_{6v})$	−5.22	$E(L_{1v})$	−3.5	−3.7	−4.25		
$E(L_{6v})$	−1.02						
$E(L_{4,5v})$	−0.95	$E(L_{3v})$	−0.3	−0.5	−0.56	−0.94	−1.4[e]
$E(L_{6c})$	4.90	$E(L_{1c})$	5.4	5.2	5.64		
$m_c(\Gamma_{6c})$	0.280				0.262		0.28[g]
m_{hh} [100]	0.546				1.036		1.76[g]
m_{hh} [111]	1.252				2.521		
m_{lh} [100]	0.258				0.302		0.23[g]
m_{lh} [111]	0.204				0.220		
m_{soh}	0.366						

[a]Stukel *et al.* (1969).
[b]Cohen and Cheleikowsky (1988).
[c]Huang and Ching (1985).
[d]Bernard and Zunger (1987).
[e]Ley *et al.* (1974).
[f]Theis (1977).
[g]Lawaetz (1971).

experimental results to form the input for the band calculation of ZnS. The fundamental gap is taken to be a direct gap of 3.8 eV (Theis, 1977). Several band structures of the zinc blende ZnS (β-ZnS) have been calculated with empirical adjustments. These include the othogonalized-plane-wave calculation (EM-OPW) (Stukel *et al.*, 1969), the nonlocal empirical pseudopotential calculation (EMP) (Cohen and Chelikowsky, 1988), and the LCAO calculation by Huang and Ching (1985). These band structures tend to give optical transition energies close to the experimental values, such as the 5.81 eV for $L_{3v} \rightarrow L_{1c}$, the 6.6 eV for $X_{3v} \rightarrow X_{1c}$ (Theis, 1977), and the 8.35 eV for $\Gamma_{15v} \rightarrow \Gamma_{15c}$ (Baars, 1967). However, the energies of the valence-band states such as X_{5v}, X_{5v}, and L_{3v}, as shown in Table 7.52, are too shallow compared to the experimental values, and a self-consistent LDA calculation (Bernard and Zunger, 1987). Because of the presence of the d bands, the x-ray photoelectric experiments (Ley *et al.*, 1974) contained large uncertainties. LDA calculations generally give accurate valence bands. The present band structure is constructed so that it has the optical transition energies in accord with experimental values and the important valence-band energies, such as X_{7v}, $L_{4,5v}$ and the X_{6v} (X_{3v}), close to the LDA calculation. The calculated band structure is plotted in Fig. 7.31. The most distinctive feature is the wide gap and relatively narrower width for the major valence band as compared to those in the III–V compounds. The calculated band energies at Γ, X, and L and the effective masses are tabulated in Table 7.52 along with results from other calculations and experiments.

7.5.2. ZnSe

The present band structure is based mainly on the results of the nonlocal empirical pseudopotential calculation (EMP) (Cohen and Chelikowsky, 1989) and the LCAO calculation by Huang and Ching (1985). Small adjustments are made to incorporate experimental information and other estimated results, which include the direct band gap (2.82 eV) (Theis, 1977) and the following optical transition energies: 4.91 eV for $L_{3v} \rightarrow L_{1c}$ (Theis, 1977), 7.80 eV for $\Gamma_{15v} \rightarrow \Gamma_{15c}$ (Walter *et al.*, 1970), and 6.0 eV for $X_{3v} \rightarrow X_{1c}$ (Petroff *et al.*, 1969). The calculated band structure is plotted in Fig. 7.32. It also has a wide direct gap. Table 7.53 lists the band energies at Γ, X, and L and the effective masses for the present band structure and those from the experiments and other calculations.

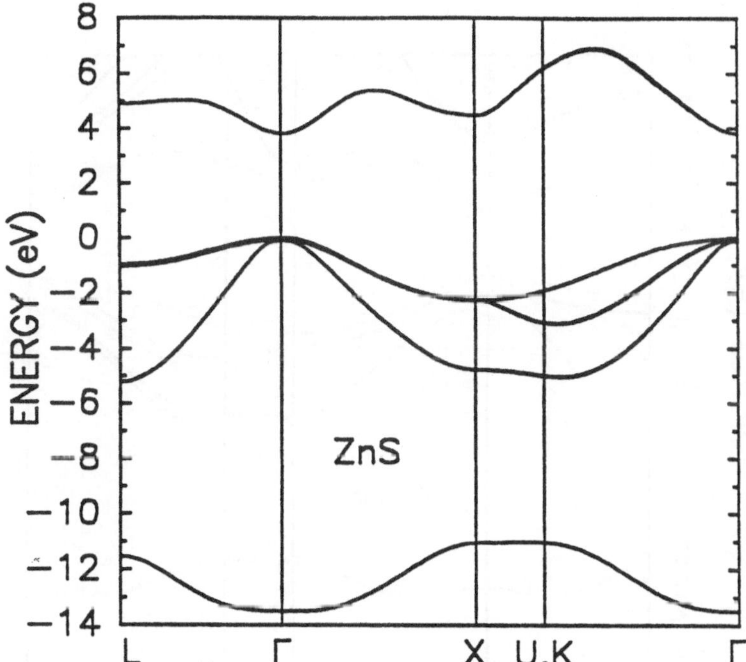

FIGURE 7.31. Calculated band structure energy in eV versus crystal wave vector of ZnS along several symmetry directions.

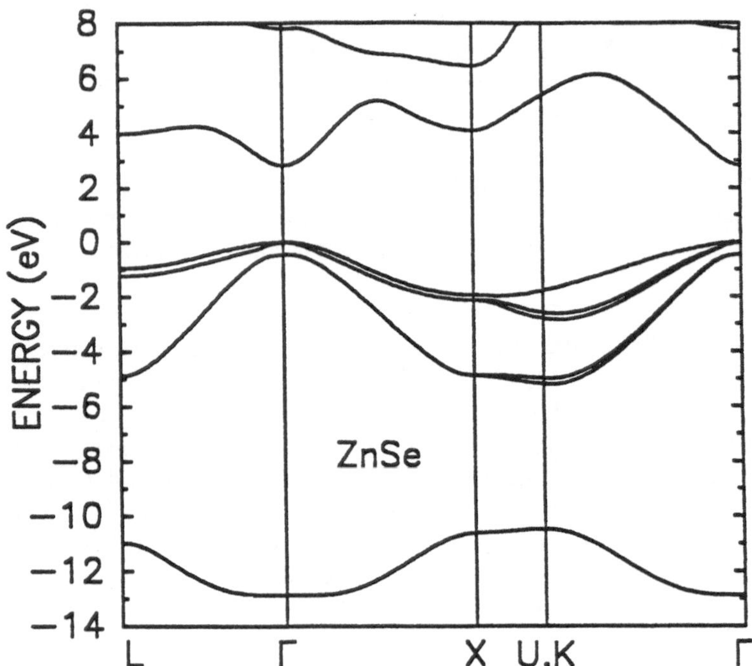

FIGURE 7.32. Calculated band structure energy in eV versus crystal wave vector of ZnSe along several symmetry directions.

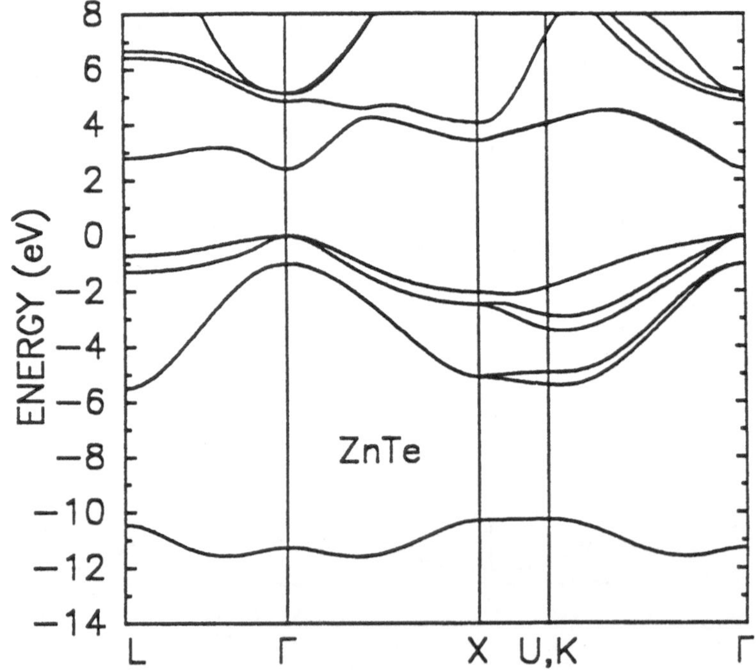

FIGURE 7.33. Calculated band structure energy in eV versus crystal wave vector of ZnTe along several symmetry directions.

TABLE 7.53. Comparison of Calculated Band Structure of ZnSe from the Present HPT Method with Experiments and Other Calculations

	Theory					Experiments	
	HPT	EMP[a]		LCAO[b]	LDA[c]	ARP[d]	
$E(\Gamma_{6v})$	−12.86	−12.25	$E(\Gamma_{1v})$	−12.27	−12.86		
$E(\Gamma_{7v})$	−0.445	−0.45				−0.6	−0.4[j]
$E(\Gamma_{8v})$	0	0	$E(\Gamma_{15v})$	0			
$E(\Gamma_{6c})$	2.82	2.76	$E(\Gamma_{1c})$	2.83			2.80[e]
$E(\Gamma_{7c})$	7.80	7.33	$E(\Gamma_{15v})$	7.38			
$E(\Gamma_{8c})$	8.06	7.42					
$E(X_{6v})$	−10.61	−10.72	$E(X_{1v})$	−10.94	−11.79		
$E(X_{6v})$	−4.90	−4.96	$E(X_{3v})$	−4.38	−4.82	−4.8	−5.3[f]
$E(X_{6v})$	−2.15	−2.17				−2.4	
$E(X_{7v})$	−1.96	−1.96	$E(X_{5v})$	−1.79	−2.2	−2.3	−2.1[g]
$E(X_{6c})$	4.07	4.54	$E(X_{1c})$	4.42			
$E(X_{7c})$	6.43	5.17	$E(X_{3c})$	6.10			
$E(L_{6v})$	−10.97	−11.08		−11.31	−12.06		
$E(L_{6v})$	−4.88	−5.08	$E(L_{1v})$	−4.67	−5.38	−5.0	
$E(L_{6v})$	−1.22	−1.04				−1.2	
$E(L_{4,5v})$	−0.94	−0.76	$E(L_{3v})$	−0.66	−0.87	−0.70	−1.3[g]
$E(L_{6c})$	4.00	3.96	$E(L_{1c})$	4.23			
$E(L_{6c})$	8.97	7.65	$E(L_{3c})$	8.72			
$E(L_{4,5c})$	9.16						
m_c (Γ_{6c})	0.170			0.173			0.17[h]
m_{hh} [100]	0.628			0.784			1.44[i]
m_{hh} [111]	1.252			1.810			
m_{lh} [100]	0.266			0.191			0.149[i]
m_{lh} [111]	0.220			0.147			
m_{soh}	0.425						

[a]Cohen and Cheleikowsky (1988).
[b]Huang and Ching (1985).
[c]Bernard and Zunger (1987).
[d]ARP = angle-resolved photoemission experiment. Qu *et al.* (1991).
[e]Cardona and Greenway (1963).
[f]Eastman *et al.* (1974).
[g]Ley *et al.* (1974).
[h]Marple (1964).
[i]Lawaetz (1971).
[j]Landolt–Bornstein (1988).

TABLE 7.54. Comparison of Calculated Band Structure of ZnTe from the Present Calculation (HPT) with Experiments and Other Calculations

	Theory					
	HPT		LCAO[c]	LDAd		Experiments
$E(\Gamma_{6v})$	−11.26	$E(\Gamma_{1v})$	−11.20	−11.16		−13.0[e]
$E(\Gamma_{7v})$	−1.004					−0.91[g] (300 K) 1.0[b]
$E(\Gamma_{8v})$	0	$E(\Gamma_{15v})$	0	0		2.39[a] 2.38[b]
$E(\Gamma_{6c})$	2.39	$E(\Gamma_{1c})$	2.39			
$E(\Gamma_{7c})$	4.82	$E(\Gamma_{15v})$	6.02			
$E(\Gamma_{8c})$	5.11					
$E(X_{6v})$	−10.28	$E(X_{1v})$	−9.81	−9.97		−11.6[e]
$E(X_{6v})$	−5.10	$E(X_{3v})$	−4.29	−5.14		−5.5[e]
$E(X_{6v})$	−2.49					
$E(X_{7v})$	−2.05	$E(X_{5v})$	−1.92	−2.27		−2.4[e]
$E(X_{6c})$	3.40	$E(X_{1c})$	3.01			−2.4[e]
$E(X_{7c})$	4.04	$E(X_{3c})$	5.07			
$E(L_{6v})$	−10.45	$E(L_{1v})$	−10.16	−10.25		
$E(L_{6v})$	−5.50	$E(L_{1v})$	−4.70			
$E(L_{6v})$	−1.28					
$E(L_{4,5v})$	−0.70	$E(L_{3v})$	−0.69	−0.94		−1.1[e]
$E(L_{6c})$	2.81	$E(L_{1c})$	3.39			
$E(L_{6c})$	6.41	$E(L_{3c})$	5.92			
$E(L_{4,5c})$	6.67					
$m_c\ (\Gamma_{6c})$	0.160		0.151			0.16[b] 0.13[g]
m_{hh} [100]	0.390		0.541		m_{hh}	1.27[f] 0.6[g]
m_{hh} [111]	1.098		1.354			
m_{lh} [100]	0.219		0.155		m_{lh}	0.154[f]
m_{lh} [111]	0.160		0.119		m_{lh}	0.154[f]
m_{soh}	0.388					

[a]Segall and Marple (1967).
[b]Cardona and Greenway (1963).
[c]Huang and Ching (1985).
[d]Bernard and Zunger (1987).
[e]Ley *et al.* (1974).
[f]Lawaetz (1971).
[g]Landolt-Bornstein (1988).

7.5.3. ZnTe

Several pieces of experimental band data are considered in the parametrization of the band structure of ZnTe. These are the 2.39-eV direct gap, the $m_c = 0.16$ mass ratio of the conduction minimum, the 1.0 eV for the spin-orbit splitting of the top of the valence band, and the following optical transition energies (Cardona and Greenway, 1963): 3.45 eV for $L_{3v} \rightarrow L_{1c}$, 4.82 eV for $\Gamma_{15v} \rightarrow \Gamma_{15c}$, and 5.45 eV for $X_{3v} \rightarrow X_{1c}$. For the valence bands, the results from the LCAO and LDA mixed-basis calculations are used, because the band energies deduced from x-ray photoemission data seem to be consistently deeper for the II–VI

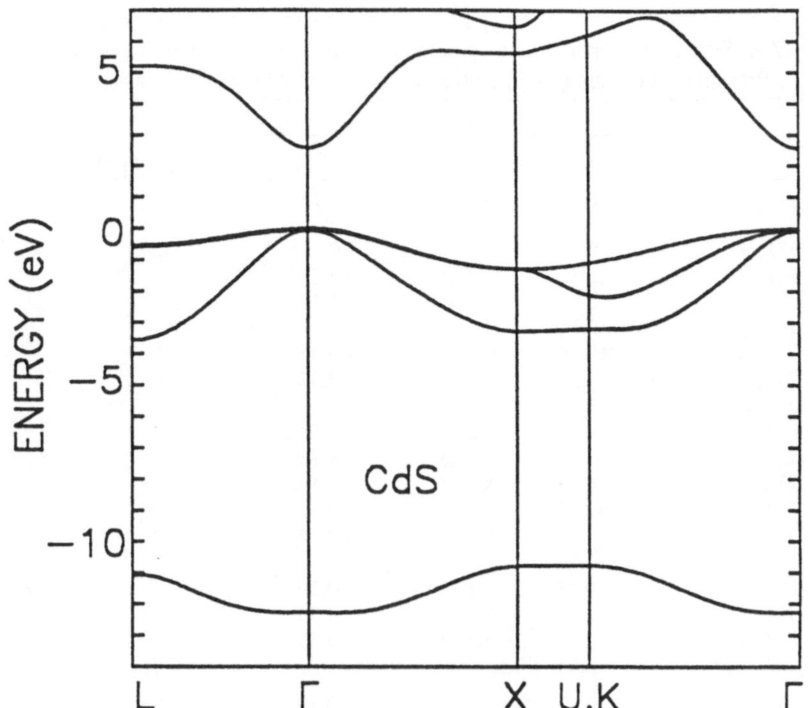

FIGURE 7.34. Calculated band structure energy in eV versus crystal wave vector of CdS along several symmetry directions.

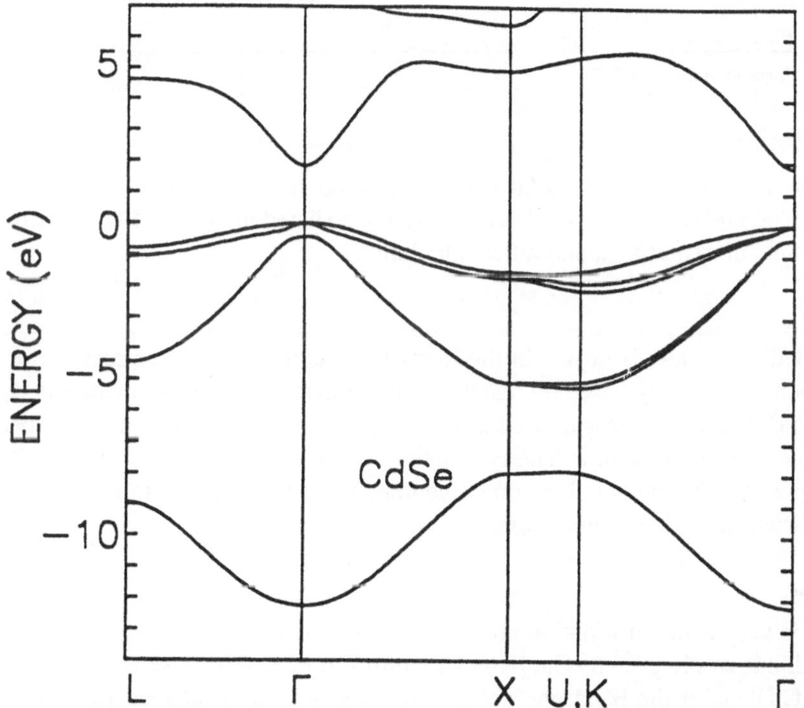

FIGURE 7.35. Calculated band structure energy in eV versus crystal wave vector of CdSe along several symmetry directions.

TABLE 7.55. Comparison of Calculated Band Structure of CdS from the Present Calculation (HPT) with Experiments and Other Calculations

	Theory			
	HPT		LCAO	Experiments
$E(\Gamma_{6v})$	−12.27	$E(\Gamma_{1v})$	−11.73	
$E(\Gamma_{7v})$	−0.075			
$E(\Gamma_{8v})$	0	$E(\Gamma_{15v})$	0	2.55[a]
$E(\Gamma_{6c})$	2.56	$E(\Gamma_{1c})$	2.51	
$E(\Gamma_{7c})$	8.20	$E(\Gamma_{15v})$	8.28	
$E(\Gamma_{8c})$	8.34			
$E(X_{6v})$	−10.78	$E(X_{1v})$	−10.46	
$E(X_{6v})$	−3.30	$E(X_{3v})$	−3.67	
$E(X_{6v})$	−1.31			
$E(X_{7v})$	−1.30	$E(X_{5v})$	−1.58	
$E(X_{6c})$	5.60	$E(X_{1c})$	4.72	
$E(X_{7c})$	6.47	$E(X_{3c})$	6.16	
$E(L_{6v})$	−11.05		−10.77	
$E(L_{6v})$	−3.57	$E(L_{1v})$	−4.02	
$E(L_{6v})$	−0.58			
$E(L_{4,5v})$	−0.53	$E(L_{3v})$	−0.59	
$E(L_{6c})$	5.20	$E(L_{1c})$	4.30	
$E(L_{6c})$	8.71	$E(L_{3c})$	9.30	
$E(L_{4,5c})$	8.80			
$m_c\ (\Gamma_{6c})$	0.173		0.209	0.14[a]
m_{hh} [100]	1.029		0.929	
m_{hh} [111]	2.321		1.935	
m_{lh} [100]	0.333		0.201	
m_{lh} [111]	0.283		0.161	
m_{soh}	0.524			

[a]Landolt–Bornstein (1988).

compounds. The calculated band structure is plotted in Fig. 7.33. The calculated band energies at Γ, X, and L and the effective masses are also listed in Table 7.54 and are compared with the results of experiment and other calculations.

7.5.4. *CdS*

Since CdS normally is grown in the wurtzite structure, the band structure for its zinc blende structure is mainly based on results obtained from the LCAO calculation by Huang and Ching (1985). The band gap used is 2.56 eV, which is inferred from the experimental value for the wurtzite structure. The calculated band structure is plotted in Fig. 7.34. The band energies at Γ, X, and L and the effective masses are tabulated in Table 7.55 along with the results from the LCAO calculation.

7.5.5. *CdSe*

There is very little information about the band structure of β-CdSe, because the cubic form of CdSe is rarely grown. The band structure shown in Fig. 7.35 is extrapolated from ZnSe and CdTe with the band gap set to 1.827 eV, which is inferred from the wurtzite structure. Table 7.56 lists the calculated band energies at Γ, X, and L and the effective masses.

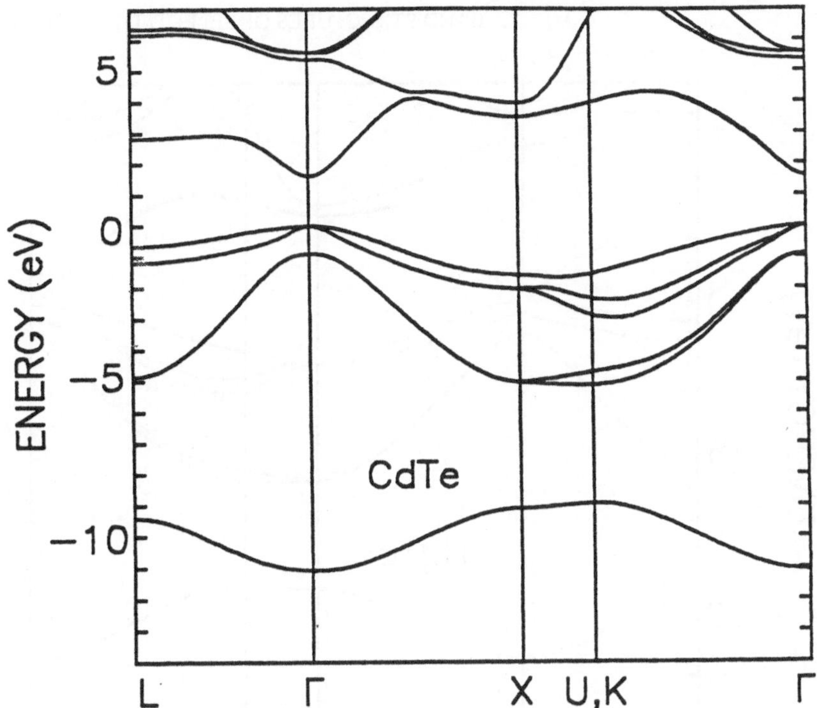

FIGURE 7.36. Calculated band structure energy in eV versus crystal wave vector of CdTe along several symmetry directions.

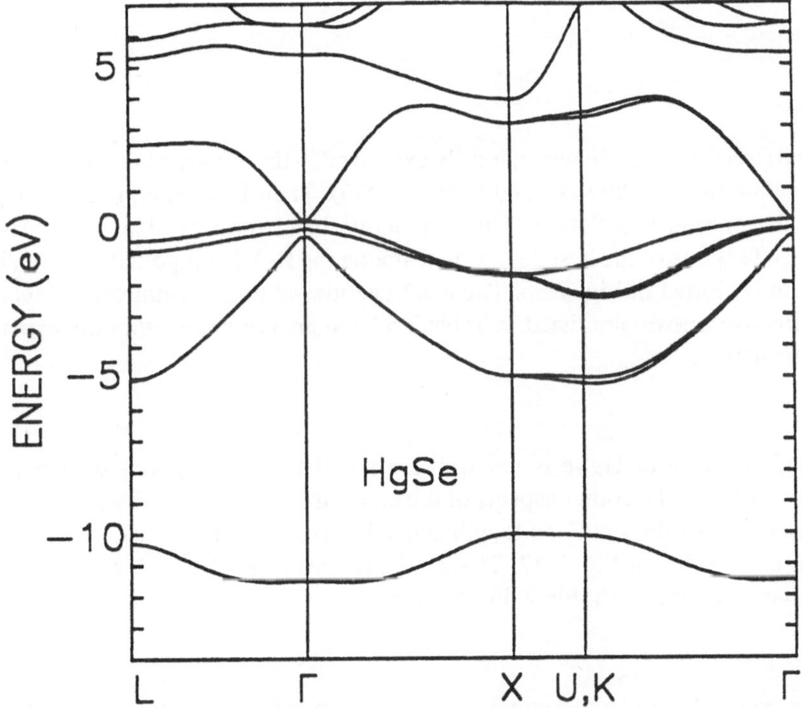

FIGURE 7.37. Calculated band structure energy in eV versus crystal wave vector of HsSe along several symmetry directions.

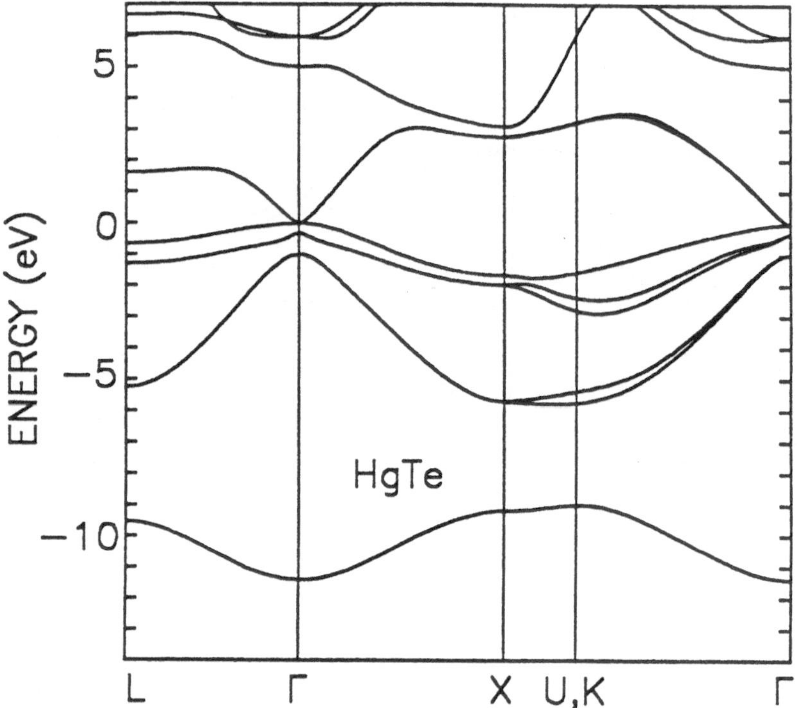

FIGURE 7.38. Calculated band structure energy in eV versus crystal wave vector of HgTe along several symmetry directions.

7.5.6. *CdTe*

The band structure of CdTe was carefully examined in the empirical nonlocal pseudopotential calculation by Chelikowsky and Cohen (1976). Their band energies are adopted as an input in the present calculation. The conduction-band mass is also fitted. The band structure of CdTe is one of the best determined among the II–VI compounds. The calculated band structure is plotted in Fig. 7.36. The band energies at three symmetry points and the important effective masses are listed in Table 7.57 and are compared with the experiments and other calculations.

7.5.7. *HgSe*

The band structure of HgSe is not well known. The band gap was deduced to be a negative gap −0.20 eV. The other aspects of the band structures are borrowed from those of HgTe modified by its different bond length and spin-orbit coupling constant. The resulting band structure is plotted in Fig. 7.37. The band energies at Γ, X, and L and the important effective masses are given in Table 7.58.

7.5.8. *HgTe*

There are several useful experimental data sets for parametrization of the band structure of HgTe. These include the negative gap $E(\Gamma_{6c}) - E(\Gamma_{8v}) = -0.304$ eV, the 0.98-eV spin-orbit splitting of the top of the valence band, the energy spacing $E(X_{6v}) - E(\Gamma_{8v}) = -5.7$ eV, and the 0.031 mass ratio to the free-electron value for the conduction-band minimum (at Γ_{8v}). In addition, the following optical transition energies deduced by Cohen and Chelikowsky (1989) are useful: 5.0 eV for $\Gamma_{8v} \to \Gamma_{7c}$, 2.25 eV for $L_{4,5v} \to L_{6c}$, 2.87 eV for $L_{7v} \to L_{6c}$, and 4.71 eV for $X_{6v} \to X_{6c}$. These results have been largely included in the parametrized band structure of HgTe shown in Fig. 7.38, with the band energies and important effective masses listed in Table 7.59.

TABLE 7.56. Comparison of Band Structure of CdSe from Present HPT Calculation with Experiments

	Theory HPT	Experiments
$E(\Gamma_{6v})$	−12.25	
$E(\Gamma_{7v})$	−0.429	−0.432[a]
$E(\Gamma_{8v})$	0	
$E(\Gamma_{6c})$	1.827	1.827[a], 1.9[b]
$E(\Gamma_{7c})$	7.50	
$E(\Gamma_{8c})$	7.68	
$E(X_{6v})$	−7.97	
$E(X_{6v})$	−5.10	
$E(X_{6v})$	−1.75	
$E(X_{7v})$	−1.56	
$E(X_{6c})$	4.90	
$E(X_{7c})$	6.42	
$E(L_{6v})$	−8.96	
$E(L_{6v})$	−4.46	
$E(L_{6v})$	−1.06	
$E(L_{4,5v})$	−0.80	
$E(L_{6c})$	4.60	
$E(L_{6c})$	7.02	
$E(L_{4,5c})$	7.16	
$m_c(\Gamma_{6c})$	0.0805	0.11[b]
m_{hh} [100]	0.831	
m_{hh} [111]	1.644	0.44[b]
m_{lh} [100]	0.120	
m_{lh} [111]	0.112	
m_{soh}	0.251	

[a]Langer *et al.* (1970) (for wurtzite structure).
[b]Landolt–Bornstein (1988).

TABLE 7.57. Comparison of Calculated Band Structure of CdTe from the Present HPT Method with Other Calculations and Experimental Data

	Theory						
	HPT	EPM[a]		LCAO[b]	ARP[c]	Experiments	
$E(\Gamma_{6v})$	−11.07	11.07	$E(\Gamma_{1v})$	−10.52			
$E(\Gamma_{7v})$	−0.89	−0.89			−0.9	−0.89[d] (293 K)	
$E(\Gamma_{8v})$	0	0	$E(\Gamma_{15v})$	0			
$E(\Gamma_{6c})$	1.60	1.59	$E(\Gamma_{1c})$	1.60		1.606[d] (2 K)	
$E(\Gamma_{7c})$	5.36	5.36	$E(\Gamma_{15v})$	4.76			
$E(\Gamma_{8c})$	5.59	5.61					
$E(X_{6v})$	−9.11	−9.12	$E(X_{1v})$	−9.13		−8.8 ± 0.3[e]	
$E(X_{6v})$	−5.05	−5.05	$E(X_{3v})$	−4.32	−4.6	−4.7 ± 0.2[e]	−5.1 ± 0.2[f]
$E(X_{6v})$	−2.05	−1.98			−2.3		
$E(X_{7v})$	−1.60	−1.60	$E(X_{5v})$	−2.05	−2.0		−1.8 ± 0.2[f]
$E(X_{6c})$	3.48	3.48	$E(X_{1c})$	2.88			
$E(X_{7c})$	3.94	3.95	$E(X_{3c})$	3.93			
$E(L_{6v})$	−9.41	−9.64		−9.46			
$E(L_{6v})$	−4.86	−4.73	$E(L_{1v})$	−4.61	−4.6		
$E(L_{6v})$	−1.18	−1.18			−1.3		
$E(L_{4,5v})$	−0.65	−0.65	$E(L_{3v})$	−0.80	−0.6	−0.7 ± 0.2[e]	−0.9 ± 0.3[f]
$E(L_{6c})$	2.82	2.82	$E(L_{1c})$	2.36			
$E(L_{6c})$	6.18	6.18	$E(L_{3c})$	5.92			
$E(L_{4,5c})$	6.37	6.35					
m_c (Γ_{6c})	0.096			0.127		0.096[g]	0.09[d]
m_{hh} [100]	0.566			0.478		0.62[h]	0.72[d]
m_{hh} [111]	1.343			1.114		0.98[h]	
m_{lh} [100]	0.128			0.125		0.092[h]	0.12[d]
m_{lh} [111]	0.113			0.095			
m_{soh}	0.479						

[a] Chelikowsky and Cohen (1976).
[b] Huang and Ching (1985).
[c] ARP = angle-resolved photoemission experiment. Qu *et al.* (1991).
[d] Landolt–Bornstein (1988).
[e] Ley *et al.* (1974).
[f] Eastman *et al.* (1974).
[g] Kanazawa and Brown (1964).
[h] Dang (1982).

TABLE 7.58. Calculated Band Structure for Cubic HgSe from Present Calculation (HPT) and Comparison with Experiments

	HPT	Experiment[a]
$E(\Gamma_{6v})$	−11.50	
$E(\Gamma_{7v})$	−0.464	−0.396 (10 K)
$E(\Gamma_{8v})$	0	
$E(\Gamma_{6c})$	−0.220	−0.205 (80 K)
$E(\Gamma_{7c})$	5.37	
$E(\Gamma_{8c})$	6.33	
$E(X_{6v})$	−10.01	
$E(X_{6v})$	−4.98	
$E(X_{6v})$	−1.78	
$E(X_{7v})$	−1.70	
$E(X_{6c})$	3.10	
$E(X_{7c})$	3.89	
$E(L_{6v})$	−10.28	
$E(L_{6v})$	−5.05	
$E(L_{6v})$	−0.94	
$E(L_{4,5v})$	−0.60	
$E(L_{6c})$	2.50	
$E(L_{6c})$	5.31	
$E(L_{4,5c})$	5.90	
$m_c\ (\Gamma_{8v})$	0.018	
m_{hh} [100]	0.774	0.78
m_{hh} [111]	2.142	
m_{lh} [100]	0.0359	
m_{lh} [111]	0.0359	
m_{soh}	0.0385	

[a]Landhold–Bornstein (1988)

TABLE 7.59. Comparison of Calculated Band Structure of HgTe from Present Calculation (HPT) with Experiments and Other Estimates

	HPT	Experiments and estimate
$E(\Gamma_{6v})$	−11.40	
$E(\Gamma_{7v})$	−0.977	−0.98[b]
$E(\Gamma_{8v})$	0	
$E(\Gamma_{6c})$	−0.304	−0.304[a] −0.3025[e] (4.4 K)
$E(\Gamma_{7c})$	5.00	
$E(\Gamma_{8c})$	5.96	
$E(X_{6v})$	−9.54	
$E(X_{6v})$	−5.70	−5.7[c]
$E(X_{6v})$	−1.98	−2.5[c]
$E(X_{7v})$	−1.68	
$E(X_{6c})$	2.74	
$E(X_{7c})$	3.08	
$E(L_{6v})$	−9.54	
$E(L_{6v})$	−5.25	
$E(L_{6v})$	−1.29	−1.2[c]
$E(L_{4,5v})$	−0.65	
$E(L_{6c})$	1.61	
$E(L_{6c})$	6.05	
$E(L_{4,5c})$	6.67	
$m_c\ (\Gamma_{8v})$	0.03	0.031[d]
m_{hh} [100]	0.545	0.378[d]
m_{hh} [111]	1.401	
m_{lh} [100]	0.040	0.028[f]
m_{lh} [111]	0.040	0.026[f]
m_{soh}	0.119	

[a]Weiler (1981).
[b]Chen and Sher (1982).
[c]Ley *et al.* (1974).
[d]Uchida (1976).
[e]Landolt–Bornstein (1988).
[f]Guldner *et al.* (1977a,b; 1983).

TABLE 7.60. Coefficients α, β, and γ (all in eV) of Eq. (7.4.1) for Several Important Band Energies of $ZnS_{1-x}Se_x$

States	α	β	γ	Temperature (K)	Source
$E(\Gamma_{6c})$	3.800	−1.602	0.622	0	Present
$E(\Gamma_{6c})$	0.703	−1.612	0.630	295	Ebina *et al.* (1974)
$E(\Gamma_{6c})$	3.822	−1.421	0.41	77	Suslina *et al.* (1977)
$E(X_{6c})$	4.450	−0.755	0.375	0	Present
$E(L_{6c})$	4.900	−1.183	0.283	0	Present
$E(\Gamma_{7v})$	−0.097	−0.404	0.055	0	Present

7.6. II–VI ZINC BLENDE PSEUDOBINARY ALLOYS

The band structures and the Hamiltonians of the II–VI constituent compounds from the preceding section are next applied to calculations of the zinc blende pseudobinary alloys using the HPT model of Section 7.3. Below we summarize the most important band features.

7.6.1. $ZnS_{1-x}Se_x$

Table 7.60 lists the coefficients α, β, and γ of Eq. (7.4.1) for the fundamental gap and other three band energies relative to the top of the valence band Γ_{8v} deduced from the present calculation. Also tabulated are the results for the gap deduced from two experiments. The calculated bowing parameter (γ or b) is comparable to one of the experimental values but is larger than the other. The experimental values for the bowing parameter of the fundamental gap of this alloy vary between 0.4 and 0.6 eV. In addition to the two values in Table 7.60, experimental values for γ include 0.6 eV (Morehead 1963), 0.63 eV (Soockindt *et al.*, 1979), 0.43 eV (Mach *et al.*, 1982) and 0.456 eV (El-Shazly *et al.*, 1985). This variation may be caused by experimental uncertainties in alloy compositions and by structural induced variations in the band gaps. The calculated values of b in Table 7.61 indicate that the quadratic approximation to the concentration variation of the band gap is only an approximation, because b decreases with x by more than 10%. The disorder contribution to the bowing parameter is small, $b_{ex} = 0.04$ eV, despite the scattering parameters being moderate (see Table 7.62). The actual concentration dependence of the gap is plotted in Fig. 7.39. The calculated value for valence-band offset is found to be well approximated by $\Delta E_v = -1.870 + 1.034x - 0.214x^2$ relative to GaAs.

7.6.2. $ZnSe_{1-x}Te_x$

Table 7.63 lists the coefficients α, β, and γ of Eq. (7.4.1) for the fundamental gap and other three band-edge states deduced from the present calculation. Experimental results are also listed. The three bowing parameters listed for the gap are in good agreement. However,

TABLE 7.61. Calculated Values of b (in eV) in Eq. (7.4.2) at Several x Values for the Fundamental Gap of $ZnS_{1-x}Se_x$, and the Total Disorder Contribution b_{ex}

x	0.1	0.25	0.5	0.7	0.9	Average	b_{ex}
$b(\Gamma)$	0.658	0.648	0.620	0.598	0.586	0.622	0.0

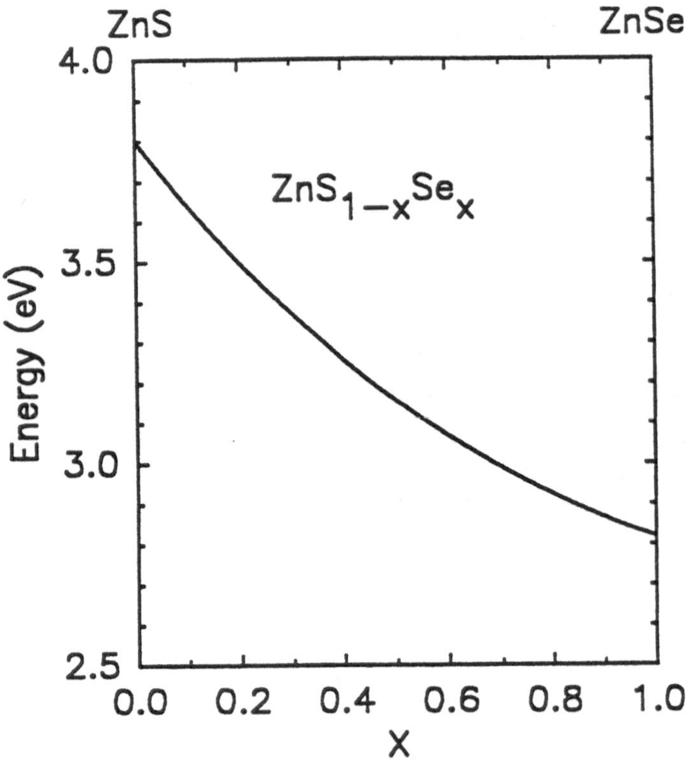

FIGURE 7.39. Band-gap energy in eV of $ZnS_{1-x}Se_x$ as a function of alloy composition x.

TABLE 7.62. Alloy Disorder Parameters Defined in Eq. (5.11.2) in the Γ-OLO Representations Centered at an Anion Site in $ZnS_{1-x}Se_x$

Representation	δ (eV)	Δ (eV)
Γ_6	−1.0228	0.0834
Γ_7	0.9406	−0.9607
Γ_8	0.5296	−0.9607

TABLE 7.63. Coefficients α, β, and γ (all in eV) of Eq. (7.4.1) for Several Important Band Energies of $ZnSe_{1-x}Te_x$

States	α	β	γ	Temperature (K)	Source
$E(\Gamma_{6c})$	2.820	−1.688	1.258	0	Present calculation
$E(\Gamma_{6c})$	2.713	−1.702	1.266	295	Ebina *et al.* (1974)
			1.28		Larach *et al.* (1957)
$E(X_{6c})$	4.070	−2.427	1.757	0	Present
$E(L_{6c})$	4.000	−2.968	1.778	0	Present
$E(\Gamma_{7v})$	−0.446	−0.756	0.198	0	Present

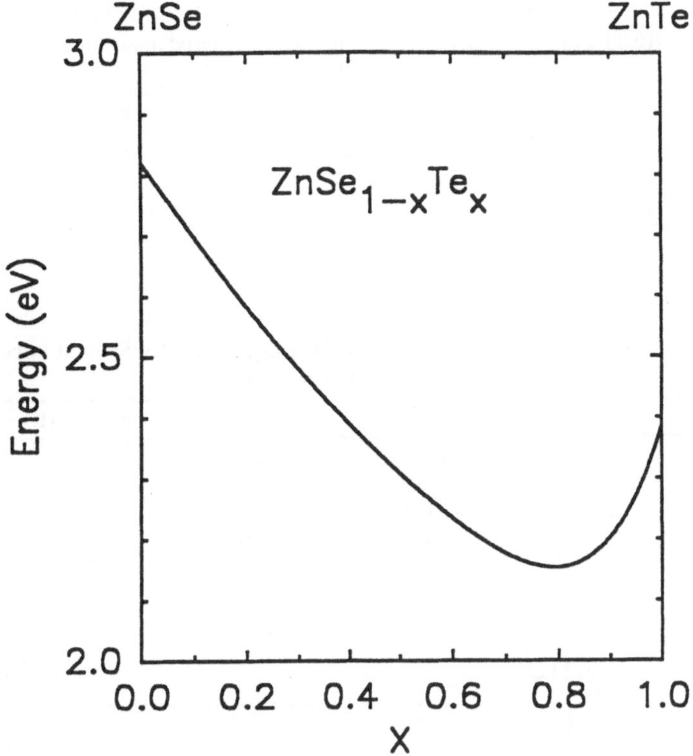

FIGURE 7.40. Band-gap energy in eV of $ZnSe_{1-x}Te_x$ as a function of alloy composition x.

the plot of band gap in Fig. 7.40 and the values of b in Eq. (7.4.2) shown in Table 7.64 clearly indicate that the quadratic form is a poor approximation for the gap. This is due to large alloy disorder, as indicated by the large fluctuations in both the diagonal and off-diagonal potential matrix elements in the Γ_6 representation centered at the anion sites (see Table 7.65). The total disorder contribution to the bowing parameter b_{ex} for this alloy is larger than the VCA value. The value of b_{ex} shown in Table 7.64 is the value calculated at $x = 0.5$. The calculated values for the valence-band offset relative to GaAs are well approximated by $\Delta E_v = -1.058 + 0.649x - 0.021x^2$. Although the calculated alloy band structures show an upward bowing for the spin-orbit splitting $\Delta_o = E(\Gamma_{8v}) - \Delta_o = E(\Gamma_{7v})$, the bowing parameter from the calculation is not as large as that deduced from a reflectance experiment (Ebina *et al.*, 1974).

7.6.3. $ZnS_{1-x}Te_x$

Table 7.66 lists the coefficients α, β, and γ of Eq. (7.4.1) for the fundamental gap and other three band-edge states deduced from the present calculation. The average bowing

TABLE 7.64. Calculated Values of b (in eV) in Eq. (7.4.2) at Several x Values for the Fundamental Gap of $ZnSe_{1-x}Te_x$, and the Total Disorder Contribution b_{ex}

x	0.1	0.25	0.5	0.7	0.9	Average	b_{ex}
$b(\Gamma)$	0.907	0.966	1.083	1.650	2.500	1.258	0.8

TABLE 7.65. Alloy Disorder Parameters Defined in Eq. (5.11.2) in the Γ-OLO Representations Centered at an Anion Site in $ZnSe_{1-x}Te_x$

Representation	δ (eV)	Δ (eV)
Γ_6	−2.109	1.211
Γ_7	0.123	−0.267
Γ_8	−0.474	−0.267

TABLE 7.66. Coefficients α, β, and γ (all in eV) of Eq. (7.4.1) for Several Important Band Energies of $ZnS_{1-x}Te_x$

States	α	β	γ	Temperature (K)	Source
$E(\Gamma_{6c})$	3.800	−4.279	2.869	0	Present
$E(X_{6c})$	4.450	−4.019	2.969	0	Present
$E(L_{6c})$	4.900	−5.171	3.081	0	Present
Δ_o	0.097	1.411	−0.504	0	Present

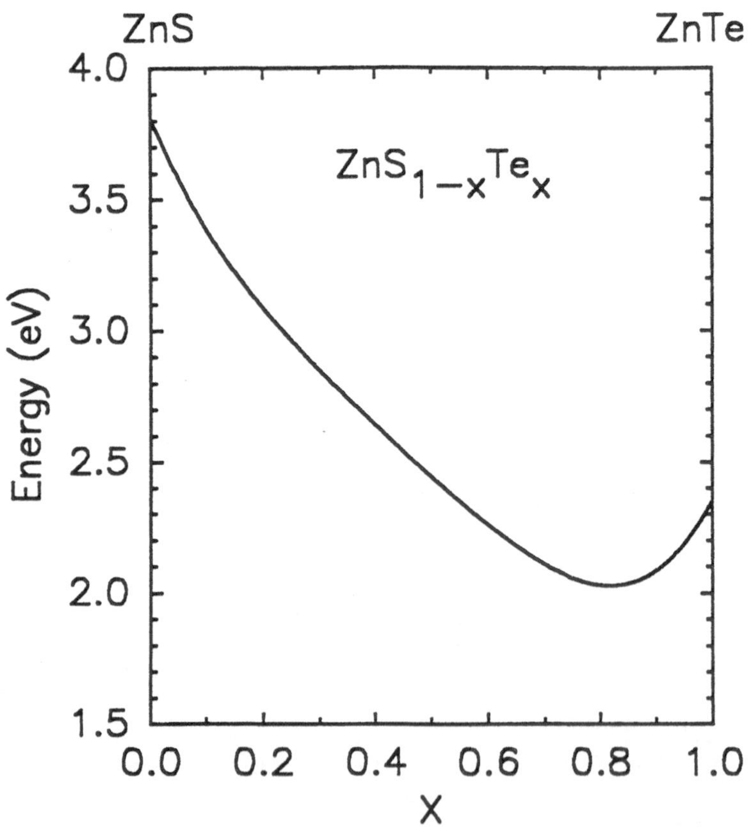

FIGURE 7.41. Band-gap energy in eV of $ZnS_{1-x}Te_x$ as a function of alloy composition x.

TABLE 7.67. Calculated Values of b (in eV) of Eq. (7.4.2) at Several x Values for the Fundamental Gap of $ZnS_{1-x}Te_x$, and the Total Disorder Contribution b_{ex}

x	0.1	0.25	0.5	0.7	0.9	Average	b_{ex}
$b(\Gamma)$	2.793	2.570	2.665	3.249	6.023	2.869	1.6

TABLE 7.68. Alloy Disorder Parameters Defined in Eq. (5.11.2) in the Γ-OLO Representations Centered at an Anion Site in $ZnS_{1-x}Te_x$

Representation	δ (eV)	Δ (eV)
Γ_6	−3.1315	1.3092
Γ_7	1.0634	−1.2182
Γ_8	0.0554	−1.2182

parameter (γ) for the calculated band gap lies between the two experimental values of 2.4 eV (Larach *et al.*, 1957) and 3.0 eV (Hill and Richardson, 1973). However, the plot of band gap in Fig. 7.41 and the values of b from Eq. (7.4.2) shown in Table 7.67 clearly indicate that the quadratic form is a poor approximation for the gap, especially around $x = 0.9$. This large deviation from the quadratic form of the gap is caused by the large alloy disorder, as indicated by the large fluctuations in both the diagonal and off-diagonal potential matrix elements in the Γ_6 representation centered at the anion sites (see Table 7.68). The total disorder contribution to the bowing parameter b_{ex} for this alloy is larger than the VCA value. The value of b_{ex} shown in Table 7.67 is the value calculated at $x = 0.5$. The calculated value for valence-band offset is well approximated by $\Delta E_v = -1.870 + 1.786x - 0.346x^2$ relative to GaAs. The calculated alloy band structures have an upward bowing for the spin-orbit splitting $\Delta_o = E(\Gamma_{8v}) - E(\Gamma_{7v})$.

7.6.4. $CdS_{1-x}Se_x$

Table 7.69 lists the coefficients α, β, and γ of Eq. (7.4.1) for the fundamental gap and the other three band-edge states deduced from the present calculation. The values of b in Eq. (7.4.2) shown in Table 7.70 indicate that the quadratic form is only a rough approximation to the concentration variation of the gap. Since the bowing is not as large as those of the three preceding alloys, the band gap as a function of x plotted in Fig. 7.42 does not vividly

TABLE 7.69. Coefficients α, β, and γ (all in eV) of Eq. (7.4.1) for Several Important Band Energies of $CdS_{1-x}Se_x$

States	α	β	γ	Temperature (K)	Source
$E(\Gamma_{6c})$	2.555	−1.123	0.395	0	Present
$E(X_{6c})$	5.600	−2.000	1.300	0	Present
$E(L_{6c})$	5.200	−2.241	1.841	0	Present
$E(\Gamma_{7v})$	−0.075	−0.178	−0.190	0	Present

TABLE 7.70. Calculated Values of b (in eV) in Eq. (7.4.2) at Several x Values for the Fundamental Gap of $CdS_{1-x}Se_x$, and the Total Disorder Contribution b_{ex}

x	0.1	0.25	0.5	0.7	0.9	Average	b_{ex}
$b(\Gamma)$	0.538	0.467	0.364	0.332	0.310	0.395	0.3

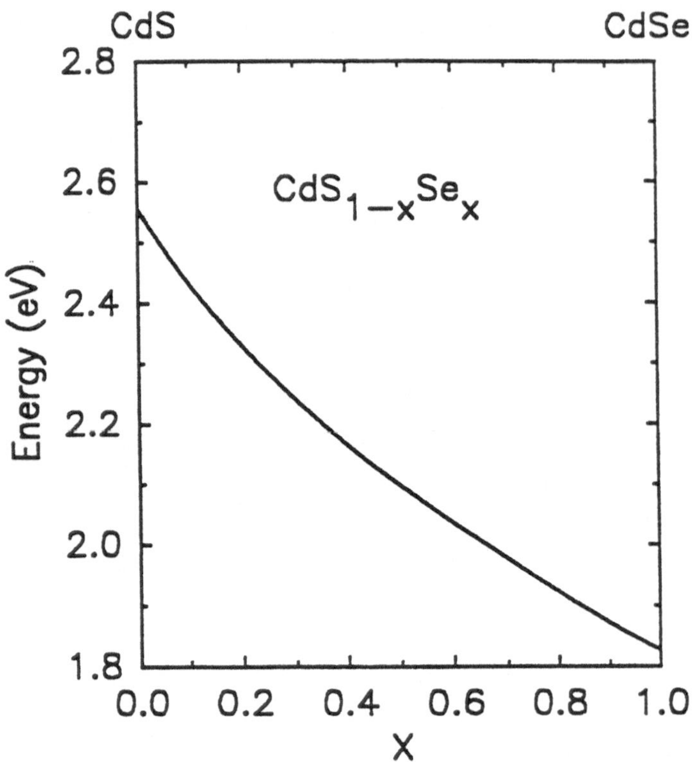

FIGURE 7.42. Band-gap energy in eV of $ZnS_{1-x}Se_x$ as a function of alloy composition x.

show this deviation. Due to moderately large disorder parameters in the Γ_6 representation (see Table 7.71), the bowing parameter is dominated by the disorder contribution, as can be inferred from the b_{ex} value in Table 7.70 calculated for $x = 0.5$. The top of the valence-band offset is found to be well approximated by $\Delta E_v = -1.390 + 0.460x - 0.050x^2$ relative to GaAs. These results are not too qualitatively different from those of the II–VI alloys so far discussed. In contrast, the experimental results of Kainthla *et al.* (1982) showed an upward bowing for the fundamental gap as a function of x, i.e., a negative value of γ (or b). However, the same experiment also showed an appreciable downward bowing for the bond length as a function of x, unlike the linear variation (Vegard's law) found for most semiconductor alloys and adopted in the calculation. Since the band gap scales as some inverse power of the lattice constant, a downward bowing in the lattice constant tends to give rise to an upward bowing

TABLE 7.71. Alloy Disorder Parameters Defined in Eq. (5.11.2) in the Γ-OLO Representations Centered at an Anion Site in $CdS_{1-x}Se_x$

Representation	δ (eV)	Δ (eV)
Γ_6	−0.7918	−1.8440
Γ_7	0.0738	0.3996
Γ_8	−0.3252	0.3996

TABLE 7.72. Coefficients α, β, and γ (all in eV) of Eq. (7.4.1) for Several Important Band Energies of $CdSe_{1-x}Te_x$

States	α	β	γ	Temperature (K)	Source
$E(\Gamma_{6c})$	1.827	–0.516	0.289	0	Present
$E(X_{6c})$	4.900	–1.760	0.340	0	Present
$E(L_{6c})$	4.800	–2.940	0.960	0	Present
$E(\Gamma_{7v})$	–0.443	–0.310	–0.138	0	Present

TABLE 7.73. Calculated Values of b (in eV) in Eq. (7.4.2) at Several x Values for the Fundamental Gap of $CdSe_{1-x}Te_x$, and the Total Disorder Contribution b_{ex}

x	0.1	0.25	0.5	0.7	0.9	Average	b_{ex}
$b(\Gamma)$	0.275	0.278	0.288	0.304	0.330	0.289	0.0

in the gap. Because these two experimental results are different from observations on all other alloy systems, it is crucial to further examine this issue experimentally. We note that both the wurtzite and zinc blende structures may coexist in a grown sample of $CdS_{1-x}Se_x$, for which the measured properties may not correspond to the present results, which assume a uniform zinc blende solid solution.

7.6.5. $CdSe_{1-x}Te_x$

Table 7.72 lists the coefficients α, β, and γ of Eq. (7.4.1) for the fundamental gap and the other three band-edge states deduced from the present calculation. The values of b in Eq. (7.4.2) shown in Table 7.73 indicate that the quadratic form is a reasonable approximation for the gap. The band gap as a function of x is plotted in Fig. 7.43. Table 7.74 shows that the disorder parameters are only moderate, so that the disorder contribution to the bowing parameter, b_{ex}, is small. The top of the valence-band offset is found to be well approximated by $\Delta E_v = -0.980 + 0.525x - 0.075x^2$ relative to GaAs. The calculated band gaps are substantially different from the experimental values of Prytkina *et al.* (1968) in many respects. First their measured gap for CdSe is smaller than that in CdTe, which is not consistent with other measured values (see Section 7.5). Second they found that the gap at 50% composition is lowest in the alloy. Again, because of the difficulty in growing a uniform zb alloy, it may not be proper to compare these measured band gaps with the present calculation, which assumes a uniform zb solid solution.

7.6.6. $CdS_{1-x}Te_x$

Table 7.75 lists the coefficients α, β, and γ of Eq. (7.4.1) for the fundamental gap and other three band-edge states deduced from the present calculation. The values of b in Eq. (7.4.2) shown in Table 7.76 indicate that the quadratic form is a reasonable approximation for the gap. The band gap as a function of x is plotted in Fig. 7.44. Table 7.77 shows that the disorder parameters are larger than those in CdSeTe, so that the disorder contribution to the bowing parameter, b_{ex}, is also larger. The valence-band offset is found to be well approximated by $\Delta E_v = -1.390 + 0.965x - 0.105x^2$ relative to GaAs.

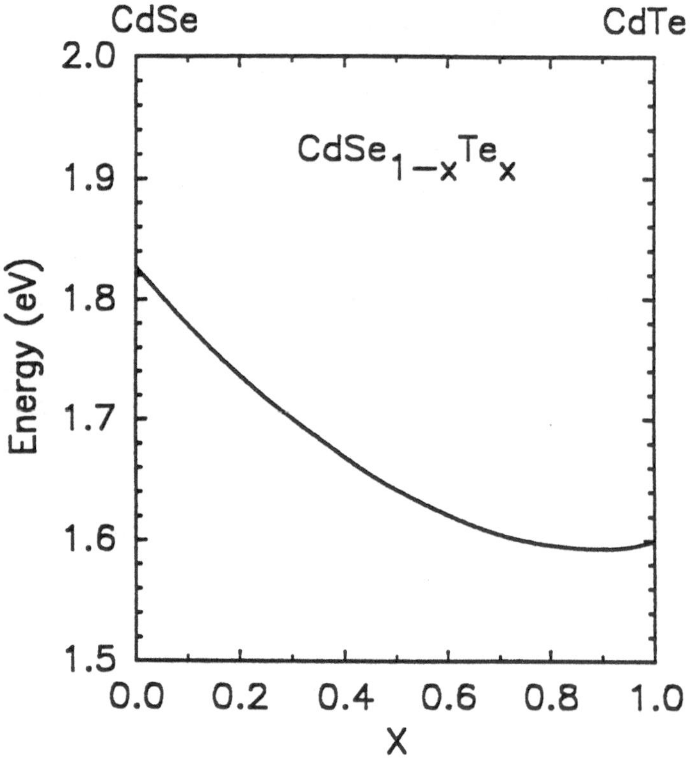

FIGURE 7.43. Band-gap energy in eV of $CdSe_{1-x}Te_x$ as a function of alloy composition x.

TABLE 7.74. Alloy Disorder Parameters Defined in Eq. (5.11.2) in the Γ-OLO Representations Centered at an Anion Site in $CdSe_{1-x}Te_x$

Representation	δ (eV)	Δ (eV)
Γ_6	–0.0514	0.7216
Γ_7	0.7229	–0.3774
Γ_8	0.2129	–0.3774

TABLE 7.75. Coefficients α, β, and γ (all in eV) of Eq. (7.4.1) for Several Important Band Energies of $CdS_{1-x}Te_x$

States	α	β	γ	Temperature (K)	Source
$E(\Gamma_{6c})$	2.555	–1.258	0.303	0	Present
$E(X_{6c})$	5.600	–2.851	0.731	0	Present
$E(L_{6c})$	5.200	–2.898	0.518	0	Present
$E(\Gamma_{7v})$	–0.075	–0.319	–0.497	0	Present γ

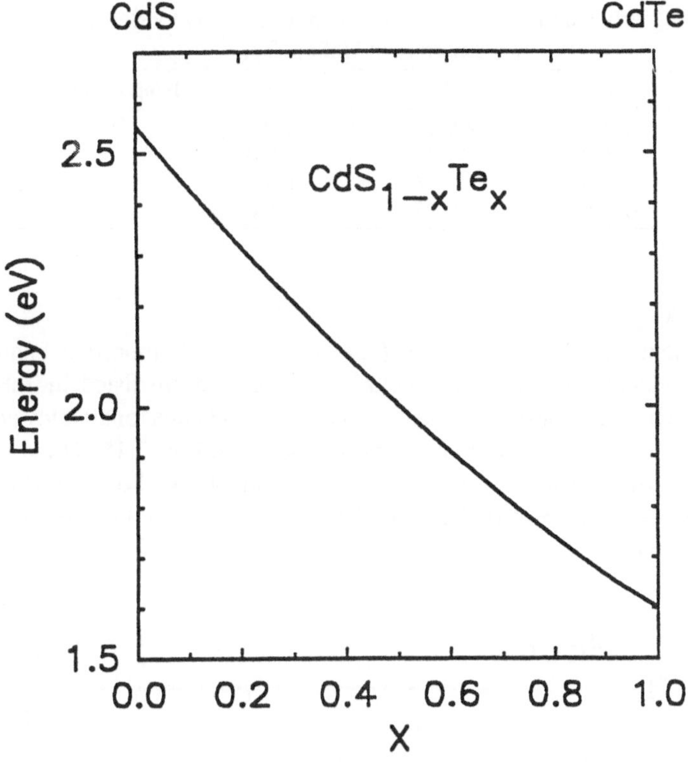

FIGURE 7.44. Band-gap energy in eV of $CdS_{1-x}Te_x$ as a function of alloy composition x.

TABLE 7.76. Calculated Values of b (in eV) in Eq. (7.4.2) at Several x Values for the Fundamental Gap of $CdS_{1-x}Te_x$, and the Total Disorder Contribution b_{ex}

x	0.1	0.25	0.5	0.7	0.9	Average	b_{ex}
$b(\Gamma)$	0.298	0.291	0.300	0.314	0.334	0.303	0.1

TABLE 7.77. Alloy Disorder Parameters Defined in Eq. (5.11.2) in the Γ-OLO Representations Centered at an Anion Site in $CdS_{1-x}Te_x$

Representation	δ (eV)	Δ (eV)
Γ_6	−0.8432	−1.0831
Γ_7	0.7967	0.0119
Γ_8	−0.1123	0.0119

TABLE 7.78. Coefficients α, β, and γ (all in eV) of Eq. (7.4.1) for Several Important Band Energies of $HgSe_{1-x}Te_x$

States	α	β	γ	Temperature (K)	Source
$E(\Gamma_{6c})$	−0.220	−0.853	0.769	0	Present
$E(X_{6c})$	3.100	−2.640	2.280	0	Present
$E(L_{6c})$	2.500	−2.753	1.863	0	Present
$E(\Gamma_{7v})$	−0.464	−0.097	−0.416	0	Present γ

7.6.7. $HgSe_{1-x}Te_x$

The coefficients α, β, and γ of Eq. (7.4.1) for several important band energies as functions of x deduced from the calculated band structure are listed in Table 7.78. The magnitudes of the bowing parameters are all large. The band gaps for all alloy concentrations are inverted, so we have negative band gaps, as shown in Fig. 7.45. There is also a large bowing of the top of the valence band, which is reasonably well described by a quadratic approximation: $\Delta E_v = -0.100 + 0.603x - 0.683x^2$. Experimental data are too sketchy for a sensible comparison.

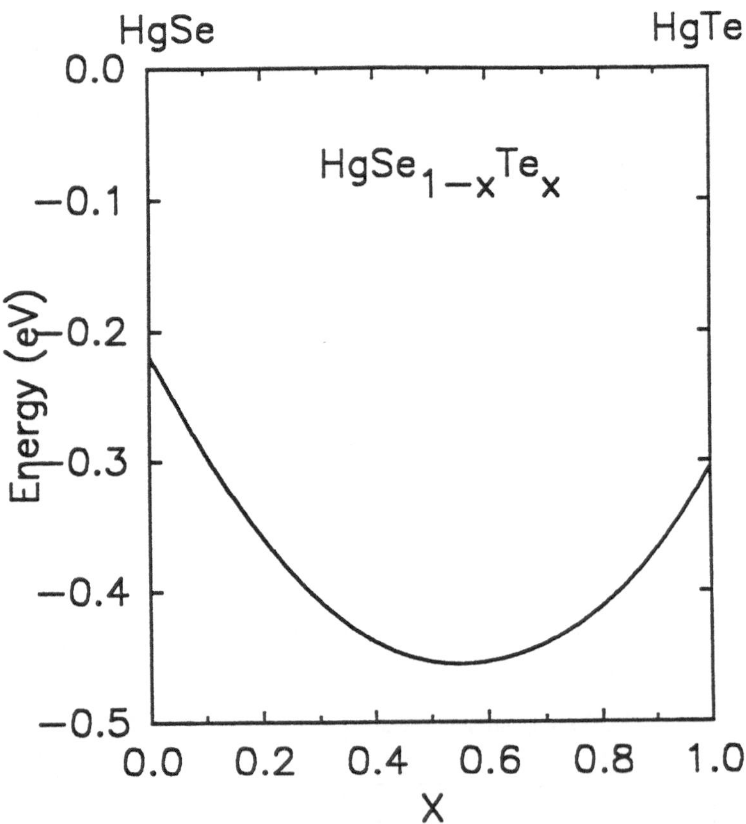

FIGURE 7.45. Band-gap energy in eV of $HgSe_{1-x}Te_x$ as a function of alloy composition x.

TABLE 7.79. Calculated Values of b (in eV) in Eq. (7.4.2) at Several x Values for the Fundamental Gap of $HgSe_{1-x}Te_x$, and the Total Disorder Contribution b_{ex}

x	0.1	0.25	0.5	0.7	0.9	Average	b_{ex}
$b(\Gamma)$	0.718	0.774	0.768	0.770	0.807	0.769	0

TABLE 7.80. Alloy Disorder Parameters Defined in Eq. (5.11.2) in the Γ-OLO Representations Centered at an Anion Site in $HgSe_{1-x}Te_x$

Representation	δ (eV)	Δ (eV)
Γ_6	1.4209	−1.3862
Γ_7	0.0304	1.1361
Γ_8	−0.4916	1.1361

7.6.8. $Cd_{1-x}Zn_xS$

The coefficients of the quadratic approximation to several important band energies as a function of x deduced from the calculated band structures of $Cd_{1-x}Zn_xS$ alloys are given in Table 7.81. The band gap as a function of alloy concentration is plotted in Fig. 7.46. The bowing is quite appreciable. Although the calculated bowing is close to the experimental value (about 0.6 eV) measured by Banerjee *et al.* (1978), the alloy sample was not a homogeneous zb solid solution. Table 7.82 shows that the quadratic form is only a rough approximation and that the alloy disorder contribution cannot be neglected. The top of the

TABLE 7.81. Coefficients α, β, and γ (all in eV) of Eq. (7.4.1) for Several Important Band Energies of $Cd_{1-x}Zn_xS$

States	α	β	γ	Temperature (K)	Source
$E(\Gamma_{6c})$	2.555	0.509	0.736	0	Present
$E(X_{6c})$	5.600	−1.406	0.256	0	Present
$E(L_{6c})$	5.200	−0.816	0.516	0	Present
$E(\Gamma_{7v})$	−0.075	0.016	−0.038	0	Present γ

TABLE 7.82. Calculated Values of b (in eV) in Eq. (7.4.2) at Several x Values for the Fundamental Band Gap and the Total Disorder Contribution b_{ex} for $Cd_{1-x}Zn_xS$

x	0.1	0.25	0.5	0.7	0.9	Average	b_{ex}
$b(\Gamma)$	0.643	0.689	0.741	0.792	0.850	0.736	0.2

TABLE 7.83. Alloy Disorder Parameters Defined in Eq. (5.11.2) in the Γ-OLO Representations at a Cation Site in $Cd_{1-x}Zn_xS$

Representation	δ (eV)	Δ (eV)
Γ_6	0.1493	1.6770
Γ_7	0.6082	0.3833
Γ_8	0.5272	0.3833

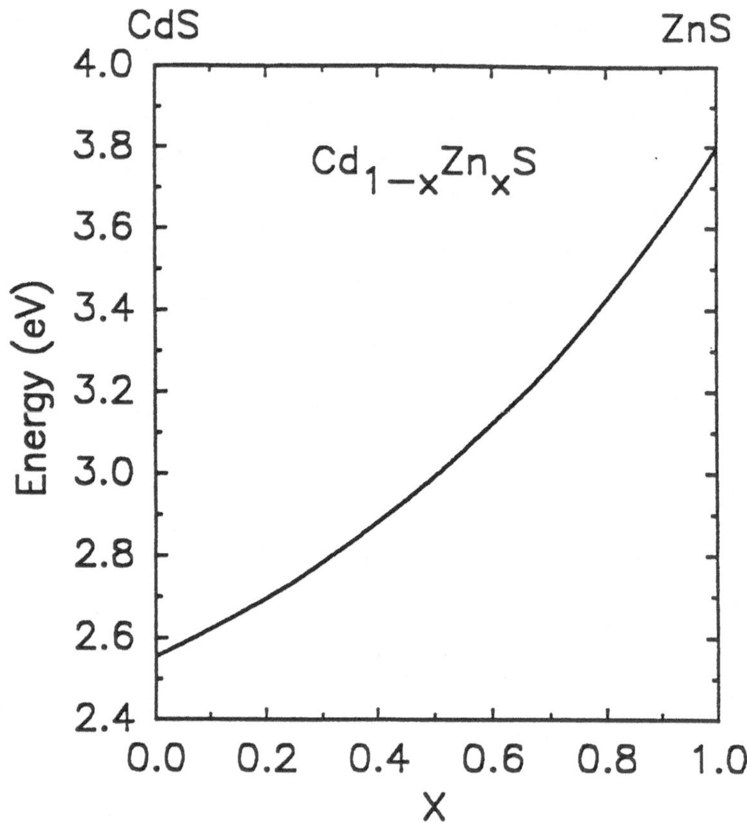

FIGURE 7.46. Band-gap energy in eV of $Cd_{1-x}Zn_xS$ as a function of alloy composition x.

valence band relative to that of GaAs can be approximated by the quadratic form: $\Delta E_v = -1.390 - 0.310x - 0.170x^2$.

7.6.9. $Cd_{1-x}Zn_xSe$

The expansion coefficients in Table 7.84, the bowing parameters in Table 7.85, the disordered parameters in Table 7.86, and the band gap as a function of x in Fig. 7.47 all show that this is a weak-scattering case with a small bowing in the gap. For the same reason, the top of the valence band is almost linear in x: $\Delta E_v = -0.980 - 0.064x - 0.006x^2$.

TABLE 7.84. Coefficients α, β, and γ (all in eV) of Eq. (7.4.1) for the Several Important Band Energies of $Cd_{1-x}Zn_xSe$

States	α	β	γ	Temperature (K)	Source
$E(\Gamma_{6c})$	1.827	0.727	0.266	0	Present
$E(X_{6c})$	4.900	−0.912	0.082	0	Present
$E(L_{6c})$	4.800	−1.481	0.681	0	Present
$E(\Gamma_{7v})$	−0.443	0.048	−0.051	0	Present γ

TABLE 7.85. Calculated Values of b (in eV) in Eq. (7.4.2) at Several x Values for the Fundamental Band Gap and the Total Disorder Contribution b_{ex} for $Cd_{1-x}Zn_xSe$

x	0.1	0.25	0.5	0.7	0.9	Average	b_{ex}
$b(\Gamma)$	0.253	0.256	0.267	0.277	0.290	0.266	0

TABLE 7.86. Alloy Disorder Parameters Defined in Eq. (5.11.2) in the Γ-OLO Representations at a Cation Site in $Cd_{1-x}Zn_xSe$

Representation	δ (eV)	Δ (eV)
Γ_6	−0.2229	−0.3620
Γ_7	−0.4083	0.3534
Γ_8	−0.4893	0.3534

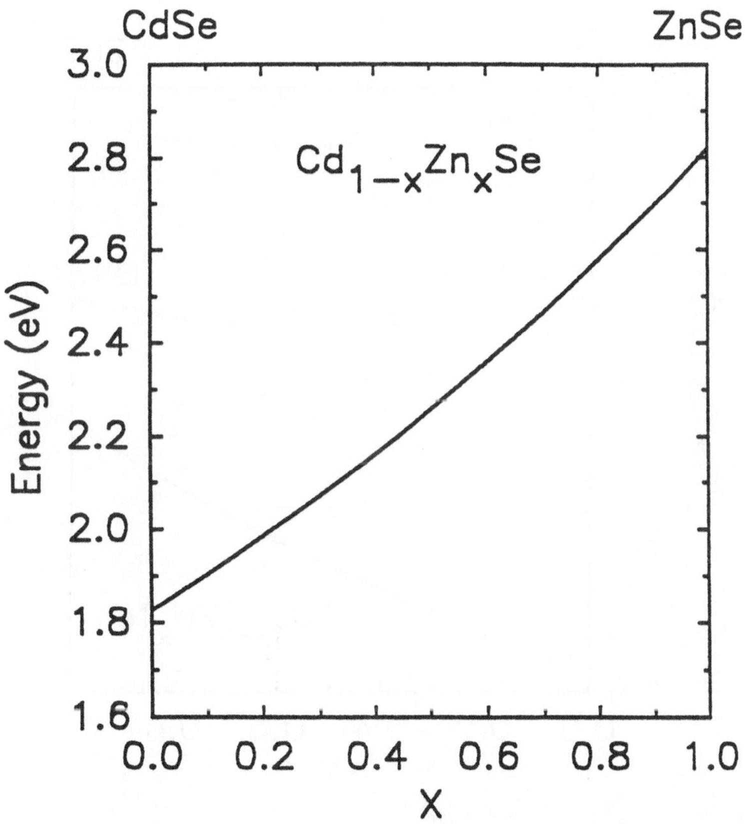

FIGURE 7.47. Band-gap energy in eV of $Cd_{1-x}Zn_xSe$ as a function of alloy composition x.

TABLE 7.87. Coefficients α, β, and γ (all in eV) of Eq. (7.4.1) for Several Important Band Energies of $Cd_{1-x}Zn_xTe$

States	α	β	γ	Temperature (K)	Source
$E(\Gamma_{6c})$	1.600	0.468	0.322	0	Present
$E(\Gamma_{6c})$	1.500	0.43	0.33	295	Ebina *et al.* (1972)
$E(X_{6c})$	3.480	−0.607	0.527	0	Present
$E(L_{6c})$	2.820	−0.452	0.442	0	Present
$E(\Gamma_{7v})$	−0.891	−0.027	−0.086	0	Present

TABLE 7.88. Calculated Values of b (in eV) in Eq. (7.4.2) at Several x Values for the Fundamental Band Gap and the Total Disorder Contribution b_{ex} for $Cd_{1-x}Zn_xTe$

x	0.1	0.25	0.5	0.7	0.9	Average	b_{ex}
$b(\Gamma)$	0.324	0.316	0.324	0.325	0.331	0.322	0.1

7.6.10. $Cd_{1-x}Zn_xTe$

Again, the expansion coefficients in Table 7.87, the bowing parameters in Table 7.88, the disordered parameters in Table 7.89, and the band energies as functions of x in Fig. 7.48 all show that this is a weak-scattering case with a moderate bowing in the gap. The calculated

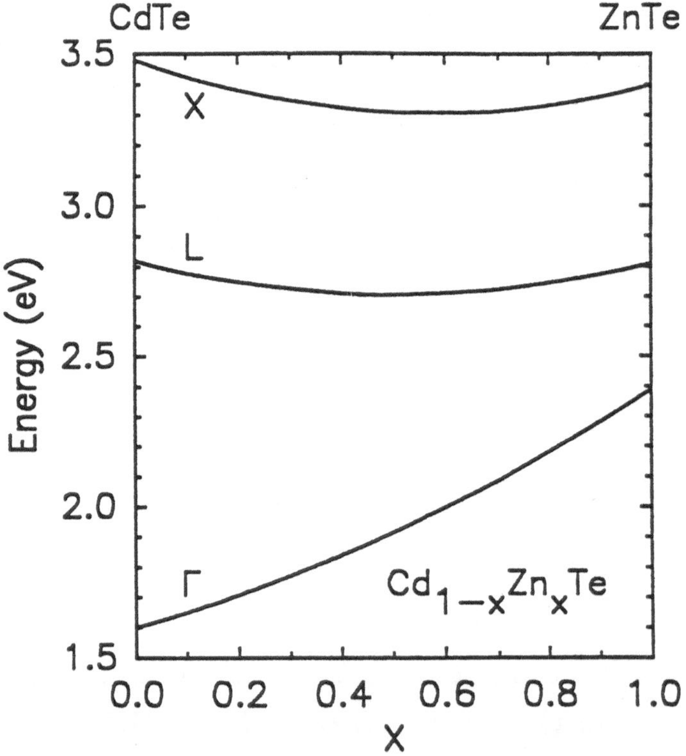

FIGURE 7.48. Conduction-band energy minima in eV at Γ, X, and L as functions of alloy composition x in $Cd_{1-x}Zn_xTe$.

TABLE 7.89. Alloy Disorder Parameters Defined in Eq. (5.11.2) in the Γ-OLO Representations at a Cation Site in $Cd_{1-x}Zn_xTe$

Representation	δ (eV)	Δ (eV)
Γ_6	0.4662	−0.0231
Γ_7	0.6432	0.5013
Γ_8	0.5622	0.5013

TABLE 7.90. Coefficients α, β, and γ (all in eV) of Eq. (7.4.1) for Several Conduction-Band-Edge States and the Spin-Orbit Spliting Δ_o for $Hg_{1-x}Cd_xSe$

States	α	β	γ	Temperature (K)	Source
$E(\Gamma_{6c})$	−0.220	1.059	0.988	0	Present
	−0.22	2.06	0.	0	Iwanowski *et al.* (1978)
$E(X_{6c})$	3.100	−0.052	1.852	0	Present
$E(L_{6c})$	2.500	0.055	2.245	0	Present
Δ_o	0.464	0.021	−0.042	0	Present

bowing parameter for the gap is in good agreement with that deduced from experiment (Ebina *et al.*, 1972). The top of the valence band is well approximated by $\Delta E_v = -0.530 + 0.174x - 0.074x^2$.

7.6.11. $Hg_{1-x}Cd_xSe$

This is an alloy which can be a semimetal (with no gap) or a semiconductor depending on the alloy composition. Figure 7.49 shows the fundamental gap as a function of concentration x deduced from the present calculation. For $x < 0.18$, the gap is negative (no gap), and the gap opens for $x > 0.18$. The coefficients of the quadratic approximations for several important energies relative to the top of the valence band derived from the calculated band structure are given in Table 7.90. The average bowing parameters for the first three energies are large, while that for the spin-orbit splitting is small. The b values in Table 7.91 indicate that the bowing parameters for the gap are smaller in the concentration range where the alloy has a negative gap than those in the positive-gap region. The average bowing parameter for the negative-gap region is 0.930 eV. Disorder contributes to about half of the bowing due to large fluctuations in all three representations, as shown in Table 7.92. The calculated valence-band edge relative to that in GaAs is well approximated by $\Delta E_v = -1.00 - 0.516x - 0.364x^2$.

Although experimental data on the band structure of this alloy are sketchy, Iwanowski *et al.* (1978) used them to conclude a linear x dependence for the gap around the zero-gap range. This linear dependence then yields a smaller x value ($\simeq 0.1$) than that in the present calculation for the transition from the negative to the positive gap.

TABLE 7.91. Calculated Values of b (in eV) in Eq. (7.4.2) at Several x Values for the Fundamental Gap and the Total Disorder Contribution b_{ex} for $Hg_{1-x}Cd_xSe$

x	0.06	0.12	0.25	0.5	0.7	0.9	Average	b_{ex}
$b(\Gamma)$	0.926	0.928	0.936	0.995	1.059	1.164	0.988	—

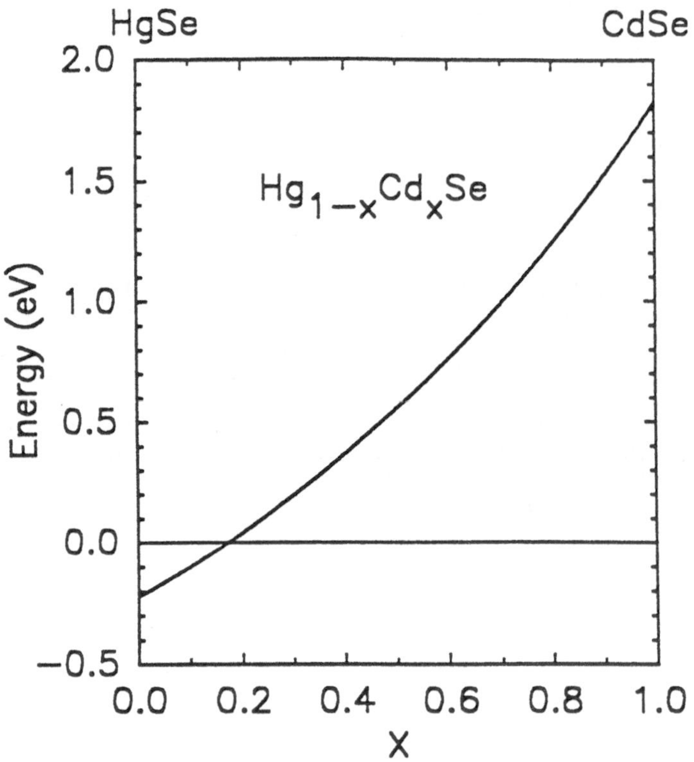

FIGURE 7.49. Band-gap energy in eV of $Hg_{1-x}Cd_xSe$ as a function of alloy composition x.

7.6.12. $Hg_{1-x}Zn_xSe$

$Hg_{1-x}Zn_xSe$, like $Hg_{1-x}Cd_xSe$, has its band gap switch from negative to positive values as the concentration x increases. Figure 7.50 shows the fundamental gap deduced from the present calculation as a function of concentration x. The curve crosses the zero-gap line at x slightly greater than 0.1. The coefficients for the quadratic approximations to several important energies relative to the top of the valence band derived from the calculated band structure are given in Table 7.93. The average bowing parameters for the first three energies are large, while that for the spin-orbit splitting is small. The b values in Table 7.94 indicate that the bowing parameters for the gap are smaller in the concentration range where the alloy has a negative gap than those in the positive-gap region. The average bowing parameter for the negative-gap region is 0.969 eV. Disorder contributes to about half of the bowing due to

TABLE 7.92. Alloy Disorder Parameters Defined in Eq. (5.11.2) in the Γ-OLO Representations at a Cation Site in $Hg_{1-x}Cd_xSe$

Representation	δ (eV)	Δ (eV)
Γ_6	−0.9181	1.0455
Γ_7	−1.5815	0.4115
Γ_8	−0.7625	0.4115

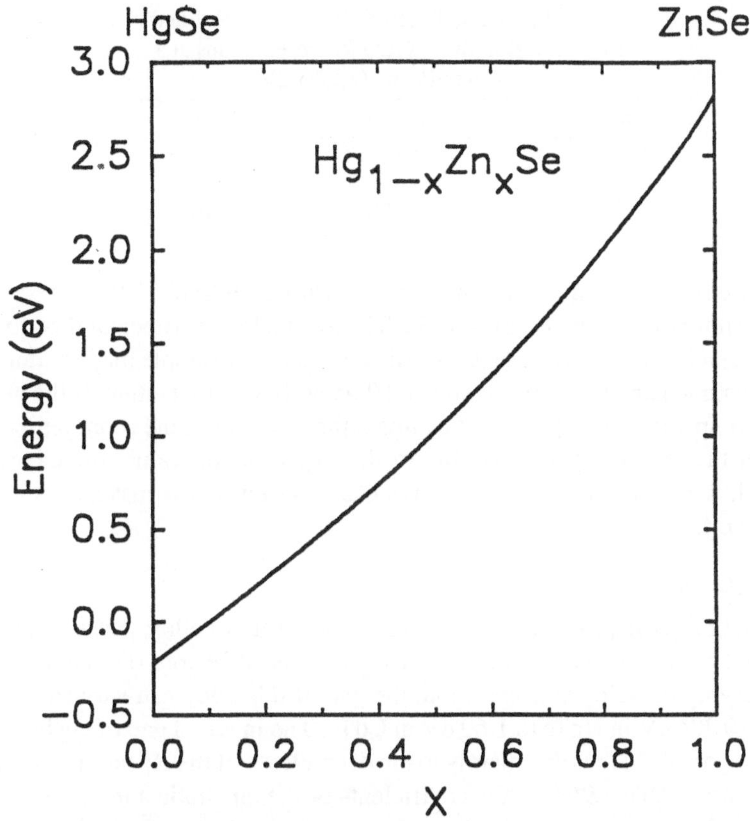

FIGURE 7.50. Band-gap energy in eV of $Hg_{1-x}Zn_xSe$ as a function of alloy composition x.

TABLE 7.93. Coefficients α, β, and γ (all in eV) of Eq. (7.4.1) for Several Important Band Energies of $Hg_{1-x}Zn_xSe$

States	α	β	γ	Temperature (K)	Source
$E(\Gamma_{6c})$	−0.220	1.928	1.112	0	Present
$E(X_{6c})$	3.100	−1.202	2.172	0	Present
$E(L_{6c})$	2.500	−0.140	1.640	0	Present
$E(\Gamma_{7v})$	−0.464	−0.041	0.059	0	Present

TABLE 7.94. Calculated Values of b (in eV) in Eq. (7.4.2) at Several x Values for the Fundamental Band Gap and the Total Disorder Contribution b_{ex} for $Hg_{1-x}Zn_xSe$

x	0.05	0.10	0.25	0.5	0.7	0.9	Average	b_{ex}
$b(\Gamma)$	0.962	0.957	1.007	1.110	1.241	1.464	1.112	—

TABLE 7.95. Alloy Disorder Parameters Defined in Eq. (5.11.2) in the Γ-OLO Representations at a Cation Site in $Hg_{1-x}Zn_xSe$

Representation	δ (eV)	Δ (eV)
Γ_6	−1.1410	0.7006
Γ_7	−1.9898	0.7708
Γ_8	−1.2518	0.7708

large fluctuations in all three representations, as shown in Table 7.95. These results are qualitatively similar to those of $Hg_{1-x}Cd_xSe$. The available experimental data for this alloy are still sketchy. Gavaleshko *et al.* (1984) indicated that their sample made a transition from negative to positive gap at x between 0.02 and 0.05 at room temperature. If the curve in Fig. 7.50 is moved upward about 0.1 eV to estimate the crossing at the room temperature, the value of x at the crossing is comparable to the experimental value just mentioned. The calculated valence-band edge relative to that of GaAs is well approximated by $\Delta E_v = -0.100 - 0.473x - 0.477x^2$.

7.6.13. $Hg_{1-x}Cd_xTe$

$Hg_{1-x}Cd_xTe$ is perhaps the most intensively studied II–VI alloy, because it has been the prime material for long-wavelength (8–12 μm) infrared detectors (Dornhaus and Nimitz, 1976). By varying the alloy concentration, the gap of this alloy can vary from an inverted band gap of −0.304 eV in HgTe to 1.60 eV in CdTe. The inverted gap of HgTe as compared to the positive gap of CdTe is due mainly to the deep s level of the Hg atom caused by a large relativistic s shift. Table 7.96 lists the coefficients of the quadratic approximations for four important energies. However, the quadratic forms are not accurate for both the fundamental gap and L gap, because the bowing parameters increase with x. This increase can be seen from Table 7.97, where b increases from 0.111 eV at $x = 0.03$ to 0.369 at $x = 0.9$. The calculated band energies of Γ_{6c}, L_{6c}, and X_{6c} are plotted against x in Fig. 7.51. The average bowing in the negative-gap region is 0.138 eV; it is 217 eV for the positive gaps. As mentioned, the Hg s-term value is deeper than the Cd s level, which becomes the main disorder parameter, as shown in Table 7.98. The disorder contributes to more than half of

TABLE 7.96. Coefficients α, β, and γ (all in eV) of Eq. (7.4.1) for Several Important Band Energies of $Hg_{1-x}Cd_xTe$

States	α	β	γ	Temperature (K)	Source
$E(\Gamma_{6c})$	−0.304	1.687	0.217	0	Present
$E(X_{6c})$	2.740	0.620	0.120	0	Present
$E(L_{6c})$	1.610	0.707	0.503	0	Present
$E(\Gamma_{7v})$	−0.977	0.160	−0.074	0	Present

TABLE 7.97. Calculated Values of b (in eV) in Eq. (7.4.2) at Several x Values for the Fundamental Band Gap and the Total Disorder Contribution b_{ex} for $Hg_{1-x}Cd_xTe$

x	0.03	0.09	0.15	0.25	0.50	0.70	0.90	Average	b_{ex}
$b(\Gamma)$	0.111	0.138	0.145	0.170	0.214	0.282	0.369	0.217	—

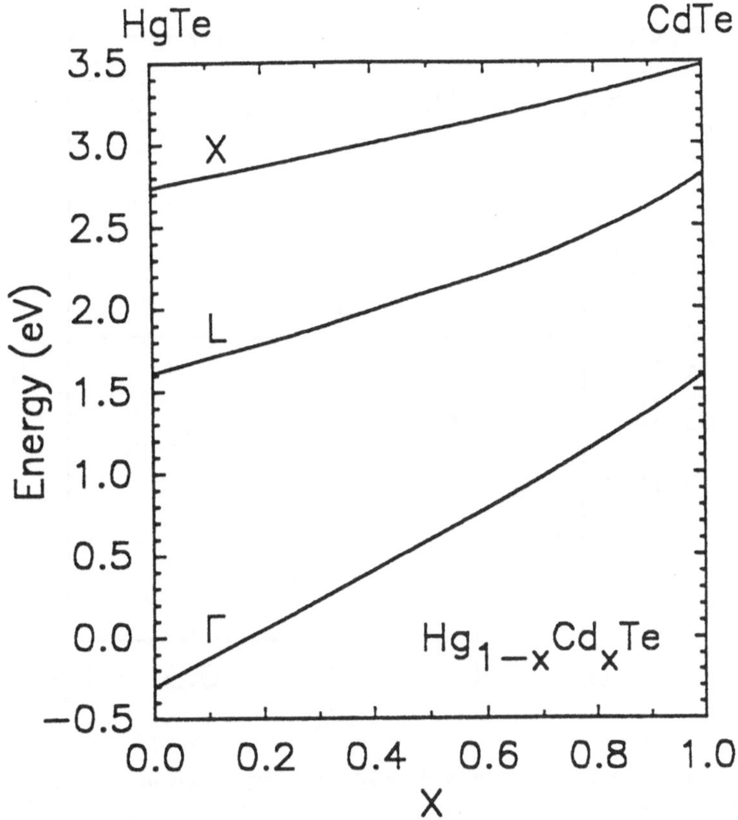

FIGURE 7.51. Conduction-band energy minima in eV at Γ, X, and L as functions of alloy composition x in $Hg_{1-x}Cd_xTe$.

the bowing parameter. The b_{ex} in Table 7.97 is the value calculated at $x = 0.5$. At $x = 0.15$, the b_{ex} value is only 0.076 eV, which contributes about the same fraction to the total bowing as b_{ex} does to the 50–50 alloy. The top of the valence band relative to that of GaAs is well described by the quadratic expression $\Delta E_v = -0.180 - 0.296x - 0.054x^2$.

Since most studies of the band structure of this alloy focused on the narrow-gap region, the calculated band gap is compared with the experimental results (see captions in Fig. 7.53) at $T = 4.2$ K in Fig. 7.52 in this range. The calculated curve lies just below the experimental values, but the overall agreement is reasonably good. There are several empirical formulas available for the band gap as a function of the temperature and the composition x. The formula

TABLE 7.98. Alloy Disorder Parameters Defined in Eq. (5.11.2) in the Γ-OLO Representations at a Cation Site in $Hg_{1-x}Cd_xTe$

Representation	δ (eV)	Δ (eV)
Γ_6	−1.4698	0.0083
Γ_7	−0.5502	−0.0653
Γ_8	0.2598	−0.0653

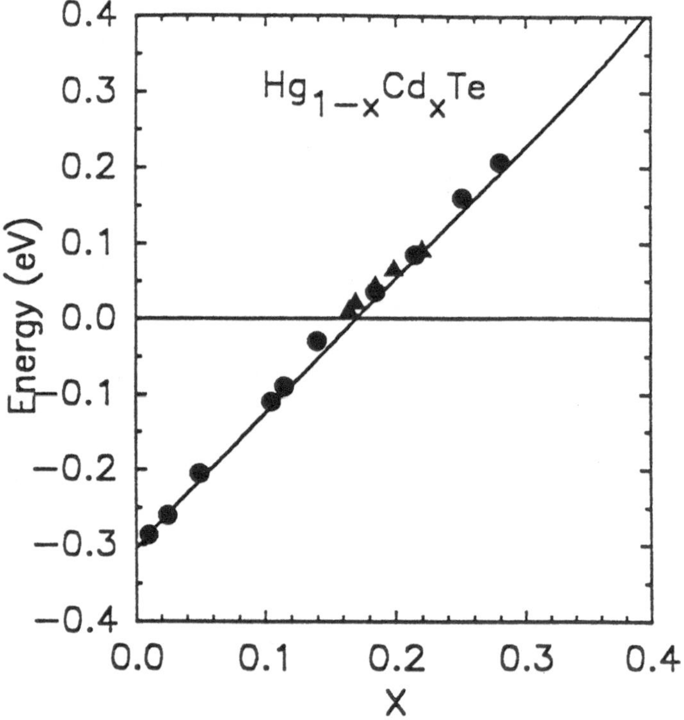

FIGURE 7.52. Calculated band energy gaps in eV (solid line) of $Hg_{1-x}Cd_xTe$ near zero-gap and experimental values (dots and triangles).

by Weiler (1981), given below, represents a good practical expression for E_g (in meV) in this region:

$$E_g = -304 + 0.63T^2 (1 - 2x)/(11 + T) + 1858x + 54x^2 \qquad (7.6.1)$$

Figure 7.53 shows a comparison of this formula and another empirical formula with experimental values. The overall deviation of Eq. (7.6.1) from experiment is small. However, we note that these empirical formulas fit the negative-gap data very well, but they consistently overestimate the positive gaps. Weiler also arrived at the following empirical formula for the conduction-band edge effective mass m_c of the alloy:

$$m/m_c = -0.6 + 19000(667 + E_g)/(1000E_g + E_g^2) \qquad (7.6.2)$$

where m is the free-electron mass and E_g is the band gap in meV.

7.6.14. $Hg_{1-x}Zn_xTe$

$Hg_{1-x}Zn_xTe$ in many respects has similar properties to $Hg_{1-x}Cd_xTe$. In electronic structure, the former has a smaller value of x for the composition to make transition from the negative gap to the positive gap, because the E_g versus x curve is steeper due to a wider band gap in ZnTe. A major difference between the two alloys is that the ZnTe lattice constant is considerably smaller (≈ 5.5%) than that of HgTe. This difference in the lattice constant introduces a larger VCA bowing parameter than that in $Hg_{1-x}Cd_xTe$. The coefficients of the

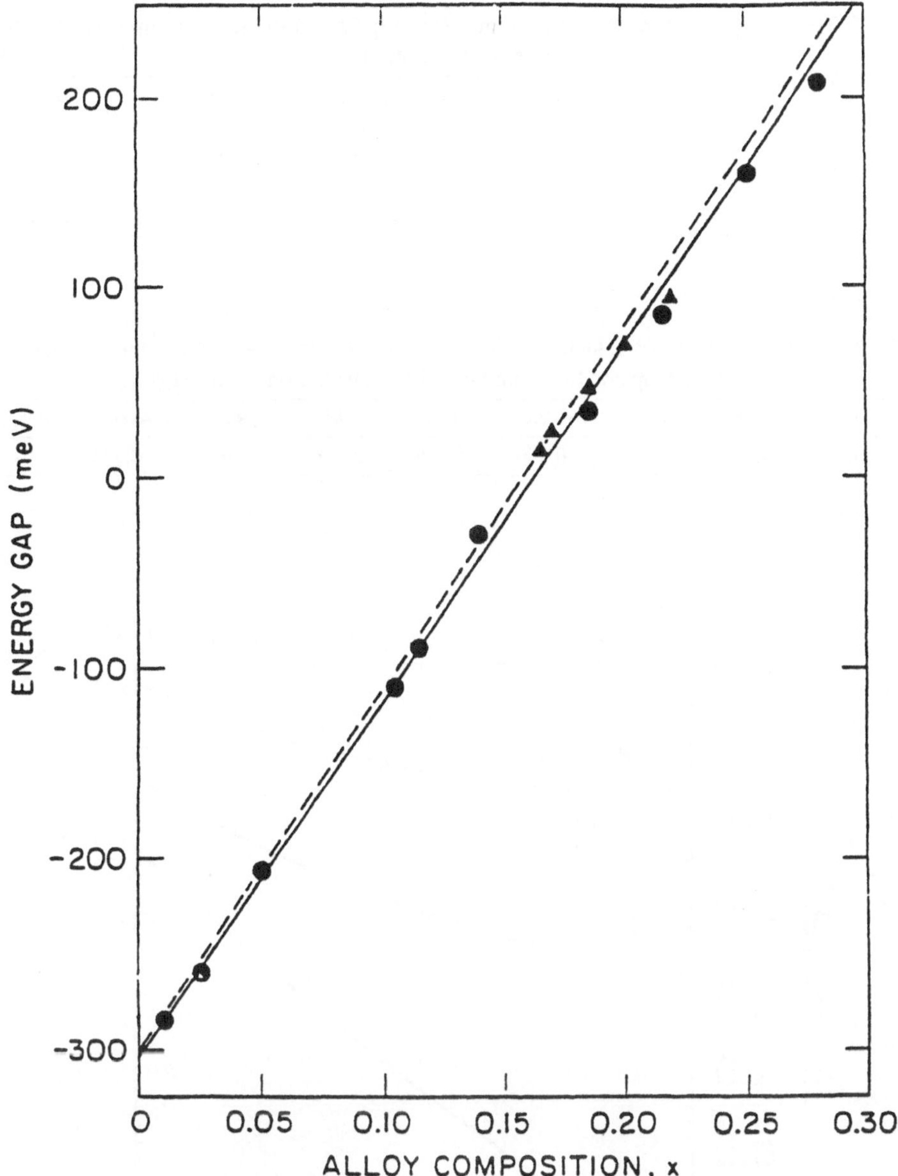

FIGURE 7.53. Energy gap of $Hg_{1-x}Cd_xTe$ for $0 \leq x \leq 0.3$ at $T = 4.2$ K, compared to the data of Guldner *et al.* (1977a,b) (●) and Dornhaus and Nimtz (1977) (▲). Solid curve: calculated from Eq. (7a); dashed curve: from Kim and Narita (1976) (Eq. (44)). (After Weiler, 1981.)

quadratic approximations to several states are given in Table 7.99. The actual x dependence of the first three lowest conduction-band energies at Γ, X, and L are displayed in Fig. 7.54. Again, the bowing parameter for the gap is smaller for the negative-gap region than that for the positive gap. The average b value is 0.335 eV for the former and 0.388 eV for the latter. The experimental results for the bowing of the gaps of these alloys are still not very consistent. For example, experimental values of the bowing parameter of the fundamental

TABLE 7.99. Coefficients α, β, and γ (all in eV) of Eq. (7.4.1) for Several Important Band Energies of $Hg_{1-x}Zn_xTe$

States	α	β	γ	Temperature (K)	Source
$E(\Gamma_{6c})$	−0.304	2.309	0.385	0	Present
$E(X_{6c})$	2.740	−0.193	0.853	0	Present
$E(L_{6c})$	1.610	0.379	0.821	0	Present
$E(\Gamma_{7v})$	−0.977	0.083	−0.110	0	Present γ

TABLE 7.100. Calculated Values of b (in eV) in Eq. (7.4.2) at Several x Values for the Fundamental Band Gap and the Total Disorder Contribution b_{ex} for $Hg_{1-x}Zn_xTe$

x	0.02	0.06	0.10	0.25	0.50	0.70	0.90	Average	b_{ex}
$b(\Gamma)$	0.315	0.332	0.339	0.356	0.396	0.423	0.476	0.385	—

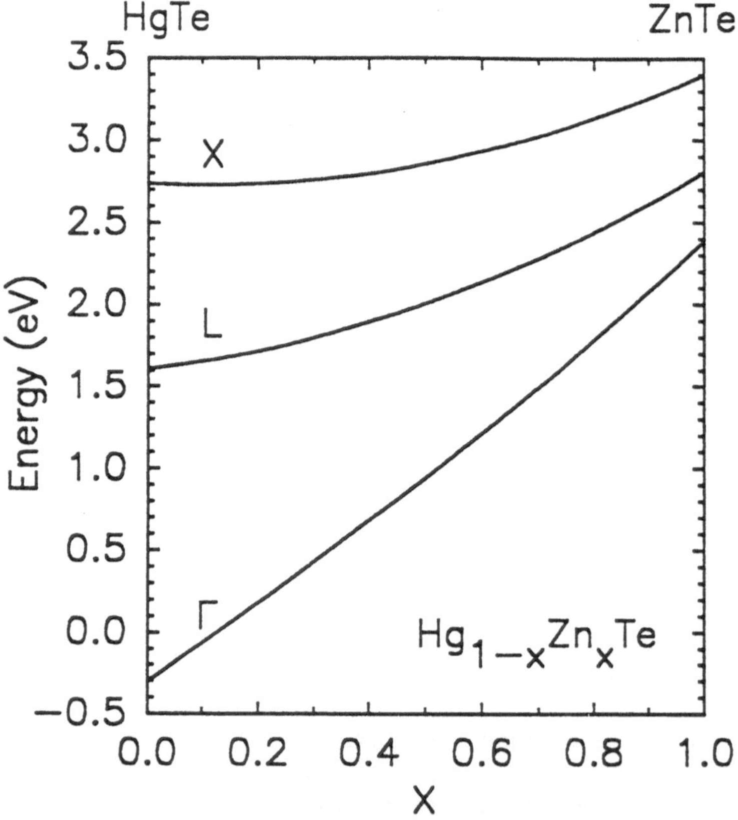

FIGURE 7.54. Conduction-band energy minima in eV at Γ, X, and L as functions of alloy composition x in $Hg_{1-x}Zn_xTe$.

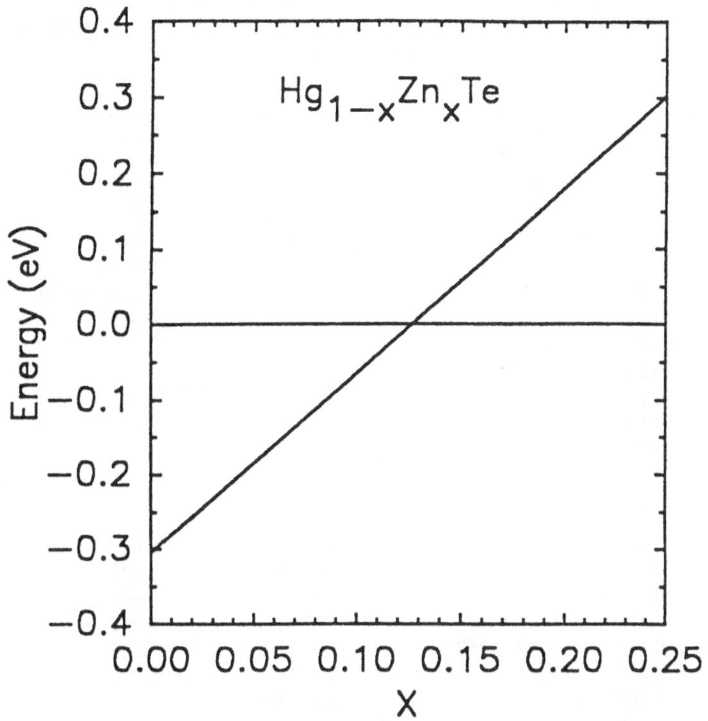

FIGURE 7.55. Band-gap at Γ as a function of concentration X for the alloy $Hg_{1-x}Zn_xTe$.

TABLE 7.101. Alloy Disorder Parameters Defined in Eq. (5.11.2) in the Γ-OLO Representations at a Cation Site in $Hg_{1-x}Zn_xTe$

Representation	δ (eV)	Δ (eV)
Γ_6	-1.0037	-0.0147
Γ_7	0.0929	0.4354
Γ_8	0.8219	0.4354

gap ranges from 0 to 0.9 eV. Sher *et al.* (1986) reported that the gap can be well described by a straight line starting with $E_g = -0.304$ at $x = 0$ and with a slope less than that of the linearly averaged gap. Since Sher *et al.* only used a few data points near the zero-gap concentrations, the curvature cannot be determined accurately from their experiments. Even with a bowing of 0.3 eV in the calculated band gap, it looks like a straight line as a function of x near the zero crossover, as shown in Fig. 7.55. Finally the valence-band top is well described by the quadratic approximation $\Delta E_V = -0.180 - 0.106x - 0.144x^2$.

7.7. CONCLUDING REMARKS

The empirical hybrid pseudopotential tight-binding model is shown to be accurate for calculating the band structures of III–V and II–VI semiconductors. The Hamiltonians and

TABLE 7.102. Bowing Parameters b for the Fundamental Energy Gaps: $E_g = \overline{E}_g - bx(1-x)$, Compiled from the Present Calculation (Cal) and Experimental Results (Exp) Quoted in Sections 7.4 and 7.6.

		Cal.	Exp.		Cal.	Exp.
GaAlAs	Γ	0.269	0.37, 0.26, 0.438	CdZnS	0.736	0.6
	(X)	0.125	0.143, 0.245, 0.16	CdZnSe	0.266	
GaAlSb	Γ	0.500	0.40, 0.48, 0.47	CdZnTe	0.322	0.33
	(X)	0.348		HgCdSe	0.988	
GaAlP	(X)	0.068		HgZnSe	1.112	
GaInP	Γ	0.517	0.758, 0.7, 0.5	HgCdTe	0.13 ($x < 0.3$)	0.054
	(X)	0.220	0.0		0.22 ($x > 0.3$)	
GaInAs	Γ	0.348	0.40	HgZnTe	0.385	0., 0.9
GaInSb	Γ	0.406	0.413			
InAlP	Γ	0.993				
	(X)	0.113				
InAlAs	Γ	0.698	0.74			
	(X)	0.200				
InAlSb	Γ	0.907				
	(X)	0.777				
AlAsP	(X)	0.400		ZnSSe	0.622	0.630, 0.41
AlAsSb	(X)	0.250		ZnSeTe	1.258	1.266, 1.28
AlPSb	(X)	0.277		ZnSTe	2.869	2.97, 3.02
GaAsP	Γ	0.116	0.174, 0.210, 0.186	CdSSe	0.395	
	(X)	0.243	0.202, 0.267	CdSeTe	0.289	
GaAsSb	Γ	0.830	1.2	CdSTe	0.303	
GaSbP	Γ	0.768		HgSeTe	0.769	
	(X)	0.252				
InAsSb	Γ	0.881	0.680			

resulting band structures of all common zinc blende compounds and alloys have been calculated and are compiled in this chapter. Following the method described and input parameters given in this chapter, the reader should be able to reproduce these band structures and calculate other band quantities as needed.

The calculated bowing parameters for the fundamental gaps of alloys are summarized in Table 102. They compare favorably with the available data. In several instances, the discrepancy between the calculated and experimental values is smaller than that between two different experiments for the same system. This information is summarized in Plate 1, a set of band gap versus lattice constant curves for all the III–V and II–IV disordered alloys, including those with direct and indirect gaps. As mentioned, the band structures of semiconductor alloys can be significantly influenced by growth conditions, which often control the state of order, long-ranged and short-ranged, of these materials. Caution must be exercised when comparing experiments to the calculated values, which assume a disordered random alloy. To extend the present model to long-range ordered semiconductor alloys, the pseudopotential form factors in Table 7.1 must be extrapolated or interpolated to additional reciprocal lattice vectors. However, the local tight-binding parameters should not be significantly altered.

This chapter does not include wide-gap semiconductors, such as SiC, GaN, AlN, and their alloys, because experimental information about the band structures of these materials is still limited and the alloy theory has yet to be extended to treat the wurtzite structure to which most of these wide-gap systems belong. The present theory should be expanded to

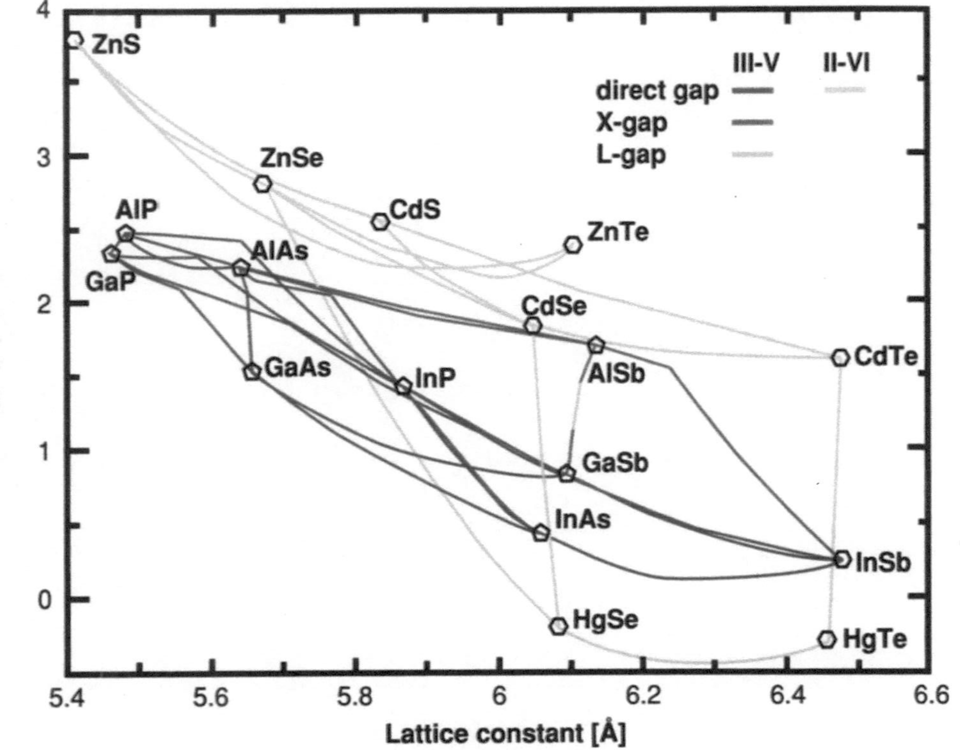

Plate 1. The fundamental energy gap versus lattice constant for many stable and metastable, pseudobinary, zinc blende-structured alloys. Both direct and indirect gap materials are represented. Direct gaps of III-V alloys are red, indirect gaps where the conduction band minimum is at the *X*-point are blue, and the one at the *L*-point is green. All the II-VI alloys have direct gaps and they are orange. Each III-V compound, designated by a pentagon, has four alloy lines connected to it. Because HgS is not on the chart, some II-VI compounds, designated by hexagons, have four and others only have three alloy lines connected to them. Some compounds have unrelated alloy lines passing through them by accident. These extra lines pass behind the compound symbol. For example, the $Al_{1-x}In_xAs$ line passes through the InP compound point.

calculate the band structures of these materials, because they have important applications in high-temperature electronics and short-wavelength electro-optical devices.

REFERENCES

Abrahams, M.S., R. Braunstein, and F.D. Rosi (1960), in *Conference on the Properties of Elemental and Compound Semiconductors,* ed. H.C. Gatos (Interscience, New York), p. 275.

Adachi, S.J. (1985), *J. Appl. Phys.* **58**, R1.

Alibert, C., G. Bordure, A. Laugier, and G. Chevallier (1972), *Phys. Rev. B* **6**, 1301.

Alibert, C., A. Joullie, A.M. Joullie, and C. Ance (1983), *Phys. Rev. B* **27**, 4946.

Aspnes, D.E. (1976), *Phys. Rev. B* **14**, 5331.

Aspnes, D.E., C.G. Olson, and D.W. Lynch (1976), *Phys. Rev. B* **14**, 4450.

Aspnes, D.E., and A.A. Studna (1973), *Phys. Rev. B* **7**, 4605.

Aubel, J.L., U.K. Reddy, S. Sundaram, W.T. Beard, and J. Comas (1985), *J. App. Phys.* **58**(1), 495.

Aulombard, R.L., L. Konczewicz, A. Kadri, A. Joullie, and J.C. Portal (1985), in *GaAs and Related Compounds*, 84, Inst. Phys. Ser. No. 74, p. 235.

Baars, J.W. (1967), in *II–VI Semiconducting Compounds*, ed. D.G. Thomas (Benjamin, New York), p. 631.

Bagguley, D.M.S., M.L.A. Robinson, and R.A. Stradling (1963), *Phys. Lett.* **6**, 143.

Banerjee, A., P. Nath, V.D. Vankar, and K.L. Chopra (1978), *Phys. St. Sol. A* **46**, 723.

Bauer, R.S., R.H. Moles, and T.C. McCill (1987), in *Semiconductor Interfaces: Formation and Properties*, eds. G. Le Lay, J. Derrin, and N. Boccara (Springer-Verlag), p. 372.

Bernard, J.E., and A. Zunger (1987), *Phys. Rev. B* **36**, 3199.

Berolo, O., and J.C. Woolley (1972), in *Proceedings of the 11th Int. Conf. on the Physics of Semiconductors* (PWN, Warsaw), p. 1420.

Biryulin, Y.F., N.V. Ganina, M.G. Milvidskii, V.V. Chaldychev, and Y.V. Shmartsev (1983), *Sov. Phys. Semicond.* (English Transl.), **17**, 68.

Biryulin, Y.F., S.P. Vul, V.V. Chaldychev, and Y.V. Shmartsev (1983), *Sov. Phys. Semicond.* (English Transl.), **17**, 65.

Blakemore, J.S. (1982), *J. Appl. Phys.* **53**, 6179.

Camassel, J., P. Merle, L. Bayo, and H. Mathieu (1980), *Phys. Rev. B* **22**, 2020.

Capizzi, M., S. Modesti, F. Martelli, and A. Frova (1981), *Solid State Commun.* **39**, 333.

Cardona, M., and D.L. Greenway (1963), *Phys. Rev.* **131**, 98.

Casey, H.C., and M.B. Panish (1978), *Heterostructure Lasers.* Part B: *Materials and Operational Characteristics* (Academic Press, New York).

Chadi, D.J. (1977), *Phys. Rev. B* **16**, 3572.

Chelikowsky, J.R., and M.L. Cohen (1976), *Phys. Rev. B* **14**, 556.

Chen, A.-B., and A. Sher (1980), *Phys. Rev. B* **22**, 3886.

Chen, A.-B., and A. Sher (1982), *Phys. Rev. B* **26**, 6603.

Cherng, M.J., R.M. Cohen, and G.B. Stringfellow (1984), *J. Electron. Mater.* **13**, 799.

Chiang, T.C., and D.E. Eastman (1980), *Phys. Rev. B* **22**, 2940.

Chiang, T.C., J.A. Knapp, M. Aano, and D.E. Eastman (1980), *Phys. Rev. B* **21**, 3513.

Chiu, T.H., W.T. Tsang, S.N.G. Chu, J. Shah, and J.A. Ditzenberger (1985), *Appl. Phys. Lett.* **46**, 408.

Cohen, M.L., and T.K. Bergstresser (1966), *Phys. Rev.* **141**, 789.

Cohen, M.L., and J.R. Chelikowsky (1989), *Electronic Structure and Optical Properties of Semiconductors*, 2nd ed. (Springer-Verlag, Berlin).

Dang, L.S. (1982), *Solid State Commun.* **44**, 1187.

Dingle, R., R.A. Logan, and J.R. Authur (1977), *Inst. Phys. Conf. Ser.* **33A**, 210.

Dornhaus, R., and G. Nimitz (1976), in *Solid State Physics*, ed. G. Hohler (Springer-Verlag, Berlin and New York).

Eastman, D.E., W.D. Grobman, J.L. Freeout, and M. Erbudak (1974), *Phys. Rev. B* **9**, 3473.

Ebina, A., E. Fukunaga, and T. Takahashi (1974), *Phys. Rev. B* **10**, 2495.

Ebina, A., K. Saito, and T. Takahashi (1972), *J. Appl. Phys.* **44**, 3659.

El Shazly, A.A., M.M.H. El Naby, M.A. Kenawy, M.M. El Nahass, H.T. El Shair, and A.M. Ebrahim (1985), *Appl. Phys. A* **36**, 51.

Gavaleshko, N.P., V.V. Khomyak, V.M. Frasunyak, L.D. Paranchinch, and S.Y. Paranchich (1984), *Sov. Phys. Semicond.* (English transl.) **18**, 967.
Goetz, K.-H., D. Bimbery, H. Jürgensen, J. Selders, A.V. Solomonov, G.F. Glinskii, and M. Razhegi (1983), *J. Appl. Phys.* **54**, 4543.
Gomyo, A., T. Suzuki, K. Kobayashi, S. Kawata, and I. Hino (1987), *Appl. Phys. Lett.* **50**, 673.
Guldner, Y., G. Bastard, J.P. Vieren, M. Voos, J.P. Faurie, and A. Million (1983), *Phys. Rev. Lett.* **51**, 907.
Guldner, Y., C. Rigaux, A. Mycielski, and Couder (1977a), *Phys. Status Solidi B* **81**, 615.
Guldner, Y., C. Rigaux, A. Mycielski, and Couder (1977b), *Phys. Status Solidi B* **82**, 149.
Harrison, W.A. (1983), *Electronic Structure* (W.H. Freeeman, San Francisco).
Heller, M.W., and R.G. Hamerly (1985), *J. Appl. Phys.* **57**, 4626.
Helm, M., W. Knap, W. Seidenbusch, R. Lassnig, and E. Gornik (1985), *Solid State Commun.* **53**, 547.
Hill, D.A., and C.F. Schwerdtfeger (1974), *J. Phys. Chem. Solids* **22**, 205.
Hill, R., and D. Richardson (1973), *J. Phys. C* **6**, L115.
Horner, G.S., A. Mascarenhas, S. Froyen, R.G. Alonso, K. Bertress, and J. Olson (1993), *Phys. Rev. B* **47**, 4041.
Huang, M., and W.Y. Ching (1985), *J. Phys. Solids* **46**, 977.
Humphreys, R.G., U. Rossler, and M. Cardona (1978), *Phys. Rev. B* **18**, 5590.
Hybertsen, M.S., and S.G. Louie (1987), *Phys. Rev. Lett.* **58**, 1551.
Iwanowski, R.J., T. Dietl, and W. Szymamska (1978), *J. Phys. Chem. Solids* **39**, 1059.
Jones, E.D., R.P. Schneider, S.M. Lee, and K.K. Bajaj (1992), *Phys. Rev. B* **46**, 7225.
Joullie, A.M., and C. Alibert (1974), *J. Appl. Phys.* **45**, 5472.
Joullie, A., B. Girault, A.M. Joullie, and A. Zien-Eddine (1982), *Phys. Rev. B* **25**, 7830.
Kainthla, R.C., D.K. Pandya, and K.L Chopra (1982), *J. Electrochem. Sci. Technol.* **Jan.**, 99.
Kanatz, T., N. Masahiko, N. Hiroshi, and N. Taneo (1992), *Phys. Rev. B* **45**, 6637.
Kanazawa, K.K., and F.C. Brown (1964), *Phys. Rev.* **135**, A1757.
Kerr, T.M., T.D. McLean, D.J. Westwood, J.D. Medland, C.E.C. Wood, and I.J. Margatroyd (1986), *J. Vac. Sci. Technol. B* **4**, 532.
Kopylov, A.A. (1985), *Solid State Commun.* **56**, 1.
Kopylov, A.A., and A.N. Pikhin (1977), *Sov. Phys. Semicond.* (English transl.) **11**, 510.
Lai-Hsu, Y.-M. (1990), *Difference-Equation Approach to the Electronic Structure of Superlattices with Applications to HgTe/CdTe and HgTe/ZnTe Systems*, Ph.D. dissertation, Auburn University.
Landolt-Bornstein (1988), *Numerical Data and Functional Relationships in Science and Technology*, ed. K.-H. Hellwidge (Springer-Verlag, Berlin), vol. 22.
Langer, D.W., R.N. Euwema, K. Era, and T. Koda (1970), *Phys. Rev. B* **2**, 4005.
Larach, S., R.E. Shrader, and C.F. Stocker (1957), *Phys. Rev.* **108**, 507.
Lawaetz, P. (1971), *Phys. Rev. B* **4**, 3460.
Lee, H.J., L.Y. Juravel, J.C Woolley, A.J. Spring Thorpe (1980), *Phys. Rev. B* **21**, 659.
Lee, M.K., R.H. Horng, and L.C. Huang (1991), *Appl. Phys. Lett.* **59**, 3261.
Leotin, J., R. Barbaste, S. Askenazy, M.S. Skolnick, R.A. Stradling, and J. Tuchendler (1974), *Solid. State Commun.* **15**, 693.
Ley, L., R.A. Pollak, F.R. McFeely, S.P. Kowalczyk, and D.A. Shirley (1974), *Phys. Rev. B* **9**, 600.
Little, C.L., D.G. Seiter, R. Kaplan, and R.J. Wagner (1983), *Phys. Rev. B* **27**, 7473.
Logothetidis, S., L. Vina, and M. Cardona (1985), *Phys. Rev. B* **31**, 947.
Lohez, D., and M. Lannoo (1983), *Phys. Rev. B* **27**, 5007.
Lorenz, M.R., and A. Onton (1970), *Proc. 10th Int. Conf. of Semiconductor Physics* (USAEC, New York), p. 444.
Lucovski, G., and M.F. Chen (1970), *Solid State Commun.* **8**, 1397.
Mach, R., P. Flogel, L.G. Suslina, A.G. Areshkin, J. Maege, and G. Voig (1982), *Phys. Stat. Sol.* **109B**, 607.
Madelung, O. (ed.) (1991), *Semiconductors Group IV Elements and III–V Compounds* (Springer-Verlag, New York).
Marciniak, H.C., and D.B. Wittey (1975), *J. Appl. Phys.* **46**, 4823.
Marple, D.T.F. (1964), *J. Appl. Phys.* **35**, 1879.
Matatagui, E., A.E. Thompson, and M. Cardona (1968), *Phys Rev.* **176**, 950.
Miura, N., G. Kido, M. Suekane, and S. Chikazumi (1983a), *J. Phys. Soc. Jpn.* **52**, 2838.
Miura, N., G. Kido, M. Suekane, and S. Chikazumi (1983b), *Physica* **117B**, **118B**, 66.
Monemar, B. (1973), *Phys. Rev. B* **8**, 5711.
Morehead, F.F. (1963), *J. Phys. Chem. Solids* **24**, 37.
Nahory, R.L., M.A. Pollack, J.C. DeWinter, and K.M. Williams (1977), *J. Appl. Phys.* **48**, 513.

Nelson, D.F., L.F. Johnson, and M. Gershenzon (1964), *Phys. Rev.* **135**, A1399.
Nelson, R.J., N. Holonyak, and W.O. Groves (1976), *Phys. Rev. B* **13**, 5415.
Nicholas, R.J., R.A. Stradling, and J.C. Ramage (1979), *J. Phys. C* **12**, 1641.
Onton, A., and R.J. Chicotka (1970), *J. Appl. Phys.* **41**, 4305.
Onton, A., and R.J. Chicotka (1971), *Phys. Rev. B* **4**, 1847.
Onton, A., R.J. Chicotka, and Y. Yacoby (1972), *11th Int. Conf. Phys. Semiconductors* (PWN, Warsaw), p. 1023.
Onton, A. and L.M. Foster (1972), *J. Appl. Phys.* **43**, 5084.
Petroff, Y., M. Balkanski, J.P. Walter, and M.L. Cohen (1969), *Solid State Commun.* **7**, 459.
Pidgeon, C.R., S.H. Groves, and J. Feinleib (1967), *Solid State Commun.* **5**, 677.
Prytkina, L.V., V.V. Volkov, A.N. Mentser, A.V. Vanyukov, and P.S. Kireev (1968), *Sov. Phys. Semicond.* (English transl.), **2**, 509.
Qu, H., J. Kanski, P.O. Nilsson, and U.O. Karlsson (1991), *Phys. Rev. B* **44**, 1762.
Roth, A.P., W.J. Keeler, and E. Fortin (1980), *Can. J. Phys.* **58**, 560.
Saxena, A.K. (1981), *Phys. Status Solidi (b)* **105**, 777.
Schwerdtfeger, C.P. (1972), *Solid State Commun.* **11**, 779.
Segall, B., and D.T.F. Marple (1967), in *Physics and Chemistry of II–VI Compounds*, eds. M. Aven and J.S. Prener (North-Holland, Amsterdam), p. 319.
Seiler, D.G., M.W. Goodwin, and A. Miller (1980), *Phys. Rev. Lett.* **44**, 807.
Shaklee, K.L., M. Cardona, and F.H. Pollak (1966), *Phys. Rev. Lett.* **16**, 48.
Sharma, A.C., N.M. Ravinda, S. Auluck, V.K. Srivastava (1983), *Phys. Stat. Solidi (b)* **120**, 715.
Sher, A., E.A. Zemel, H. Feldstein, and A. Raizman (1986), *J. Vac. Sci. Technol. A* **4**(4), 2024.
Skromme, B.J., and G.E. Stillman (1984), *Phys. Rev. B* **29**, 1982.
Slater, J.C., and G.F. Koster (1954), *Phys. Rev.* **94**, 1498.
Solal, F., G. Jezequel, F. Houzay, A. Barski, and R. Pinchaux (1984), *Solid State Commun.* **52**, 37.
Soockindt, L., D. Etienne, J.P. Marchand, and L. Lassabatere (1979), *Surf. Sci.* **86**, 378.
Stukel, D.J., R.N. Euwema, T.C. Collins, F. Herman, and R.L. Kortum (1969), *Phys. Rev.* **179**, 740.
Suslina, L.G., D.L. Fedorov, S.G. Konnikov, F.F. Kodzhespirov, A.A. Andreev, and E.G. Sharial (1977), *Sov. Phys. Semicond.* (English transl.), **11**, 1132.
Takayama, J., K. Shimomae, and C. Hamaguchi (1981), *J. Jpn. Appl. Phys.* **20**, 1265.
Takizawa, T. (1983), *Phys. Soc. Jpn.* **52**, 1057.
Temkin, H., and V.G. Keramidas (1980), *J. Appl. Phys.* **51**, 3269.
Theis, D. (1977), *Solid State Sol. (b)* **79**, 125.
Uchida, S. (1976), *J. Phys. Soc. Jpn.* **40**, 118.
Wakefield, B., M.A.G. Halliwell, T. Kerr, D.A. Andrews, G.J. Davis, and D.R. Wood (1984), *Appl. Phys. Lett.* **44**, 341.
Walter, J.P., M.L. Cohen, Y. Petroff, and M. Balkanski (1970), *Phys. Rev. B* **1**, 2662.
Williams, E.W., and V. Rhen (1968), *Phys. Rev.* **172**, 172.
Williams, G.P., F. Cerrina, J. Anderson, G.J. Lapeyre, R.J. Smith, J. Hermanson, N.J.A. Knapp (1983), *Physica* **117B**, **118B**, 350.
Varfolomeeve, A.V., R.P. Seisyan, and R.N. Yokimova (1975), *Sov. Phys. Semicond.* (English transl.), **9**, 530.
Voigt, T., W. Schmid, K.W. Benz, and M.H. Pilkulin (1982), in *GaAs and Related Compounds,* **81**, Inst. Phys. Conf. Ser. No. 63, p. 77.
Wei, S.-H., and A. Zunger (1990), *Appl. Phys. Lett.* **56**, 662.
Weiler, M.H. (1981), *Semicond. Semimet.* **16**, 119.
Zengin, D.M. (1983), *J. Phys. D* **16**, 653.
Zunger, A., and S. Mahajan (1993), in *Handbook on Semiconductors*, Vol. 3, 2nd ed. (Elsevier, Amsterdam).

APPENDIX 7A: BAND STRUCTURE CALCULATION USING HPT

Although the text contains sufficient information for calculating the band structures of all the systems considered in this chapter, here explicit instructions are provided to help readers perform these calculations.

The Hamiltonian for a given zinc blende semiconductor in the present HPT consists of three parts:

$$H = H_0 + H_1 + H_{so} \tag{7A.1}$$

where H_0 is calculated from the pseudopotential form factors in Table 7.1, H_1 is a second-neighbor tight-binding Hamiltonian, and H_{so} is the spin-orbit interaction. Both H_1 and H_{so} are treated as adjustable. There are 17 parameters in H_1, including 4 term value $\delta\varepsilon$'s, 5 first-neighbor matrix element h's, and 8 second-neighbor two-center integral v's. For a given compound, there are only two parameters λ^A and λ^C in H_{so}. These parameters are listed in Table 7.2 for the III–V compounds and in Table 7.51 for the II–VI systems. All three parts may be thought of as expanded in a set of orthonormal local orbitals (OLO) with eight orbitals per atom (one s, three p's, and then doubled with spins). We describe how to use these results to calculate H at each $\mathbf{k}$ inside the Brillouin zone, so it can be diagonalized to obtain the band structure.

H_{so} is the simplest of the three. Let $|\text{L}j\alpha\rangle$ be an OLO of the type α (s, p_x, etc.) belonging to the jth atom (anion or cation) in the unit cell labeled by a lattice vector $\mathbf{L}$. These OLOs may be denoted as $|\text{L}j\alpha\sigma\rangle$, with σ being up or down to include the spin orientations. A simple way to write H_{so} is to use OLO in the Γ representations, $|\text{L}j\gamma\rangle$, which are related to $|\text{L}j\alpha\sigma\rangle$ through a unitary transformation defined in Eq. (5.10.10) for each Lj. In this Γ-representation basis H_{so} is diagonal:

$$H_{so} = \sum_{\text{L}} \sum_{j} \sum_{\gamma} |\text{L}j\gamma\rangle \lambda^j_\gamma \langle \text{L}j\gamma| \tag{7A.2}$$

where $\lambda^j_\gamma = 0$, $-2\lambda^j$, and λ^j, for $\gamma = \Gamma_6$, Γ_7, and Γ_8, respectively, and the superscript j denotes either anion or cation. Since the other two parts of the Hamiltonian are usually expressed in the basis $|\text{L}j\alpha\sigma\rangle$, H_{so} can then be transformed to this basis from Eq. (7A.2) using the transformation defined in Eq. (5.10.10). H_{so} in $|\text{L}j\alpha\sigma\rangle$ is still block diagonal for each L, each block is a 16×16 matrix consisting of two 8×8 diagonal blocks belonging to anion and cation, respectively. Since H_{so} is site diagonal, in the Bloch basis $|\mathbf{k}j\alpha\sigma\rangle$ (see Eq. 7.1.1) H_{so} has the same 16×16 matrix for every $\mathbf{k}$.

Next consider H_1. It is spin-independent; i.e., all the matrix elements between different spins are zero, and the matrix element between two up spins is the same as that between two down spins. Hence we need only consider matrix elements of the kind

$$\langle \mathbf{k}j\alpha|H_1|\mathbf{k}j'\alpha'\rangle = \sum_{\text{L}} e^{-(\mathbf{L}+\tau_{j'}-\tau_j)\cdot\mathbf{k}} \langle 0j\alpha|H_1|\text{L}j'\alpha'\rangle \tag{7A.3}$$

In computing Eq. (7A.3), we can define $\mathbf{d} = \mathbf{L} + \tau_j' - \tau_j$ and group the sum into shells of distances $|\mathbf{d}|$. Harrison (1983, p. 77) has given the expressions of $\langle \mathbf{k}j\alpha|H_1|\mathbf{k}j'\alpha'\rangle$ for the diagonal and first-shell terms. Note that the tabulated values of the first-neighbor parameters (h's) are four times the actual local matrix elements, but the second-neighbor parameters (v's) are the actual two-center integrals. For example, h^{AC}_{ss} is $4\langle 0_{As}|H_1|\mathbf{d}_{cs}\rangle$ with $\mathbf{d} = \tau_c = (1/4, 1/4, 1/4)a$. The second-shell expression can be done in a similar way by summing up the 12 second-neighbor atoms of the same kind ($j = j'$) and using the expressions in Eqs. (2.4.3) through (2.4.6) to relate the two-center integrals (v's) in Tables 7.5 and 7.51 to the local matrix elements $\langle 0j\alpha|H_1|\text{L}j'\alpha'\rangle$ that are needed.

Finally to compute H_0 we need to use the wave functions explicitly. All the calculations in this chapter are done with Slater orbitals of the form $u_s(\mathbf{r}) = re^{-\gamma r}$, $u_x(\mathbf{r}) = xre^{-\gamma r}$, $u_y(\mathbf{r}) = yre^{-\gamma r}$, and $u_z(\mathbf{r}) = zre^{-\gamma r}$ for each atom. Two different exponent parameters $\gamma \equiv \beta(2\pi/a)$, one for the anions and the other for the cations, are used. Their values are $\beta_A = 1.8$ and $\beta_C = 1.5$. These nonorthogonal local orbitals are denoted $|u_{Lj\alpha}\rangle$, and their corresponding Bloch basis functions are denoted $|\phi_{kj\alpha}\rangle$ in Section 7.1. We first compute the overlap matrix $S_{j\alpha,j'\alpha'}(\mathbf{k}) = \langle\phi_{kj\alpha}|\phi_{kj'\alpha'}\rangle$, the kinetic matrix $T_{j\alpha,j'\alpha'}(\mathbf{k}) = \langle\phi_{kj\alpha}|p^2/2m|\phi_{kj'\alpha'}\rangle$, and potential matrix $V_{j\alpha,j'\alpha'}(\mathbf{k})$. Here V is the pseudopotential given in Table 7.1:

$$V(r) = \sum_{\mathbf{g}} V_{\mathbf{g}}\, e^{i\mathbf{g}\cdot\mathbf{r}} \tag{7A.4}$$

The computation of these matrix elements may be facilitated by expanding $|\phi_{kj\alpha}\rangle$ in plane waves $|\mathbf{k}+\mathbf{g}\rangle$:

$$|\phi_{kj\alpha}\rangle = \sum_{g} C_{j\alpha}(\mathbf{k}+\mathbf{g})|\mathbf{k}+\mathbf{g}\rangle \tag{7A.5}$$

The expansion coefficients can then be calculated explicitly:

$$C_{j\alpha}(\mathbf{k}+\mathbf{g}) = \langle\mathbf{k}+\mathbf{g}|\phi_{kj\alpha}\rangle \equiv (N/\Omega)^{1/2} e^{-i\mathbf{g}\cdot\boldsymbol{\tau}_j} A_{j\alpha}(\mathbf{k}+\mathbf{g}) \tag{7A.6}$$

where Ω/N is the volume per unit cell, $\boldsymbol{\tau}_j$ is the position of the atom with respective to the lattice points with $\boldsymbol{\tau}_A = (0,0,0)$ and $\boldsymbol{\tau}_C = (1/4,1/4,1/4)a$, and the $A_{j\alpha}(\mathbf{q})$ is the integral

$$A_{j\alpha}(\mathbf{q}) = \int e^{-i\mathbf{q}\cdot\mathbf{r}} u_{j\alpha}(\mathbf{r})\, d^3r \tag{7A.7}$$

Using the Slater orbitals in the preceding paragraph, we obtain

$$A_s(\mathbf{q}) = 8\pi(3\gamma^2 - q^2)/(\gamma^2 + q^2)^3 \tag{7A.8}$$

$$A_x(\mathbf{q}) = -32i\pi q_x(5\gamma^2 - q^2)/(\gamma^2 + q^2)^4 \tag{7A.9}$$

and similarly for $A_y(\mathbf{q})$ and $A_z(\mathbf{q})$. In terms of the expansion coefficients the overlap matrix is given by

$$S_{j\alpha,j'\alpha'}(\mathbf{k}) = \sum_{\mathbf{g}} C^*_{j\alpha}(\mathbf{k}+\mathbf{g})C_{j'\alpha'}(\mathbf{k}+\mathbf{g}) \tag{7A.10}$$

The kinetic energy matrix in the crystal units is

$$T_{j\alpha,j'\alpha'}(\mathbf{k}) = \sum_{\mathbf{g}} C^*_{j\alpha}(\mathbf{k}+\mathbf{g})C_{j'\alpha'}(\mathbf{k}+\mathbf{g})|\mathbf{k}+\mathbf{g}|^2 \tag{7A.11}$$

Finally the potential matrix is given by

$$V_{j\alpha,j'\alpha'}(\mathbf{k}) = \sum_{\mathbf{g}}\sum_{\mathbf{G}} C^*_{j\alpha}(\mathbf{k}+\mathbf{g}+\mathbf{G})C_{j'\alpha'}(\mathbf{k}+\mathbf{G})V_{\mathbf{g}} \tag{7A.12}$$

Reasonably converged values for all these matrices can be obtained with $\mathbf{g}$ truncated at the sixth shell, with a total number of 89 $\mathbf{g}$ vectors.

Having calculated the overlap matrix $S(\mathbf{k})$ and the Hamiltonian matrix $H_N(\mathbf{k}) = T(\mathbf{k}) + V(\mathbf{k})$, we can obtain $H_0(\mathbf{k})$ from

$$H_0(\mathbf{k}) = B^+(\mathbf{k})H_N(\mathbf{k})B(\mathbf{k}) \tag{7A.13}$$

with the B matrix given by $B = US_0^{-1/2}U^+$, where U is a unitary matrix that transforms $S(\mathbf{k})$ into a diagonal S_0 by $U^+SU = S_0$.

APPENDIX 7B: VCA HAMILTONIAN, ALLOY DISORDER AND MOLECULAR ATA CALCULATION

7B.1. The Alloy Hamiltonian in HPT

In this appendix, we describe the alloy Hamiltonian and the VCA and ATA calculations. Let us start with the three parts of $\hat{H}$ in Eq. (7.3.1):

$$\hat{H} = \hat{H}_0 + \hat{H}_1 + \hat{H}_{so} \tag{7B.1}$$

and consider them to be expanded in the OLO basis $|Lj\alpha\sigma\rangle$ constructed from Slater orbitals. These orbitals are for a VCA crystal in a pseudobinary alloy $A_{1-x}B_xC$ with a lattice constant obeying Vegard's law:

$$a = (1-x)a_{AC} + xa_{BC} \tag{7B.2}$$

$\hat{H}_0$ is the "universal" part of H_0 of Eq. (7A.1) now extended to the VCA alloy. Since the A and B atoms carry different potentials with them and we want to represent the disorder potential by the diagonal and first-neighbor interactions, we remove the diagonal part D from H_0 to form $\hat{H}_0$:

$$\hat{H}_0 = H_0 - D \tag{7B.3}$$

The four independent values of the diagonal matrix elements in crystal units are $D_s^A = 0.616$, $D_p^A = 1.971$, $D_s^C = 2.214$, and $D_p^C = 0.616$. These values are evaluated from Brillouin zone integrations of the diagonal matrix elements of $H_0(\mathbf{k})$ of Eq. (7A.13)

$\hat{H}_1$ is constructed by adding the diagonal D matrix to the constituents' H_1 matrices. To be specific, let us consider $Ga_{1-x}In_xAs$ as an example. We assume that the diagonal term values of A and B preserve their pure-crystal values, and those of C atoms take their concentration averaged values. The term values of each atom in the pure constituent compounds, denoted E_α^j (e.g., E_s^A), are also given in Table 7.2 for the III–V compounds and in Table 7.51 for the II–VI systems. These values were obtained by adding the D_α^j to the $\delta\varepsilon_\alpha^j$,

$$E_\alpha^j = D_\alpha^j + \delta\varepsilon_\alpha^j \tag{7B.4}$$

and then converting it to eV from crystal units. Thus, for $Ga_{1-x}In_xAs$, we take from Table 7.2 the values $E_s^{Ga} = 0.464$ eV and $E_s^{In} = 0.341$ eV for any composition x, but use the concentration average $E_s^{As} = -6.26(1-x) - 6.36x$ for the s-term value (in eV) for each As atom.

The first-neighbor matrix elements h in $\hat{H}_1$ in crystal units (defined in terms of the VCA lattice constant a) is calculated from the constituent's value h_0 by the following scaling:

$$h = h_0(\bar{d}/d)^2 \tag{7B.5}$$

where $\bar{d}$ is the VCA bond length and d is the actual bond length between the two orbitals under consideration. For example, for an A–C bond in an $A_{1-x}B_xC$ alloy, d is taken to be

$$d = d_{AC} + x(d_{BC} - d_{AC})/4 \tag{7B.6}$$

where d_{AC} and d_{BC} are respectively the bond lengths in the pure AC and BC constituent compounds. This ratio $\bar{d}/d$ is a good approximation to that between the VCA bond length and the bimodal bond length shown in Fig. 1.5. Finally the two-center integral v's in H_1 for two second-neighbor atoms are approximated in VCA as the concentration averages.

The spin-orbital Hamiltonian $\hat{H}_{so}$ for an $A_{1-x}B_xC$ alloy, in analogy to the pure-constituent compounds, is best represented in the Γ-representation OLO basis, given in Eq. (7A.2). However, the values of λ now vary from site to site, depending on the atomic occupation. The values of λ for atoms A and B are taken directly from Tables 7.2 or 7.51, but those for the C atom are approximated by concentration weighted averages.

7B.2. The VCA Hamiltonian

Having defined the alloy Hamiltonian, the VCA Hamiltonian is simply the concentration average:

$$\bar{H} = \langle\hat{H}\rangle = \hat{H}_0 + \langle\hat{H}_1\rangle + \langle\hat{H}_{so}\rangle \tag{7B.7}$$

$\hat{H}_0$ is universal in crystal units. $\langle\hat{H}_1\rangle$, in addition to the second-neighbor two-center integrals, involves the average of the term values

$$\langle E^j_\alpha\rangle = (1-x)E^j_\alpha(\mathrm{AC}) + xE^j_\alpha(\mathrm{BC}) \tag{7B.8}$$

and the first-neighbor matrix elements

$$\langle h\rangle = (1-x)h(\mathrm{AC}) + xh(\mathrm{BC}) \tag{7B.9}$$

where h(AC) and h(BC) are the quantities defined in Eq. (7B.5). Similarly $\langle\hat{H}_{so}\rangle$ involves the average of the spin-orbit coupling constants

$$\langle\lambda^j\rangle = (1-x)\lambda^j(\mathrm{AC}) + x\lambda^j(\mathrm{BC}) \tag{7B.10}$$

With all the matrix elements defined, one can transform $\bar{H}$ into the **k** representation in the Bloch basis $|\mathbf{k}j\alpha\sigma\rangle$, as is done for the pure-constituent compounds in band structure calculations.

7B.3. Disorder Hamiltonian and ATA Calculation

In the molecular CPA and ATA calculations, the disorder Hamiltonian is assumed to be the sum of molecular contributions:

$$U = \sum U_L \tag{7B.11}$$

Each molecule contains 8 hybrid orbitals (16 if spins are included) surrounding the center alloying atom A or B, as shown in Fig. 5.15. These hybrid orbitals are then transformed into the A_1 and T_2 basis functions according to Eq. (5.10.4). Finally, we construct the Γ

representation from these s- and p-like functions according to Eq. (5.10.10). In this Γ-representation basis, the molecular contribution U_L becomes block diagonal into eight 2×2 matrices: two for Γ_6, two for Γ_7, and four for Γ_8 representations. These 2×2 matrices are denoted $U_\gamma(L)$, where γ specifies the representation and L indicates atom type A or B. The values of these matrix elements depend on whether the center atom is A or B. These disorder matrices are conveniently defined in terms of the scattering potential

$$W_\gamma(L) \equiv U_\gamma(A) - U_\gamma(B) \tag{7B.12}$$

such that

$$U_\gamma(A) = xW_\gamma \quad U_\gamma(B) = -(1-x)W_\gamma \tag{7B.13}$$

The form of these W_γ matrices is given in Eq.(5.11.2); i.e.,

$$W\gamma = \begin{pmatrix} \delta & \Delta \\ \Delta & 0 \end{pmatrix} \tag{7B.14}$$

We describe the calculation of the disorder parameters δ and Δ for the three representations.

For the Γ_6 representation, the value for δ is just the difference between the s-term values of the A and B atoms. For example, in $Ga_{1-x}In_xAs$, $\delta = E_s^{Ga} - E_s^{In} = 0.464 - 0.341 = 0.123$ eV. The 0.124 eV value tabulated in Table 7.25 is calculated this way with better precision. The value of Δ is the difference in the matrix elements connecting a cation s state at the origin and a Γ_6 state associated with the second-shell hybrids (labeled 5 through 8 in Fig. 5.15). This difference is caused by the fact that these matrix elements depend on which atom is at the central site. One can show that this Δ for $Ga_{1-x}In_xAs$ is given by

$$\Delta = [h_{ss}^{AC}(GaAs) - h_{ss}^{AC}(InAs)]/4 - 3[h_{xs}^{AC}(GaAs) - h_{xs}^{AC}(InAs)]/4 \tag{7B.15}$$

Remember that in this equation the values for h should be computed from the tabulated values h_0 according to Eq. (7B.5), and that the tabulated h_0 values are four times the matrix elements between first-neighbor OLOs. The value for Δ thus calculated for $Ga_{1-x}In_xAs$ at $x = 0.5$ is $\Delta = 0.42$ eV, which is the value given in Table 7.25.

For both the Γ_7 and Γ_8 representations, the off-diagonal scattering potential matrix element Δ has the same value -0.163 eV at $x = 0.5$, which is calculated from

$$\begin{aligned}\Delta = -[&h_{sx}^{AC}(GaAs) - h_{sx}^{AC}(InAs) + h_{xx}^{AC}(GaAs) \\ &- h_{xx}^{AC}(InAs) + 2h_{xy}^{AC}(GaAs) - 2h_{xy}^{AC}(InAs)]/4\end{aligned} \tag{7B.16}$$

Again, these h values should be computed according to Eq. (7B.5). The diagonal matrix elements are different, due to different spin-orbit shifts. For Γ_7, δ in $Ga_{1-x}In_xAs$ is given by

$$\delta(\Gamma_7) = E_p^{Ga} - E_p^{In} - 2(\gamma^{Ga} - \gamma^{In}) \tag{7B.17}$$

so $\delta = 3.762 - 4.296 + 2(0.05 - 0.092) = -0.45$ eV. For Γ_8 it is given by

$$\delta(\Gamma_8) = E_p^{Ga} - E_p^{In} + (\lambda^{Ga} - \lambda^{In}) \tag{7B.18}$$

with a calculated value of $\delta = -0.576$ eV.

Although the scattering potential matrices have a slight x dependence, because h in Eq. (7B.5) also has a slight x dependence, calculations using these tabulated values of δ and Δ for all alloy concentrations introduce only negligible errors.

7B.4. Band Calculation Using the Molecular ATA

Now that both the VCA and disorder Hamiltonians are defined, the molecular ATA calculation can be carried out in a straightforward manner. The ATA self-energy Σ can be written as the sum of molecular contributions:

$$\Sigma = \sum \Sigma_{\mathrm{L}} \tag{7B.19}$$

because the disorder Hamiltonian in Eq. (7B.11) is also block diagonal into molecular contributions. Each Σ_{L}, in the Γ representation defined in the preceding subsection, is further block diagonalized into eight 2×2 matrices, denoted generally as Σ_{γ}. These self-energies are invariant under lattice translation operations. They each satisfy the ATA equation

$$\Sigma_{\gamma} = \langle t_{\gamma}\rangle[1 + g_{\gamma}\langle t_{\gamma}\rangle]^{-1} \tag{7B.20}$$

where the average t matrix is given by

$$\langle t_{\gamma}\rangle = (1-x)t_{\gamma}(\mathrm{A}) + xt_{\gamma}(\mathrm{B}) \tag{7B.21}$$

with $t_{\gamma}(\mathrm{A})$ given by

$$t_{\gamma}(\mathrm{A}) = U_{\gamma}(\mathrm{A})[1 - g_{\gamma}U_{\gamma}(\mathrm{A})\rangle]^{-1} \tag{7B.22}$$

and similarly for $t_{\gamma}(\mathrm{B})$. The g_{γ} in Eqs. (7B.20) and (7B.22) are the corresponding Green functions associated with the VCA Hamiltonian in the Γ representation. They are also 2×2 matrices, as defined in Eq. (5.10.11) and are calculated according to the Brillouin zone integration in Eq. (5.10.12). After the calculation of these Σ_{γ}, we transform the self-energy operators into **k**-space, denoted $\Sigma(\mathbf{k})$. Then the shift of the band energy away from its VCA value is calculated from the expectation value

$$\delta\varepsilon_n(\mathbf{k}) = \langle n\mathbf{k}|\Sigma(\mathbf{k})|n\mathbf{k}\rangle \tag{7B.23}$$

where $|n\mathbf{k}\rangle$ is the eigenket of $\overline{H}$ with a corresponding eigenvalue $\varepsilon_n(\mathbf{k})$.

PROBLEMS

CHAPTER 1

1. Bond Lengths and Relaxation Parameters of Pseudobinary Alloys

Consider a simple lattice model for a 50–50 pseudobinary zinc blende alloy $A_{0.5}B_{0.5}C$. Start with a virtual crystal with an average bond length $d = (d^0_{AC} + d^0_{BC})/2$, in which the C atoms occupy a fcc sublattice and the averaged atoms (of A and B) occupy another sublattice. Each atom is connected to its nearest-neighboring atoms by springs with a force constant κ. There is no strain energy in this virtual crystal. Now replace a virtual atom by a real A atom, as illustrated in Fig. 3.3. Because the average bond length d is not the same as the natural bond length d^0_{AC}, strain is built in the four bonds surrounding the A atom. If we only allow these four C atoms to relax while the rest of the atoms are held in their original positions, show that when the strain energy is minimized the relaxation parameter Γ (defined in Eq. 1.3.2) for these four AC bonds is exactly 75%.

2. Crystal Diffraction and Miller Indices (hkl)

The scattered wave associated with a plane wave $e^{i\mathbf{q}\cdot\mathbf{r}}$ scattered by an atom at the origin is given by $\psi = f(\theta)e^{iqr}/r$, where $f(\theta)$ is the scattering amplitude.

(a) Show that the scattered wave associated with $e^{i\mathbf{q}\cdot\mathbf{r}}$ scattered by the same atom located at $\mathbf{R}$ is given by $\psi = e^{-i\Delta\mathbf{q}\cdot\mathbf{R}}f(\theta)e^{iqr}/r$, where $\Delta\mathbf{q} = (\mathbf{q}' - \mathbf{q})$ with $\mathbf{q}'$ defined by $\mathbf{q}' = q\mathbf{r}/r$. Extend this to the scattering of $e^{i\mathbf{q}\cdot\mathbf{r}}$ by a collection of atoms and show that the sum of the scattered waves is given by $\psi = Fe^{iqr}/r$, where the effective scattering amplitude is given by $F = \Sigma f_n \exp(-i\Delta\mathbf{q}\cdot\mathbf{R}_n)$ with n summed over all atoms.

(b) For a crystal, the atomic positions can be written as $\mathbf{R}_n = \mathbf{l} + \boldsymbol{\tau}_j$, where $\mathbf{l}$ is a lattice vector and $\boldsymbol{\tau}_j$ is the position relative to $\mathbf{l}$ of the jth atom belonging to a unit cell. Then the effective scattering amplitude can be written as

$$F(\Delta\mathbf{q}) = \sum_l e^{-i\Delta\mathbf{q}\cdot\mathbf{l}} \sum_j f_j e^{-i\Delta\mathbf{q}\cdot\boldsymbol{\tau}_j}$$

The condition for constructive diffraction is $\Delta\mathbf{q} = \mathbf{G}$, where $\mathbf{G}$ is a reciprocal lattice vector, because by definition $\mathbf{G}\cdot\mathbf{l}$ is 2π times an integer. Then F becomes $F(\mathbf{G}) = Nf$ with f given by

$$f(\mathbf{G}) = \sum_j f_j e^{-i\mathbf{G}\cdot\boldsymbol{\tau}_j} \tag{A}$$

where N is the number of unit cells. Show that the allowed G vectors for a zinc blende semiconductor, written as

$$\mathbf{G} = (h, k, l)2\pi/a \tag{B}$$

are those with all odd or all even integers h, k, and l. What is the expression for f for a given set of h, k, and l?

(c) For any cubic crystal, which includes simple cubic (sc), body-centered cubic (bcc), and fcc, one can choose the unit cell to be a simple cube. Then the **G** given by Eq. (B) for any integer set (hkl) is allowed. However, not all these (hkl) will have a nonzero scattering amplitude. Show that the f of Eq. (A) for a given set of (hkl) for a zinc blende semiconductor AC is given by

$$f = (f_A + f_C e^{-i(h+k+l)\pi/2})(1 + e^{-i(h+k)\pi} + e^{-i(h+l)\pi} + e^{-i(k+l)\pi})$$

Then show that f is not zero only when all (hkl) are even or odd.

3. *Diffraction Patterns for Ordered and Disordered Alloys*

(a) Consider an ABC_2 alloy ordered in the CuAuI structure described in Section 1.4.1 with a c/a ratio $\beta = 2$. Show that the basic reciprocal lattice vectors are given by $\mathbf{g}_1 = (1,1,0)2\pi/a$, $\mathbf{g}_2 = (1,-1,0)2\pi/a$,and $\mathbf{g}_3 = (0,0,1)2\pi/a$. Then show that the allowed (hkl) of **G** given by Eq. (B) have a one-to-one correspondence to the diffraction patterns shown in Fig. 1.8a.

(b) What is the effective $f\,(hkl)$ for this alloy when the unit cell is taken to be a simple cube?

(c) Show that for those (hkl) forbidden for the zinc blende but allowed for the CuAuI structure, the effective f is given by $f = f_A - f_B$.

(d) Show that the average f of part (c) for a partially ordered CuAuI structure is given by $f = S(f_A - f_B)$. Thus, the intensities of these "forbidden" spots are related to the order parameter S. For a disordered alloy, $S = 0$, and the diffraction patterns are the same as that of the constituent zinc blende compounds as shown in Fig. 1.2.

CHAPTER 2

1. *Single-Particle Schrödinger Equation in the Local Density Approximation*

Show that by taking partial derivative with respect to ϕ_ν^*, the variational calculation of Eq. (2.2.7) yields the Schrödinger equation (2.2.8) with V given by Eqs. (2.2.9) through (2.2.12).

2. *Tight-Binding Matrix Elements in the Two-Center Approximation*

Show that Eqs. (2.4.3) through (2.4.6) hold in the two-center approximation.

3. The Bond-Orbital Model (BOM)

(a) Use the interaction parameter η's in Eq. (2.5.1) and the definition of V_2 to obtain the result $V_2 = -24.5/d^2$.
(b) Show that the bonding energy for a homopolar semiconductor is given by Eq. (2.5.10).
(c) Follow steps in Section 2.5 and use Table 2.1 to check the results in Table 2.2 for several systems (e.g., Si, GaAs, CdTe, and CuBr).

4. Application of BOM to Impurity Formation Energies

(a) Compute the energy required to substitute a Ga atom in GaAs by an In atom using BOM and assuming no lattice relaxation.
(b) Repeat part (a) except now allow the four As atoms surrounding the impurity to relax with a lattice relaxation parameter of 75% (see Problem 1 of Chapter 1).

5. Excess Energy of Ordered Alloys

Calculate the excess energy of $GaInAs_2$ in the CuAuI structure using BOM and the bond lengths in Table 1.6.

CHAPTER 3

1. Valence-Force-Field Model (VFF)

Show that the three independent elastic constants are related to the VFF force constants α and β by $B = (\alpha + \beta/3)/a$ (Eq. (3.3.2)), $C_{11} - C_{12} = 4\beta/a$ (Eq. (3.3.3)), and $C_{44} = 2\alpha\beta/(\alpha+\beta)a$ (Eq. (3.3.6)).

2. Elastic Constants in the Bond-Orbital Model

Show that in BOM the three independent elastic constants collapse to a simple relation $9/C_{44} = 6/(C_{11} - C_{12}) + 4/B$ following the steps described in Section 3.5.

3. Bulk Modulus in an Ordered Alloy

Show that the bulk modulus for a chalcopyrite structure in the simple spring model described in Section 3.7.1 deviates from the concentration-weighed average by a negative value ΔB given in Eq. (3.7.2). Show that in the VFF ΔB is given by Eq. (3.7.3).

4. Bulk Modulus in a Disordered Alloy

Derive the expressions for the effective spring constant k in Eq. (3.7.10) and effective bond length d given in Eq. (3.7.11), following the model and steps described in Section 3.7.2.

CHAPTER 4

1. Quasi-Chemical Approximation (QCA)

(a) Show that the pair probability that minimizes the alloy free energy of (4.2.11) in QCA is given by r in Eq. (4.2.12).
(b) Show that Eq. (4.2.12) is also a solution of Eq. (4.2.15).

(c) Show that the critical temperature in QCA is $T_c = \Omega/\lambda R$ with λ given by Eq. (4.2.17).

2. *Common Tangent Line and Equal Chemical Potentials in Equilibrium*

Consider the phase equilibrium between two phases designated by α and β of a binary alloy $A_{1-x}B_x$. Show that the equilibrium concentrations x_α and x_β of these phases at a given temperature and given by a common tangent line (e.g., Fig. 4.3b) can also be obtained from the condition that the chemical potentials of A and B in the two different phases are equal: $\mu_A^\alpha(x_\alpha) = \mu_A^\beta(x_\beta)$ and $\mu_B^\alpha(x_\alpha) = \mu_B^\beta(x_\beta)$.

3. *Composition Diagram for Ternary Alloys*

Show that the concentrations a, b, and c, defined in Fig. 4.8, satisfy the condition $a + b + c = 1$.

4. *Liquidus Curves for III–V Compounds*

Use Eqs. (4.4.8) and (4.4.9) and the data in Table 4.2 to generate the liquidus curves for GaAs, InAs, and InSb shown in Fig. 4.5.

5. *Mixing Energy in Pseudobinary Alloy*

Show that the mixing energy based on Eq. (4.6.13) is one-fourth of the value for the case where lattice relaxation is forbidden (refer to Problem 1 in Chapter 1).

6. *Entropy of Mixing for Pseudobinary Alloys in GQCA and CVM*

Show that the mixing entropy based on CVM tetrahedral clusters actually containing the bonds, i.e., Eq. (4.10.10), is the same as that using GQCA.

7. *Special Solution for GQCA*

Show that $\eta = 1$ is the physical solution for Eq. (4.7.15) if the reduced mixing energy parameters satisfy the symmetrical relation $\Delta_j = \Delta_{J-j}$.

CHAPTER 5

1. *Band Gaps in Semiconducting Polymers—The SSH Model*

The origin of band gaps and the nature of excitations in semiconducting polymers are distinctly different from those in the common semiconductors emphasized in this book. A tight-binding model by Su, Schrieffer, and Heeger (1979, the SSH model) for polyacetylene $(CH)_n$ provides a convenient framework for discussing these quantities. This model only deals with the π-orbitals of the C atoms. Each carbon atom has one π-electron. The TB Hamiltonian is given by

$$H = \sum_n \sum_m |n\rangle h_{nm} \langle m|$$

where n and m are the site indices. If the carbon atoms are equally spaced, the matrix elements are $h_{nm} = \varepsilon_0$ for $n = m$, and $h_{nm} = -\gamma$ for $m = n + 1$ or $m = n - 1$. ε_0 can be set to zero without losing generality. Using the periodic boundary condition, we obtain the energy dispersion

$$\varepsilon(k) = -2\gamma \cos(ka)$$

for the π-electrons with wave vector lying inside the Brillouin zone (BZ), $-\pi/a \le k \le \pi/a$.

(a) Use this expression to calculate the band structure energy per atom for the occupied states.

(b) We shall show that the energy in (a) is not the ground state energy. Consider that, instead of being equally spaced, the chain undergoes "dimerization," i.e., the C atoms shift alternatively with displacements u and $-u$ along the chain direction. Then the first-neighbor TB matrix elements take two values: $-\gamma - \delta$ for closer pairs and $-\gamma + \delta$ for the farther pairs. Show that these changes split the bands into two given by

$$\varepsilon_t(k) = \pm 2(\gamma^2 \cos^2(ka) + \delta^2 \sin^2(ka))^{1/2}$$

where k now lies within the reduced BZ: $-\pi/(2a) \le k \le \pi/(2a)$. There is a gap $E_g = 4\delta$ separating the valence and conduction bands.

(c) The dimerization in (b) increases the energy of the σ bonds not considered explicitly. A simple way to represent this increase in energy is $E_R = 2NKu^2$, where k is an effective spring constant and N is the total number of C atoms. Note that the change in the TB matrix elements δ can also be related to the atomic displacements u by $\delta = 2\alpha u$ for small u. Use Harrison's universal TB parameters to deduce an expression for α.

(d) Evaluate the total energy per atom for this case by summing up the band structure energy in (b) and the strain energy in (c), and show that this energy is lower than that in (a). Minimize this energy with respect to u to find the equilibrium value of u, the energy gained per atom due to dimerization, and the band gap. Find the numerical values of these quantities using the following parameters given by SSH:

$$\gamma = 2.5 \text{ eV}, \alpha = 4.1 \text{ eV/Å}, \text{ and } K = 21 \text{ eV/Å}^2.$$

Note in part (d), one encounters the elliptical function of the second kind

$$E(\zeta) = \int_0^{\pi/2} (1 - \zeta \sin^2 \theta)^{1/2} \, d\theta$$

It is sufficient to use the approximation $E(1 - z^2) = 1 + z^2[\ln(4/z) - 0.5]/2$ for positive $z << 1$ in the above calculation.

2. **k·p** *Theory and Effective Mass*

(a) The effective mass formulas in Eq. (5.4.8) are approximations to the **k·p** theory. This theory is based on an expansion of the energy eigenfunction associated with a wave vector **k** in terms of those of a neighboring wave vector $\mathbf{k}_0$; i.e.,

$$\psi_k(\mathbf{r}) = \sum_n C_n \psi_{nk_0}(\mathbf{r}) \exp(i\delta\mathbf{k}\cdot\mathbf{r})$$

where $\mathbf{k} = \mathbf{k}_0 + \delta\mathbf{k}$ and n is the band index. Use this expression in the Schrödinger equation $[p^2/2m + V]\psi = E\psi$ and take the dot product with $\psi_{n'k_0}$ to obtain

$$\sum_n \left(\left[\frac{\hbar^2(\delta k)^2}{2m} + \varepsilon_n(k_0) - E \right] \delta_{n'n} + \frac{\hbar}{m} \delta\mathbf{k}\cdot\mathbf{p}_{n'n} \right) C_n = 0$$

where $\mathbf{p}_{n'n} = \langle \psi_{n'k_0} | \mathbf{p} | \psi_{nk_0} \rangle$.

(b) For small δk and for a nondegenerate energy level, the $\delta\mathbf{k}\cdot\mathbf{p}$ term can be treated by perturbation theory. Show that the energy level is given by Eq. (5.4.2); i.e.,

$$E_n(k) = \varepsilon_n(\mathbf{k}_0) + \hbar\delta\mathbf{k}\cdot\mathbf{v}_n(\mathbf{k}_0) + \frac{1}{2}\hbar^2 \sum_{\alpha\beta}\sum \left(\frac{1}{m^*}\right)_{\alpha\beta} \delta k_\alpha\, \delta k_\beta$$

with $\mathbf{v}_n$ being the group velocity. Also show and that the inverse effective tensor is given by

$$\left(\frac{1}{m^*}\right)_{\alpha\beta} = \frac{1}{m}\left(\delta_{\alpha\beta} + \frac{2}{m} \mathrm{Re} \left[\frac{\sum_n (p^{\alpha}_{n'n})^* p^{\beta}_{n'n}}{\varepsilon_n(k_0) - \varepsilon_{n'}(k_0)} \right] \right)$$

Note that Eq. (5.4.8) corresponds to the case when only two states at the band edge are considered.

3. *Green Function Applied to an Impurity*

(a) Consider a single impurity in an otherwise perfect crystal. The one-electron Hamiltonian is given by $H = H_0 + V$, where H_0 is the Hamiltonian in the perfect crystal and V is the impurity potential. Show that the change in the density of states introduced by the impurity is given by

$$\Delta\rho(E) = -\frac{d}{dE} \mathrm{Im}\,(\ln[\det(1 - gV)])/\pi$$

where $g \equiv 1/(E + i0 - H_0)$ is the Green function in the perfect crystal.

(b) Note that in the calculation of the determinant in (a) the size of the matrix is governed by that of the impurity potential. In the single-band model with a site-diagonal impurity potential $V = |0\rangle\delta\langle 0|$, the g needed is simply $F = \langle 0|g|0\rangle$. For a localized impurity level E to form outside the band, the condition is that $1 - F\delta = 0$. Find a criterion for δ for having a localized impurity state in the one-dimensional model with H_0 given by Eq. (5.2.7).

(c) Use the potential in (b) and the band model specified by Eqs. (5.9.6) and (5.9.7) to investigate the change of density of states as a function of energy E for several values of $\delta = 0.5, -1.0$, and 2.0.

4. *CPA Calculation for Binary Alloys*

Use the model in Section 5.9 to perform CPA calculations for a 50–50 binary alloy with the following scattering strengths: $\delta = 0.5, 1.0, 1.5$, and 2.0. In each case plot both the real and imaginary parts of the self-energy and the density of states as a function of energy.

5. Tight-Binding Band Structures and Spin-Orbit Splitting

(a) Perform band structure calculations for Si and GaAs using the first-neighbor TB Hamiltonian with the TB parameters given by Eqs. (2.4.3) through (2.4.6), Harrison's universal expressions in Eq. (2.5.1), and the term values in Table 2.1. Plot the bands as a function of the wave vector similar to those shown in Fig. 2.3. Find the values of the band gaps.

(b) Treat the spin-orbit Hamiltonian as diagonal in the local orbital basis described in Eq. (5.10.10). Note that only the p levels have a splitting. The coupling constants are taken to be the atomic values given in Table 2.1. Add this spin-orbit Hamiltonian to (a) and then perform the band structure calculation for Si and GaAs. Plot the bands again, and find the effects of the spin-orbit coupling on the states near the band edge.

CHAPTER 6

1. Master Equation, H-Theorem, and a Priori *Distribution*

Consider the master equation (6.1.4) and define a quantity $h \equiv \Sigma P_n \ln P_n$.

(a) Show that $dh/dt \le 0$ in general. This is Boltzmann's H-theorem. Since the entropy S is related to h by $S = -kh$, this result shows that the entropy of a supersystem never decreases.

(b) Show that the equality in (a) holds only when all P_n's are equal. This result implies the *a priori* equilibrium distribution.

2. Time-Dependent Solution of the Supersystem Master Equation

(a) Show that Eq. (6.1.4) can be written as a matrix equation, $d\mathbf{p}/dt = -\mathbf{W}\mathbf{p}$. Here $\mathbf{p}$ is a column matrix with the nth matrix element defined as $p_n = P(n,t)$; $\mathbf{W}$ is a symmetrical stoachastic matrix with $W_{nm} \equiv W(n/m)$ for $n \neq m$, and $W_{nn} = 1/\tau_n \equiv \Sigma_m' \, W(m/n)$, where the sum over m excludes n.

(b) Show that the equation in (a) has a general solution given by

$$\mathbf{p}(t) = \sum_{\nu=1}^{N} c_\nu \mathbf{p}_\nu e^{-\Omega_\nu t}$$

where c_ν is a constant and $\mathbf{p}_\nu$ is an eigenvector of $\mathbf{W}$ with a corresponding eigenvalue Ω_ν. Here N is the total number of states involved.

(c) Find an expression for c_ν in terms of the initial condition $\mathbf{p}(0)$.

(d) Given the ordering of the eigenvalues $\Omega_1 < \cdots < \Omega_N$, demonstrate that $\Omega_1 = 0$ exists; i.e., there exists an equilibrium solution (see Problem 1).

(e) Suppose $N = 2$. Write out the solution for $P(1,t)$ and $P(2,t)$ given $P(1,0) = 0$ and $P(1,0) = 1$. What is the structure of the $\mathbf{W}$ matrix? Plot $P(1,t)$ and $P(2,t)$ as a function of t and find their values at large t.

3. Brooks' Formula for Alloy Limited Mobility

Derive Brooks' formula, Eq. (6.4.1), following the steps from Eq. (6.4.3) to (6.4.5). Then show that this formula is modified to become Eq. (6.4.9) when the conduction electrons have two anisotropic masses m_l and m_t, and there are alloy fluctuations in both the s and p term values.

4. Particle and Momentum Relaxation Time by Optical Phonon Scattering

Derive the expressions for τ_R and τ_m given by Eqs. (6.3.20) and (6.3.21) respectively.

5. Quantum Mechanical Basis of Transport Theory

Suppose that there is an electric field **E** applied to a crystal. An electron will move under the influence of **E**. The Hamiltonian for the electron is $H_E = H + e\mathbf{E}\cdot\mathbf{r}$, where H is the Hamiltonian when **E** is absent. The eigenfunction of H is denoted $\phi(\mathbf{k},\mathbf{r};t) = u(\mathbf{k},\mathbf{r})\exp(i[\mathbf{k}\cdot\mathbf{r} - \varepsilon(\mathbf{k})t/\hbar])$, where $\varepsilon(\mathbf{k})$ is the eigen (band) energy.

(a) Show that

$$\psi(\mathbf{k},\mathbf{r};t) = u[\mathbf{k}(t),\mathbf{r};t]\exp\{i[\mathbf{k}(t)\cdot\mathbf{r}] - \frac{i}{\hbar}\int_0^t \varepsilon\,[\mathbf{k}(t')]\,dt'\}$$

where $\mathbf{k}(t) = \mathbf{k} - e\mathbf{E}t/\hbar$ is an approximate solution to the time-dependent Schrödinger equation when **E** is present, if $|\partial u(\mathbf{k},\mathbf{r})/\partial\mathbf{k} << u(\mathbf{k},\mathbf{r})/k|$ and $|e\mathbf{E}t/\hbar| << k$ (either E is small or for short time).

(b) Show that $\langle\mathbf{v}\rangle \equiv \langle\phi(\mathbf{k})|\mathbf{p}/m|\phi(\mathbf{k})\rangle = \nabla_k\varepsilon(\mathbf{k})/\hbar$.

(c) Then demonstrate that

$$\langle\mathbf{v}(t)\rangle \equiv \int \psi^*(\mathbf{k},\mathbf{r};t)(\mathbf{p}/m)\psi(\mathbf{k},\mathbf{r};t)\,d^3r = \nabla_{k(t)}\,\varepsilon[\mathbf{k}(t)]/\hbar$$

and so

$$\langle\mathbf{a}(t)\rangle = d\langle\mathbf{v}(t)\rangle/dt = \nabla_{k(t)}\,\langle\mathbf{v}(t)\rangle\cdot(-e\mathbf{E}/\hbar)$$

$$= \frac{1}{\hbar^2}\nabla_{k(t)}\,\nabla_{k(t)}\varepsilon[\mathbf{k}(t)]\cdot(-e\mathbf{E}) \equiv (1/m^*)\cdot(-e\mathbf{E})$$

where the equality defines the mass tensor m^*:

$$(1/m^*)_{ij} = \frac{1}{\hbar^2}\frac{\partial^2\varepsilon(\mathbf{k})}{\partial k_i\,\partial k_j}$$

6. Transient Solution to the Boltzmann Equation

Show how the expansion method in Eq.(6.7.3) would be generalized to solve the transient Boltzmann equation rather than its steady-state form.

Index

Page numbers in bold indicate tables; underlined page numbers refer to figures.